# SOIL
# PHYSICS
# COMPANION

Edited by

## A. W. Warrick, Ph.D.

## CRC Press
Taylor & Francis Group
Boca Raton  London  New York

CRC Press is an imprint of the
Taylor & Francis Group, an **informa** business

CRC Press
Taylor & Francis Group
6000 Broken Sound Parkway NW, Suite 300
Boca Raton, FL 33487-2742

First issued in paperback 2019

© 2002 by Taylor & Francis Group, LLC
CRC Press is an imprint of Taylor & Francis Group, an Informa business

No claim to original U.S. Government works

ISBN-13: 978-1-4665-7984-2 (hbk)
ISBN-13: 978-1-138-74837-8 (pbk)

| Library of Congress Cataloging-in-Publication Data |
| --- |
| Soil physics companion / edited by Arthur W. Warrick. |
| p. cm. |
| Includes bibliographical references. |
| ISBN 0-8493-0837-2 (alk. paper) |
| 1. Soil physics. I. Warrick, Arthur W. |
| S592.3 .S676 2001 |
| 631.4′3—dc21          2001043627 |

Library of Congress Card Number 2001043627

**Visit the Taylor & Francis Web site at**
**http://www.taylorandfrancis.com**

**and the CRC Press Web site at**
**http://www.crcpress.com**

# Preface

The *Soil Physics Companion* contains nine chapters addressing most significant topics of contemporary soil physics. The chapter authors are leading scientists and recognized authorities on their respective topics. Information from the *Handbook of Soil Science* is augmented with example boxes to add flavor and interest. These additions include problems, further descriptions and historical backgrounds. In cases where more information is desired, current references are provided for an entry into the vast literature available.

Soil physics includes the descriptions of the physical aspects of the soil system and of transport processes. The usual physical setting is at or just below the soil surface, but most of the concepts and descriptions are valid to all depths and for all similar geological and extraterrestrial materials. Included are descriptions of new devices useful for measurements at the Earth's surface and reaching into the vadose zone.

Chapters 1, 2, and 7 emphasize the soil solids. Included are descriptions of the matrix as well as necessary definitions to describe both static and dynamic aspects of the soils. At the start, the soil is considered as static to facilitate quantification of mass, particle size and surface areas. Later the dynamics of tillage and temporal variations due to natural and human actions are examined.

Soil water is the primary theme of Chapters 3 and 4, a major part of Chapters 5 and 6 and important for the other chapters as well. Along with general principles, measurement methodology and instruments are discussed for determining both soil water content and potential.

Energy balance and the thermal regime are the topics of Chapter 5. Appropriate definitions, measurement techniques and the transport of energy are given. Included is a detailed description of the soil–plant–atmospheric interface, which represents a common convergence point for many of the world's problems of food production, water resources and environmental pollution, including global warming.

Separate chapters are devoted to solute transport and soil–gas movement. Solute transport is fundamental in terms of environmental pollution, nutrient management and soil quality. Gas movement historically emphasized soil aeration relevant for cultural practices and microorganisms; today, it is also a major consideration for soil remediation and global gases.

Spatial variability is treated separately in Chapter 9 in recognition of the importance of the heterogeneity of all soil properties; in fact, of all measurable natural quantities. In soil physics, the development of the quantitative aspects of variation is necessary for both site characterization and for predictions into the future. This has been an active area for connecting soil systems to varied disciplines, including remote sensing, hydrology and resource management.

# About the Editor

**Dr. Arthur W. Warrick** received a B.S. in mathematics followed by M.S. and Ph.D. degrees in soil physics at Iowa State University. He is currently Professor of Soil Science in the department of Soil, Water and Environmental Science at the University of Arizona, Tucson, with an Adjunct Appointment in Hydrology and Water Resources. He is a Fellow of the American Society of Agronomy, the Soil Science Society of America and the American Geophysical Union. His general areas of research are in the movement of water and solutes in soil and in the quantification of the variability of soil physical properties.

# Contributors

**D. A. Angers**
Soil and Crops Research and Development
  Center
Agriculture and Agri-Food Canada
Ste-Foy, Quebec, Canada

**Thomas Baumgartl**
Institute of Soil Science
University of Kiel
Kiel, Germany

**Steven R. Evett**
USDA/ARS/CPRL
Bushland, Texas

**Rainer Horn**
Institute of Soil Science and Plant Nutrition
Kiel, Germany

**B. D. Kay**
Department of Land Resource Science
University of Guelph
Guelph, Ontario, Canada

**Feike J. Leij**
U.S. Salinity Laboratory
Riverside, California

**Joel W. Massmann**
University of Washington
Seattle, Washington

**Alex B. McBratney**
Department of Agricultural Chemistry
  & Soil Science
University of Sydney
Sydney, Australia

**D. J. Mulla**
University of Minnesota
St. Paul, Minnesota

**Jean Phillippe Nicot**
Duke Engineering and Services
Austin, Texas

**Dani Or**
Plants, Soils, & Biometerology
Utah State University
Logan, Utah

**David E. Radcliffe**
Department of Crop and Soil Sciences
University of Georgia
Athens, Georgia

**Todd C. Rasmussen**
School of Forest Resources
University of Georgia
Athens, Georgia

**Bridget R. Scanlon**
Bureau of Econ. Geology
University of Texas
Austin, Texas

**Joseph M. Skopp**
Department of Agronomy
University of Nebraska
Lincoln, Nebraska

**Jon M. Wraith**
Land Resources Department
Montana State University
Bozeman, Montana

**Martinus Th. van Genuchten**
U.S. Salinity Laboratory
Riverside, California

# Table of Contents

**Chapter 5**    Water and Energy Balances at Soil–Plant–Atmosphere Interfaces

*Steven R. Evett*

**Chapter 6**    Solute Transport

*Feike J. Leij and Martinus Th. van Genuchten*

**Chapter 7**    Soil Structure

*B. D. Kay and D. A. Angers*

**Chapter 8**    Soil Gas Movement in Unsaturated Systems

*Bridget R. Scanlon, Jean Phillippe Nicot, and Joel W. Massmann*

**Chapter 9**    Soil Spatial Variability

*D. J. Mulla and Alex B. McBratney*

# 1 Physical Properties of Primary Particles

*Joseph M. Skopp*

## 1.1 INTRODUCTION

This chapter discusses the following physical properties of the primary particles: particle density, particle shape, particle size distribution and surface area. In addition, two soil properties related to packing are also presented: bulk density and porosity. The definitions and ideas behind these properties have built into them the concept or assumption of discrete primary particles as the primary soil constituent. If organic matter or amorphous materials and cementing agents are abundant, then the importance of primary particles is reduced.

## 1.2 PARTICLE DENSITY ($\rho_p$)

### 1.2.1 DEFINITION

The particle density represents one of the fundamental soil physical properties. Particle density is defined as the mass of soil particles divided by the volume occupied by the solids (i.e., excluding voids and water). Typical values for soils range from 2.5–2.8 Mg/m$^3$, with 2.65 Mg/m$^3$ being representative of many soils. Particle density provides few insights into soil physical processes. Consequently, one frequently overlooks the errors associated with particle density. However, its value enters into calculations of more useful soil properties, such as porosity and particle size distribution.

### 1.2.2 TYPICAL VALUES — MINERAL, WHOLE SOIL

Each individual soil mineral has a characteristic particle density. Values for different minerals can be found in Klein and Hurlbut (1985). Quartz, a predominant soil mineral, has a value of about 2.65 Mg/m$^3$ which is why this value is frequently given as representing all soils. In contrast, gypsum has a value of 2.32 Mg/m$^3$, biotite a value of 2.80–3.20 Mg/m$^3$ and hematite a value of 4.80–5.30 Mg/m$^3$. The particle density of a soil is an average for the distribution of soil minerals present.

A determination of particle density may be made for individual minerals, size fractions or the soil as a whole. Organic matter removal takes place in order to reduce variation. With standard procedures and removal of organic matter, a propagated error of less than ± .01 Mg/m$^3$ is possible. (Propagated error is the combination of all instrument errors when making a determination of a soil physical property.) Generally, determinate errors (i.e., biases) play a greater role in the analysis of particle density than indeterminate errors (i.e., random errors).

Perhaps the most interesting question in the determination of soil particle density is the role of organic matter (with a typical density of 1.0 Mg/m$^3$) in surface horizons. Most standard methods remove organic matter; thus the particle density reflects only the mineral phase. This value is the best value for use in particle size analysis, but may not be the best value for calculating porosity. Including organic matter means that changes in soil management may change this particle density.

### 1.2.3  METHODS — WATER PYCNOMETER, AIR PYCNOMETER, OTHER

Three methods of determining particle density will be examined. The most common determination of soil particle density uses a pycnometer or some variation. A pycnometer is any device which can be made to retain a reproducible or measurable volume. The soil sample is introduced into the pycnometer and the displaced volume of a fluid is determined. Water is typically the displaced fluid, but air can also be used.

The water pycnometer method generally requires the removal of organic matter prior to use. This avoids problems with trapped air and increased variability. The water pycnometer method requires good temperature control.

An alternative air pycnometer procedure uses a gas as the displacing fluid and the ideal gas law to calculate the volume of solids. The principles of an air pycnometer are straightforward, but care must be taken to prevent temperature changes. The air pycnometer does not require the removal of organic matter, which is particularly valuable when the total porosity (or total void space) or air-filled porosity is required without knowing soil bulk density or water content.

A less common method of determining particle density uses a vibrating tube which is filled with a solution or suspension. The resonant frequency of the vibrating tube provides a very precise means for determining the density of the suspension.

---

**EXAMPLE 1.1**   Particle density of whole soil. Is the particle density obtained from standard techniques suitable for determining total porosity?

The standard equation to calculate porosity requires the particle density. Standard methods require the removal of organic matter to carry out this procedure. The effective particle density including organic matter can easily be 10% less, which can result in a significant error.

---

## 1.3  PARTICLE SHAPE

### 1.3.1  DEFINITIONS — SHAPE FACTORS, FRACTALS

Particle shape influences specific surface, as well as particle size, analysis. It is also expected that particle shape has a strong influence on the packing of particles and soil strength, as well as transport properties. Unfortunately, particle shape is difficult to measure and few determinations exist in the literature.

A variety of terms exist to define particle shape. Some of the definitions require reference to a figure of an ideal particle (Figure 1.1). Let L, B and T represent the length, breadth and thickness of a particle. Then Heywood (1947) defines the following:

$$\text{Flatness Ratio} = B/T$$

$$\text{Elongation Ratio} = L/B$$

Other dimensionless terms are:

$$\text{Sphericity} = \frac{\text{surface area of equivalent sphere}}{\text{actual surface area}}$$

$$\text{Circularity} = \frac{\text{circumference of circle with the same area}}{\text{actual perimeter}}$$

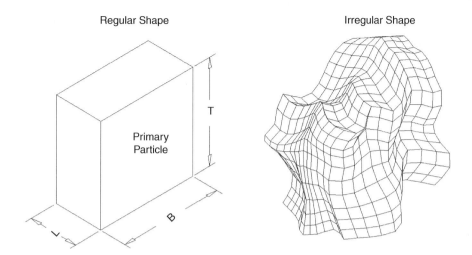

Regular Shape      Irregular Shape

**FIGURE 1.1** Ideal individual soil particle.

**TABLE 1.1**
**Shape factors for three-dimensional physical models with L = d, where d is a diameter or edge length, and h = disc height or rod length.**

| Geometry | Area | $C_2$ | Volume | $C_3$ |
|---|---|---|---|---|
| Sphere | $\pi d^2$ | $\pi$ | $\pi d^3/6$ | $\pi/6$ |
| Disc or cylinder | $(\pi dh + \pi d^2/2)$ | $(\pi h/d + \pi/2)$ | $\pi d^2 h/4$ | $\pi h/4d$ |
| Cube | $6d^2$ | $6$ | $d^3$ | $1$ |
| Square rod | $2d^2 + 4dh$ | $2 + (4h/d)$ | $d^2 h$ | $h/d$ |

$$\text{Rugosity} = \frac{\text{actual perimeter}}{\text{circumference of circumscribing circle}}$$

These last two terms work best for a two-dimensional image or the projection of the particle outline onto a flat surface.

A microscopic picture of soil describes either the solids or the voids. A simple picture of the solids might be a sphere or cube, while that of the voids might be a cylindrical tube or a slit between two flat surfaces. These pictures, or physical models, of the soil make it possible to deduce relations describing surface area, packing, water retention or water movement.

It is possible to apply these ideas to models that are not simple geometrical figures. Here, a characteristic length replaces the edge length of a cube and a dimensionless shape factor ($C_k$) describes the deviation of the shape from simple geometries (Table 1.1). These two factors are used to describe the area and volume as:

$$A = C_2 L^2 \quad \text{and} \quad V = C_3 L^3 \qquad [1.1]$$

where L is the characteristic length and $C_2$ or $C_3$ is the shape factor for area or volume, respectively. For a cube, if L equals d (the length of an edge), find: $C_2 = 6$ and $C_3 = 1$. Examples of $C_2$ and $C_3$ for other shapes are in Table 1.1. Two particles differing in size may or may not have the same shape. Typically, weathering or size reduction changes not only the total dimension but also the shape factor.

### 1.3.2 METHODS

At least three methods exist to examine particle shape. First is direct observations under a microscope or using an image analysis system. Commercial image analysis systems exist that automatically provide shape factors or similar properties.

The second method is an indirect technique from the variation of viscosity with the concentration of suspended particles. Increasing the solids concentration results in deviations from a pure liquid viscosity. These deviations are dependent on particle geometry as well as concentration of the suspension. The viscosity of the suspension ($\eta$) and the viscosity of pure solvent ($\eta_s$) usually behave as follows:

$$\eta/\eta_s = 1 + Kf \qquad [1.2]$$

where f is the volume fraction of suspended material and K is an empirical constant, which is a shape parameter (Einstein, 1906). For spheres, $K = 2.5$ is a constant and changes with the shape of the particle. Kahn (1959) applied this technique to examine the shape of clay particles. Similar techniques (Egashira and Matsumoto, 1981) provide estimates of a/b (the ratio of major to minor axis) for montmorillonite (200–300), kaolinite (15–25) and mica (10–20).

The third method of particle shape analysis uses the scattering characteristics of light passing through a soil suspension and relies on measurements of the angles of scattered light. Instruments commercially available for particle size analysis can be used.

---

**EXAMPLE 1.2**  Shape factor for perimeter. Define the shape factor ($C_1$) for perimeter (P) of a particle. Show two ways to estimate $C_1$ from the examination of microscopic images.

By analogy with Equation [1.1], $P = C_1L$ where L is the characteristic length or diameter. The shape factor is:

$$C_1 = P/L$$

The perimeter may be estimated directly. Or the area (A) of the particle's image is used to get the perimeter, $P = A/L$, and then:

$$C_1 = A/L^2$$

L must be carefully defined (e.g., minimum diameter, maximum diameter or some average). A can be a projected area on the image, or the three-dimensional surface of a particle. Note that calculations using these two equations will not necessarily yield the same shape factor.

---

## 1.4  PARTICLE SIZE DISTRIBUTION (PSD)

### 1.4.1 DEFINITIONS (INCLUDING CONCEPT OF TEXTURE)

Particle size distribution (PSD) is the most fundamental physical property of a soil and defines the soil texture. The particle sizes present and their relative abundance sharply influence most physical properties.

Soil particle size (or effective diameters) provide the basis for a classification system. A range of diameters may be given a special designation (e.g., 2.0 mm to 1.0 mm is very coarse sand). Typically the ranges form a logarithmic scale with particles in a given size range termed soil separates. The size boundaries vary with country or discipline. Comparisons of the names given to a size range are given elsewhere (Gee and Bauder, 1979; Sheldrik and Wang, 1993).

The phrase, equivalent diameters, is used to emphasize the role of the measurement technique in determining particle size. If identical particles are measured by different techniques, they may appear to have different diameters. It is conceivable that two soils with identical PSDs (as determined by a single method) will show differences in other physical properties resulting from distinct particle shapes.

Defining the diameter of an irregularly shaped particle is not a trivial task. A single parameter, the diameter, characterizes a smooth sphere. The symmetry of the sphere and its smoothness mean that no other information need describe it. As soon as a distortion of the sphere occurs (i.e., into a jelly bean) then at least three diameters are possible. Some particle size analysis methods may orient the particle into a preferred direction (e.g., settling of a particle in a liquid). Other methods (e.g., image analysis) may observe several possible orientations.

## 1.4.2 TYPICAL DISTRIBUTIONS

Typical data for a variety of soils are presented in Table 1.2. Interpretation of particle size analysis data requires either the drawing of graphs or the calculation of summary coefficients, which are discussed in Section 1.4.7.

Graphs of a PSD typically select the dependent variable as either the cumulative fraction up to a size or incremental fraction of soil in a size interval. The incremental fraction ($F_i$) is usually the mass of particles within a size interval ($X_{i-1}$ to $X_i$) divided by the total mass of solids with the index i specifying the size interval. The cumulative fraction ($G_i$) is the sum of all fractions for particle sizes less than the $X_i$ value.

A typical graphical expression of PSD uses the logarithm of particle diameter (Figures 1.2 and 1.3). The shape and position of the graph provides qualitative information about the soil texture. Soils frequently show a log normal distribution of particle sizes so that a graph of fraction versus the logarithm of particle diameters appears to be normally distributed.

## 1.4.3 DISPERSION AND FRACTIONATION

Particle size analysis (or mechanical analysis) consists of isolating various particle sizes or size increments and then measuring the abundance of each size. Most methods accomplish this in two steps. First, the soil is dispersed, or separated into individual primary particles. Second, the dispersed sample is fractionated, or the amounts of each size interval are measured.

There are three objectives of dispersion: 1) removal of cementing agents, 2) rehydration of clays, and 3) the physical separation of individual soil particles. It is sufficient to recognize that organic matter and amorphous minerals are the primary cementing agents. When either of these are present in large amounts (e.g., Histosols or Oxisols), then dispersion may be difficult or meaningless. Generally, soil dispersion occurs using a combination of chemical and mechanical means.

It is important to recognize that the fraction of soil that is a single size cannot be determined. What is detected is the fraction of soil within a particular particle size interval or the cumulative fraction of all particles less than a given size. The use of sieves typically determines the mass fraction (mass of particles in a size interval divided by total mass). Microscopic counting results in a number fraction (number of particles in a size interval divided by total number of particles). Photometric techniques typically determine the area fraction. Other methods result in volume or line fractions, depending on the sensing procedure. Thus, while all the methods are capable of observing PSDs, not all the methods provide results that are equal or directly comparable.

**TABLE 1.2**
**Particle size distribution of soil samples representing a variety of soil types from the United States. Mass percent of total sample for the indicated size class.**

| Soil* | Sand | | | | | Silt | | | | Clay |
|---|---|---|---|---|---|---|---|---|---|---|
| | Very Coarse 2–1 mm | Coarse 1–0.5 mm | Medium 0.5–0.25 mm | Fine 0.25–0.10 mm | Very Fine 0.1–0.05 mm | Coarse 50–20 µ | Medium 20–10 µ | Fine 10–5 µ | Fine 5–2 µ | 2–0 µ |
| Anthony | 18.05 | 13.71 | 17.68 | 12.93 | 8.92 | 7.41 | 2.69 | 2.20 | 1.37 | 15.04 |
| Ava | 0.53 | 0.56 | 1.25 | 0.82 | 0.67 | 12.80 | 21.69 | 15.71 | 9.63 | 30.63 |
| Chalmers | 0.74 | 0.62 | 1.67 | 1.38 | 2.52 | 19.42 | 20.18 | 11.18 | 7.44 | 34.89 |
| Davidson | 0.71 | 2.38 | 6.52 | 6.02 | 3.39 | 3.32 | 4.83 | 4.08 | 7.43 | 61.32 |
| Fanno | 8.45 | 4.87 | 2.40 | 9.96 | 9.06 | 5.92 | 4.27 | 4.05 | 5.56 | 46.46 |
| Kalkaska | 0.19 | 1.79 | 47.99 | 36.26 | 5.19 | 1.34 | 1.01 | 1.33 | 0.18 | 4.67 |
| Mohave | 15.25 | 11.30 | 12.40 | 8.02 | 5.42 | 30.36 | 5.00 | 1.34 | 0.43 | 10.45 |
| Molokai | 1.29 | 2.64 | 4.57 | 6.64 | 7.91 | 5.78 | 8.30 | 6.08 | 4.88 | 52.00 |
| Nickolson | 0.67 | 0.31 | 0.44 | 0.42 | 1.18 | 12.90 | 13.47 | 15.27 | 5.41 | 49.89 |
| Wagram | 7.48 | 20.70 | 32.06 | 21.81 | 5.84 | 2.00 | 1.37 | 1.59 | 3.31 | 3.84 |

* USDA Taxonomic Names are: Anthony = torrifluvents, Ava = fragiudalf, Chalmers = endoaquolls, Davidson = paleudult, Fanno = haplustalf, Kalkaska = haplargid, Mohave = calciargid, Molokai = eutrostox, Nickolson = fragiudlaf, Wagram = paleudult.

*Source:* From Hendricks, D.M., personal communication, 1997.

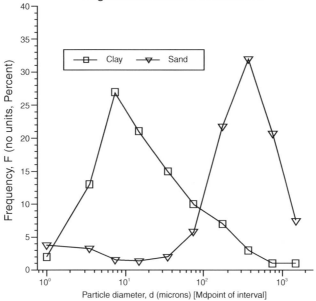

FIGURE 1.2 Frequency graph of particle size distribution.

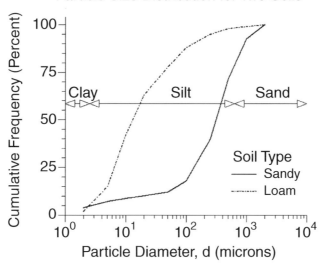

FIGURE 1.3 Cumulative frequency graph of particle size distribution.

### 1.4.4 Sieving

The process of sieving is that of placing the particles on a pattern of holes. Small particles may fall through while the sieve retains the larger particles. Either air or water may be the fluid to support the particles as they sort on the sieve. Dry sieving has a lower practical limit of 50 μm, while wet sieving can separate somewhat smaller particle sizes. Sieve holes may be square (using a wire cloth mesh) or round, although square holes are most common. The use of sieves with square openings will not result in measurements equivalent to those using sieves with round holes.

The use of words such as effective or nominal diameters with sieves is in recognition of the imperfect separation that may occur. Placement of a soil sample on a sieve does not result in instantaneous separation. Several factors influence the time to achieve a fixed level of separation. These factors include: sample size, shaking intensity, particle shape, particle size and hole geometry. Since samples vary in their sieving characteristics, it is best to run a trial sample. Errors on a single set of sieves typically are less than 1%, while comparisons between sieves show random errors of about 4%.

Many of the standard sieve sizes correspond to the class limits for USDA soil separates. Surprisingly, no standard sieve is available for the 50 μm cutoff between the sand and silt separates. Consequently, sieving cannot distinguish this class boundary using standard sieves.

### 1.4.5 Sedimentation

Below 50 μm, sieving is an inefficient and difficult procedure. For soil samples, sedimentation in water is one alternative procedure. A suspension of the dispersed sample settles in water, and at preselected times a measurement is made of the density of particles (mass of particles in a volume of liquid) at a specified depth within the sedimentation cylinder. Variations in the method occur as to the determination of suspension density. In all cases, Stokes' Law is central to the derivation of an equation which relates the time of settling to the size of particle sampled.

Two classic means of determining the density of a suspension exist: hydrometer method and pipet method. In the hydrometer method, the influence of density on a floating object (the hydrometer) is observed. As density decreases (due to settling out of soil particles), the hydrometer sinks. A calibration scale converts the depth of the float (i.e., hydrometer) to the suspension density (expressed as gL). The pipet method directly removes a sample from the suspension. The concentration of solids is determined by drying the pipeted sample. Nondestructive gamma ray absorption provides a commercially available alternative to both pipet and hydrometer. This method has the advantages of undisturbed and repeated sampling.

The hydrometer method uses higher concentrations of soil and may be less accurate than the pipet method. However, the hydrometer method allows repeated sampling at many points of the distribution (since no sample is withdrawn). Problems exist with the pipet method due to convection currents near the tip of the pipet, effect of sample removal and greater potential for operator error. These problems suggest that the hydrometer method may be preferable in some circumstances.

Stokes' Law for the viscous drag on a sphere is combined with buoyant and gravitational forces to obtain a settling equation. The particle shape is also assumed to be a sphere in evaluating the other forces. Combining forces and solving for v (the particle velocity) gives rise to the settling equation:

$$v = 2r^2g(\rho_p - \rho_w)/9\eta \qquad\qquad [1.3]$$

where r is the equivalent radius, g is the gravitational acceleration, $\rho_w$ is the fluid density, and $\eta$ is the viscosity.

The particle velocity is the distance traveled divided by the time, or x/t. The settling equation allows a particle radius to be calculated at a particular x and t, if the particle density and solution viscosity are known. This equation is basic to all gravity sedimentation procedures but requires a number of assumptions.

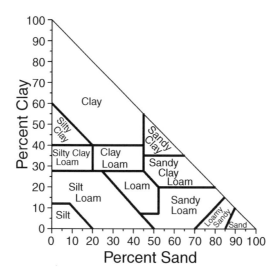

**FIGURE 1.4** Alternative texture triangle using clay (%) and sand (%) to determine texture class name. For example, a soil with 50% clay and 30% sand-sized particles would be assigned a textural class name of clay (from the Proceedings of the Soil Science Society of America, 1962, vol. 26, pg. 613. With permission.)

### 1.4.6 OTHER METHODS

Alternatives exist to fractionation by settling under the influence of gravity in water. Settling can be speeded up, through the use of a centrifuge. Settling can occur in air, called air elutriation. The settling equation for air elutriation is the same as for water except that the velocity is that of the air floating particles upward. Complete avoidance of settling may occur using microscopic methods. Image analysis in conjunction with microscopic methods can determine particle shape or geometry parameters. Other techniques include: conductometric (e.g., Coulter counter) and light scattering.

### 1.4.7 INTERPRETATION OF RESULTS

A graph can clearly present the elements of a PSD. However, it is desirable to have a more compact means to express the properties of the distribution. This is the origin of soil textural class which summarizes the particle size properties in a single phrase. The phrase chosen depends on the relative abundance of sand, silt and clay, irrespective of variations within these size ranges, through the use of a textural triangle (Figure 1.4). Typical textural triangles and the textural class names are given in Gee and Bauder (1986) and Loveland and Whalley (1991). Note that while the names of three soil separates are similar to the names of textural classes, a textural class name does not limit the sizes that may be present. In other words, a clay texture class contains sand, silt and clay separates while the clay separate contains only clay-sized particles.

The use of only the total amount of sand, silt and clay to describe texture results in a bias which is the assumption that all particles within the range of the sand (or silt or clay) size are equivalent. For the clay fraction, this is a particularly misleading assumption and is partly responsible for the low correlations frequently observed when using clay in regression analysis. Two soils with the same % clay may differ in the amounts of fine clay versus coarse clay. Additional differences in the shape and mineralogy of particles can also cause variations in soil behavior.

A more interesting approach determines the moments of the size distribution. This gives a quantitative measure of the mean, spread and asymmetry of a PSD. More importantly, it increases the power of correlative studies relating texture and any other physical, chemical or biological factors of interest.

Other measures exist in the engineering literature and elsewhere to describe a PSD. One approach defines the grain diameter ($D_n$) at which n% passes through a sieve. Therefore, $D_{40}$ represents that size for which 40% of the particles are smaller and 60% are larger. Various combinations of these $D_n$ characterize the distribution. One example is the uniformity coefficient:

$$C_u = D_{60}/D_{10} \qquad\qquad [1.4]$$

The uniformity coefficient provides information as to how narrow the distribution is, with 1. being the minimum when only a single size is present. The second ratio presented is the coefficient of gradation ($C_g$):

$$C_g = (D_{30})^2/D_{60}D_{10} \qquad\qquad [1.5]$$

Indirect descriptions of the particle size distribution are possible by using an equation or model. These models contain parameters which in turn characterize the distribution. The problem in using these models is first to determine the parameters, second, to determine the appropriateness of the model, and third, to interpret the parameters.

One model is of particular interest: the power function. The use of a power function suggests an underlying fractal process. Where this applies, the exponent ($\delta$) relates to the fractal dimension (n) as $\delta = 3 - n$, and n is between 0 and 3. Tyler and Wheatcraft (1992) apply this technique to several materials and report n values between 2 and 3. The log normal model and parameters have also been useful. Campbell (1985) uses these to estimate the water-holding properties of soil.

---

**EXAMPLE 1.3**   Particle size analysis be sedimentation. How long does a particle of diameter 0.1 mm take to fall 10 cm in water at 25°C?

Use the formula: $v = 2 (\rho_p - \rho_f)gr^2/9\eta$
Assume rapid steady state or: $v = x/t$ Solve for $t$:

$$t = 9\eta x/2(\rho_p - \rho_f)gr^2$$

Substitute for variables and assume: $\rho_p = 2.65 \ Mg/m^3$

**Viscosity and density from *CRC Handbook of Chemistry and Physics*.**

$$\eta = .8904 \ cp \text{ at } 25°C \quad \text{and} \quad \rho_f = .997 \ Mg/m^3 \text{ at } 25°C$$

$$g = 9.80 \ m/sec^2 \quad \text{and} \quad r = .05 \ mm$$

$$t = 9 \ (.8904 \ cp) \ (.10 \ m)/2(2.65 \ Mg/m^3 - 1.00 \ Mg/m^3)$$

$$* \ (9.80 \ m/sec^2) \ (.00005 \ m)^2$$

Convert units (this may be done prior to substitution):

$$t = 9 \ (.890 \ g/m\text{-}sec)(Mg/10^6 \ g)(.10 \ m)/2(2.65 \ Mg/m^3 - 1.00 \ Mg/m^3)$$

$$* \ (9.80 \ m/sec^2) \ (.00005 \ m)^2$$

Calculate and cancel units to obtain: $t = 9.91 \ sec$

---

**TABLE 1.3**
**Ranges of specific surfaces of clay minerals, selected soil components and whole soils compiled from a variety of sources.**

| Component | Specific Surface ($m^2/g$) |
|---|---|
| Kaolinite | 15–20 |
| Illites | 80–100 |
| Bentonite | 115–260 |
| Montmorillonite | 280–500 |
| Organic Matter | 560–800 |
| Calcite | .047 |
| Crystalline Iron Oxides | 116–184 |
| Amorphous Iron Oxides | 305–412 |
| **Soils** | |
| Sands | <10 |
| Sandy loams and silt loams | 5–20 |
| Clay loams | 15–40 |
| Clay | >25 |

## 1.5 SPECIFIC SURFACE AREA

### 1.5.1 DEFINITIONS

The surface area of the individual particles is an important factor in nutrient or pesticide adsorption, water absorption, soil strength and soil transport properties. The surface area of a soil has contributions from primary particles, amorphous mineral coatings and organic matter. These individual contributions may overlap and cancel. Further, the surface area of some expanding minerals may change with the water content and chemical composition of the soil solution.

The surface area is an extensive quantity (i.e., depends on how much soil is present). A more satisfying alternative is the introduction of an intensive quantity, the specific surface, which is either the surface area per mass ($S_m$) or per volume ($S_v$). The specific surface per volume changes with soil compaction.

### 1.5.2 TYPICAL VALUES — SEPARATES, WHOLE SOIL ($S_m$)

Table 1.3 shows typical values for specific minerals as well as values for whole soils. Amorphous materials and soil organic matter can greatly affect soil values. The whole soil values include the effects of particle size distribution as well as typical soil organic matter and amorphous mineral levels for temperate zone soils.

### 1.5.3 METHODS

Determination of surface area occurs in a number of ways. Direct measurements usually rely on the adsorption of either a gas (typically nitrogen or argon under high vacuum) or a liquid (historically ethylene glycol; but more recently, ethylene glycol monoethylether-EGME or water).

Either a multi- or monomolecular film is deposited or a gas adsorption isotherm is determined. A critical point in the use of all these procedures is the means by which we ensure a multi- or monomolecular layer. Control of the adsorbed phase occurs through regulation of the gas phase. More commonly for liquid adsorption, the use of a desiccator helps to fix the total pressure and partial pressure of the adsorbed component. In the EGME method a mixture of EGME and $CaCl_2$ regulates the vapor pressure of EGME. In water absorption a separate saturated lithium nitrate solution regulates the relative humidity. The lithium nitrate and EGME methods are convenient procedures for soils; however, both methods are sensitive to variations in temperature. Values for specific surface vary with the method used.

Another direct method examines photomicrographs of primary particles. The classical approach determines the probability of a needle (randomly placed on the photo) either falling within a pore or intersecting the pore edge. Current image analysis instrumentation allows this evaluation (through the particle size distribution) as part of many software packages. Unfortunately, most particle size distributions are not detailed enough in the smallest size range to accurately estimate specific surface.

### 1.5.4 MODELS

Models are used to describe how particle size influences specific surface and the general relation of particle geometry to specific surface. Starting with the specific surface $S_m = A/\rho_p V$, where A is the surface area of solids and V is the volume of solids, A and V are expressed in terms of the shape factors $C_2$ and $C_3$ with a characteristic dimension L:

$$S_m = C_2/\rho_p C_3 L \qquad [1.6]$$

This relation states that specific surface is the reciprocal of the characteristic length. The shape factors also influence surface area. Using Table 1.1, for spheres: $S_m = 6/\rho_p d$; while for a disc or flat plate: $S_m = [4 + 2(d/h)]/\rho_p d$. A large d/h ratio corresponds to a flat plate, while a small d/h ratio corresponds to a prism or needle-like shape. A ratio of d/h = 1 is identical to the result for spheres (Figure 1.5). The specific surface of fine clays may be one or more orders of magnitude greater than for coarse clays.

---

**EXAMPLE 1.4**   Specific surface area from particle shape factors. Derive a relation for the specific surface area of rod-shaped particles.

Start with the specific surface $S_m = A/\rho_p V$, where A is the surface area of solids and V is the volume of solids. Then express A and V in terms of the shape factors $C_2$ and $C_3$ from Table 1.1 with a characteristic dimension L:

$$S_m = C_2 L^2/\rho_p C_3 L^3$$

$$S_m = C_2/\rho_p C_3 L$$

$$S_m = [2+(4h/d]/\rho_p h$$

where h is the rod length (and characteristic length) and d rod edge length.

---

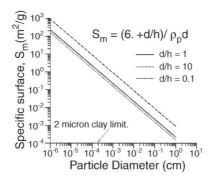

**FIGURE 1.5** Effect of particle diameter on specific surface for materials of a single size.

## 1.6  BULK DENSITY ($\rho_B$) AND POROSITY ($\varphi$)

### 1.6.1  Definitions

To quantify the state of compaction and the amount of pore space in a soil, the volume and mass of solids which are extensive quantities are related by the intensive term, bulk density ($\rho_B$):

$$\rho_B = \text{(mass of solids)/(volume of solids and voids)} \qquad [1.7]$$

The voids are the pore spaces that may hold either air, water or other liquids.

A related term to density is specific gravity. Specific gravity is defined as the ratio of the density (particle, bulk, fluid or any other density) to the density of water at 4°C and standard pressure. The specific gravity is dimensionless and the reference density of water is exactly 1.00 Mg/m$^3$.

### 1.6.2  Typical Values

The bulk density is a key physical property of any porous material and changes in response to disturbance or soil management practices. Yet there is a limit to any modification of bulk density. The particle size distribution along with packing controls the range of possible values.

Bulk density varies with the packing of the soil particles. Typically, sands pack more closely and values range from 1.4 to 1.9 Mg/m$^3$. Clays tend to bridge and cannot pack as tightly, giving values from 0.9 to 1.4 Mg/m$^3$. Textures between sands and clay vary in their bulk density accordingly. The wide range in bulk densities for a particular texture indicates that other factors (such as organic matter and compactive history) have an important influence on this property.

Field determinations of bulk density have relatively low precision (typically ± .05 Mg/m$^3$), which means that only about 10–20 different states of bulk density can be distinguished. An increase in measurement precision is inadequate because sampling bias and natural variability are similar in magnitude to typical measurement errors. Bulk density typically has coefficients of variations in the range of 10–40%.

Bulk density is highly dependent on soil conditions at the time of sampling. Changes in soil swelling due to changes in water content alters the bulk density. Thus, comparisons of bulk density must control or compensate for water content where swelling is significant. Other factors such as traffic patterns can also influence the bulk density. Consequently, a determination of bulk density may require the use of the above definition and the specification of the conditions at the time of sampling.

If the bulk density of a soil is fixed, then the relative amount of pore space is also fixed. To make this concept more precise, a term is needed to describe the amount of pore space (or voids). With pore space, volume rather than mass dimensions are more appropriate. Porosity as an intensive quantity is defined as:

$$\text{Porosity} = \varphi = \text{volume of voids/volume soil} \qquad [1.8]$$

Alternately, it can be shown that this definition is equivalent to the following (if the particle density includes organic matter):

$$\varphi = 1. - (\rho_B/\rho_P) \qquad [1.9]$$

This relation is the result of definitions and is not empirical. Using typical bulk densities, the total porosity of sandy soils is less than that of finer textured soils. It implies that for every value of bulk density, a given soil has only one possible value of the porosity. However, it is not true that a soil has only one possible value of the bulk density.

### 1.6.3 RELATED TERMS

A number of other expressions which characterize the amount of air in a soil are listed with their names, symbols and definitions:

$$\text{Air Filled Porosity} = \varphi_a = \text{volume air/volume soil} = \varphi - \theta_v \qquad [1.10]$$

where $\theta_v$ is the volumetric water content of a soil and defined in a later chapter.

$$\text{Void Ratio} = \text{volume of voids/volume of soils} = \varphi/(1 - \varphi) \qquad [1.11]$$

$$\text{Wet Density} = \rho_M = \text{(mass of solids + water)/volume of soil} \qquad [1.12]$$

Note that "wet" in the last expression refers to the inclusion of water in the numerator and not to how much water is present. It is possible to calculate the wet density of a dry soil and to convert between bulk density and wet density using the relation:

$$\rho_B = \rho_M - \theta_v \rho_W \qquad [1.13]$$

Wet density is not a preferred means of expressing the packing density of a soil.

### 1.6.4 FURTHER USES FOR BULK DENSITY

Another use for bulk density is the conversion of any gravimetric quantity (i.e., an intensive quantity that we express on a per gram basis) to a volumetric basis. For example, in Section 1.5 there are two kinds of specific surface: $S_v$ and $S_m$. These are surface areas per quantity of volume or mass. The explicit relation between the two is:

$$S_v = S_m \rho_B \qquad [1.14]$$

Some researchers try relating bulk density to factors like root penetration, soil strength and compaction (Table 1.4; SCS, 1981). These attempts generally meet with mixed success.

**TABLE 1.4**
**Approximate bulk densities that restrict root penetration.**

| Texture | Critical Bulk Density for Soil Resistance | |
|---|---|---|
| | High (Mg/m³) | Low (Mg/m³) |
| Sandy | 1.85 | 1.60 |
| Coarse-loamy | 1.80 | 1.40 |
| Fine-loamy | 1.70 | 1.40 |
| Coarse, Fine-Silty | 1.60 | 1.30 |
| Clayey | (Depends on both clay % and structure) | |

## 1.6.5 METHODS

Typically, soil bulk density is determined by inserting a ring into the soil. The ring is of known volume and upon extraction the soil core within the ring is dried to determine mass of solids and water present at the time of sampling. The major difficulties are first, the presence of stones or organic matter (possibly alive) and second, the compaction of the core while sampling may bias the volume. The effect of stones and compaction can be minimized by using a larger sampling ring.

There will be sites where a ring technique is not feasible. For example, coarse-textured soils may not remain in the ring or it may be difficult to drive the sampling ring to the desired depth. Further, since soil compressibility depends on water content, there will be times when sampling a particular site may show bias because the soil is too wet. The presence of live or dead roots also poses a problem, particularly near the soil surface and in soils managed with reduced tillage.

An alternative field method relies on hand-excavating soil. This results in an irregularly shaped hole whose volume we must determine. The hole is filled with a measurable volume (sand, an air-filled balloon or a water-filled balloon) which allows accurate calibration. Gamma ray attenuation methods have been developed for field use. Details can be found in references by Campbell and Henshall (1991) and Culley (1993).

Laboratory columns allow for greater precision in the direct determination of bulk density. All packed columns show systematic variations with depth which depend on the packing technique used. One alternative laboratory procedure uses gamma ray attenuation. Other techniques that may have application include computer assisted tomography (CT) and sensing of soil dielectric properties.

## REFERENCES

Anderson, S.H., R.L. Peyton and C.J. Gantzer. 1990. Evaluation of constructed and natural soil macropores using x-ray computed tomography. *Geoderma,* 46:13–29.
Baver, L.D., W.H. Gardner and W.R. Gardner. 1972. *Soil Physics,* 4th ed., John Wiley and Sons, New York.
Blake, G.R. and K.H. Hartge. 1986. *Methods of Soil Analysis,* pt. 1, chap. 14, A. Klute (ed.), ASA, Madison, WI.
Borggaard, O.K. 1982. The influence of iron oxides on the surface area of soil, *J. Soil Sci.,* 33:443–449.
Bower, C.A. and F.B. Gschwend. 1952. Ethylene glycol retention by soils as a measure of surface area and interlayer swelling, *SSSA Proc.,* 16:342–345.
Campbell, D.J. and J.K. Henshall. 1991. Bulk density in soil analysis physical methods, chap. 7, K.A. Smith and C.E. Mullins (eds.), M. Dekker, Inc., New York.
Campbell, G.S. 1985. Soil physics with basic, Elsener, Amsterdam.
Carter, D.L., M.M. Mortland and W.D. Kemper. 1986. Specific surface, *in Methods of Soil Analysis,* pt. 1, chap. 16, Physical and mineralogical methods, ASA, Madison, WI.

Culley, J.L.B. 1993. Density and compressibility in soil sampling and methods of analysis, chap. 50, M.R. Carter (ed.), *Canadian Society of Soil Sci. Lewis Publ.,* Boca Raton, FL, p. 823.

Davies, R. 1984. Particle size measurement: experimental techniques, *in Handbook of Powder Science and Technology,* chap. 2, M.E. Fayel and L. Otten (ed.), Van Nostrand Reinhold Co., New York.

Egashira, K. and J. Matsumoto. 1981. Evaluation of the axial ratio of soil clays from gray lowland soils based on viscosity measurements, *Soil Sci. Plant Nutr.,* 27:273–279.

Einstein, A. 1906. A new determination of molecular dimensions, *Annalen der Physik.,* 19:289–306, translated in: *Investigations on the Theory of Brownian Movement,* Dover Press.

Elder, J.P. 1979. Density measurements by the mechanical oscillator, *in Methods in Enzymology,* vol. 61, Academic Press, 12–25.

Elghamry, W. and M. Elashkar. 1962. Simplified textural classification triangles, *Soil Sci. Soc. Am. Proc.,* 26:612–613.

Folk, R.L. 1966. A review of grain-size parameters, *Sedimentology,* 6:73–93.

Gee, G.W. and J.W. Bauder. 1979. Particle size analysis by hydrometer: a simplified method for routine textural analysis and a sensitivity test of measurement parameters, *Soil Sci. Soc. Am. Proc.,* 43:1004–1007.

Gee, G.W. and J.W. Bauder. 1986. Particle size analysis, *in Methods of Soil Analysis,* pt. 1, chap. 15, Second ed., A. Klute (ed.), ASA, Madison, WI. 383–411.

Heywood, H. 1947. Symposium on particle size analysis, *Trans. Inst. Chem. Eng.,* 22:214.

Jensen, E. and H.M. Hansen. 1961. An elutriator for particle-size fractionation in the sub-sieve range, *Soil Sci.,* 92:94–99.

Kahn, A. 1959. Studies on the size and shape of clay particles in aqueous suspensions, *Clays Clay Miner,* 6:220–236.

Karsten, J.H.M. and W.A.G. Kotze. 1984. Soil particle size analysis with the gamma attenuation technique, *Commun, Soil Sci., Plant Anal.,* 15:731–739.

Klein, C. and C.S. Hurlbut, Jr. 1985. *Manual of Mineralogy* (after James D. Dana), 20th edition, Wiley, New York.

Loveland, P.J. and W.R. Whalley. 1991. Soil analysis physical methods, *in* K.A. Smith and C.E. Mullins, (eds.), M. Dekker, Inc., New York, 620.

Nelson, R.A. and S.B. Hendricks. 1943. Specific surface of some clay minerals, soils, and soil colloids, *Soil Sci.,* 56:285–296.

Orchiston, H.D. 1955. Adsorption of water vapor, III. Homoionic montmorillonites at 25°C, *Soil Sci.,* 79:71–78.

Santo, L.T. and G.Y. Tsuji. 1977. Soil bulk density and water content measurements by gamma-ray attenuation techniques, *Tech. Bull. 98,* Hawaii Agric. Expt. Stn.

Scheidegger, A.E. 1960. *The Physics of Flow through Porous Media,* revised edition, University of Toronto Press.

SCS. 1981. Tables for discussion of estimation of bulk density, Approximation 3, In-house report of NSSL, R.B. Grossman, Lincoln, NE.

Sheldrick, B.H. and C. Wang. 1993. Particle size distribution in soil sampling and methods of analysis, M.R. Carter, (ed.), chap. 47, *Canadian Society of Soil Sci. Lewis Publ.,* Boca Raton, FL, 823.

Shirazi, M.A. and L. Boersma. 1984. A unifying quantitative analysis of soil texture, *Soil Sci. Soc. Am. J.,* 48:142–147.

Streeter, V.L. and E.B. Wylie. 1975. Fluid mechanics, sixth edition, McGraw-Hill Inc., New York.

Suarez, D.L. and J.D. Wood. 1984. Simultaneous determination of calcite surface area and content in soils, *Soil Sci. Soc. Am. J.,* 48:1232–1235.

Svedberg, T. and J.B. Nichols. 1923. Determination of size and distribution of size of particles by centrifugal methods, *J. Am. Chem. Soc.,* 45:2910–2917.

Tanner, C.B. and S.J. Bourget. 1952. Particle-shape discrimination of round- and square-holed sieves, *Soil Sci. Soc. Am. Proc.,* 16:88.

Tyler, S.W. and S.W. Wheatcraft. 1992. Fractal scaling of soil particle-size distributions: analysis and limitations, *Soil Sci. Soc. Am. J.,* 56:362–369.

Uhland, R.E. 1949. Physical properties of soils as modified by crops and management, *SSSA Proc.,* 14:361–366.

# 2 Dynamic Properties of Soils

*Rainer Horn and Thomas Baumgartl*

## 2.1 INTRODUCTION

Soils undergo intensive changes in their physical, chemical and biological properties during natural soil development and as a result of anthropogenic processes such as plowing, sealing, erosion by wind and water, amelioration, excavation and reclamation of devastated land. In agriculture, soil compaction as well as erosion by wind and water are classified as the most harmful processes which not only end in a reduction of the productivity of the site, but are also responsible for groundwater pollution, gas emissions and higher energy requirements to obtain a comparable yield. In forestry, normal plant and soil management, tree harvesting and clear cutting affect site-specific properties including organic matter loss, groundwater pollution and gas emissions, which have the potential to cause global changes. Furthermore, soil amelioration, especially by deep tillage prior to replanting, often causes irreversible changes in properties and functions. These interrelationships have been described by Soane and van Ouverkerk (1994) and quantified by Hakansson et al. (1987). Oldeman (1992) showed that about 33 million ha of arable land are already completely devasted by soil compaction in Europe alone while the total area of degraded land worldwide exceeds about 2 billion ha. Physical (soil erosion and deformation) and chemical processes are responsible for about 1.6 and 0.4 billion ha of degraded soils, respectively. Worldwide population growth will reduce the average area per person for food and fiber production from 0.27 ha today to < 0.14 ha within 40 years. Consequently, a more detailed analysis of soil and site properties is needed to manage soils in accordance with their potential properties.

With respect to physical processes and soil degradation, Voorhees and Allmaras (1998), Morras and Piccolo (1998), and Horn (1998) postulated that soil erosion can be linked to soil tillage processes, because soil in seedbeds is susceptible and easily transported by wind or water. When soils become saturated, transport as sheet or gully erosion is inititiated, often down to the plow pan. If the tilled soil dries out, transport by wind may occur resulting in severe reductions in potential site productivity. Preparation of a seedbed leads to abrupt changes in the transport of gas, water, ions and heat between the tilled and deeper soil layers (Boone and Veen, 1994; Lipiec and Simota, 1994; Stepniewski et al., 1994). This is especially true in terms of preferential flow through structured soils. Such anthropogenic changes make the discussion of dynamics within the soil profile relevant. In addition, the interactions between soil structure, water status, and aeration of structured soils in relation to root growth and compressibility of arable land have been described by Emerson et al. (1978) and Horn et al. (2001).

Information on dynamic soil properties including the process of structure formation is important in classifying, using and sustaining soils in accord with their physical, mechanical, chemical or biological capacity.

## 2.2 PROCESSES IN AGGREGATE FORMATION

Soils containing more than 15% clay tend to form structural units known as aggregates by both physical and biological processes (Hillel, 1980; McKenzie, 1989). Aggregates may vary in size from crumbs (< 2 mm) to polyhedrons or subangular blocks (0.005–0.02 m) to prisms and columns

**TABLE 2.1**
**Methods for determining soil physical properties.**

| Method | Dimension | Soil Condition |
|---|---|---|
| Atterberg consistency test | water content (%, w/w) | homogenized soil |
| Proctor test | water content (%, w/w) | homogenized soil |
|  | bulk density (g cm$^{-3}$) | single aggregate |
| Mean weight diameter | (–) | partly homogenized soil samples |
| Wet sieving | length (cm) | single aggregates |
| Percolation |  |  |
| Irrigation |  |  |

> 0.1 m (Horn et al., 1989a; Becher 1992). Aggregates have either sharp rectangular edges or are defined by nonrectangular shear planes (Hartge and Horn, 1977; Hartge and Rathe, 1983; Babel et al., 1995). Aggregated soils are always stronger than homogenized material. Physical, chemical and biological processes vary between the inter- and intra-aggregate volumes. Thus the dynamics of hydraulic, mechanical, biological and chemical processes affect soil intensity properties strongly (Goss and Reid, 1979; Tippkötter, 1988, Dexter et al., 1988; Haynes and Swift, 1990, Horn et al., 1994). Furthermore, when strength is defined mechanically and/or hydraulically, one cannot extrapolate beyond the imposed limits which must be specified.

Structural properties are always in a dynamic equilibrium and influence nearly all site properties and functions including aeration, water infiltration, capillary rise, accessibility of particle surfaces for exchange, sorption reactions and biological activity (Junkersfeld and Horn, 1997). If leaching of nutrients or organic substances alters the properties of pore walls or the outer surfaces of aggregates, chemical changes (ionic strength) have to be considered in addition to mechanical properties to adequately deal with dynamic processes in structured soils. More detailed information on aggregate formation and changes in chemical properties is presented by Horn et al. (1994) and Kay (1990).

Mechanical processes and their dynamic aspects will be described and compared in homogenized and structured soils in order to highlight the effect of soil structure, tillage and timber harvesting on properties and processes in an ecosystem. Both capacity and intensity parameters and their measurements will be discussed in relation to mechanical properties. Thereafter, the effects of structure on ion sorption and desorption, and hydraulics will be discussed and modeled.

## 2.3   DETERMINATION OF MECHANICAL PARAMETERS

Soils consist of solid, liquid and gaseous phases; chemical, physical and biological processes determine site properties and functions depending on the degree of soil development. Capacity and intensity parameters and functions will be used to either define basic material properties or to quantify material functions. The latter include the definition of well defined limits (validity of the material properties, i.e., whether or not they are elastic or plastic changes due to mechanical loading) and the derivation of induced changes in physical, chemical and biological functions.

### 2.3.1   Capacity Measurements

Capacity methods quantify material properties which are constant but which differ by location in the field. The methods that can be used to compare mechanical properties can be found in Burke et al. (1986) and are summarized in Table 2.1.

### 2.3.1.1  Soil Consistency

Atterberg limits of homogenous soil material are defined as a function of water content on a mass basis ($\theta_m$) and are related to soil strength properties. The results are also used to predict soil workability.

*Liquid Limit*

The liquid limit is defined as the water content ($\theta_m$) at a certain amount of energy applied to the soil in the Casagrande apparatus (e.g., 25 strokes). Each set of samples is homogenized, placed in a special bowl and V shape trenched from top to bottom. Thereafter, the sample will be pushed up and down by the special equipment continuously and the number of strokes counted. As soon as the trench has been closed within 1 cm, the water content ($\theta_m$) is determined. This test is repeated for at least four water content values. The liquid limit increases with clay and organic matter contents, ionic strength, cation valency and proportion of 2:1 clay minerals in the soil.

*Plastic Limit*

The plastic limit is defined as the water content ($\theta_m$) when homogenized soil samples start to crack when rolled to a diameter < 4 mm.

*Plasticity Value*

The difference between the water content ($\theta_m$) at the liquid and plastic limits is the plasticity value, often used as an index of soil workability. Sensitivity to plastic deformation increases with increasing plasticity value; the smaller the value, the sooner the soil can be trafficked without further soil deformation. The higher the plasticity value, the smaller the angle of internal friction for sandy soils. Many attempts have been made to correlate the plasticity value to soil strength (Kezdi, 1969; Hartge and Horn, 1992; Kretschmer, 1997). In principle, this test only gives information on mimimum strength values.

### 2.3.1.2  Proctor Test

The Proctor test evaluates the effects of water content ($\theta_m$), and organic and mineral composition of homogenized soil on compactability. Optimum bulk density ($\rho_{opt}$) and water content ($\theta_{opt}$) for maximum compactability is determined at a given applied energy (e.g., $3 \times 25$ hammer strokes of known amplitude and weight), after a series of soil tests have been performed at different water contents. The coarser the soil sample, the higher the Proctor bulk density value at a smaller water content ($\theta_{opt}$). Sandy soils have higher values than silty, or clayey, soils, while the latter require a higher water content ($\theta_{opt}$) to reach the optimum bulk density ($\rho_{opt}$ = Proctor density). The more heterogeneous the grain size distribution, the higher the Proctor density at higher water content.

Strongly aggregated soils behave like coarser soil materials (the Proctor density is higher at a lower optimal water content). If aggregates themselves are destroyed during the test, the Proctor density gets smaller but the water content increases compared to originally completely homogenized material.

### 2.3.1.3  Mean Weight Diameter

Aggregate stability is determined by wet sieving. The soil samples are prewetted to a given pore water pressure and then sieved through a set of sieves from 8 to 2 mm diameter. The difference between the aggregate size distribution at the beginning and end of sieving under water for a given time is calculated as the mean weight diameter (MWD). This value is qualitatively related to aggregate strength and increases with increasing aggregate stability.

#### 2.3.1.4  Penetration Resistance

Soil resistance to any kind of deformation is determined by various types of penetrometers. The most simple is a thin metal rod (< 0.5 cm diameter) with a defined tip shape. Frequently, the tip angle is 30° to simulate root properties or earthworm shapes. The penetrometer can be either pushed into the ground by the constant weight of a falling hammer or it can be driven by a motor at constant speed. The output readings can be either penetration depth per hammer stroke, or depth stress depletions which have to be overcome by the penetrating body in more sophisticated models. Penetration resistance is correlated with root growth, earthworm activity and tillage effects. When penetration resistance exceeds 2 MPa, root growth is often reduced by half, while values > 3 MPa often prevent root growth. Tillage may increase the critical stress value of a hardpan to > 3.5 MPa depending on the nature of the pore system and the type of soil structure. Because the penetrometer needle is not as flexible as a root, which can choose planes of weakness for growth, penetrometer readings quantify resistances mainly in the vertical direction. In addition, the penetrometer readings only integrate impeding effects and cannot identify the causes. Despite a voluminous literature on the effect of increasing bulk density and/or water content on penetration resistance, extrapolation to land management situations is limited. Penetrometer readings can be used to create maps of derived properties (e.g., definition of sites with a given strength irrespective of its origin). Such data can be interpreted using statistical variograms, fractal analyses or simply by stating that values are spatially different.

#### 2.3.1.5  Other Methods

*Bulk Density*

In completely homogenized and/or structureless sandy soils, bulk density can be used as a first approximation of soil strength. However, because the dimensions of bulk density are mass per unit volume and not mass in a given constant volume, the interpretation is limited when dealing with structured and/or unsaturated soils. No relationship between strength and bulk density exists in aggregated soils nor is it possible to derive other properties from such data except in the case of a seedbed or newly disturbed sites.

*Structural Stability in Alcohol/Water Mixtures*

More detailed information on this very qualitative method is presented by Hartge and Horn (1992) and Burke et al. (1986).

#### 2.3.2  Intensity Parameters

Soil formation including aggregate development involves changes in both physical and mechanical properties and therefore requires the exact definition of the limits within which properties are quantified. This is true, because *in situ* soil formation processes have to be linked to internal and external conditions (climatic, mechanical, physical, hydrological or chemical aspects) for a particular situation. For example, all properties such as soil strength, stress attenuation, changes in soil structure or pore distribution, water and ion fluxes, gas exchange and accessibility of exchange sites for cations are material functions (with well-defined and quantified limits). Consequently, in order to deal with dynamic soil properties, stress, strain and strength definitions are initially required to later define the limits of the material functions with respect to the application of external stresses.

---

**EXAMPLE 2.1**

1. Explain the main processes in aggregate formation and specify their influences on hydraulic, pneumatic, thermal, and chemical properties.
2. How does structure formation in soils change soil mechanical parameters?
3. Why express capacity and intensity parameters in soil physics various soil states?

### 2.3.2.1  Stress Theory

*Definitions*

Before discussing methods of field and laboratory stress measurement as well as factors influencing compaction, one needs to differentiate between several terms used to define dynamic compressive properties. These definitions have been taken from Kezdi (1969), Bradford and Gupta (1986), Hartge and Horn (1991), and Fredlund and Rahardjo (1993).

*Force* applied to a soil per unit area is defined as stress.

*Stresses* working along the surface will also induce stresses in the soil, which may result in a three-dimensional deformation of the soil volume or will be transmitted as a rigid body. The mechanical behavior of a soil (volume change and shear strength) can be described in terms of the soil stress state.

*Strength* quantifies mechanically based material functions and depends on internal parameters such as particle size distribution, type of clay mineral, nature and amount of adsorbed cations, content and type of organic matter, aggregation induced by swelling and shrinking, stabilization by roots and humic substances, bulk density, pore size distribution and pore continuity of the bulk soil and single aggregates, water content and/or water suction (Horn, 1981).

*Stress State*

The number of stress state variables required to define the stress state depends primarily upon the number of phases involved. The effective stress $\sigma'$ can be defined as a stress variable for saturated conditions and is the difference between the total ($\sigma$) and the neutral stress ($u_w$) which is equal to the pore water pressure:

$$\sigma' = \sigma \pm u_w \qquad\qquad [2.1]$$

where $\sigma'$ is transmitted by solid and ($u_w$) by the liquid phase, respectively. In unsaturated soils, stresses are transmitted by the solid, liquid and gaseous phases. Thus, Equation (2.1) becomes:

$$\sigma' = (\sigma - ua) + X\,(u_a - u_w) \qquad\qquad [2.2]$$

where $u_a$ and $u_w$ are pore air and water pressures, respectively, and X is a factor which depends on the degree of saturation. At 0 kPa water suction ($\psi m = 0$) X = 1, while at $\psi m = 10^6$ kPa, X = 0.

For sandy, less compressible and nonaggregated soils, X can be calculated as:

$$X = 0.22 + 0.78\ Sr \qquad\qquad [2.3]$$

where Sr is degree of saturation. For silty and clayey soils, the values of the parameters in Equations (2.1) and (2.2) depend on soil aggregation, pore arrangement and strength, and hydraulic properties. Thus the material function of the components in structured soils is only valid as long as the internal soil strength is not exceeded by the externally applied stresses. It changes if, for example, aggregates are destroyed during soil deformation and the structure properties are reduced to those which depend on texture.

In current discussions on sustainability, soil compaction is repeatedly mentioned as one of the main threats to agriculture and forestry which should be avoided (Soane and van Ouwerkerk, 1994; Horn et al., 1995). The extent of soil deformation can be predicted by stress strain processes and by their relative proportions. In the absence of gravitational and other applied forces, stresses in three-phase soil systems are divided into three normal stresses and six shearing stresses acting on a cube; at equilibrium the shearing forces reduce to three. Therefore, three normal stresses ($\sigma_x$, $\sigma_y$, $\sigma_z$) and three shearing stresses ($\tau_{xy}$, $\tau_{xz}$, $\tau_{yz}$) must be determined to define the stress state at a point (Nichols et al., 1987; Horn et al., 1992; Harris and Bakker, 1994).

The stress state can be desribed completely by a symmetric matrix:

$$\begin{bmatrix} \sigma_x & \tau_{xy} & \tau_{xz} \\ \tau_{xy} & \sigma_y & \tau_{yz} \\ \tau_{xz} & \tau_{yz} & \sigma_z \end{bmatrix}$$

[2.4]

For symmetric matrices it is always possible to find a coordinate system, in which the matrix becomes diagonal. In this principal axis system the stress matrix simplifies to

$$\begin{bmatrix} \sigma_1 & 0 & 0 \\ 0 & \sigma_2 & 0 \\ 0 & 0 & \sigma_3 \end{bmatrix}$$

[2.5]

with principal stresses $\sigma_1$, $\sigma_2$, and $\sigma_3$. For a simpler characterization of the stress state two invariants of the stress matrix are often used, the mean normal stress MNS and the octahedral shear stress OCTSS (Koolen 1994):

$$MNS = \tfrac{1}{3}\left(\sigma_1 + \sigma_2 + \sigma_3\right)$$

[2.6]

$$OCTSS = \tfrac{2}{3}\sqrt{\left(\sigma_1 - \sigma_2\right)^2 + \left(\sigma_2 - \sigma_3\right)^2 + \left(\sigma_3 - \sigma_1\right)^2}$$

[2.7]

*Stress Propagation*

Each applied stress is transmitted into the soil in three dimensions and can alter the physical, chemical and biological properties if internal mechanical strength was exceeded. The type of external force applied, time dependency and number of compaction events can either change properties to depths by divergent processes or destroy a given structure by shear forces such as kneading. The latter case may result in complete homogenization and normal shrinkage behavior. (Horn 1988).

Stress propagation theories are rather old and have been often modified and adapted to *in situ* situations. The fundamental theories of Boussinesque (1885) are only valid for completely elastic material, while Fröhlich (1934) or Söhne (1953) (all cited in Horn, 1988) included elastoplastic properties through the introduction of concentration factor values. More complete descriptions of these models are given by Koolen and Kuipers (1985), Bailey et al. (1986, 1992), and Johnson and Bailey (1990).

### 2.3.2.2  Strain Theory

Every change in stress state results in soil deformation. If stress exceeds the internal soil strength, the plastic (irreversible) portion increases strongly. Analogous to stress, strain can be described by normal ($\varepsilon_x$, $\varepsilon_y$, $\varepsilon_z$) and shear strain ($\varepsilon_{xy}$, $\varepsilon_{xz}$, $\varepsilon_{yz}$), being described completely by the symmetric strain matrix:

$$\begin{bmatrix} \varepsilon_x & \varepsilon_{xy} & \varepsilon_{xz} \\ \varepsilon_{xy} & \varepsilon_y & \varepsilon_{yz} \\ \varepsilon_{xz} & \varepsilon_{yz} & \varepsilon_z \end{bmatrix}$$

[2.8]

In the corresponding principal axis system the strain matrix reduces to:

$$\begin{bmatrix} \varepsilon_1 & 0 & 0 \\ 0 & \varepsilon_2 & 0 \\ 0 & 0 & \varepsilon_3 \end{bmatrix}$$  [2.9]

The strain matrix completely describes the local soil deformation. For example, the volumetric strain ($\varepsilon_{vol}$) can be calculated as:

$$\varepsilon_{vol} = \varepsilon_1 + \varepsilon_2 + \varepsilon_3 = \varepsilon_x + \varepsilon_y + \varepsilon_z$$  [2.10]

(Note that the trace of a matrix is invariant under coordinate transformations.) Furthermore, the strain components and their proportions depend on internal and external parameters and require the determination of all components in a three-dimensional volume.

### 2.3.2.3 Stress/Strain Processes

Generally, mechanical processes in soils are described by the stress-strain-equation:

$$\begin{bmatrix} \sigma_x(\vec{x},t) & \tau_{xy}(\vec{x},t) & \tau_{xz}(\vec{x},t) \\ \tau_{xy}(\vec{x},t) & \sigma_y(\vec{x},t) & \tau_{yz}(\vec{x},t) \\ \tau_{xz}(\vec{x},t) & \tau_{yz}(\vec{x},t) & \sigma_z(\vec{x},t) \end{bmatrix} = f \begin{bmatrix} \varepsilon_x(\vec{x},t) & \varepsilon_{xy}(\vec{x},t) & \varepsilon_{xz}(\vec{x},t) \\ \varepsilon_{xy}(\vec{x},t) & \varepsilon_y(\vec{x},t) & \varepsilon_{yz}(\vec{x},t) \\ \varepsilon_{xz}(\vec{x},t) & \varepsilon_{yz}(\vec{x},t) & \varepsilon_z(\vec{x},t) \end{bmatrix}$$  [2.11]

The function f defines soil properties. In the case of nonlinear isotropic elastic properties, the matrix notation reads as follows:

$$\begin{bmatrix} \sigma_x & \tau_{xy} & \tau_{xz} \\ \tau_{xy} & \sigma_y & \tau_{yz} \\ \tau_{xz} & \tau_{yz} & \sigma_z \end{bmatrix} = \frac{E}{1+v} \cdot \begin{bmatrix} \varepsilon_x - \varepsilon_m & \varepsilon_{xy} & \varepsilon_{xz} \\ \varepsilon_{xy} & \varepsilon_y - \varepsilon_m & \varepsilon_{yz} \\ \varepsilon_{xz} & \varepsilon_{yz} & \varepsilon_z - \varepsilon_m \end{bmatrix} + \frac{E}{3 \cdot (1 - 2 \cdot v)} \cdot \begin{bmatrix} \varepsilon_m & 0 & 0 \\ 0 & \varepsilon_m & 0 \\ 0 & 0 & \varepsilon_m \end{bmatrix}$$  [2.12]

where
$E$ = Young's modulus
$n$ = Poisson's ratio
$\varepsilon_m = \frac{1}{3} \cdot \left( \varepsilon_x + \varepsilon_y + \varepsilon_z \right) = \frac{1}{3} \cdot \varepsilon_{vol}$

### 2.3.2.4 Definition of Soil Deformation

Considering the soil as a continuum soil movement is described as a translation field $\vec{d}(\vec{x},t)$ whose local properties are usually characterized by the three components of its spatial derivative:

$$\nabla \circ \vec{d} = \begin{bmatrix} \frac{\partial}{\partial x} d_x & \frac{\partial}{\partial x} d_y & \frac{\partial}{\partial x} d_z \\ \frac{\partial}{\partial y} d_x & \frac{\partial}{\partial y} d_y & \frac{\partial}{\partial y} d_z \\ \frac{\partial}{\partial z} d_x & \frac{\partial}{\partial z} d_y & \frac{\partial}{\partial z} d_z \end{bmatrix}$$  [2.13]

1. rotation

$$rot\left(\vec{d}\right) = \nabla \times \vec{d} = \begin{bmatrix} \frac{\partial}{\partial y} d_z - \frac{\partial}{\partial z} d_y \\ \frac{\partial}{\partial z} d_x - \frac{\partial}{\partial x} d_z \\ \frac{\partial}{\partial x} d_y - \frac{\partial}{\partial y} d_x \end{bmatrix}$$ 

[2.14]

2. compaction/decompaction = divergence = volumetric strain

$$\varepsilon_{vol} = div\left(\vec{d}\right) = \nabla \cdot \vec{d} = tr\left(\nabla \circ \vec{d}\right) = \frac{\partial}{\partial x} d_x + \frac{\partial}{\partial y} d_y + \frac{\partial}{\partial z} d_z$$ 

[2.15]

3. shearing

$$\varepsilon_s = \frac{1}{2}\left(\nabla \circ \vec{d} + \left(\nabla \circ \vec{d}\right)^T\right) - \frac{1}{3} div\left(\vec{d}\right) \cdot \begin{bmatrix} 1 & 0 & 0 \\ 0 & 1 & 0 \\ 0 & 0 & 1 \end{bmatrix}$$ 

[2.16]

While rotation does not result in any deformation or change in soil volume (except the rotation of the principal axis system of any anisotropic material property), volumetric strain and shearing result in a deformation (Figures 2.1 and 2.2). Thus, soil deformation is a process more complex than a volume reduction which is the same as volumetric strain expressing only one degree of freedom, whereas deformation at constant volume (shearing) summarizes 5 degrees of freedom. Although it is useful to distinguish between shear and compaction/decompaction processes as they typically exhibit very different effects, all deformations in soils are combinations of both utilizing the full range of 6 degrees of freedom.

Compression refers to a process that describes the increase in soil mass per unit volume (increase in bulk density) under an externally applied load or under changes of internal pore water pressure. Examples of externally applied static or dynamic loads are vibration, rolling, trampling, etc., while internal forces per unit area include such things as pore water pressure or suction caused by a hydraulic gradient.

In saturated soils, compression is called consolidation, while in unsaturated soils, it is called compaction. Consolidation, therefore, depends on the drainage of excess soil water determined by hydraulic conductivity and gradient. However, during compaction less compressible air will be expelled as a function of air permeability, pore continuity and water saturation in the profile. Consolidation tests are therefore mainly used in civil engineering (e.g., road construction) and have only limited application to agricultural soils.

Soil compressibility described by the shape of the stress strain curve, is defined as a resistance to a volume decrease, when the soil is subjected to a mechanical load. Compaction tests are used both for laboratory and for field soil compression characterizations.

In the laboratory, soil compaction refers to the compression of small soil samples, whereas in the field three-dimensional changes in bulk density, pore size distributions and strength of an elemental soil volume to depth and laterally are involved (Gräsle and Nissen, 1996). In laboratory compaction, the optimum diameter-to-height ratio of the test cylinder should be ≥ 5 in order to avoid friction between cylinder wall and sample. Compaction tests to determine soil strength can be carried out on homogenized, undisturbed or bulk soil samples or single aggregate samples at different pore water pressure (i.e., water suction) (Baumgartl, 1991).

Compactability is the difference between the initial and maximum densities to which a soil can be compacted by a given amount of energy at a defined water content.

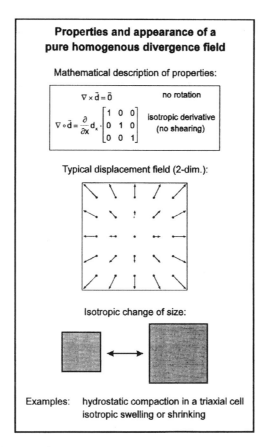

**FIGURE 2.1** Processes of compaction and decompaction and their mathematical characterization including some examples.

### 2.3.3 STRESS MEASUREMENTS *IN SITU*

*Stress Distribution*

Stress distribution measurements on undisturbed soil profiles under traffic with different speeds, loads and contact areas describe the type and intensity of soil strength, stress attenuation and soil deformation. In unsaturated soils, pore water pressure at the time of loading further affects these parameters. One of the major problems in validating compaction models is sensor installation at different depths and distances from the perpendicular line. Because of the structural disturbance due to excavation and installation of the sensors, the data obtained only describe the stress patterns for homogenized or artificially mixed soil. Thus, only in nonaggregated sandy soils can one expect to find minimal effects due to installation. In aggregated soils, sensors must be installed in undisturbed soil from the side to minimize structural disturbance so that octahedral normal and shear stresses can be obtained. The pressure sensor is a foreign body with deformation properties different than those of the soil. If the pressure sensor itself is weaker than the soil, the registered stress will underestimate the real stress at that depth. If, however, the pressure sensor stiffness exceeds that of the surrounding soil, stress concentrates on the more rigid transducer body and, therefore, the real soil stress is overestimated. The different types of stress transducers available are described in Table 2.2.

Plastic bodies in the form of pneumatic or hydraulic cylinders, balls or discs made of silicone or rubber change their volume in response to the applied stress. Prior to measurement, pressure cells must be filled with water or air and prestressed to 80 kPa which affects the stress/strain

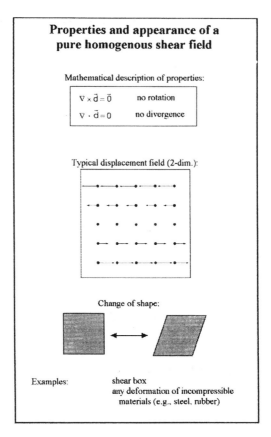

**FIGURE 2.2** Process of shearing including its mathematical characterization. Pure shearing results in no change in density, but changes in angles between particles. Consequently, there is no divergence and no rotation.

**TABLE 2.2**
**Different types of stress state transducers.**

| Principle | Material | Size | Deformation | Measured Values | Calculated Values | Author |
|---|---|---|---|---|---|---|
| Pneumatic | rubber | cylinder | plastic | soil stiffness | — | Kögler (1933) |
| Hydraulic | rubber | disk | plastic | 1 defined stress | — | Söhne (1951) |
| Hydraulic | steel | disk | elastic | 1 defined stress | — | Franz (1958) |
| Hydraulic | rubber | sphere | plastic | mean normal stress | — | Blackwell (1978) |
| Hydraulic | silicon | cylinder | plastic | mean normal stress | — | Bolling (1984) |
| Strain gauge | silicon | sphere | plastic | mean normal stress | — | Verma (1975) |
| Strain gauge | steel | disk | elastic | 1 defined normal stress | — | Cooper (1975) |
| Strain gauge | aluminum | disk | elastic | 1 defined normal stress | — | Horn (1980) |
| Strain gauge | steel | cube | elastic | 3 defined normal stresses | — | Prange (1960) |
| Strain gauge | aluminum | half sphere | elastic | 6 defined normal stresses | $\sigma_{1,2,3}$ $\sigma_{xy,yz,xz}$ | Nichols et al. (1987) Horn et al. (1992) Harris and Bakker (1994) |
| Strain gauge | aluminum | sphere | elastic | 6 defined normal stresses | $\sigma_{1,2,3}$ $\sigma_{xy,yz,xz}$ | Kühner (1997) |

*Source:* From Bolling, I., 1986, modified.

modulus of the sensor and the measured pressure value (Horn et al., 1991). Theoretically, the sensor elasticity should be the same as that of the surrounding undisturbed soil, which is almost impossible to achieve. Generally, plastic sensors are weaker than the soil and consequently, stresses will be underestimated. A plastic stress transducer measures an average normal stress. The direction of stresses cannot be identified if cylindrical or spherical transducers are used and shear stresses cannot be determined.

When rigid bodies are used as stress transducers, piezoelectric materials or strain gauges are placed on diaphragm material made of aluminum or steel. In comparison to plastic stress transducers, the stress/strain modulus of a rigid transducer cannot be matched to the surrounding soil. Peattie and Sparrow (1954) have suggested that the optimum ratio of the transducer stress-strain modulus to that of the soil should be 10.

The determination of the stress distribution in normal soils using strain gauges has been described by Horn (1981), Burger et al. (1987), van den Akker et al. (1994) and Blunden et al. (1994). Ellies et al. (1995,1996) and Ellies and Horn (1996) tested the applicability of such techniques for volcanic ash soils which behave in a manner different from that of normal mineral soils. Horn (1995) summarized the physical/mechanical properties of deep tilled soils and quantified the effect of traffic on stress distribution. Kühner et al. (1994) and Kühner (1997) have quantified the stress distribution under various *in situ* conditions and Horn (1998) defined stress and strength conditions for soils under different land use systems. Blunden et al. (1992,1994), Kirby et al. (1997), Trein (1995), Kühner (1997) and van den Akker et al. (1994) describe various other aspects of stress distribution and attenuation in soils leading to soil protection strategies and engineering solutions. The stress distribution in partially deep tilled soils (slit plow) is also described by Horn et al. (1998). Olson (1994) found that even a slight increase in moisture content or contact area always resulted in a large increase in the concentration factor value. Because measurements and calculations by Blunden et al. (1992,1994), Kirby et al. (1997), Trein (1995) and van den Akker et al. (1994) were based on measurements of only the vertical component of applied stress, the missing data was obtained by modeling in order to predict stress state in soils. However, such an approach can only verify processes which are assumed to be dominant in homogenous soil systems (seedbed), while analysis of stress distribution in unsaturated structured soils is a more complex problem with a multitude of processes operating in an interactive fashion. This is particularly true for unsaturated, nonlinear, hysteretic soils whose composition changes with time or during tillage, thus altering their physical and chemical properties. Consequently, for the determination of stress paths, a Stress State Transducer (SST) system (6 normal stresses on three mutually orthogonal and three nonorthogonal planes) can be used to measure all stress components at a single point. Based on continuum mechanics theory, octahedral principle and shear stresses can be partly measured and partly calculated for a cube cut from the continuum. It should be stressed that such stress propagation cannot be handled in the same way as for soil parent material (i.e., completely homogenized or stiff material as in soil mechanics).

### 2.3.3.1 Strain Determination

The principle involved in the determination of soil strain has been described by Koolen and Kuipers (1985). One of the earliest systems to determine the volumetric strain path *in situ* during traffic was that of Gliemeroth (1953), while those of McKibben and Green (1936), Gill and van den Berg (1967), van den Akker and Stuiver (1989) and Okhitin et al. (1991) could only be used to determine position changes of inserted particles (colored sticks, spheres, etc.) relative to their original positions. The calculation of the strain path under *in situ* conditions is not possible if the adjacent soil is disturbed during installation and if one is unable to characterize sensor movement during deformation or if sensors are missing in various positions; consequently, physical properties cannot be predicted (Erbach et al., 1991). Subsequently, Kühner et al. (1994) developed a new Deformation Transducer System (DTS), which when installed in the undisturbed soil volume prior to traffic,

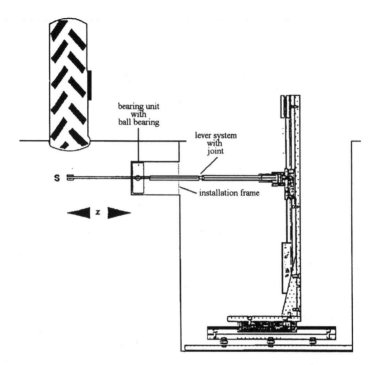

**FIGURE 2.3** Combined stress/strain apparatus to determine volumetric stresses and biaxial strain during traffic (Kühner, 1997).

records the nature of stress-induced movements of soil particles. If four DTSs are installed at various positions and distances from the wheel rut, both the strain path and volumetric strain matrix can be quantified.

#### 2.3.3.2  Stress/Strain Determination

The stress/strain apparatus described by Kühner (1997) (Figure 2.3), consists of a SST sensor block, which is connected to a mobile measuring device in order to determine movements in the x and z directions. A data logging system with a frequency of approximately 40 observations $s^{-1}$ is used to record stresses and sensor displacement in both directions.

### 2.3.4  MEASUREMENTS OF SOIL STRENGTH

The determination of soil strength parameters requires measurements under well-defined laboratory conditions (Table 2.3).

#### 2.3.4.1  Uniaxial Compression Test

The uniaxial compression test is used to define the pressure at which soil begins to fail at a given water content. A vertical normal stress ($\sigma_1$) is applied to the soil sample, while the stresses on the planes mutually perpendicular to the $\sigma_1$ direction ($\sigma_2 = \sigma_3$) are zero. The uniaxial compression test is often used to determine the tensile strength of single aggregates (crushing test).

#### 2.3.4.2  Confined Compression Test

The soil strength relationships of undisturbed, homogenized soils and single aggregates are quantified in the confined compression test. In contrast to the uniaxial compression test, stresses in the

**TABLE 2.3**
**Methods of determining soil strength.**

| Method | Dimension | Derived | Soil Condition |
|---|---|---|---|
| Uniaxial compression | pressure (Pa) | — | homogenized soil single aggregates structured bulk soil |
| Confined compression test | pressure (Pa) | precompression stress (Pa) | homogenized soil structured bulk soil |
| Triaxial test | pressure (Pa) | cohesion (Pa) angle of internal friction (°) | homogenized soil structured bulk soil single aggregates |
| Direct shear test | pressure (Pa) | cohesion (Pa) angle of internal friction (°) | homogenized soil structured bulk soil single aggregates |

$\sigma_2$ and $\sigma_3$ direction are undefined (rigid wall of the soil cylinder). Both the time and load dependent changes in soil deformation are measured; the slope of the virgin compression line (i.e., the compression index), and the transition from the overconsolidated to the virgin compression line (i.e., the precompression stress) can be determined. The precompression stress is defined as the stress value at the intersection of the less steep recompression curve and the virgin compression line. The latter straight line portion has a steeper slope if plotted on a semilog scale. Many methods are available to determine the precompression stress but that of Casagrande is most frequently used (Bölling, 1971).

### 2.3.4.3 Triaxial Test

Undisturbed cylindrical soil samples are loaded with an increasing vertical principal stress ($\sigma_1$), while the horizontal principal stresses ($\sigma_2 = \sigma_3$) are defined and kept constant. Shear stresses occur in any plane other than those of the principal stresses. The shear parameters (cohesion and angle of internal friction) can be determined from the slope of the Mohr's circles failure line. Due to aggregate deterioration and prevented drainage of excess soil water, the Mohr Coulomb failure line is bent toward smaller slope values. However, the number of contact points, strength per contact point and pore geometry affect triaxial test results. There are three types of triaxial tests:

1. In the consolidated drained test (CD), the soil sample is equilibrated with the mean normal stresses prior to an increase in the vertical stress ($\sigma_1$); the pore water drains off when the decrease in volume exceeds the air-filled pore space. Therefore, the applied stresses are assumed to be transmitted as effective stresses via the solid phase. However, Baumgartl (1989) found an additional change in the pore water pressure during extended CD triaxial tests depending on soil hydraulic properties. Thus, shear speed and low hydraulic conductivity, high tortuosity and small hydraulic gradients affect the drainage of excess soil water and the effective stresses (Horn et al., 1995).
2. In the consolidated undrained triaxial test (CU), pore water cannot be drained off the soil as vertical stress increases. Thus, high hydraulic gradients occur and the pore water reacts as a lubricant with a low surface tension. Thus, in the CU test, shear parameters are much smaller and pore water pressure values are much greater than those in the CD test.
3. The highest neutral stresses and therefore the lowest shear stresses are measured in the unconsolidated undrained test (UU), where neither the effective nor the neutral stresses are equilibrated with the applied principal stresses at the beginning of the test.

Thus, cohesion and angle of internal friction are strongly dependent on the compression and drainage conditions during triaxial tests. In terms of strength of agricultural soils under traffic, texture, pore water pressure (i.e., water suction), nature of aggregates, and soil structure determine the preference for a particular test.

### 2.3.4.4  Direct Shear Test

In the direct shear test, the type and direction of the shear plane, which is assumed to be affected only by normal and shear stresses, is fixed. Normal stress is applied in vertical and shear stress in the horizontal direction. As in the triaxial test, cohesion and angle of internal friction are influenced by shear speed and drainage conditions for a given soil.

## 2.4  EFFECT OF SOIL STRUCTURE AND DYNAMICS ON STRENGTH AND STRESS/STRAIN PROCESSES

### 2.4.1  Soil Strength

As stress is applied, soil deformation occurs at the weakest point in the soil matrix and further increase in stress results in the formation of failure zones. Therefore, the strength of the failure zone is equal to the energy required to create a new unit of surface area or to initiate a crack (Skidmore and Powers, 1982) and is called the apparent surface energy (Hadas, 1987). Consequently, soil stability is related to strength distribution in the failure zones. In principle, soil structure will be stable if the applied stress is smaller than the strength of the failure zone, i.e., if the bond strength at the points of contact exceeds the external stress.

### 2.4.1.1  Homogenized Soils

If only mechanical properties of homogenized soils are compared, gravelly soils are less compressible than sandy or finer textured soils (Horn, 1988), and any deviation of the particles from the spherical form results in an increase in the shearing resistance and soil strength (Gudehus, 1981; Hartge and Horn, 1991). Ellies et al. (1995) found that the angle of internal friction for medium and coarse sands at comparable bulk density and water content but of different origin (quartz, basalt, volcanic ash and shells) ranged from 25 to 50°. Soil strength also depends on the type and content of clay and exchangable cations. At the same bulk density and pore water pressure, compressibility of homogenized soil increases with clay content and decreases with organic matter content. At the same clay content, soils are more readily compressed for a lower bulk density or wetter soil. Cohesive forces between illite, smectite and vermiculite are greater than for kaolinite which has a larger angle of internal friction because of its size and shape. Additionally, the higher the valency of the adsorbed cations, and the higher the ionic strength, the greater the shearing resistance under a given condition (Yong and Warkentin, 1966; Mitchell, 1994).

Shearing resistance increases depending both on the nature and the content of organic matter. In agricultural soils with the same physical and chemical properties, the mechanical parameters (precompression stress value, angle of internal friction and cohesion) are greater when the relative contents of carbohydrates, condensed lignin subunits, bound fatty acids and aliphatic polymers are higher (Hempfling et al., 1990).

### 2.4.1.2  Effect of Aggregation

Mechanical strength of structured soils with comparable internal parameters depends on aggregation, actual and maximum predrying, and composition and arrangement of the pore system. For

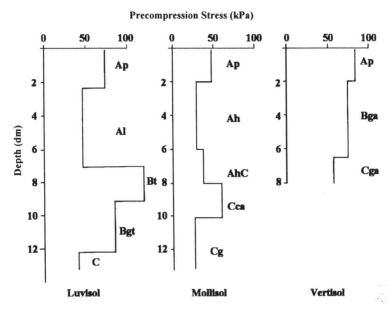

**FIGURE 2.4** Precompression stress values for 3 soil profiles at ψm = – 6 kPa.

comparable grain size distribution and pore water pressure, soil strength increases with aggregation (i.e., coherent < prismatic < blocky < crumbly, platy < subangular blocky). If the relative reduction in water-filled pores (the multiplication factor X in the effective stress equation) is smaller than the actual decrease in pore water pressure (more negative), strength increases. Conversely, drier soils get weaker. Nevertheless, dynamic changes in mechanical properties also depend on the frequency of swell/shrink and wet/dry events and on the actual pore water pressure. Soil becomes stronger if it is redried and rewetted and if pore water pressure gradients over longer distances promote particle movement and rearrangement until entropy is minimized (Semmel et al., 1990). In addition, soil strength is promoted by two different mechanisms. The increase in strength results from either an increase in the total number of contact points between single particles (i.e., an increase in effective stress) or in shear resistance per contact point (Hartge and Horn, 1984). Thus, even if soil bulk densities are similar, strength may be quite different.

Pedogenic effects on normal mechanical strength of three soils are illustrated in Figure 2.4. Luvisols derived from loess are characterized by clay illuviation, which results in decreased strength in the Al horizon but increased strength in the Bt horizon due to aggregation. Calcium precipitation in the corresponding horizon of the Mollisol also leads to a strength increase. In all three soil types, the parent material was weakest. Processes such as annual plowing and tractor traffic create very strong plow pans and layers with precompression stress values similar to the contact pressure of a tractor tire or even higher due to lug effects (up to 300%). Strength increases due to compression by glacier movements have been detected by the determination of stress/strain behavior of undisturbed soil samples. The transition from overconsolidation stress to the virgin compression line depends to a great extent on internal parameters and applied stresses in the plow pan. In addition, strength decreases in the A horizon due to plowing and seedbed preparation can be followed until texture dependent values are reached.

Marked strength changes when crossing horizons are observed for soil profiles which have been deep-loosened or plowed (60 cm) or loosened and/or mixed with other soil material. In the case of a partial deep loosening by a trenching slit plow (Blackwell et al., 1985; Reich et al., 1985), traffic must be restricted by use of smaller machinery or controlled tracks perpendicular to the homogenized soil volume. Freezing and thawing affect soil strength, because aggregates become

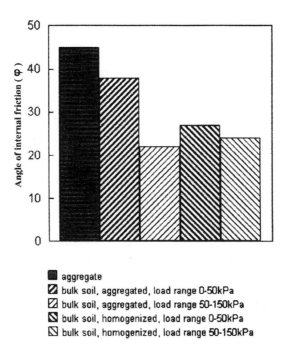

■ aggregate
▨ bulk soil, aggregated, load range 0-50kPa
▨ bulk soil, aggregated, load range 50-150kPa
◪ bulk soil, homogenized, load range 0-50kPa
◪ bulk soil, homogenized, load range 50-150kPa

**FIGURE 2.5** Changes in shear strength for various applied stress ranges for single aggregates, bulk soil and homogenized material.

denser or are destroyed by ice lens formation during freezing. Aggregates may act as rigid bodies which allow the exfoliation processes to start from the outer skin because of ice lens formation. Both effects are called soil curing, but result in completely different strength values and physical properties of the bulk soil (Horn, 1985; Kay, 1990).

Soils with a pronounced vertical pore system are stronger than those with randomized or extremely horizontal pore systems because the vertical (bio-) pores are equilibrated with the major principal stress ($\sigma_1$). Thus, untilled or minimum tilled soils are stronger than conventionally tilled soils (Horn, 1986).

However, each soil deformation requires air-filled pore space and/or high hydraulic conductivity to drain off the released pore water. The smaller the hydraulic conductivity, gradient and pore continuity, the more stable the soil during short-term loading. In sandy soils, this effect is small, because the initial settlement equals the total strain. With increasing clay content, however, the time dependent soil settlement (proportion of the initial to the primary and secondary settlement) is reduced. This results in an increase in the precompression stress value for short-term loading. Again, these differences are smaller in stronger aggregated loamy and clayey soils than in sandy soils.

The development of structure always results in an increase in soil strength. Secondary large pores can only be created if the aggregates become denser and stronger so as to carry the same stresses over fewer contact points. These strength differences can also arise from changes in the angle of internal friction, cohesion and stress dependent changes in shear strength for various applied stress ranges for single aggregates, undisturbed and homogenized material (Figure 2.5).

Increased aggregation increases soil strength under comparable hydraulic conditions. Relative to the very high value of the angle of internal friction for a single aggregate, that for bulk soil is smaller and decreases when a certain stress range is exceeded. As applied stresses increase, the angle of internal friction resembles that of homogenized material which emphasizes that each type of structure is only valid for a well-defined stress (mechanical or hydraulic) range. When this value is exceeded, only texture-dependent properties remain.

## 2.4.2 STRESS DISTRIBUTION IN SOILS

Any load applied at the soil surface is transmitted to the soil in three dimensions by the solid, liquid and gas phases. If air permeability is high enough to allow immediate deformation of the air filled pores, soil settlement is mainly affected by fluid flow.

However, fluid flow may be delayed because changes in water content or pore water pressure depend on the hydraulic conductivity, gradient and pore continuity. Thus, the intensity and form of pressure transmission are again affected by soil strength. In the following discussion, stress distribution in both homogenized soils and in aggregated systems will be defined. Based on the theory of Boussinesq (1885), who assumed that all soils behave in a completely elastic manner, Froehlich (1934) (both cited in Horn, 1981) described a more realistic form of the equipotential lines in terms of textural dependent concentration values. Under saturated conditions, these values should range between 3 for very strong material and 9 for soft soil, and should be higher for increasing clay content. In weak soils with high concentration factor values, stresses applied are transmitted to depth but remain concentrated around the perpendicular center line. On the other hand, in strong soils with low values of the concentration factor, pressure is transmitted more horizontally in a shallower soil layer.

At the same contact pressure, stresses are transmitted deeper when the contact area is larger. Furthermore, stress distribution patterns in the soil are not only different for tire lugs and the intervening area, but are also affected by the stiffness of the carcass (Horn et al., 1987). Thus, there are no well-defined equipotential stress lines in soils (Horn et al., 1989b) but the concentration factor values must be related to precompression stresses in relation to applied stress, and contact area for a given texture (DVWK, 1995). The effect of wheel speed on stress distribution must also be considered.

### 2.4.2.1 Effect of Internal Parameters

At a given bulk density and pore water pressure, applied stress at the soil surface will be transmitted deeper as silt and clay increase while stress attenuation will be greater in a smaller volume as the soil dries. At a given particle size distribution, water content and bulk density, stress attenuation will increase with increasing organic matter content. Based on many stress distribution measurements in the field and in monoliths (Horn et al., 1989b), a general correlation scheme involving texture, precompression stress and tire contact area to derive the concentration factor value has been used to approximately predict the stress distribution in soils (DVWK, 1995).

### 2.4.2.2 Effect of Structure on Stress Attenuation and Strain

The effect of aggregation on stress distribution and its consequences for ecological parameters is understood well, while the effect of traffic on stress differentiation is still controversial. Under *in situ* conditions stress attenuation in soils with the same internal parameters is greater if the soils are more aggregated. Concentration factor values expressed as precompression stress values are smaller for better aggregated and drier soils (Burger et al., 1988, DVWK, 1995). Not only the pressure but also the size and shape of the contact area affect stress distribution in unsaturated structured soils (Figure 2.6).

If internal strength values are smaller than the external forces applied, repeated traffic results in increased soil strength. For example, if a loessial Luvisol is repeatedly traversed at constant water content, horizontal minor stresses decrease while the major vertical stress increases (Figure 2.7).

These soil strength changes increase the concentration factor values due to the more pronounced vertical stress distribution, and each loading consequently results in a smaller effective stress relative to neutral stress (i.e., negative pore water pressure). During soil loading, stress components can vary as well as the ratio between mean normal and octahedral shear stress. Shear stresses create

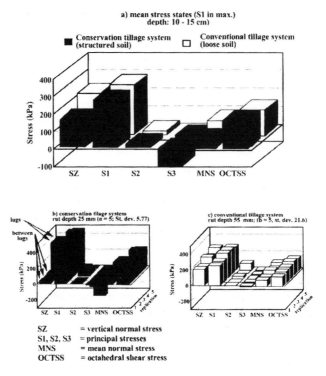

FIGURE 2.6  Stress distribution under a tire with lug effects (Kühner, 1997).

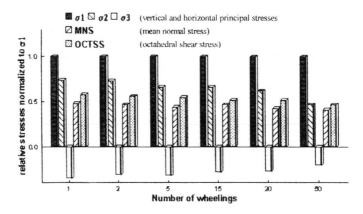

FIGURE 2.7  Stress distribution in a loessial Luvisol derived due to repeated traffic for 1 day at constant water content.

volumetric strain by soil particle displacement and rearrangement of pores, formation of gas bubbles and increased pore water pressure resulting from reduced hydraulic conductivity. With increasing traffic frequency, measured stresses in the upper soil horizons become smaller due to a more pronounced stress attenuation, while the opposite is true for the deeper soil horizons, where even an increase in the measured stress can be detected (Horn et al., 1998). This additional increase in soil strength caused by the consecutive traffic events can be explained by the elasticity of the topsoil, which induces a plastic deformation at depth in the still weaker subsoil which irreversibly becomes compacted. Because soil amelioration to depth by ripping, deep or slit plowing only loosens compacted soil if it is dry enough to suffer brittle failure, such methods degrade those structural

---

**TABLE 2.4**

**Changes in air capacity (AC) due to traffic or plowing in the consecutive soil horizons at given initial soil physical properties.**

traffic, topsoil (seedbed): clayey, silty loam (Ut3) at $\psi$m –6kPa

$a_0 = 53.73$;   $a_1 = -23.38$;   $b = 7$

plowing, deeper soil horizons: prism, Lt3 (very clayey loam), at $\psi$m = –6 kPa

$a_0 = 29.04$;   $a_1 = 15.27$;   $b = 5$

---

strengths which are dependent on texture. Such strength declines require a complete change in tillage and agricultural practices. Such traffic experiments also provide information on soil structure deterioration due to kneading. If soil water content is high compared to that to be drained off or hydraulic conductivity is low, homogenization and structure deterioration can be obtained from particle displacement measurements and the change in the proportions of octahedral shear strength relative to the principal stresses during traffic. The vertical and shear components become more important, which weakens the total system. Such effects are even more likely when slip and vibration effects of machinery are considered (Kühner et al., 1994). Deep ruts, platy structure, tractors stuck in the mud, and completely homogenized pore systems are all indicative of inappropriate soil or site management which will result in yield decreases or increased erosion. Additional applications of deterministic models are described by Veenhof and McBride (1996), Hakansson (1994), McBride and Joosse (1996), Nissen and Gräsle (1996). The magnitude of changes in the pore system due to soil deformation in the virgin compression stress range have been predicted by multiple regression analysis based on data from more than 80 soil profiles (Nissen, 1998; DVWK, 1997). If the effect of plowing or traffic on a weak seedbed or on changes in air space in the plow pan are considered as functions of internal strength and applied stress exceeding the precompression stress value, the following equation can be used:

$$AC = b \cdot \log_{10}\left(10^{c/b} + 1\right) \text{ with } c = a_0 + a_1 \cdot \log_{10}\left(\frac{\sigma_n}{1\,kPa}\right) \qquad [2.17]$$

in which AC is the air capacity and the empirical parameters (Table 2.4) depend on particle size distribution, aggregation and pore water pressure.

## 2.4.3 EFFECT OF STRESS APPLICATION AND ATTENUATION ON SOIL STRAIN

If external stress is smaller than internal soil strength, no deformation results and *vice versa*. The latter can either result in constant displacement of volume, or soil compaction (Kühner et al., 1994; Pytka et al., 1995). In order to distinguish between these processes, both the rut depth and the vertical movement of a given soil volume below the rut must be known. The extent to which soil strain occurs during traffic and the extent to which various tillage implements (conventional/conservation) deform a soil at a given pore water pressure, is shown in Figure 2.8. In the conventional tillage treatment in a loessial Luvisol, passage of a tractor (front/rear wheel) results in a pronounced vertical (up to 8 cm) and horizontal forward and backward (up to 2 cm) displacement. Under conservation tillage, these soil deformations are less because of a higher internal soil strength leading to a maximum vertical displacement of < 4 cm after two traffic events and a much less pronounced horizontal displacement. With increasing aggregate development, soil strength increases and aggregate deterioration is less pronounced during displacement and alteration of the pore system due to the infilling of interaggregate pores by smaller particles. Nevertheless, all stresses which are not attenuated to levels below soil strength result in volume alterations, even if the applied stresses vary for different soil types, land uses, tillage systems and environmental conditions.

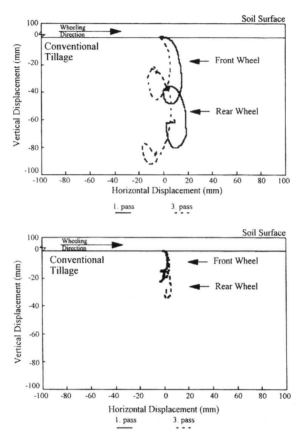

**FIGURE 2.8** Strain distribution in structured soils due to traffic. Particle movement or soil volume is more pronounced in the conventionally tilled soil while soil deformation is less intensive in the conservation tillage plot.

---

**EXAMPLE 2.2**

1. Specify three kinds of compression tests and explain why there are different results. Specify the boundary conditions.
2. Explain the physical meaning of precompression stress and which factors are of main influence with respect to soil strength.
3. How far can stress strain behavior be applied to predict the sustainable soil amelioration intensity by deep loosening?
4. Specify the possibilities of mass movement and deformation of soil volumes (2 each).
5. Which stress variables define the stress state (1-dimensional in a 3-phase system; 3-dimensional in a 1-phase system)?

---

## 2.5 FURTHER DYNAMIC ASPECTS IN SOILS

### 2.5.1 HYDRAULIC ASPECTS

#### 2.5.1.1 Water Retention Curve

With increasing aggregation, the interaggregate pore volume becomes coarser and water retention curves start to differ from those dependent only on texture. Aggregates are normally very dense and strong, containing mostly fine with very few coarse pores and less plant available water, while

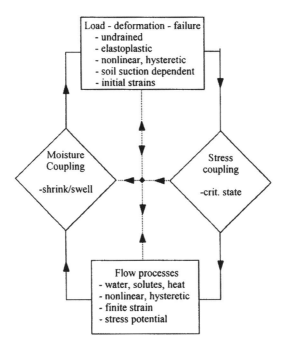

**FIGURE 2.9** Schematic diagram to define limits for the validity of data sets for hydraulic and mechanical parameters.

their bulk densities can exceed 2 g cm$^{-3}$ (Hartge and Horn, 1991). Thus, structure development always coincides with a change in pore system and function. As is true for stress/strain functions which can be subdivided into a virgin compression line and an overconsolidated stress range, water retention curves can be divided in a virgin dessication range and a rewetting or drying suction range (Junkersfeld and Horn, 1997). The effective stress equation shows that soil strength can be gained from either mechanical or pore water pressure which affect structure and pore size distribution. Junge et al. (1997) verified these relationships (Figure 2.9). Exceeding the degree of predryness or prestress by either mechanical or negative pore water pressure ($\overset{\wedge}{=}$ ψm) always results in a further soil deformation by normal shrinkage, rearrangement of particles, additional formation of aggregates or complete deterioration of existing pores.

### 2.5.1.2  Darcy's Law

*Hydraulic Conductivity*

Given laminar flow and a homogeneous pore system, one-dimensional water flux is described by Darcy's law. Values of the hydraulic conductivity range between 10$^{-4}$ and 10$^{-13}$ ms$^{-1}$ depending on water potential, texture and structure. Under saturated conditions, hydraulic conductivities range between 10$^{-4}$ and 10$^{-9}$ m s$^{-1}$ in clay soils, respectively. Since hydraulic conductivity is also affected by structure and texture, it will be higher in highly fractured or aggregated compared to compacted and dense soils. The hydraulic conductivity value (K) depends on volume and continuity of conducting pores. In structured soils with large cracks, (K) for the bulk soil is high but low inside single aggregates (Horn, 1990). Depending on aggregate density and pore continuity, (K) may be 4 orders of magnitude lower in single aggregates than bulk soil except where the aggregate contains mostly sand, in which case there may be no difference. The effects of structure on hydraulic conductivity persist under unsaturated conditions where any changes in structure directly affect hydraulic conductivity. At less negative pore water pressures, unsaturated hydraulic conductivity in single aggregates decreases with compaction of the structural elements (prisms less than polyhedral or subangular blocks) compared to fluxes in bulk soil. Only in weak aggregates are differences and ranges smaller (Gunzelmann et al., 1987). After exceeding the crossover suction at very negative pore water pressures (Hillel, 1980), hydraulic conductivity is higher in aggregates than in the bulk system. The heterogeneity of flow paths in aggregates compared to bulk soil is further enhanced, since outer surfaces of aggregates contain more clay than within where the pores are larger (Horn, 1987). Consequently, water and/or air flow in or out of single aggregates is further reduced, which can also be explained by the increasing tortuosity of the pore system at different positions in the aggregate (Glinski and Stepniewski, 1985).

*Hydraulic Gradient*

Generally, hydraulic gradients (m m$^{-1}$) vary by only half an order of magnitude, depending on pore water pressure, particle and pore size distributions (Hartge and Horn, 1977). Differences in hydraulic gradients should be large in aggregated dense soils particularly under conservation tillage, at least initially which may result in reduced plant water uptake efficiency. Differences in root length density which can be correlated with hydraulic properties can be used as indicators of rhizosphere function (Vetterlein and Marschner, 1993).

*Water Flux Density*

As differences in hydraulic properties between bulk soil and single structural elements increase, fluid flow becomes more pronounced in interaggregate pores reducing the possibility of obtaining an equilibrated pore water pressure profile. Pore continuity in macropores and a more tortuous pore system in single aggregates induces preferential flow especially near saturation (Beven and Germann, 1982). Youngs and Leeds-Harrison (1990) pointed out that a pore water pressure gradient causes water to flow preferentially in macropores with little flow within aggregates when both the macro- (interaggregate) and micropores (intraaggregate) are saturated. However, when saturated macropores surround unsaturated aggregates, solutes will be transported by diffusion into or from the aggregates depending on the concentration gradient. If macropores are empty, pores inside the aggregates become isolated, severely reducing the possibility for redistribution of water and solutes between them. Booltink and Bouma (1991), Edwards et al. (1993), and Tolchel'nikov et al. (1991) showed that hydraulic properties determined for bulk soil do not account for or explain pore continuity and/or pore accessibility with respect to water flux. Methods for macropore and intraaggregate flux measurements are given by Gunzelmann (1990), Plagge (1991) and Jardine et al. (1990). Jury et al. (1991), Kutilek and Nielsen (1994), and Nielsen and Kutilek (1995) deal with effects of soil aggregation in structured soils on changes in the hydraulic properties.

### 2.5.2  CHEMICAL ASPECTS

Brusseau and Rao (1990) pointed out that solute transport in aggregated soils is often characterized by nonideal phenomena, usually ascribed to the presence of immobile domains within porous medium, which influence the dynamics of physical processes.

### 2.5.2.1  Ion Transport

Soil ion transport rate (J) (mmol m$^{-2}$ d$^{-1}$) includes mass flow and diffusion for both liquid and solid phases, source/sink terms (exchange processes), ion precipitation, redox reactions and biological decay processes. Assuming laminar solute flow pattern in pores, water flow near the particle surface is retarded relative to that in the pore center because the soil solution becomes more viscous and concentrated as the particle surface is approached. Consequently, with decreasing pore size, ion diffusion in the liquid phase is reduced. With increasing tortuosity, which depends on both the volume and geometry of the water pathway, the ion concentration gradient is further reduced. In unsaturated soils with a high clay content, f is further increased by reduced thickness and discontinuity of water films, increased density of the adsorbed cation swarm, anion exclusion and increased viscosity.

### 2.5.2.2  Effect of Structure on Ion Exchange and Transport

Inter- and intraaggregate porosity affects exchange site accessibility and adsorption/desorption. At constant moisture, convective ion transport (e.g., Ca and Mg) is much smaller in single aggregates than in bulk soil, especially homogenized material (Horn et al., 1989a). In well-structured soils, acidic cations are more concentrated in the soil solution from bulk soil than single aggregates where basic cations tend to accumulate (Taubner, 1993). Ion diffusion from single aggregates is lower compared to bulk soil at a given time because (1) the bulk densities of single aggregates are larger than those of bulk soil; (2) the number of accessible reaction sites depends on the ratio of outer surface to sample mass which increases with decreasing aggregate diameter (spherical aggregate); (3) flow length increases with aggregate size; and (4) the average pore size and continuity are much smaller due to a higher clay content at the outer surface of a single aggregate.

Soil solution from undisturbed soil always contains more acidic cations than homogenized material with the difference becoming more pronounced as macroporosity increases. As soil becomes less structured and/or saturated, the differences become smaller. However, even in less aggregated soils under unsaturated conditions, the values would never be the same as the equilibrium soil solution (Hantschel et al., 1988; Kaupenjohann, 1991; Hartmann et al., 1998). This chemical disequilibrium was also verified by Hantschel and Pfirrmann (1990) under *in situ* conditions. They showed that the base status of the equilibrium soil solution from undisturbed forest soils was significantly higher than that from aggregated soil samples that released essentially $Al^{3+}$ and $H^+$ during percolation. Taubner (1993) showed that the soil solution from single aggregates contains less $H^+$ and more basic cations than the corresponding solution from interaggregate pores of the bulk soil. These differences are more pronounced as soils become more structured and heavier textured. Exchange sites show the same heterogeneity between aggregates and the bulk soil. In general, $Al^{3+}$ and $H^+$ are enriched on the outer surfaces of aggregates, while in the center the base saturation is higher. Hartmann et al. (1998) developed a model to predict the ion release under unsaturated soil conditions. The release of pore water pressure as dependent on changes in base saturation were plotted as a function of the dessication and/or flux times. Starting with a completely saturated undisturbed soil, cation release and/or accessibility of exchange sites decrease initially. Only after exceeding the crossover suction value does ion release increase due to greater hydraulic conductivity of the aggregates at that pore water pressure.

The extent to which ion diffusion is affected by the accessibility of particle surfaces inside the aggregates, and concentration gradient between soil solution and the bulk soil itself has been described by Horn and Taubner (1989). In their experiments to determine cumulative potassium release rates from single aggregates (polyhedral; texture: loamy clay) of structured bulk and homogenized soil (< 2 mm) under saturated conditions, the release rates per unit mass were highest for homogenized material. In aggregates, however, K release from internal sites was retarded. The larger the aggregates, the smaller were the release rates. Bhadoria et al. (1991a, 1991b) furthermore quantified the impedance factor f for Cl diffusion in soils as affected by bulk density and water content. An increase in bulk density from 1.38 to 1.76 Mg m$^{-3}$ at constant gravimetric moisture content of 7% decreased the impedance by a factor of 3, while at water contents greater than 10%, it increased linearly with increasing bulk density.

Gisi et al. (1996) and Buchter et al. (1992) defined the consequences of such chemical disequilibrium for ecosystem modeling approaches. Augustin (1992) also considered the microbiological differences at various positions inside the aggregate in terms of complexation and exchange processes.

## 2.6  MODELING DYNAMIC COUPLED PROCESSES

Hydraulic, thermal or gaseous transport processes in unsaturated structured soils must be mutually linked. The mathematical equations for water, gas and heat transport in soils, and the effects of

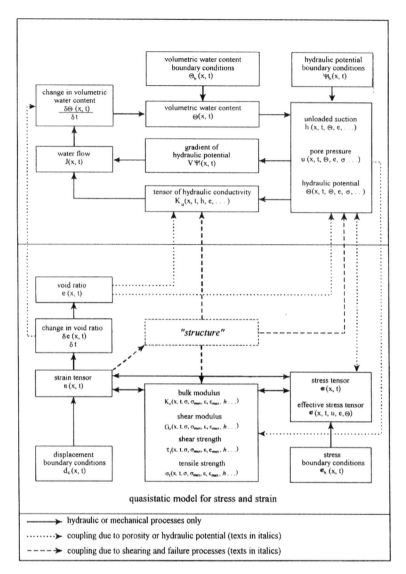

**FIGURE 2.10** Schematic representation of a coupled hydraulic elastoplastic soil model. Conceptual model to define and to quantify coupled processes in soils.

tillage on changes in structure are available (Jury et al., 1991). However, a schematic picture of a coupled hydraulic-elasto-plastic soil model (Gräsle et al., 1996) will be illustrated (Figure 2.10) as one possibility for coupling processes in order to demonstrate their complexicity on potential use as on environmental impact model (O'Sullivan and Simota, 1995).

### 2.6.1 COMPONENTS OF A MODEL TO PREDICT COUPLED PROCESSES

The main constituents of such a model for coupled processes can be defined as follows:

#### 2.6.1.1 Load-Deformation-Failure

This process describes the load-deformation-failure response of soil to a change in stress, load, strain and/or displacement, with all other factors held constant. It includes: (1) nonlinear elastic behavior of the material as a function of stress and its history, soil water suction and stress (Richards,

1978), (2) changes in initial stress or strain as a result of swell/shrink behavior caused by changes in soil water suction, solute content or temperature (Richards, 1986), (3) stress/strain path dependency or hysteretic behavior in load deformation response (Richards, 1979) and (4) shear and tensile failure with dilatance or compression, and with strain softening.

### 2.6.1.2 Coupling Processes for a Change of Stress or Strain

During any kind of soil deformation or shearing, the changes in hydraulic properties and functions must be considered as pointed out in the effective stress equation. In addition, changes in soil stress due to swelling or shrinkage will also cause changes in soil water potential. This effect of stress on water potential is sometimes referred to as the stress potential (Croney and Coleman, 1961) or the stress component of field measurements of soil water suction (Richards, 1986). Load deformation analysis enables changes in soil water potential to be calculated for inclusion in subsequent water flow analyses. The displacements can also be used to calculate the new geometry and material velocities for the analysis of water flow in soils undergoing finite strain. Such analysis can be also extended to three-dimensional flow problems, which are of interest in the plow pan and in well-structured horizons below.

### 2.6.2 Validation of Coupled Processes by a Finite Element Model (FEM)

In order to demonstrate the capabilities of a modeling approach, Stress State Transducer (SST) data and the mechanical parameters for three soil horizons from a traffic experiment on a conservation tillage plot (shallow chiselling up to 8 cm, Luvisol derived from glacial till) were used to compare the measured stress and strain with those which predicted by the FEM technique (Richards et al., 1997). If only the mechanical parameters of the three soil horizons at given rut depth and applied stress are considered, no good agreement could be obtained between the measured and the calculated data with increasing depth (Figure 2.11). However, if mechanical data are included in the calculation for a depth of 20 to 40 cm where previously plowing (8 years ago) had been carried out and where the shear strength and bulk modulus data were different from those of the adjacent soil horizons, then a reasonably good agreement between the measurements and model predictions was obtained. Corresponding measurements and modeling of coupled ion and water fluxes through unsaturated structured soils can be used to quantify aggregation effects on ion diffusion, dispersion and mass flow and to predict possible changes in pore systems and functions induced by stresses. They can be also used to estimate time dependent effects on nutrient, heavy metal or pesticide transport through structured soils and to include the effect of drying or wetting on these processes.

---

**EXAMPLE 2.3**

1. Why do mechanical considerations require the knowledge of the stress state as well as the strain state?
2. How do hydrological properties influence the stress state and why must hydraulic and mechanical properties have to be always linked?

---

## 2.7 CONCLUSIONS

1. Mechanical soil properties can be determined by capacity or intensity methods, which result in relative or in actual strength values. However, the development of soil structure presents a problem because of the required homogenization during soil preparation.
2. The determination of soil strength by intensity parameters requires the measurement of volumetric stress and strain.

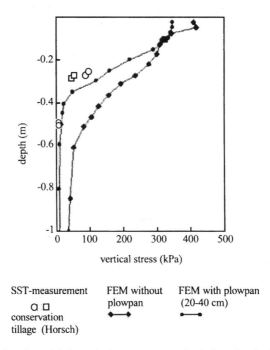

**FIGURE 2.11** Determined and modeled vertical stress versus depth in a Luvisol derived from glacial till under conservation tillage (type Horsch). The corresponding data sets are taken from 3 depths (10, 30, 50 cm). The properties of a plowpan layer were included, which had been created up to 1989 and which had a thickness of 20 cm (from 20–40 cm) (Kühner, 1997).

3. Structure development in arable and forest soils always results in increased soil strength. With increased aggregation, strength increases and the characteristics of the water retention curve become more pronounced.

4. In structured soils, hydraulic properties of inter- and intraaggregate pores are different with the differences in hydraulic conductivity, plant available water and air capacity becoming more pronounced as structure development increases. In addition, the root penetrability is reduced.

5. The increase in strength with decreasing pore water pressure depends on the pattern of the water retention curve for single aggregates and bulk soil.

6. With increasing degree of aggregation, pore water pressure at the same water content in the bulk soil becomes more negative. Single aggregates are always stronger than the bulk soil or the homogenized material. This is borne out by smaller increases in pore water pressure during loading and by higher aggregate bulk density. The shear strength parameters (angle of internal friction and cohesion) are always higher for single aggregates than bulk soil for a given applied stress. As soon as the applied stress exceeds internal soil strength, the aggregate or bulk aggregated soil will become homogenized. Thus, the pattern of the Mohr Coulomb failure line resembles that of the homogenized material after exceeding this stress value.

7. Soil strength is strongly affected by spatial differences in pore water pressure. Hydraulic properties of single aggregates and their arrangement in the bulk soil are therefore most important if soil strength and the corresponding stress distribution is to be predicted. Stress attenuation and stress distribution (intensity and spatial direction) due to loading under running wheels can vary to a great extent depending on soil and machinery parameters.

8. Structure development always involves reduced accessibility to exchange sites for soil solutions, retarded exchange processes and stronger chemical disequilibrium. Only at the lowest entropy (at an increased accessibility) do structured soils become macroscopically homogeneous. With reduced pore water potential (reduced flux rate), the exchange process becomes more complete. At higher saturations, anaerobic soil volumes in aggregates become more extensive for longer periods. Thus, soil physicochemical properties vary intensely throughout the year.

9. Depending on the objective, either empirical or mechanistic models can be used with varying degrees of certainty and reliability. The analysis of unsaturated structured soils is a very complex problem, with many multidisciplinary processes operating in an interactive fashion. Long-term soil stability involves materials which are unsaturated, nonlinear hysteretic and the soil composition changes with time. During tillage operations, many physical and mechanical properties are altered or at least affected, requiring a more precise definition of the actual conditions and the expected changes.

10. Within this conceptual framework, almost any practical problem in soils and soft rocks can be modeled using established theories or, where they are not applicable, experimentally based material responses. Each process can be experimentally simulated in laboratory or field tests. This can be done quickly and cheaply by simple tests carried out under the predicted field conditions of stress, moisture, density, etc. Alternately, complex nonlinear and hysteretic functions of the load-deformation-failure process with or without the failure component included can be implemented as a linear elastic problem. The prediction of simple seepage through soils may only require a simple linear water flow process. Where contaminant flow is also considered, then the coupled water and solute flow processes must be considered.

## REFERENCES

Augustin, S. 1992. Mikrobielle Stofftransformationen in Bodenaggregaten, 152 S., Dissertation, Universität Göttingen, Germany.

Babel, U., H. Vogel, M. Krebs, G. Leithold, and C. Hermann. 1995. Micromorphological investigation on soil aggregates, *in* K.H. Hartge and R. Stewart (eds.), *Soil structure: its development and function,* Lewis Publishers, 11–30, Boca Raton, FL.

Bailey, A.C., C.E. Johnson, and R.L. Schafer. 1986. A model for agricultural soil compaction, *J. Agric. Eng. Res.,* 33:257–262.

Bailey, C., R.L. Raper, C.E. Johnson, and E.C. Burt. 1992. An integrated approach to soil compaction modelling, *Proc. Int. Agric. Eng. Conf.,* 213–226, Uppsala, Sweden.

Baumgartl, T. 1991. Spannungsverteilung in unterschiedlich texturierten Böden und ihre Bedeutung für die Bodenstabilität. Diss, Christian-Albrechts Universität, Kiel, Germany.

Becher, H.H. 1992. Die Bedeutung der Festigkeitsverteilung in Einzelaggregaten für den Wasser- und Stofftransport im Boden. Z. Pflanzenernähr. Bodenk, 155:361–366.

Beven, K.J. and P. Germann. 1982. Macropores and water flow in soils, *Water Resour. Res.,* 18:1311–1325.

Bhadoria, P.B.S., J. Kaselowsky, N. Claassen, and A. Jungk. 1991a. Phosphate diffusion coefficients in soil as affected by bulk density and water content, *Z. Pflanzenernähr. Bodenk,* 154:53–57.

Bhadoria, P.B.S., J. Kaselowsky, N. Claassen, and A. Jungk. 1991b. Impedance factor for chloride diffusion in soil as affected by bulk density and water content, *Z. Pflanzenernähr. Bodenk,* 154:69–72.

Blackwell, J., R. Horn, N. Jayawardane, R. White, and P. Blackwell. 1985. Vertical stress distribution under tractor wheeling in a partially deep loosened Typic Paleustalf, *Soil Till. Res.,* 13:1–12.

Blunden, B.G., McBride, R.AS., H. Daniel, and P.S. Blackwell. 1994. Compaction of an earthy sand by rubber tracked and tyred vehicles, *Aust. J. Soil Res.,* 32:1095–1108.

Blunden, B.G., McLachlan, and J.M. Kirby. 1992. A recording system for measuring *in situ* soil stresses due to traffic, *Soil Till. Res.,* 25:35–42.

Bölling, W.M. 1971. Zusammendrückung und Scherfestigkeit von Böden, Springer Verlag, Berlin, Germany.

Bolling, I. 1986. How to predict soil compaction from agricultural tires, *J. Terramech.,* 22:205–223.

Booltink, H.W.G. and J. Bouma. 1991. Physical and morphological characterisation of bypass flow in a well structured clay soil, *Soil Sci. Soc. Am. J.,* 55:1249–1254.

Boone, F.R. and B.W. Veen. 1994. Mechanisms of crop responses to soil compaction, B.D. Soane and C.van Ouwerkerk (eds.), *Soil Compaction in Crop Production,* Elsevier Publishers, 237–264, Amsterdam, Netherlands.

Bradford, J.M. and S. Gupta. 1986. Soil compressibility, A. Klute. (ed.), Methods of soil analysis, Part I. 2nd ed., *Am. Soc. Agron.,* 479–492, Madison, WI.

Brusseau, M.L. and P.S.C. Rao. 1990. Modeling solute transport in structured soils, a review, *Geoderma,* 46:169–192.

Buchter, B., G. Richner, R. Schulin, and H. Flühler. 1992. Übersicht über bestehende Stofftransportmodelle zur Analyse von Mobilisierung und Auswaschungsprozessen von Schwermetallen im Boden, Schriftenreihe Umwelt 165, Bundesamt f. Umwelt, Wald und Landschaft, Bern, Switzerland.

Burger, N., M. Lebert, and R. Horn. 1987. Druckausbreitung unter fahrenden Traktoren im natürlich gelagerten Boden, Mitt. Dtsch. Bodenk. Ges., 55:135–141.

Burger, N., M. Lebert, and R. Horn. 1988. Prediction of the compressibility of arable land, *Catena Suppl.,* 11:141–151.

Burke, W., D. Gabriels, and J. Bouma. 1986. Soil structure assessment, Balkema Publisher, Rotterdam, The Netherlands.

Canarache, A. 1990. PENETR: A generalized semi-empirical model estimating soil resistance to penetration, *Soil Till. Res.,* 16:51–70.

Croney, D. and D. Coleman. 1961. Pore pressure and suction in soil, *in Pore pressure and suction in soils,* 31–37, Butterworths, London.

Dexter, A.R. 1988. Advances in characterisation of soil structure, *Soil Till. Res.,* 11:199–239.

Dexter, A.R., R. Horn, and W. Kemper. 1988. Two mechanisms of age hardening, *J. Soil Sci.,* 39:163–175.

Drescher, J., R. Horn, and M. de Boodt. 1988. Impact of water and external forces on soil structure, *Catena Suppl.,* 11:175.

DVWK. 1995. Merkblätter 234. Gefügestabilität ackerbaulich genutzter Mineralböden. Teil I: Mechanische Belastbarkeit. Wirtschafts- und Verlagsges. Gas und Wasser mbH, Bonn, Germany.

DVWK. 1997. Merkblätter zur Wasserwirtschaft 235. Gefügestabilität ackerbaulich genutzter Mineralböden. Teil II: Auflastabhängige Veränderung von bodenphysikalischen Kennwerten. Wirtschafts- und Verlagsges. Gas und Wasser mbH, Bonn, Germany.

Edwards, W.M., M.J. Shipitalo, and L.B. Owens. 1993. Gas, water and solute transport in soils containing macropores: a review of methodology, *Geoderma,* 57:31–49.

Ellies, A. and R. Horn. 1996. Stress distribution in Hapludands under different use, *Z. Pflanzenernähr. Bodenk,* 159:113–120.

Ellies, A., R. Horn, and R. Smith. 1996. Transmision de presiones en el perfil de algunos suelos, *Agro. Sur.,* 24:4–12.

Ellies, A., R. Grez, and Y. Ramirez. 1995. Cambios en las propiedades humectantes en suelos sometidos a diferentes manejos, *Turrialba,* 45:42–48.

Emerson, W.W., R.D. Bond, and A.R. Dexter. 1978. Modification of soil structure. John Wiley & Sons, Chichester, UK.

Erbach, D.C., G.R. Kinney, A.P. Wilcox, and A.E. Abo-Abda. 1991. Strain gauge to measure soil compaction, *Trans. ASAE,* 34:123–128.

Fredlund, D.G. and H. Rahardjo. 1993. Soil mechanics for unsaturated soils, J.Wiley and Sons, Chichester, UK.

Ghodrati, M. and W.A. Jury. 1990. A field study using dyes to characterize preferential flow of water, *Soil Sci. Soc. Am. J.,* 54:1558–1563.

Gill, W.R. and G.E. van den Berg. 1967. Soil dynamics in tillage and traction, *USDA Agric. Man.,* 316.

Gisi, U., R. Schenker, R. Schulin, X.F. Stadelmann, and H. Sticher. 1990. Bodenökologie. Thieme, Stuttgart, Germany.

Gliemeroth, G. 1953. Untersuchungen über Verfestigungs- und Verlagerungsvorgänge im Ackerboden unter Rad- und Raupenfahrzeugen, Z. Acker. Pflanzenbau, 96:219–234.

Glinski, J. and W. Stepniewski. 1985. Soil aeration and its role for plants, CRC Press, Boca Raton, FL.

Goss, M. and F.B. Reid. 1979. Influence of perennial ryegrass roots on aggregate stability, *Agric. Res. Council Letcombe Lab. Ann. Rep.,* 1978:24–25.

Gräsle,W. and B. Nissen. 1996. Bestimmung der Vorbelastung bei verhinderter und zugelassener Seitendehnung, Teil I: Theoretische Grundlagen und Auswertungsverfahren. Mitt. Dtsch. Bodenkde. Ges., 80:327–330.

Gräsle, W., B.G. Richards, T. Baumgartl, and R. Horn. 1996. Interaction between soil mechanical properties of structured soils and hydraulic processes: Theoretical fundamentals of a model, E.E. Alonso and P. Delage (eds.), *Unsaturated Soils.* vol. II, 719–725, Balkema Publishers, Paris.

Gudehus, G. 1981. Bodenmechanik, *Enke Verlag,* Stuttgart, Germany.

Gunzelmann, M. 1990. Die Quantifizierung und Simulation des Wasserhaushalts von Einzelaggregaten und strukturierten Gesamtböden unter besonderer Berücksichtigung der Wasserspannungs-/Wasserleit-fähigkeitsbeziehung von Einzelaggregaten, Ph.D. thesis, Bayreuther Bodenk, Ber., 129.

Gunzelmann, M., U. Hell, and R. Horn. 1987. Die Bestimmung der Wasserspannungs-/Wasserleitfähigkeits-Beziehung von Bodenaggregaten. Z. *Pflanzenernähr. Bodenk,* 150:400–402.

Hadas, A. 1987. Dependence of true surface energy of soils on air entry pore size and chemical constituents, *Soil Sci. Soc. Am. J.,* 51:187–191.

Hakansson, I. 1994. Subsoil compaction by high axle load traffic, *Soil Till. Res.,* 29:105–306.

Hakansson, I., W.B. Voorhees, P. Elonen, G.S.V. Raghavan, B. Lowery, A.L.M. van Wyik, K. Rasmussen, and H. Riley. 1987. Effect of high axle load traffic on subsoil compaction and crop yield in humid regions with annual freezing, *Soil Till. Res.,* 10:259–268.

Hantschel, R., and T. Pfirrmann. 1990. Bodenkundliche Untersuchungen am Forschungsschwerpunkt Wank — Bedeutung für die Waldschäden in den Kalkalpen, *Forstl. Centralbl.,* 153:25–38.

Hantschel, R., M. Kaupenjohann, J. Gradl, R. Horn, and W. Zech. 1988. Ecologically important differences between equilibrium and percolation soil extracts, *Geoderma,* 43:213–227.

Harris, H.D. and D.M. Bakker. 1994. A soil stress transducer for measuring *in situ* soil stresses, *Soil Till. Res.,* 29:35–48.

Hartge, K.H. and R. Horn. 1977. Spannungen und Spannungsverteilungen als Entstehungsbedingungen von Aggregaten, Mitteilgn Deutsch. Bodenkd. Gesell., 25:23–33.

Hartge, K.H. and R. Horn. 1984. Untersuchungen zur Gültigkeit des Hooke'schen Gesetzes bei der Setzung von Böden bei wiederholter Belastung, Z. Acker- Pflanzenbau, 153:200–207.

Hartge, K.H. and R. Horn. 1991. Einführung in die Bodenphysik, 2nd ed., Enke Verlag, Stuttgart, Germany.

Hartge, K.H. and R. Horn. 1992. Bodenphysikailsches Praktikum, 3rd ed., Enke Verlag, Stuttgart, Germany.

Hartge, K.H. and I. Rathe. 1983. Schrumpf- und Scherrisse-Labormessungen, *Geoderma,* 31:325–336.

Hartmann, A., W. Graesle, and R. Horn. 1998. Cation exchange processes in structured soils at various hydraulic properties, *Soil Till. Res.,* 47:196–205.

Haynes, R.J. and R.S. Swift. 1990. Stability of soil aggregates in relation to organic constituents and soil water content, *J. Soil Sci.,* 41:73–83.

Hempfling, A., H. Schulten, and R. Horn. 1990. Relevance of humus composition for the physical/mechanical stability of agricultural soils: a study by direct pyrolysis-mass spectrometry, *J. Anal. Appl. Pyrol.,* 17:275–281.

Hillel, D. 1980. Fundamentals of soil physics. Academic Press, London, UK.

Horn, R. 1981. Die Bedeutung der Aggregierung von Böden für die mechanische Belastbarkeit in dem für Tritt relevanten Auflastbereich und deren Auswirkungen auf physikalische Bodenkenngrößen, Schrift-enreihe FB 14 TU Berlin, Germany.

Horn, R. 1985. Die Bedeutung der Trittverdichtung durch Tiere auf physikalische Eigenschaften alpiner Böden, Z. *Kulturtech. Flurber,* 26:42–51.

Horn, R. 1986. Auswirkung unterschiedlicher Bodenbearbeitung auf die mechanische Belastbarkeit von Ack-erböden, Z. *Pflanzenernähr. Bodenk.,* 149:9–18.

Horn, R. 1987. The role of structure for nutrient sorptivity of soils, Z. *Pflanzenernähr. Bodenk.,* 150:13–16.

Horn, R. 1988. Compressibility of arable land, *Catena Suppl.,* 11:53–71.

Horn, R. 1989. Strength of structured soils due to loading — a review of the processes on macro- and microscale; European aspects, W.E. Larson, G.R. Blake, R.R. Allmaras, W.B. Voorhees, and S. Gupta (eds.), *Mechanics and Related Processes in Structured Agricultural Soils,* 9–22, Kluwer Academic Publishers, Dordrecht, Netherlands.

Horn, R. 1990. Aggregate characterization as compared to soil bulk properties, *Soil Till. Res.,* 17:265–289.

Horn, R. 1995. Stress transmission and recompaction, *in* Tilled and segmently disturbed subsoils under trafficking, *in* Jayawardane, S. (ed.), Subsoil management techniques, *Advances in Soil Science,* 187–210, CRC Press, Boca Raton, FL.

Horn, R. 1998. The effect of static and dynamic loading on stress distribution, soil deformation and its consequences for soil erosion, H.P. Blume et al. (ed.), Towards Sustainable Land Use, *Advances in Geoecology,* 31:233-256, 1625, Catena Verlag, Cremlingen, Germany.

Horn, R. and A.R. Dexter. 1989. Dynamics of soil aggregation in a homogenized desert loess, *Soil Till. Res.,* 13:254–266.

Horn, R. and H. Taubner. 1989a. Effect of aggregation on potassium flux in a structured soil, *Z. Pflanzenernähr. Bodenk.,* 152:99–104.

Horn, R., N. Burger, M. Lebert, and G. Badewitz. 1987. Druckfortpflanzung in Böden unter fahrenden Traktoren, *Z. Kulturtech. Flurber.,* 28:94–102.

Horn, R., H. Taubner, and R. Hantschel. 1989a. Effect of structure on water transport, proton buffering and nutrient release, *Ecol. Stud.,* 77:323–340.

Horn, R., M. Lebert, and N. Burger. 1989b. Vorhersage der mechanischen Belastbarkeit von Böden als Pflanzenstandort auf der Grundlage von Labor- und in situ-Messungen, Abschlußbericht Bayr. SMLFU, Munich.

Horn, R., T. Baumgartl, S. Kühner, M. Lebert, and R. Kayser. 1991. Zur Bedeutung des Aggregierungsgrades für die Spannungsverteilung in strukturierten Böden, *Z. Pflanzenernähr. Bodenk.,* 154:21–26.

Horn, R., C. Johnson, H. Semmel, R. Schafer, and M. Lebert. 1992. Stress measurements in undisturbed unsaturated soils with a stress state transducer (SST)-theory and first results, *J. Plant Nutr. Soil Sci.,* 155:269–274.

Horn, R., H. Taubner, M. Wuttke, and T. Baumgartl. 1994. Soil physical properties related to soil structure, *Soil Till. Res.,* 35:23–36.

Horn, R., W. Stepniewski, T. Wlodarczyk, G. Walensik, and E.F.M. Eckhardt. 1994. Denitrification rate and microbial distribution within homogeneous soil aggregates, *Int. Agrophy.,* 8:65–74.

Horn, R., T. Baumgartl, R. Kayser, and S. Baasch. 1995. Effect of aggregate strength on changes in strength and stress distribution in structured bulk soils, Hartge, K.H. and R. Stewart (eds.), *Soil Structure: its Development and Function,* 31–52, CRC Press, Boca Raton, FL.

Horn, R., W. Gräsle, and S. Kühner. 1996. Einige theoretische Überlegungen zur Spannungs- und Deformationsmessung in Böden und ihre meßtechnische Realisierung, *Z. Pflanzenernähr. Bodenk.,* 159:137–142.

Horn, R., H. Kretschmer, T. Baumgartl, K. Bohne, A. Neupert, and A.R. Dexter. 1998. Soil mechanical properties of a partly reloosened (slit plough system) and a conventionally tilled over consolidated gleyic luvisol derived from glacial till, *Int. Agrophys.,* 12:143–154.

Horn, R., J.J.H. van den Akker, and J. Arvidsson. 2000. Subsoil compaction. Distribution, processes, consequences, *Advances in Geoecology,* Catena Verlag, Reiskirchen, 462.

Jardine, P.M., G.V. Wilson, and R.J. Luxmoore. 1990. Unsaturated solute transport through a forest soil during rainstorm events, *Geoderma,* 46:103–118.

Johnson, C.E. and A.C. Bailey. 1990. A shearing strain model for cylindrical stress states, *Proc. ASAE Pap.,* 90–1085.

Junge, T., J. Thienemann, W. Gräsle, T. Baumgartl, and R. Horn. 1997. Zum Einfluss von Spannungen und Porenwasserdrücken auf die mechanische Stabilität von Basisabdichtungen. Teil II: Theoretishche Überlegungen und Konsequenzen für die gängige Einbautechnik, *Müll und Abfall,* 8:475–479.

Junkersfeld, J. and R. Horn. 1997. Über die räumliche und zeitliche Variabilität scheinbar fixer Wasserhaushaltsgrößen am Beispiel von Bodenaggregaten, Z. Pflanzenernähr, Bodenk, 160:179–186.

Jury, W.A. and W.R. Gardner. 1991. *Soil Physics,* 5th ed., John Wiley & Sons, New York.

Kaupenjohann, M. 1991. Chemischer Bodenzustand und Nährelementversorgung immisionsbelasteter Fichtenbestände in NO-Bayern, Ph.D. thesis, Universitat Bayreuth, Germany.

Kay, B.D. 1990. Rates of change of soil structure under different cropping systems, *Adv. Soil Sci.,* 12:1–41.

Kezdi, A. 1969. *Handbuck der Bodenmechanik,* VEB Verlag, Berlin.

Kirby, J.M., B.G. Blunden, and C.R. Trein. 1997. Stimulating soil deformation using a critical state model. II. Soil compaction, *Europ. J. Soil Sci.,* 48:59–70.

Kirby, M.J. and M.F. O'Sullivan. 1997. Critical state soil mechanics analysis of the constant cell volume triaxial test, *Europ. J. Soil Sci.,* 48:71–79.

Koolen, A.J. 1994. Mechanics of soil compaction, *in* Soane, B.D. and C. van Ouwerkerk (eds.), Soil compactionin crop production, 23–44, Elsevier, Amsterdam.

Koolen, A.J. and H. Kuipers. 1985. Agricultural soil mechanics, Springer-Verlag, Berlin.

Kretschmer, H. 1997. Körnung und Konsistenz, *in* H.P. Blume, P. Felix-Henningsen, W. Fischer, H.G. Frede, R. Horn, and K. Stahr (eds.), *Handbuch der Bodenkunde*, Ecomed Verlag, Landsberg, Germany.

Kühner, S. 1997. Simultane Messung von Spannungen und Bodenbewegungen bei statischen und dynamischen Belastungen zur Abschätzung der dadurch induzierten Bodenbeanspruchung, Ph.D. thesis, Christian-Albrechts University, Kiel, Germany.

Kühner, S., T. Baumgartl, W. Gräsle, T. Way, R. Raper, and R. Horn. 1994. Three dimensional stress and strain distribution in a loamy sand due to wheeling with different slip, *Proc. 13th Int. Soil Till. Res. Org. Conf.*, 591–597.

Kutilek, U. and D. Nielsen. 1994. *Soil Hydrology*, Catena Verlag, Cremlingen, Germany.

Larson, W.E., G.R. Blake, R.R. Allmaras, W.B. Vorhees, and S. Gupta. 1989. Mechanics and related processes in structured agricultural soils, Kluwer Academic, Dordrecht, The Netherlands.

Lipiec J. and C. Simota. 1994. Role of soil and climate factors in influencing crop responses in central and eastern Europe, *in* Soane, B.D. and C. van Ouwerkerk (eds.), *Soil Compaction in Crop Production*, 365–390, Elsevier, Amsterdam.

McBride, R.A. and P.J. Joosse. 1996. Overconsolidation in agricultural soils, II. Pedotransfer functions for estimating preconsolidation stress, *Soil Sci. Soc. Am. J.*, 60:373–380.

McKenzie, B. 1989. Earthworms and their tunnels in relation to soil physical properties, Ph.D. thesis, University of Adelaide, Australia.

McKibben, E.G. and R.L. Green. 1936. Transport wheels for agricultural machinery, VII: Relative effects of steel wheels and pneumatic trires on agricultural soils, *Agric. Eng.*, 21:183–185.

Mitchell, J.K. 1976. *Fundamentals of Soil Behavior*, John Wiley & Sons, New York.

Morras, H. and G. Piccolo. 1998. Biological recuperation of degraded ultisols in the province of Misiones, 1345–1361. H.P. Blume et al. (ed.), Towards Sustainable Land Use, *Advances in Geoecology*, 31, 1625, *Catena Verlag*, Cremlingen.

Nichols, T.A., A.C. Bailey, C.E. Johnson, and D. Grisso. 1987. A stress-state transducer for soil, *Trans. ASAE*, 30:1237–1241.

Nielsen, D. and M. Kutilek. 1993. Soil hydrology, *Catena Verlag*, Cremlingen, Germany.

Nissen, B. 1998. Gefügestabilität ackerbaulich genutzter Mineralböden — auflastabhängige Veränderung von bodenphysikalischen Kennwerten, Ph.D. thesis, Christian Albrechts University Kiel, in press.

Nissen, B. and W. Gräsle. 1996. Bestimmung der Vorbelastung bei verhinderter und zugelassener Seitendehnung. Teil II: Experimentelle Ergebnisse. Mitt. Dtsche. Bodenkde. Ges., 80:331–334.

O'Sullivan, M.F. and C. Simota. 1995. Modelling the environmental impacts of soil compaction: a review, *Soil Till. Res.*, 35:69–84.

Okhitin, A.A., J. Lipiec, S.Tarkiewiecz, and A.V. Sudakov. 1991. Deformation of silty loam soil under a tractor tyre, *Soil Till. Res.*, 19:187–1995.

Oldeman, L.R. 1992. Global extent of soil degradation, *Proc. Symp. Soil Resilience and Sustainable Land Use*, Budapest, Hungary.

Olson, H.J.1994. Calculation of subsoil compaction, *Soil and Tillage Res.*, 29:105-111.

Peattie, K.R. and R.W. Sparrow. 1954. The fundamental action of earth pressure cells, *J. Mech. Phys. Solids*, 2:141–155.

Plagge, R. 1991. Bestimmung der ungesättigten hydraulischen Leitfähigkeit im Boden. Bodenökologie und Bodengenese, Heft 3. dissertation, TU, Berlin, Germany.

Pytka, J, R. Horn, S. Kühner, H. Semmel, and D. Blazejczak. 1995. Soil stress state determination under static load, *Int. Agrophys.*, 9:219–227.

Reich, J., H. Unger, H. Streitenberger, C. Mäusezahl, C. Nussbaum, and P. Steinert. 1985. Verfahren und Vorrichtung zur Verbesserung Verdichteter Unterböden, Agrartechnik Berlin, 41:57–62.

Richards, B.G. 1978. Application of an experimentally based non-linear constituative model to soils in the laboratory and field tests, *Aust. Geomech. J.*, G8:20–30.

Richards, B.G. 1979. The method of analysis of the effects of volume change in unsaturated expansive clays on engineering structures, *Aust. Geomech. J.*, G9:27–41.

Richards, B.G. 1986. The role of lateral stresses on soil water relations in swelling soils, *Aust. J. Soil Res.*, 24:457–467.

Richards, B.G. 1992. Modelling interactive load-deformation and flow processes in soils, including unsaturated and swelling soils, *Proc. 6th. Aust-NZ Conf. Geomech.*, 18–37.

Richards, B.G., Baumgartl, T., Horn, R., Gräsle, W. 1997. Modelling the effects of repeated wheel loads on soil profiles, *Int. Agrophysics,* 11:71–87.

Semmel, H., R. Horn, U. Hell, A.R. Dexter, and E.D. Schulze. 1990. The dynamics of soil aggregate formation and the effect on soil physical properties, *Soil Tech.,* 3:113–129.

Skidmore, E.L. and D.H. Powers. 1982. Dry soil aggregate stability: Energy based index, *Soil Sci. Soc. Am. J.,* 46:1274–1279.

Soane, B.D. and C. van Ouwerkerk. 1994. Soil compaction in crop production, Elsevier Publishers, Amsterdam, Netherlands.

Stepniewski, W., J. Glinski, and B.C. Ball. 1994. Effects of soil compaction on soil aeration properties, B.D. Soane and C.van Ouwerkerk (eds.), *Soil Compaction in Crop Production,* 167–190, Elsevier Publishers, Amsterdam, Netherlands.

Taubner, H. 1993. Stoffdynamik unterschiedlich strukturierter immissionsbelasteter Böden — Vergleichende Untersuchungen an Gesamtboden und Aggregaten, dissertation, Christian-Albrechts University, Kiel, Germany.

Tippkötter, R. 1988. Aspekte der Aggregierung, Habilitationsschrift, University of Hannover, Germany.

Tolchel'nikov, Yu.S., E.M. Samoylova, A.M. Grebennikov, E.A. Kondrashkin, and A.V. Mazur. 1991. Influence of fissuring on water regime of southern chernozems of Western Siberia, *Sov. Soil. Sci.,* 11:24–32.

Trein, C.R. 1995. The mechanisms of soil compaction under wheels, Ph.D. thesis, Silsoe College, Crainfield University, Silsoe, UK.

van den Akker, J.J.H., and H.J. Stuiver. 1989. A sensitive method to measure and visualize deformation and compaction of the subsoil with a photographed point grid, *Soil Till. Res.,* 14:209–217.

van den Akker, J.J.H., W.B.M. Arts, A.J. Koolen, and H.J. Stuiver. 1994. Comparison of stresses, compactions and increase of penetration resistances caused by a low ground pressure tyre and a normal tyre, *Soil Till. Res.,* 29:125–134.

Veenhof, D.W. and R.E.A. McBride. 1996. Overconsolidation in agricultural soils, I. Compression and consolidation behaviour of remoulded and structured soils, *Soil Sci. Soc. Am. J.,* 60:362–373.

Vetterlein, D. and H. Marschner. 1993. Use of a microtensiometer technique to study hydraulic lift in a sandy soil planted with pearl millet (*Pennisetum americanum* [L.]Leeke), *Plant Soil,* 149:275–282.

Voorhees, W.B. and R.R. Allmaras. 1998. Longterm soil degradation in the United States by compaction from heavy agricultural machinery, H.P. Blume et al. (ed.), Towards Sustainable Land Use, *Advances in Geoecology,* 31, 1625, *Catena Verlag,* Cremlingen, Germany.

Yong, E. and B. Warkentin. 1966. *Introduction to Soil Behavior,* Macmillan, New York, NY.

Youngs, E.G. and P.B. Leeds-Harrison. 1990. Aspects of transport processes in aggregated soils, *J. Soil Sci.,* 41:665–675.

# 3 Soil Water Content and Water Potential Relationships

*Dani Or and Jon M. Wraith*

## 3.1 INTRODUCTION

Water in soil occupies pore spaces that arise from the physical arrangement of the particulate solid phase, competitively and often concurrently with the soil gas phase (Chapters 1, 7, 8). While hidden from casual view, highly substantial volumes of water are commonly stored in soils. For example, one ha of medium textured soil 1 m deep and having field capacity water content of 20% by volume may store sufficient water to fill 4,000 200-L barrels. This reservoir serves as a substantial buffer, thus enabling consistent plant growth in areas having scattered or sporadic precipitation. Other soil organisms, many of which are beneficial, also rely heavily on the water holding characteristics of soils for their existence. On the other hand, soil water is a highly dynamic entity, exhibiting substantial variation in both time and space. This is particularly true near the soil surface, and in the presence of active plant roots. Changes in soil water content and its energy status affect many soil mechanical properties including strength, compactibility and penetrability, and may cause changes in the bulk density of swelling soils. The liquid phase characteristics affect the soil gaseous phase and the rates of exchange between these phases, as well as other important soil properties such as the hydraulic conductivity which governs the rate of water and soluble chemical flow.

The purpose of this chapter is to introduce basic concepts related to the amount and energy state of water in soil. These concepts are prerequisite to quantify and manage soil water storage, to obtain predictions concerning rates and directions of water flow and solute transport, to utilize soils as building or foundation materials, and for many other purposes. The term soil water is used here to represent the soil liquid phase which is typically a water solution containing dissolved salts, organic substances and gases.

The water status in soils is defined by: (1) the amount of water in the soil, or soil water content ($\theta$), and (2) the force by which water is held in the soil matrix, soil energy content or soil water potential ($\psi$). These soil water attributes are related to each other through a function known as the soil water characteristic (SWC).

## 3.2 SOIL WATER CONTENT

Many agronomic, hydrologic and geotechnical practices require knowledge of the amount of water contained in a particular soil volume. Some of the most common methods used to characterize and determine soil water content, on a mass or volume basis, in the laboratory or *in situ*, will be described.

### 3.2.1 DEFINITIONS

#### 3.2.1.1 Soil Water Content on Mass Basis (Gravimetric)

Mass or gravimetric soil water content is expressed relative to the mass of oven dry soil according to:

$$\theta_m = \frac{\text{mass of water}}{\text{mass of dry soil}} = \frac{(\text{mass wet soil})-(\text{mass oven dry soil})}{\text{mass oven dry soil}} \qquad [3.1]$$

and has units of kg kg$^{-1}$ or other consistent mass units.

0-8493-0837-2/02/$0.00+$1.50
© 2002 by CRC Press LLC

### 3.2.1.2  Soil Water on Volume Basis

It is often desirable to express water content on a volume basis. The volumetric water content ($\theta_v$) is defined as the volume of water per bulk volume of soil:

$$\theta_v = \frac{\text{volume of water}}{\text{bulk volume of soil}} = \frac{(\text{mass of water/density of water})}{\text{bulk soil volume}} \qquad [3.2]$$

It also represents the depth ratio of soil water (i.e., the depth of water per unit depth of soil). The conversion between gravimetric and volumetric water contents requires knowledge of the soil dry bulk density ($\rho_b$) which is the ratio of oven dry soil mass to its original volume, and the density of water. The conversion formula is given by:

$$\theta_v = \theta_m \frac{\rho_b}{\rho_w} \qquad [3.3]$$

where $\rho_w$ is the density of water (1000 kg m$^{-3}$ at 20°C). Alternatively, a soil sample of known original volume may be processed as in the gravimetric method, and $\theta_v$ determined as: [(mass water/water density)]/sample volume.

### 3.2.1.3  Water Content on Relative Saturation Basis

An additional means of characterizing the soil water content is in terms of the degree of saturation:

$$\Theta = \frac{(\text{volume of water filled pore space})}{\text{total volume of soil pore space}} = \frac{\theta_v}{\theta_{vs}} \qquad [3.4]$$

where $\theta_{vs}$ is volumetric soil water content under completely water-saturated conditions. This index ranges from zero in completely dry soil to unity in a saturated soil. The degree of saturation is also commonly termed effective saturation or relative water content.

### 3.2.1.4  Soil Water Storage

It is often convenient to express the quantity of soil water in a specific soil depth increment in terms of soil water storage or equivalent depth of soil water (units of length). Equivalent depth of soil water, $D_e$ (m), is calculated as $D_e = \theta_v * D$, where D is the soil depth increment (m) having volume water content $\theta_v$. This quantity is useful to relate aboveground water dimensions of rainfall, irrigation or evaporation (L) to belowground dimensions ($\theta_v$, m$^3$ water m$^{-3}$ soil). For example, one may wish to quantify changes in soil water content arising from addition of water by rainfall or loss of water by evaporation (Chapter 8). For suitable accuracy it is often necessary to sum the equivalent depth relationship over discrete soil depth layers having distinct water contents:

$$D_e = \sum_{i=1}^{n} \theta_{vi} D_i \qquad [3.5]$$

where i denotes depth increments.

## 3.2.2 Relationships to Soil Solid and Gaseous Phases

Because the soil pore spaces are shared between soil water and soil air, a few basic interrelationships among the amount of pore space and the soil water and air contents will be addressed. These concepts are treated in greater detail in Chapter 1.

The total volume of spaces in a soil not occupied by the solid phase is termed the porosity ($\phi$). The porosity indicates pore volume in the soil relative to total soil volume and its value generally lies in the range 0.3 to 0.6. Coarse-textured soils tend to have less total pore space than fine-textured soils, because of the nature of intraparticle packing arising from different solid particle sizes (Chapter 1). Typical porosity values range from about 0.3 in coarse (sandy) soils to about 0.6 in fine (clayey) soils. The porosity may be quantified as $\phi = 1 - \rho_b/\rho_p$, where the ratio of soil bulk density ($\rho_b$) to density of the soil solids ($\rho_p$) characterizes the relative proportion of soil solids; thus, soil voids (pores) occupy the remaining volume. On average, volumetric, areal and lineal porosities are assumed to be equal.

A related index is the air-filled porosity ($\phi_a$) which is a complementary quantity to $\theta_v$, and describes the relative volume of total soil volume occupied by soil air. Because $\phi$ may be occupied by soil water and/or air, $\phi_a$ may be conveniently characterized as: $\phi_a = \phi - \theta_v$. The expected ranges for $\phi_a$ in soils of varying texture are thus similar to those for $\theta_v$.

## 3.2.3 Measurement of Soil Water Content

### 3.2.3.1 Thermogravimetry

This is a direct and destructive method whereby a soil sample is obtained by augering or coring into the soil; its volume need not be known. The sample is weighed at its initial wetness and then dried to remove interparticle absorbed water, but not structural water trapped within clay lattices known as crystallization water. The conventional protocol is to oven dry samples at 105°C until the soil mass becomes stable; this usually requires 24 to 48 h or more, depending on the sample size, wetness and soil characteristics (texture, aggregation, etc.). The difference between the wet and dry weights is the mass of water held in the original soil sample (Section 3.2.1). Despite the somewhat arbitrary specification of a standard oven temperature (105°C) and the variable drying period (depending on specific conditions), the gravimetric method is considered the standard against which many indirect techniques are calibrated. The method is not without bias and/or error, however, and Gardner (1986) discussed some sources of these including the potential for water loss between sampling time and initial weighing, precision of the three weights involved (wet, dry and tare), and the unknown amount of soil texture-dependent residual water associated with the clay fraction.

The primary advantages of gravimetry are the direct and relatively inexpensive processing of samples. The shortcomings of this method are its labor- and time-intensive nature, the time delay required for drying (although this may be shortened by use of a microwave rather than conventional oven, the methodology has not yet been standardized), and the fact that the method is destructive, thereby prohibiting repetitive measurements within the same soil volume.

### 3.2.3.2 Neutron Scattering

This is a nondestructive but indirect method commonly used for repetitive field measurement of volumetric water content. It is based on the propensity of H nuclei to slow (thermalize) high energy fast neutrons. A typical neutron moisture meter consists of: (1) a probe containing a radioactive source that emits high energy (2–4 MeV) fast (1600 km s$^{-1}$) neutrons, as well as a detector of slow neutrons; (2) a scaler to electronically monitor the flux of slow neutrons; and optionally (3) a datalogger to facilitate storage and retrieval of data (Figure 3.1). The radioactive source commonly contains a mixture of [241]Am and Be at 10 to 50 mCi. The [241]Am emits $\beta$ particles which strike the Be and cause emission of fast neutrons.

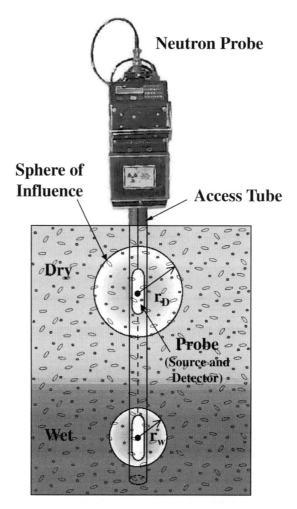

**FIGURE 3.1** An illustration of a neutron probe device for measuring soil water content $r_D$ and $r_w$ represent different radii of measurement in dry and wet soils.

When the probe is lowered into an access tube, fast neutrons are emitted radially into the soil where they collide with various atomic nuclei. Collisions with most nuclei are virtually elastic, causing only minor loss of kinetic energy by the fast neutrons. Collisions with H nuclei, which have similar mass to neutrons, cause a significant loss of kinetic energy and slow down the fast neutrons [consider a marble (neutron) colliding with a similarly sized ball bearing (H nucleus) versus a bowling ball (larger atomic nucleus)]. When, as a result of repeated collisions, the speed of fast neutrons diminishes to those at ambient temperature (about 2.7 km s$^{-1}$), with corresponding energies of about 0.03 eV, they are called thermalized or slow neutrons. Thermalized neutrons rapidly form a cloud of nearly constant density near the probe, where the flux of the slow neutrons is measured by the detector. The average loss of the neutrons' kinetic energy, thus the relative number of slow neutrons, is therefore proportional to the amount of H nuclei in the surrounding soil. The primary source of H in soil is water; other sources of H in a given soil are assumed to be constant and are accounted for during calibration. Although several non H substances including C, Cd, Bo, Cl and Li which may be present in trace amounts in some soils may also thermalize fast neutrons, these may generally also be effectively compensated through soil specific calibration.

Calibration of the neutron probe is thus required to account for background H sources and other local effects (soil bulk density, trace neutron attenuators), and is conveniently achieved by

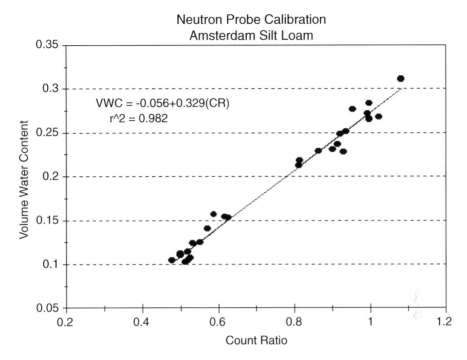

**FIGURE 3.2** A neutron moisture meter calibration relationship.

paired measurements of soil water content and neutron probe counts. The calibration curve (Figure 3.2) is usually linear and relates $\theta_v$ to slow neutron counts or count ratio (CR):

$$\theta_v = a + b(CR) \qquad [3.6]$$

where CR is the ratio of slow neutron counts at a specific location in the soil to a standard count obtained with the probe in its shield. For many soils, the calibration relation is approximately the same. Use of the count ratio rather than raw slow neutron counts compensates for the slow decay of the radioactive source over time.

The sphere of influence about the radiation source varies between about 15 cm (wet soil) to perhaps 70 cm (very dry soil), depending on how far fast neutrons must travel in order to collide with a requisite number of H nuclei. An approximate equation for the radius of influence (r, in cm) as a function of soil wetness is (IAEA, 1970):

$$r = 15\left(\theta_v\right)^{-1/3} \qquad [3.7]$$

Thus, the neutron scattering method is unsuitable for measurement near the soil surface because a portion of the neutrons may escape the soil. Advantages of this method include the ability to repeatedly measure $\theta_v$ at the same locations, averaging of the measured water content over a substantial soil volume, moderate equipment cost (generally about $4,000 to $6,000), and ability to measure soil water content at multiple depths and locations using the same equipment. Limitations or disadvantages include the radiation hazard and attendant licensing requirements, relatively poor (and uncertain) spatial resolution, unsuitability for near-surface measurements, and the soil-specific calibration requirement.

### 3.2.3.3 Electric and Dielectric Methods

Measurement methods based on changes in soil electric properties due to changes in soil water content have been used for decades (Smith-Rose, 1933; Babb, 1951), mostly in the area of exploration geophysics. Background information concerning such applications may be found in a number of sources, including Olhoeft (1985), who addressed low frequency electrical properties of porous media, Selig and Mansukhani (1975), who discussed early attempts to use resistance and capacitance techniques for soil water content measurements, and Hoekstra and Delaney (1974) concerning soil dielectric properties at very high frequencies. Presently, the most common electric approaches for soil water content measurement may be grouped according to: (1) electrical resistance techniques (Spaans and Baker, 1992), (2) capacitance methods (Dean et al., 1987), (3) time domain reflecto-metry (TDR) methods (Topp et al., 1980; Dalton et al., 1984), and (4) combinations of frequency domain, resonance and capacitive techniques (von Hippel, 1954; Hilhorst and Dirksen, 1994).

### 3.2.3.4  Electric Resistance Methods

Changes in the electrical resistivity of soils with changes in water content (and with soluble ionic constituents) have been used to develop simple and inexpensive sensors to infer soil water status. These sensors usually consist of concentric or flat electrodes embedded in a porous matrix and connected to lead wires for measurement of electrical resistance within the sensor's porous matrix. The commonly used term gypsum block arises from early models which were in fact made of gypsum (Bouyoucos and Mick, 1940), and from the practice of saturating the matrix of many sensors made from alternative materials with gypsum to buffer local soil ionic effects. The sensor is embedded in the soil and allowed to equilibrate with the soil solution. The matric potential of water in the sensor is determined from the measured electrical resistance through previously determined calibration of the sensor itself (electrical resistance versus matric potential). Under equilibrium conditions, the sensor matric potential is equal to the soil water matric potential (below); however, the sensor water content may be different than the soil. Hence, these measurements are often used to infer soil water matric potential from which the soil water content may be estimated based on a known relationship between these quantities (Gardner, 1986). With proper calibration for a particular soil, the sensor could be used to infer soil water content directly (Kutilek and Nielsen, 1994). The main advantages of electrical resistance sensors are their low cost and simple measurement requirements. Measurements may be obtained using a simple resistance meter, or more conveniently acquired automatically using a data logger. On the other hand, the usual require-ment for specific calibration of each sensor and for each soil to obtain acceptable accuracy, and lack of sensitivity under wet conditions, render this measurement method appropriate mostly as a qualitative indicator of soil water status (Spaans and Baker, 1992).

### 3.2.3.5  Capacitance Methods

The electrical capacitance of two electrodes inserted in the soil is dependent upon the soil dielectric constant. The bulk soil dielectric constant is dominated by the dielectric constant of soil water ($\varepsilon_w \approx 80$) because of its large magnitude relative to that of soil solids ($\varepsilon_s \approx 5$) and soil air ($\varepsilon_a = 1$). The basic relationship between the soil dielectric constant and the electrical capacitance between two parallel plates of area A and spacing d is:

$$C = \frac{A\varepsilon * \varepsilon_0}{d} \qquad [3.8]$$

where $\varepsilon *$ is the relative complex dielectric constant (or permittivity) of the soil which contains both real ($\varepsilon'$) and imaginary ($\varepsilon''$) components, with $\varepsilon * = \varepsilon' - i\varepsilon''$, $i = \sqrt{-1}$, and $\varepsilon_0$ the dielectric permittivity of free space = $8.854 \times 10^8$ F m$^{-1}$. Note that the relative dielectric permittivity ($\varepsilon *$) is the ratio of the complex dielectric permittivity to the permittivity of free space = $\varepsilon */\varepsilon_0$. In most applications, only the real part of the dielectric is considered and the subscript 'b' is commonly used to denote the relative complex dielectric constant of the bulk soil. Capacitance devices for field water content

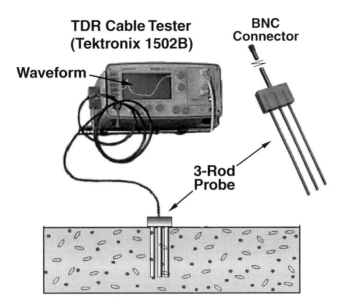

**FIGURE 3.3** TDR cable tester with 3-rod probe embedded vertically in surface soil layer.

measurement are often based on annular conductor design rather than parallel plates (Kutilek and Nielsen, 1994) to facilitate depth measurements through boreholes (Dean et al., 1987).

Commercial capacitance soil water gauges often use a resonant LC circuit relating changes in resonant frequency of the circuit to changes in the soil capacitance (Dean et al., 1987; Evett and Steiner, 1995). Some gauges use a more direct capacitance bridge method to determine the unknown soil capacitance. Among the common capacitance-based measurement systems capable of soil water content profiling are the Sentry 200-AP (Troxler, NC, USA) evaluated by Evett and Steiner (1995) which uses access tubes, and a permanently installed modular bank of sensors (EnviroScan, Sentek, Australia) evaluated by Paltineanu and Starr (1997).

Advantages of capacitance methods include their lack of radiation hazard, and lower expense than transmission line approaches such as TDR (below). They share the neutron probe's variable and uncertain measurement volume, and annular air gaps around sensors that utilize access tubes can cause substantial measurement errors. Hence, buried probe designs seem to perform more reliably at present than those inserted into soil access tubes. Finally, capacitance methods share similar issues of relating the measured dielectric constant to soil water content as do other dielectric-based approaches.

### 3.2.3.6 Time Domain Reflectometry

Progress in devices for accurate measurement of reflected electromagnetic signals traveling in transmission cables and waveguides led to the introduction of several measurement methods for soil dielectric properties using waveguides embedded in the soil, such as time domain reflectometry (TDR) (Figure 3.3).

TDR measures the apparent dielectric constant of the soil surrounding a waveguide, at microwave frequencies of MHz to GHz. The propagation velocity (v) of an electromagnetic wave along a transmission line (waveguide) of length L embedded in the soil is determined from the time response of the system to a pulse generated by the TDR cable tester. The propagation velocity (v = 2L/t) is a function of the soil bulk dielectric constant according to:

$$\varepsilon_b = \left(\frac{c}{v}\right)^2 = \left(\frac{ct}{2L}\right)^2$$

[3.9]

where c is the velocity of electromagnetic waves in vacuum ($3 \times 10^8$ m s$^{-1}$), and t is the travel time for the pulse to traverse the length of the embedded waveguide in both directions (down and back). The soil bulk dielectric constant is governed by the dielectric constant of liquid water ($\varepsilon_w \approx 80$), as the dielectric constants of other soil constituents are much smaller, e.g., soil minerals ($\varepsilon_s \approx 3$ to 5), frozen water (ice $\varepsilon_i \approx 4$) and air ($\varepsilon_a = 1$). This large disparity of the dielectric constants makes the method relatively insensitive to soil composition and texture (other than organic matter and some clays), and thus, a good method for liquid soil water measurement. Also, because the dielectric constant of frozen water is much lower than for liquid, the method may be used in combination with a neutron probe or other technique which senses total soil water content, to separately determine the volumetric liquid and frozen water contents in frozen or partially frozen soils (Baker and Allmaras, 1990).

Two basic approaches have been used to establish the relationships between $\varepsilon_b$ and $\theta_v$. The first approach is empirical, whereby mathematical expressions are simply fitted to observed data without using any particular physical model. For example, Topp et al. (1980) found good agreement using a third order polynomial between $\varepsilon_b$ and $\theta_v$ for multiple soils. The second approach uses a model of the dielectric constants and the volume fractions of each of the soil components to derive a predictive relationship between the composite (bulk) dielectric constant and soil water (i.e., a specific component). Such a physically based approach, called a dielectric mixing model, was adopted by Dobson et al. (1985), Roth et al. (1990), and others.

TDR calibration establishes the relationship between $\varepsilon_b$ and $\theta_v$. For this discussion, calibration is conducted in a fairly uniform soil without abrupt changes in soil water content along the waveguide. The empirical relationship for mineral soils as proposed by Topp et al. (1980):

$$\theta_v = -5.3\times10^{-2} + 2.92\times10^{-2}\varepsilon_b - 5.5\times10^{-4}\varepsilon_b^2 + 4.3\times10^{-6}\varepsilon_b^3 \qquad [3.10]$$

provides adequate description for many soils and for the water content range $\theta < 0.5$ (which covers the entire range of interest in most mineral soils), with an estimation error of about 0.013 for $\theta_v$. However, Equation [3.10] fails to adequately describe the $\varepsilon_b - \theta_v$ relationship for water contents exceeding 0.5, and for organic rich soils, mainly because Topp's calibration was based on experimental results for mineral soils and concentrated in the range of $\theta_v < 0.5$.

Roth et al. (1990) proposed a physically based calibration model which considers dielectric mixing of the constituents and their geometric arrangement. According to this mixing model, $\varepsilon_b$ of a three-phase system may be expressed as:

$$\varepsilon_b = \left[\theta_v\varepsilon_w^\beta + (1-\varphi)\varepsilon_s^\beta + (\varphi-\theta_v)\varepsilon_a^\beta\right]^{\frac{1}{\beta}} \qquad [3.11]$$

where $\varphi$ is soil porosity, $-1 < \beta < 1$ summarizes the geometry of the medium in relation to the axial direction of the wave guide ($\beta = 1$ for an electric field parallel to soil layering, $\beta = -1$ for a perpendicular electrical field, and $\beta = 0.5$ for an isotropic two-phase mixed medium), $1 - \varphi$, $\theta_v$ and $\varphi - \theta_v$ are the volume fractions and $\varepsilon_s$, $\varepsilon_w$ and $\varepsilon_a$ are the dielectric constants of the solid, aqueous and gaseous phases, respectively. Note that these components sum to unity. Rearranging Equation [3.11] and solving for $\theta_v$ yields:

$$\theta_v = \frac{\varepsilon_b^\beta - (1-\varphi)\varepsilon_s^\beta - \varphi\varepsilon_a^\beta}{\varepsilon_w^\beta - \varepsilon_a^\beta} \qquad [3.12]$$

which determines the relationship between $\varepsilon_b$ as measured by TDR and $\theta_v$. Many have used $\beta = 0.5$ which is shown by Roth et al. (1990) to produce a calibration curve very similar to the third

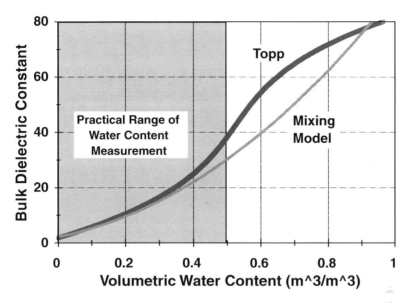

**FIGURE 3.4** Relationships between bulk soil dielectric constant and $\theta_v$ expressed by two commonly used TDR calibration approaches.

order polynomial in Equation [3.10] for the water content range $0 < \theta_v < 0.5$. Introducing into Equation [3.12] common values for each component such as $\beta = 0.5$, $\varepsilon_w = 81$, $\varepsilon_s = 4$, and $\varepsilon_a = 1$ yields the following simplified form:

$$\theta_v = \frac{\sqrt{\varepsilon_b} - (2 - \varphi)}{8}$$ [3.13]

A comparison between Equation [3.10] and a calibration curve based on Equation [3.12] with $\varphi = 0.5$ is depicted in Figure 3.4.

The main advantages of TDR over other methods for repetitive soil water content measurement are: (1) superior accuracy to within 1 or 2% of $\theta_v$, (2) minimal calibration requirements (in many cases soil specific calibration is not needed), (3) no radiation hazards associated with neutron probe or gamma attenuation techniques, (4) excellent spatial and temporal resolution, and (5) simple to obtain continuous soil water measurements through automation and multiplexing. Limitations of the TDR method include relatively high equipment expense, limited applicability under highly saline conditions due to signal attenuation, and the soil-specific calibration required for soils with high clay or organic matter contents. Severe attenuation of TDR waveforms in the presence of high salinity and/or some clays having high surface area and surface charges may interfere with or even preclude water content measurements.

### 3.2.3.7 Frequency Domain and Other Methods

Several new sensors and measurement methods are based on combinations of capacitive, reflective and frequency-shift principles, all of which are governed by the soil dielectric properties. This trend appears highly promising for the development of accurate and cost effective sensors for soil water content measurement and will likely dominate future developments in this area. An example of such a stand-alone sensor (Figure 3.5) is the water content reflectometer (Campbell Scientific Inc., Logan, UT) which provides an indirect measurement of soil volumetric water content based on changes in soil dielectric permittivity. High speed electronic components are configured in an oscillator circuit which is connected to parallel rods acting as a waveguide. The rods are inserted

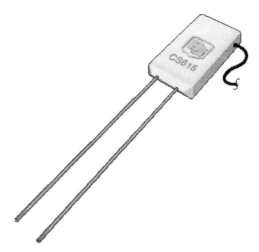

**FIGURE 3.5** A water content reflectometer sensor. [Source: Campbell Scientific, Inc., Logan, UT, with permission.]

in the monitored soil volume (typical rod length is 0.3 m). As soil water content changes, the resultant dielectric property causes a shift in the oscillation frequency of the circuit. A calibration relationship is established between the output frequency of the circuit and $\theta_v$. The time required for the actual measurement is less than 20 milliseconds. The method is sensitive to soil electrical conductivity and an adjustment must be made to the calibration when soil solution conductivity exceeds about 2 dS m$^{-1}$. Another commercially available stand-alone sensor based on frequency domain (FD) measurements is described in detail by Hilhorst and Dirksen (1994).

Other methods for soil water measurement include gamma ray attenuation techniques using dual probe apparatus for bulk density and water content, x-ray computed tomography, methods based on fiber optics, nuclear magnetic resonance (NMR) and an array of geophysical methods (including ground penetrating radar and electrical resistivity sounding). Partial dependence of porous media thermal properties on water content can form the basis for indirect inference of water content or matric potential based on measured thermal responses. A few of these are discussed in Section 3.3.4.4. Additional information on these and other methods may be found in Klute (1986), Baker (1990), Campbell and Mulla (1990), and Carter (1993).

### 3.2.4 APPLICATIONS OF SOIL WATER CONTENT INFORMATION: THE WATER BALANCE

A primary use of soil water content information is for evaluation of the hydrologic water balance described by:

$$P + I = ET + D + R - \Delta W \qquad [3.14]$$

where P is precipitation, I is irrigation, ET is evapotranspiration (evaporative water loss from the soil and from plants), D is drainage or deep percolation, R is surface runoff, and $\Delta W$ is change in water storage within the profile (soil water depletion). W is defined as the equivalent depth of water (De) stored in the soil profile under consideration, and $\Delta W = (W_{initial} - W_{final})$ (Figure 3.6). These parameters are all associated with a given specific time interval. The convention used here is that inputs to the soil profile are taken as positive, and outputs negative. The concept is based on conservation of mass (water), and is similar to the familiar exercise of balancing inputs and outlays from a checking account, for example. Under typical conditions, $\Delta W$ is fairly significant over the short term (weeks to months), but generally evens out to about zero over one to several years.

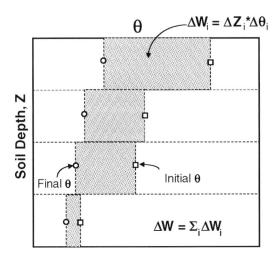

**FIGURE 3.6** Schematic of soil water depletion calculations for a soil profile divided into four depth increments. Total change in water storage is the sum of depletion in each layer.

### 3.2.4.1 Field Capacity, Wilting Point and Plant-Available Soil Water

Observations of water content changes in the soil profile following wetting by irrigation or rainfall show that the rate of change decreases in time. In some cases, water content attains a nearly constant value within 1 to 2 days after wetting, following soil water redistribution in response to internal drainage. Field capacity is defined as the water content at which internal drainage becomes essentially negligible. The attainment of a near constant water content at field capacity ($\theta_{vFC}$) is not always assured. It is dependent on: (1) the depth of wetting and the antecedent (initial) water content of the soil profile (for a soil which is moist at the onset of wetting and for deep wetting, the rate of redistribution is slower and the apparent value of $\theta_{vFC}$ is higher), and (2) the presence of impeding layers or a watertable which affect the rate and extent of water redistribution.

Another often misunderstood soil water content-related index is the permanent wilting point, which is defined as the water content at which plants can no longer extract soil water at a rate sufficient to meet physiological demands imposed by loss of water to the atmosphere, and thus irreversibly wilt and die. This water content ($\theta_{vWP}$) is primarily dependent on the soil's ability to transmit water, but also to some degree on the plant's ability to withstand or mitigate drought. Though commonly taken as $\theta_v$ at −1.5 MPa (−15 bars) matric potential, there is substantial variation among plant species in their abilities (and strategies) to resist soil drought, with some plants surviving to well below this standard wilting point index. The permanent wilting point should not be confused with the phenomenon of transient wilting, which is commonly observed during the afternoon when evaporative demand is greatest. In this case plants are able to rehydrate to some extent at night.

A primary practical use of field capacity and wilting point concepts is the determination of a plant available soil water range (PASW) (Figure 3.7). Soil water storage available for plant use is generally calculated as being between field capacity and wilting point ($\theta_{vFC} - \theta_{vWP}$), as water contents higher than $\theta_{vFC}$, while usually plant-available, are generally not sustained for long times except under specific circumstances. Plant-available soil water storage is an important factor in the determination of irrigation amounts for a cropped field or other soil-plant system. For practical purposes, irrigation amounts in excess of field capacity are lost to deep percolation, and thus, should be avoided in the interests of water resource efficiency as well as potential leaching of soluble chemicals (Chapter 6).

A useful rule of thumb is to estimate $\theta_{FC}$ as $\theta_s/2$, and $\theta_{WP}$ as $\theta_{FC}/2$; in other words, a soil exhibiting this property will have lost 50% of its saturated water content at field capacity, and another 50% of the remaining water by the wilting point. Richards (1954) was probably the first

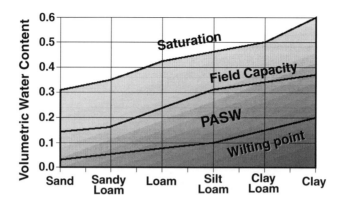

**FIGURE 3.7** Schematic of estimated plant-available soil water content (PASW) for a range of soil textural classes.

to make this observation. Banin and Amiel (1970) and Dahiya et al. (1988) have also demonstrated that the assumed ratio of 1:1/2:1/4 is a good approximation, based on measured correlations on many soils. Banin and Amiel (1970) found strong relationships between clay content or surface area (which are strongly related) and the water content at these indices.

### 3.2.4.2  Sources of Additional Information on Soil Water Content Indices

The information on these indices is fragmentary due to their dependency on other conditions in the soil profile (field capacity) and the plant in question (wilting point). The work of Slater and Williams (1965) provides a reasonable framework for estimating field capacity, and more recent information may be found on the Internet. A study by Waters (1980) is a good source of information regarding the wilting point. Dahiya et al. (1988) presented strong correlations among the saturation percentage, field capacity, wilting point and available water based on data for 438 soils, with correlation coefficients in excess of 0.92. Banin and Amiel (1970) provided similar information concerning relationships among clay content, surface area and water content at several index points. Petersen et al. (1996) also related soil surface area and other properties to their water holding capacity.

## 3.3  SOIL WATER ENERGY

As previously stated, water status in soils is characterized by both the amount of water present and by its energy state. Soil water is subjected to forces of variable origin and intensity, thereby acquiring different quantities and forms of energy. The two primary forms of energy of interest here are kinetic and potential. Kinetic energy is acquired by virtue of motion and is proportional to velocity squared. However, because the movement of water in soils is relatively slow (usually $< 0.1$ m h$^{-1}$), its kinetic energy is negligible. The potential energy which is defined by the position of soil water within a soil body and by internal conditions is largely responsible for determining soil water status under isothermal conditions.

Like all other matter, soil water tends to move from where the potential energy is high to where it is lower, in its pursuit of a state known as equilibrium with its surroundings. The magnitude of the driving force behind such spontaneous motion is a difference in potential energy across a distance between two points of interest. At a macroscopic scale, one can define potential energy relative to a reference state. The standard state for soil water is defined as pure and free water (no solutes and no external forces other than gravity) at a reference pressure, temperature and elevation, and is arbitrarily given the value of zero (Bolt, 1976).

### 3.3.1 TOTAL SOIL WATER POTENTIAL AND ITS COMPONENTS

Soil water is subject to several force fields whose combined effects result in a deviation in potential energy relative to the reference state called the total soil water potential ($\psi_T$) and is defined as:

> The amount of work that an infinitesimal unit quantity of water at equilibrium is capable of doing when it moves (isothermally and reversibly) to a pool of water at similar standard (reference) state (similar pressure, elevation, temperature and chemical composition).

It should be emphasized, however, that there are alternative definitions of soil water potential using concepts of chemical potential or specific free energy of the chemical species water (which is different than the soil solution termed soil water here). Some of the arguments concerning the proper definitions and their scales of application are presented by Corey and Klute (1985), Iwata et al. (1988) and Nitao and Bear (1996).

Recognizing that these fundamental concepts are subject to an ongoing debate, only simple and widely accepted definitions of these quantities which are applicable at macroscopic scales, and which yield an appropriate framework for practical applications, will be presented.

The primary forces acting on soil water held within a rigid soil under isothermal conditions can be conveniently grouped as (Day et al., 1967): (1) matric forces resulting from interactions of the solid phase with the liquid and gaseous phases, (2) osmotic forces owing to differences in chemical composition of the soil solution, and (3) body forces induced by gravitational and other inertial force fields (centrifugal).

The thermodynamic approach whereby potential energy rather than forces are used is particularly useful for equilibrium and flow considerations. Equilibrium would require the vector sum of these different forces acting on a body of water in different directions to be zero; this is an extremely difficult criterion to deal with in soils. On the other hand, potential energy defined as the negative integral of the force over the path taken by an infinitesimal amount of water, when it moves from a reference location to the point under consideration, is a scalar quantity. Consequently, the total potential can be expressed as the algebraic sum of the component potentials corresponding to the different fields acting on soil water:

$$\psi_T = \psi_m + \psi_s + \psi_p + \psi_z \qquad [3.15]$$

$\psi_m$ is the matric potential resulting from the combined effects of capillarity and adsorptive forces within the soil matrix. Dominating mechanisms for these effects include: (1) adhesion of water molecules to solid surfaces due to short-range London-van der Waals forces and extension of these effects by cohesion through H bonds formed in the liquid, (2) capillarity caused by liquid-gas and liquid-solid-gas interfaces interacting within the irregular geometry of soil pores, and (3) ion hydration and binding of water in diffuse double layers. There is some disagreement regarding the definition of this component of the total potential. Some consider all contributions other than gravity and solute interactions (at a reference atmospheric pressure). Others use a tensiometer (Section 3.3.4) to measure and provide a practical definition of the matric potential in a soil volume of interest (Hanks, 1992). The value of $\psi_m$ ranges from zero when the soil is saturated to negative numbers when the soil is dry (note that $\psi_m = 0$ mm is $> \psi_m = -1000$ mm). Because it is often more convenient or intuitive to work with positive than negative quantities, the terms matric suction or tension are commonly used. Each of these represents the absolute value of $\psi_m$.

$\psi_s$ is the solute or osmotic potential determined by the presence of solutes in soil water, which lower its potential energy and its vapor pressure. The effects of $\psi_s$ are important in the presence of: (1) appreciable amounts of solutes, and (2) a selectively permeable membrane or a diffusion barrier which transmits water more readily than salts. The effects of $\psi_s$ are otherwise negligible when only liquid water flow is considered and no diffusion barrier exists. The two most important diffusion barriers in the soil are (1) soil-plant root interfaces (cell membranes are selectively permeable), and (2) soil-water-air interfaces; thus, when water evaporates, salts are left behind. In dilute solutions the solute potential (also called the osmotic pressure) is proportional to the concentration and temperature according to:

$$\psi_s = -RTC_s \qquad [3.16]$$

where $\psi_s$ is in kPa, R is the universal gas constant [$8.314 \times 10^{-3}$ kPa m$^3$ mol$^{-1}$ K$^{-1}$)], T is absolute temperature (K), and $C_s$ is solute concentration (mol m$^{-3}$). A useful approximation which may be used to estimate $\psi_s$ in kPa from the electrical conductivity of the soil solution at saturation ($EC_s$) in dS m$^{-1}$ is

$$\psi_s = -36EC_s \qquad [3.17]$$

$\psi_p$ is pressure potential defined as the hydrostatic pressure exerted by unsupported water (i.e., saturating the soil) overlying a point of interest. Using units of energy per unit weight provides a simple and practical definition of $\psi_p$ as the vertical distance from the point of interest to the free water surface (unconfined watertable elevation). The convention used here is that $\psi_p$ is always positive below a watertable, or zero if the point of interest is at or above the watertable. In this sense, nonzero magnitudes of $\psi_p$ and $\psi_m$ are mutually exclusive; either $\psi_p$ is positive and $\psi_m$ is zero (saturated conditions), or $\psi_m$ is negative and $\psi_p$ is zero (unsaturated conditions), or $\psi_p = \psi_m = 0$ at the free watertable elevation. Another definition that is used in some quarters is to combine $\psi_m$ and $\psi_p$ as used here into a single component that adopts negative magnitude under unsaturated conditions and positive magnitude under saturated conditions.

$\psi_z$ is gravitational potential which is determined solely by the elevation of a point relative to some arbitrary reference point, and is equal to the work needed to raise a body against the Earth's gravitational pull from a reference level to its present position. When expressed as energy per unit weight, the gravitational potential is simply the vertical distance from a reference level to the point of interest. The numerical value of $\psi_z$ itself is thus not important (it is defined with respect to an arbitrary reference level); what is important is the difference or gradient in $\psi_z$ between any two points of interest. This value will not change with different reference point locations.

Soil water is at equilibrium when the net force on an infinitesimal body of water equals zero everywhere, or when the total potential is constant in the system. Though the last statement is a logical consequence of the definitions above, it is not strictly true as pointed out by Corey and Klute (1985). They argue that constant total potential is a necessary but not a sufficient condition, and for thermodynamic equilibrium to prevail, three conditions must be met simultaneously: (1) thermal equilibrium or uniform temperature, (2) mechanical equilibrium meaning no net convection producing force, and (3) chemical equilibrium meaning no net diffusional transport or chemical reaction. In most practical applications, however, the macroscopic definition of the total potential and equilibrium conditions based on it is completely adequate (Kutilek and Nielsen, 1994).

The difference in chemical and mechanical potentials between soil water and pure water at the same temperature is known as the soil water potential ($\psi_w$):

$$\psi_w = \psi_m + \psi_s + \psi_p \qquad [3.18]$$

Note that the gravitational component ($\psi_z$) is absent in this definition. Soil water potential is thus the result of inherent properties of soil water itself, and of its physical and chemical interactions with its surroundings, whereas the total potential includes the effects of gravity (an external and ubiquitous force field).

**TABLE 3.1**
**Units, dimensions, and common symbols for potential energy of soil water.**

| Units | Symbol | Name | Dimensions | SI units | cgs units |
|---|---|---|---|---|---|
| Energy/mass | $\mu$ | Chemical potential | $L^2 t^{-2}$ | $J\ kg^{-1}$ | $erg\ g^{-1}$ |
| Energy/volume | $\psi$ | Soil water potential, suction or tension | $M\ (Lt^2)^{-1}$ | $N\ m^{-2}$ (Pa) | $erg\ cm^{-1}$ |
| Energy/weight | $h$ | Pressure head | $L$ | m | cm |

Total soil water potential and its components may be expressed in several ways depending on the definition of a unit quantity of water. Potential may be expressed as (1) energy per unit of mass, (2) energy per unit of volume or (3) energy per unit of weight. A summary of the resulting dimensions, common symbols and units are presented in Table 3.1.

Only $\mu$ has actual units of potential; $\psi$ has units of pressure, and $h$ head of water. However, the above terminology (potential energy versus units of potential) is widely used in a generic sense in the soil and plant sciences. The various expressions of soil water energy status are equivalent, with:

$$\mu = \psi/\rho_w = gh \qquad [3.19]$$

where $\rho_w$ is density of water (1000 kg m$^{-3}$ at 20°C) and g is gravitational acceleration (9.81 m s$^{-2}$).

### 3.3.2 INTERFACIAL FORCES AND CAPILLARITY

The matric potential is often the largest component of the total potential in partially saturated soils. To better understand the origins of this important potential which attains nonzero values only under partial saturation when all three phases (liquid, gas and solid) are present, some of the properties of water in relation to porous media which give rise to this phenomenon must be discussed.

#### 3.3.2.1 Surface Tension

At the interface between water and solids or other fluids such as air, water molecules are exposed to different forces than are molecules within the bulk water. For example, water molecules inside the liquid are attracted by equal cohesive forces to form H bonds on all sides, whereas molecules at the air-water interface feel a net attraction into the liquid because the density of water molecules on the air side of the interface is much lower and all H bonds are toward the liquid. The result is a membrane-like water surface having a tendency to contract; thus energy is stored in the form of surface tension (as in a stretched spring). Different liquids vary in their liquid-gas (LG) surface tension ($\sigma_{LG}$) expressed as energy per unit area = force per unit length. For example, water at 20°C: 72.7 mN m$^{-1}$ = mJ m$^{-2}$; ethyl alcohol: 22 mN m$^{-1}$ ( = dynes cm$^{-1}$ = erg cm$^{-2}$); and mercury: 430 mN m$^{-1}$.

#### 3.3.2.2 Contact Angle

If liquid is placed in contact with a solid in the presence of a gas (three-phase system), the angle measured from the solid-liquid (SL) interface to the liquid-gas (LG) interface is the contact angle ($\gamma$). For a drop resting on a solid surface at equilibrium, the vector sum of the forces acting to spread the drop (outward) is equal to the opposing forces. This relationship is summarized by Young's equation (Adamson, 1990):

$$\sigma_{LG}\cos\gamma + \sigma_{SL} - \sigma_{GS} = 0 \qquad [3.20]$$

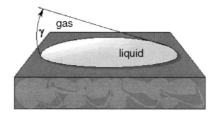

Small Contact Angle
Liquid "wets" the Solid

Large Contact Angle
Liquid "repels" the Solid

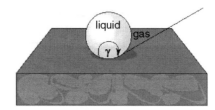

**FIGURE 3.8** Liquid-solid-gas contact angles.

where $\sigma$ is the respective interfacial surface tension. The equilibrium contact angle is therefore:

$$\gamma = \cos^{-1}\left[\frac{\sigma_{GS} - \sigma_{SL}}{\sigma_{LG}}\right] \qquad [3.21]$$

When liquid is attracted to the solid (adhesion) more than to other liquid molecules (cohesion), the angle is small and the solid is said to be wettable by the liquid. Conversely, when the cohesive exceeds the adhesive force, the liquid repels the solid and $\gamma$ is large (Figure 3.8). The contact angle of water on clean glass is very small, and is commonly taken as $0°$. The contact angle of soil water on soil minerals is also commonly assumed $\approx 0°$. The different contact angles of water and other liquids with soil solids, soil air and with each other, are important contributors to the behavior of multiphase organic liquid/water/soil/air mixtures.

### 3.3.2.3 Curved Surfaces and Capillarity

Surface tension is associated with the phenomenon of capillarity. When the liquid-gas interface is curved rather than planar (flat), the resultant surface tension force normal to the liquid-gas interface creates a pressure difference across the interface. The pressure is greater at the concave side of the interface by an amount that is dependent on the radius of curvature and the surface tension of the fluid. For a hemispherical liquid-gas interface having radius of curvature R, the pressure difference is given by the Young-Laplace equation:

$$\Delta P = 2\sigma/R \qquad [3.22]$$

where $\Delta P = P_{liq} - P_{gas}$ when the interface curves into the gas (water droplet in air), or $\Delta P = P_{gas} - P_{liq}$ when the interface curves into the liquid (air bubble in water, water in a small glass tube). In many instances a bubble may not be spherical, or an element of liquid may be confined by irregular solid

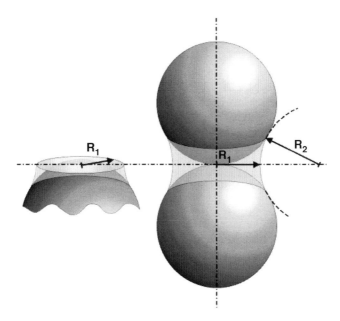

**FIGURE 3.9** Radii of curvature for pendular ring of water between two spherical solids.

surfaces resulting in two (or more) different radii of curvature such as water in pendular rings between two spherical solid particles (Figure 3.9). The Young-Laplace equation for this case is given by:

$$\Delta P = \sigma \left( \frac{1}{R_1} + \frac{1}{R_2} \right)$$

[3.23]

Note that (1) this equation reduces to Equation [3.22] for $R_1 = R_2$; and (2) the sign of $R_i$ is negative for convex interfaces ($R_2 < 0$) and positive for concave interfaces ($R_1 > 0$).

### 3.3.3 THE CAPILLARY MODEL

#### 3.3.3.1 Capillary Rise

When a small cylindrical glass tube (capillary) is dipped in free water, a meniscus is formed as a result of the contact angle between the water and the walls of the tube and from consideration of minimum surface energy. The smaller the tube the larger the degree of curvature, resulting in a larger pressure difference across the air-water (gas-liquid) interface. The pressure at the water side ($P_w$) will be lower than atmospheric pressure ($P_0$). This pressure difference will cause water to rise into the capillary tube until the upward force across the water-air interface is balanced by the weight of water in the tube (Figure 3.10). Because the radius of meniscus curvature $R = r/\cos\gamma$ where r is the tube radius, the height that water will rise in a capillary tube of radius r with $\gamma$ contact angle is:

$$h = \frac{2\sigma \cos o}{\rho_w g r}$$

[3.24]

Combining all the constants in Equation [3.24] (using typical values at room temperature) yields a simple and useful approximation:

$$h(m) = \frac{14.84}{r(\mu m)}$$

[3.25]

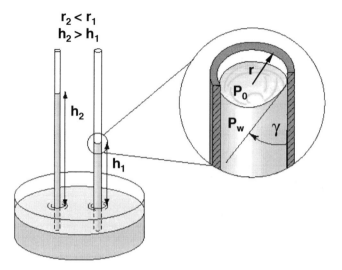

**FIGURE 3.10** Capillary rise in cylindrical glass tubes.

When expressed as potential energy per unit weight, $\psi_m = -h$ and thus may be related to equivalent pore radius through Equation [3.24].

### 3.3.3.2 Conceptual Models for Water in Soil Pore Space

The complex geometry of soil pore space creates numerous combinations of interfaces, capillaries and wedges around which films of water are formed, resulting in a variety of air-water and solid-water contact angles. Water is thus drawn into and/or held by these interstices in proportion to the resulting capillary forces. In addition, water is adsorbed onto solid surfaces with considerable force at close distances. Due to practical limitations of present measurement methods, no distinction is made between the various mechanisms affecting water in porous matrices (i.e., capillarity and surface adsorption). Common conceptual models for water retention in porous media and matric potential rely on a simplified picture of soil pore space as a bundle of capillaries. The primary conceptual steps made in such models are illustrated in Figure 3.11. The representation of soil pores as equivalent cylindrical capillaries greatly simplifies modeling and parameterization of soil pore space. The roles of water films at very low saturation levels, and the unique contribution of surface adsorption to the matric potential are beyond the scope of this chapter. Interested readers are referred to reviews by Nitao and Bear (1996), and Parker (1986).

### 3.3.3.3 The Bubbling Pressure

Application to soils of the conceptual bundle of cylindrical capillaries model discussed in Section 3.3.3.2 yields a distribution of pore sizes (pore radii) as will be elaborated in Section 3.4, and in Example 3.3. The idea is that we can calculate the properties (numbers and radii distribution) of capillary tubes that make up a theoretical bundle that obeys the capillary rise equation (Equation [3.24]) and which will have the same water retention properties as measured for a given body of soil. Thus, soil volume is represented as an equivalent bundle of capillaries from which useful information may be inferred about soil pore space. While clearly nonintuitive at the pore scale (e.g., round tubes compared with irregular soil pores formed by the intersections of angular soil solids), the model works well in many cases for modeling the behavior of water in soil volumes. The lack of consideration of thin liquid water films surrounding soil solids, and of partially-filled pores (the capillary model explicitly categorizes all pores as either water-filled or drained [gas-filled]), are perhaps the largest shortcomings of this conceptual approach.

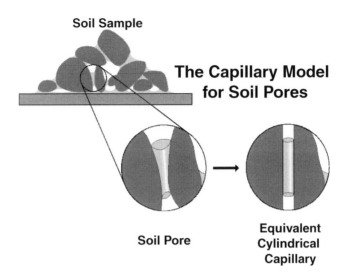

**Soil Sample**

**The Capillary Model for Soil Pores**

**Soil Pore**

**Equivalent Cylindrical Capillary**

**FIGURE 3.11** Concept of equivalent cylindrical capillary to represent soil pore spaces.

The largest pore in the distribution is an important identifier/characteristic of the soil. To gauge the size of the largest pore in the system, ask the following question: given that the (equivalent) pores are cylindrical and liquid filled (according to the capillary rise equation), what would be the positive pressure necessary to evacuate the largest pore in the system? In other words, what would be the pressure required at the atmospheric side of the pore to offset the negative pressure at the liquid side of the meniscus formed in the largest pore? The pressure required to evacuate the largest pore is called the *bubbling pressure* and it signifies the minimum pressure required to start the desaturation process of a soil sample. The term bubbling pressure thus refers to the fact that this critical pressure will force the gas phase (a bubble) through the previously water-filled pore.

The concept of bubbling pressure is also important for design of porous materials that are required to remain saturated to a specified pressure. An example is the tensiometer cup described next. The design question is often posed as follows: what would be the largest allowable pore size in the material if it is supposed to remain saturated (thereby preventing air entry) up to a certain pressure x? The solution is obtained by simply rearranging the capillary rise Equation [3.24] (or [3.25] for water with zero contact angle) and solving for r given the critical pressure value x for h.

### 3.3.4 MEASUREMENT OF SOIL WATER POTENTIAL COMPONENTS

#### 3.3.4.1 Tensiometer for Measuring Soil Matric Potential

A tensiometer consists of a porous cup, usually made of ceramic having very fine pores, connected to a vacuum gauge through a water-filled tube (Figure 3.12). The porous cup is placed in intimate contact with the bulk soil at the depth of measurement. When the matric potential of the soil is lower (more negative) than the equivalent pressure inside the tensiometer cup, water moves from the tensiometer along a potential energy gradient to the soil through the saturated porous cup, thereby creating suction sensed by the gauge. Water flow into the soil continues until equilibrium is reached and the suction inside the tensiometer equals the soil matric potential. When the soil is wetted, flow may occur in the reverse direction, i.e., soil water enters the tensiometer until a new equilibrium is attained. The tensiometer equation is:

$$\psi_m = \psi_{gauge} + \left( z_{gauge} - z_{cup} \right) \qquad [3.26]$$

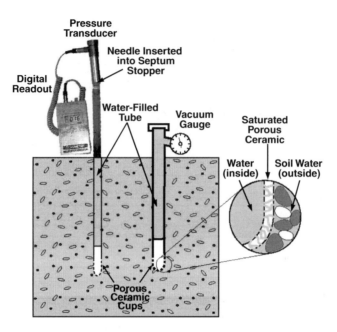

**FIGURE 3.12** Pressure transducer and vacuum gauge tensionmeters showing porous ceramic cup.

The vertical distance from the gauge plane ($z_{gauge}$) to the cup ($z_{cup}$) must be added to the matric potential measured by the gauge (expressed as a negative quantity) to obtain the matric potential at the depth of the cup, when potentials are expressed per unit of weight. This accounts for the positive head at the depth of the ceramic cup exerted by the overlying tensiometer water column.

Electronic sensors called pressure transducers often replace the mechanical vacuum gauges. The transducers convert mechanical pressure into an electric signal which can be more easily and precisely measured. In practice, pressure transducers can provide more accurate readings than other gauges, and in combination with data logging equipment are able to supply continuous measurements of soil matric potential.

The standard tensiometer range is limited to suction values (absolute value of the matric potential) < 100 kPa (1 bar or 10 m head of water) at sea level, and this value decreases proportionally with elevation. Thus, other means are needed to measure or infer soil matric potential under drier conditions.

Peck and Rabbidge (1969) described an osmotic tensiometer that relied on a confined aqueous solution, rather than pure free water as the reference state. A membrane highly impermeable to the confined solution allowed their device to cover the entire range of 0 to –1.5 MPa, unless soil solutes were excluded from the instrument by a vapor gap. Portable tensiometers that may be extended into boreholes for use at depths to several hundred meters have also been developed (Hubbell and Sisson, 1996). Achieving and maintaining adequate hydraulic contact between the porous cup and the soil at these depths was identified as an important issue, but these sensors may be highly useful in some mining, engineering, deep recharge and hazardous waste applications.

**EXAMPLE 3.1**   Historical Note: Who Invented the Tensiometer?

One of the most useful devices for monitoring soil water status is the tensiometer. It consists of a porous cup (usually made of ceramic with very fine pores) connected to a vacuum gauge (mechanical or electronic transducer) through a rigid water-filled tube (Figure 3.12). The porous cup is placed in intimate contact with the bulk soil at the depth of measurement. When the matric potential of the soil is lower (more negative) than the equivalent pressure inside the tensiometer cup, water moves from the tensiometer along a potential energy gradient to the soil through the saturated porous cup, thereby creating suction sensed by the gauge. Water flow into the soil continues until equilibrium is reached and the suction inside the tensiometer equals the soil matric potential. When the soil is wetted, flow may occur in the reverse direction, i.e., soil water enters the tensiometer until a new equilibrium is attained.

In their textbook *Soil Physics*, Marshall et al. (1996) attribute the introduction of the tensiometer to Richards (1928, whose original drawing is shown in Diagram 3.1.1). Cassel and Klute (1986) attribute the first complete description of the tensiometer to an abstract published by Gardner et al. (1922). The description in Gardner's abstract reads: "*The apparatus used consists of a porous cup closed with a water-tight joint and connected through a tall tube to an exhaust pump. The cup is surrounded by a thin layer of soil in the outer vessel and atmospheric pressure is maintained on the soil side. The pressure is then reduced on the water side and measured, and by means of a glass tube the amount of water in the soil is determined.*"

While it has been accepted for quite some time that Willard Gardner (Utah Agricultural Experiment Station, Logan, Utah) invented the tensiometer, Haines (1927) showed a design of a modern tensiometer (Diagram 3.1.2) attributed to Livingston (1918). Inspection of work published by Livingston (1908, 1918) and Livingston and Hawkins (1915) convincingly shows that the tensiometer was invented more than a decade before the description by Gardner et al. (1922).

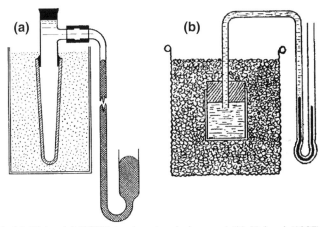

**DIAGRAM 3.1.1** (a) Richards' (1928) tensiometer design; and (b) Haines' (1927) design attributed to Livingston (1915).

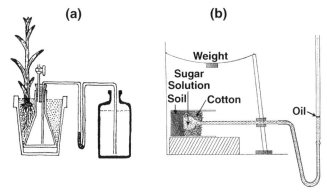

**DIAGRAM 3.1.2** (a) Livingston's (1908, 1918) auto-irrigator for maintaining constant matric potential in potted plant root zone; (b) tensiometer-osmometer designed by Hawkins and Livingston (1915) to measure the "water supplying power of the soil" (notation of original figure was enhanced for clarity).

### 3.3.4.2  Piezometer for Measuring Hydrostatic Pressure Potential

In a saturated soil such as below a watertable, soil water is under positive hydrostatic pressure. The pressure potential ($\psi_p$) equals the vertical distance from a point in the soil to the surface of the free watertable (recall that one expresses potential in terms of distance, length or head when potential energy is expressed per unit of weight). The piezometer is a hollow tube placed in the soil to depths below the watertable. It extends to the soil surface and is open to the atmosphere. The bottom of the piezometer is perforated to allow for soil water under positive hydrostatic pressure to enter the tube. Water enters the tube and rises to a height equal to that of the free watertable. The water level within the piezometer may be determined using a variety of manual or automated measurement techniques.

### 3.3.4.3  Psychrometry for Measuring Water Potential

Under equilibrium conditions the soil water potential is equal to the potential of water vapor in the surrounding soil air. A psychrometer measures the relative humidity (RH) of the water vapor which is related to the water potential of the vapor ($\psi_w$) through the Kelvin equation:

$$RH = e/e_o = \exp^{\left[(M_w \psi_w)/(\rho_w RT)\right]}$$
[3.27]

where e is water vapor pressure, $e_o$ is saturated vapor pressure at the same temperature, $M_w$ is the molecular weight of water (0.018 kg mol$^{-1}$), R is the ideal gas constant (8.31 J K$^{-1}$ mol$^{-1}$ or 0.008314 kPa m$^3$ mol$^{-1}$ K$^{-1}$), T is absolute temperature (K), and $\rho_w$ is the density of water (1,000 kg m$^{-3}$ at 20°C). Rearranging and taking a log-transformation of Equation [3.27] yields:

$$\psi_w = \frac{RT\rho_w}{M_w} \ln\left(e/e_o\right)$$
[3.28]

Equation [3.28] can be further simplified for the range of $e/e_o$ near 1 often encountered in soils (the entire range of plant-available water is between $e/e_o = 1$ and 0.98):

$$\psi_w = \frac{RT\rho_w}{M_w}\left(\frac{e}{e_o} - 1\right) \approx 462T\left(\frac{e}{e_o} - 1\right)$$
[3.29]

for $\psi_w$ in kPa. Note that because most salts are nonvolatile, the psychrometric measurement of $\psi_w$ is the sum of the osmotic and matric potentials ($\psi_w = \psi_m + \psi_s$). The soil-air interface acts as a diffusion barrier in allowing only water molecules to pass from liquid to vapor state.

A psychrometer measures the difference between dry bulb and wet bulb temperatures to determine the relative humidity. The dry bulb is the temperature of the ambient air (nonevaporating surface), and the wet bulb is the temperature of an evaporating surface (generally lower than the dry bulb temperature). The relative humidity determines the rate of evaporation from the wet bulb junction, and thus the extent of temperature depression below ambient.

A thermocouple psychrometer consists of a fine-wire chromel-constantan or other bimetallic thermocouple. A thermocouple is a double junction of two dissimilar metals. When the two junctions are subjected to different temperatures, they generate a voltage difference (Seebeck effect). Conversely, when an electrical current is applied through the junctions, it creates a temperature difference between the junctions by heating one while cooling the other, depending on the current's direction. For typical soil use, one junction of the thermocouple psychrometer is suspended in a thin-walled porous ceramic or stainless screen cup buried in the soil (Figure 3.13), while another is embedded in an insulated plug to measure the ambient temperature at the same location. In

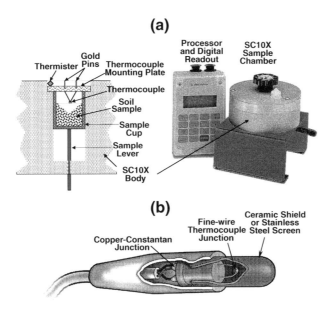

**FIGURE 3.13** (a) Model SC 10X sample chamber for psychrometric laboratory measurements of soil water potential [source: Decagon Devices, Inc., Pullman, WA, with permission]; and (b) a field psychrometer with porous ceramic shield [figure courtesy of Wescor Inc, Logan, Utah].

psychrometric mode, the suspended thermocouple is cooled below the dew point by means of an electrical current (Peltier cooling) until pure water condenses on the junction. The cooling current then stops, and as water evaporates, it draws heat from the junction (heat of vaporization), depressing it below the temperature of the surrounding air until it attains a wet bulb temperature. The warmer and drier the surrounding air, the higher the evaporation rate and the greater the wet bulb depression. The difference in temperatures between the dry and wet bulb thermocouples is measured and used to infer the relative humidity (or relative vapor pressure) using the psychrometer equation:

$$\frac{e}{e_o} = 1 - \left[\frac{s + \gamma}{e_o}\right]\Delta T \qquad [3.30]$$

where s is the slope of the saturation water vapor pressure curve ($s = de_o/dT$), $\gamma$ is the psychrometric constant (about 0.067 kPa $K^{-1}$ at 20°C), and $\Delta T$ is the temperature difference (K). The slope (s), and $e_o$ are functions of temperature only and can be approximated by closed-form expressions (Brutsaert, 1982).

A typical temperature depression measurable by a good psychrometer is on the order of 0.000085 C $kPa^{-1}$. This means that any errors in measuring wet bulb depression can introduce large errors into psychrometric determinations. Thermal equilibrium is, therefore, a prerequisite to obtaining reliable readings, as any temperature difference between wet and dry sensors resulting from thermal gradients will be (erroneously) incorporated into the relative humidity calculation.

Summarizing, psychrometric measurements of soil water potential are based on equilibrium between liquid soil water and water vapor in the ambient soil atmosphere. The drier the soil, the fewer water molecules escape into the ambient atmosphere, resulting in lower relative humidity (lower vapor pressure). When the osmotic potential is negligible, the soil water potential measured by a psychrometer is nearly equal to the soil matric potential. In principle, soil psychrometers may be buried in the soil and left for long periods, although corrosion is a problem in some environments.

## Line-Source Heat Dissipation Sensor

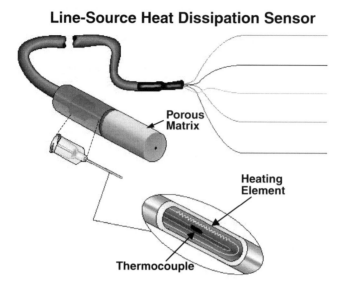

**FIGURE 3.14** Schematic of CSI 229 heat dissipation sensor. [Source: Campbell Scientific, Inc., Logan, UT, with permission.]

### 3.3.4.4   Heat Dissipation in Rigid Porous Matrix for Measuring Matric Potential

The rate of heat dissipation in a porous medium is dependent on specific heat capacity, thermal conductivity and density. The heat capacity and thermal conductivity of a porous matrix are affected by its water content, and hence related to its matric potential. Heat dissipation sensors contain line- or point-source heating elements embedded in a rigid porous matrix with fixed pore space. The measurement is based on applying a heat pulse by passing a constant current through the heating element for a specified time, and analyzing the temperature response measured by a thermocouple fixed at a known distance from the heating source (Phene et al., 1971; Bristow et al., 1993). Sensors are individually or uniformly calibrated in terms of heat dissipation versus sensor wetness (i.e., matric potential). With the heat dissipation sensor buried in the soil, changes in soil matric potential result in a gradient between the soil and the porous matrix that induces a water flux between the two materials until a new equilibrium is established. The water flux changes the water content of the porous matrix which, in turn, changes the thermal conductivity and heat capacity of the sensor. In this manner, the measured thermal response of the sensor may be related to soil wetness. A typical useful matric potential range for such sensors is –10 to –1,000 kPa.

A similar line-source sensor with a fine-wire heating element axially centered in a cylindrical ceramic matrix having a radius of 1.5 cm and length of 3.2 cm is depicted in Figure 3.14. A thermocouple is located adjacent to the heating element at mid-length. Both the heating wire and the thermocouple are contained in the shaft portion of a hollow needle. Because the thermocouple is located adjacent to the heating element, as the soil dries and water moves out of the ceramic, the magnitude of temperature change during a given period under constant heating current and duration will increase due to the reduced thermal conductivity of the porous matrix. The magnitude of the measured temperature increase and/or decrease is often linearly related to the natural logarithm of matric potential.The accuracy and operational limits of these sensors were evaluated by Reece (1996), who also presented an improved calibration method for between-sensor variability, based on normalizing sensor readings by oven dry sensor thermal conductivity. A primary advantage of these sensors is their wide range of applicability from about –0.1 to –12.0 MPa.

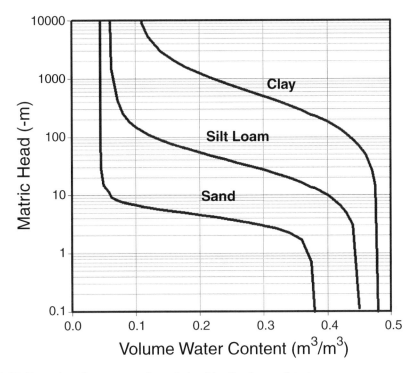

**FIGURE 3.15** Example soil water retention relationships for three soil textures.

New or improved techniques based on relating sensor thermal responses to soil matric potential continue to be developed. Accuracy, repeatability and spatial resolution of specific sensors or methods are important considerations to their potential applications and in the analysis of soil water measurements.

## 3.4 SOIL WATER CONTENT–ENERGY RELATIONSHIPS

### 3.4.1 SOIL WATER CHARACTERISTIC

A soil water characteristic (SWC) curve describes the functional relationship between soil water content ($\theta_m$ or $\theta_v$) and matric potential under equilibrium conditions. The SWC is an important soil property related to the distribution of pore space (sizes, interconnectedness), which is strongly affected by texture and structure, as well as related factors including organic matter content. The SWC is a primary hydraulic property required for modeling water flow, for irrigation management and for many additional applications related to managing or predicting water behavior in the porous system. A SWC is a highly nonlinear function and is relatively difficult to obtain accurately. Because the matric potential extends over several orders of magnitude for the range of water contents commonly encountered in practical applications, the matric potential is often plotted on a logarithmic scale. Several SWC curves for soils of different textures demonstrating the effects on porosity (saturated water content) and on the slope of the relationships resulting from variable pore size distributions are depicted in Figure 3.15.

### 3.4.2 MEASUREMENT OF SWC RELATIONSHIPS

Several methods are available to obtain measurements needed for SWC estimation. The basic requirement is for pairs of $\psi_m - \theta$ measurements over the wetness range of interest. Among the primary experimental problems in determining a SWC are: (1) the limited functional range of the tensiometer, which is often used for *in situ* measurements, (2) inaccurate $\theta$ measurements in some cases, (3) the difficulty in obtaining undisturbed samples for laboratory determinations, and (4) a slow rate of equilibrium under low matric potential (i.e., dry soils).

*In situ* methods are considered the most representative for determining SWCs, particularly when a wide range of $\psi_m$ – $\theta$ values are obtained. An effective method to obtain simultaneous measurements of $\psi_m$ and $\theta_v$ utilizes TDR probes installed in the soil at close proximity to transducer tensiometers, with the changing values of each attribute monitored through time as the soil wetness varies. Large changes in $\psi_m$ and $\theta_v$ can be induced under highly evaporative conditions near the soil surface, or in the presence of active plant roots.

### 3.4.2.1  Laboratory Estimation Using Pressure Plate and Pressure Flow Cell

The pressure plate apparatus consists of a pressure chamber enclosing a water saturated porous plate, which allows water, but prevents air flow, through its pores (Figure 3.16). The porous plate is open to atmospheric pressure at the bottom, while the top surface is at the applied pressure of the chamber. Sieved soil samples (usually < 2 mm) are placed in retaining rubber rings in contact with the porous plate and left to saturate in water. After saturation is attained, the porous plate with the saturated soil samples is placed in the chamber and a known gas (commonly $N_2$ or air) pressure is applied to force water out of the soil and through the plate. Flow continues until equilibrium between the force exerted by the air pressure and the force by which soil water is being held by the soil ($\psi_m$) is reached.

Soil water retention in the low suction range of 0 to 10 m ( = 1 bar = 0.1 MPa) is strongly influenced by soil structure and its natural pore size distribution. Hence, undisturbed intact soil samples (cores) are preferred over repacked samples for the wet end of the SWC. The pressure flow cell (Tempe cell) can hold intact soil samples encased in metal rings. The operation of the cell follows that of the pressure plate, except the pressure range is usually lower (0 to 10 m). The porous ceramic plates for both the pressure plate and the flow cell must be completely saturated, a process which may take a few days to achieve. Following equilibrium between soil matric potential and the applied air pressure, the soil samples are removed from the apparatus, weighed and oven dried for gravimetric determination of water content. An estimate of the sample bulk density must be provided to convert $\theta_m$ to $\theta_v$ in the case of disturbed samples. Because of differences in pore sizes and geometry, the water content of repacked soils at a given matric potential should not be used to accurately infer $\theta$ of intact soils at the same $\psi_m$.

Several pressure steps may be applied to the same samples when using flow cells. The cells may be disconnected from the air pressure source and weighed to determine the change in water content from the previous step, then reconnected to the air pressure and a new (greater) pressure step applied. Water outflow from the cells may also be monitored and related to the change in sample water content.

### 3.4.2.2  Field Measurement Methods–Sensor Pairing

Despite the paramount importance of SWC determination *in situ,* suitable measurement techniques are severely lacking at present. The most common approach is to use paired sensors such as neutron moisture meter access tubes or TDR waveguides and tensiometers to determine water content and matric potential simultaneously and in the same soil volume. Single probes that combine TDR with tensiometry have also been proposed (Baumgartner et al., 1994). The limitations of most sensor pairing techniques stem from: (1) differences in the soil volumes sampled by each sensor (e.g., large volume averaging by a neutron moisture meter versus a small volume sensed by a heat dissipation sensor or a psychrometer), (2) the time required for matric potential sensors to reach equilibrium and (3) limited ranges and deteriorating accuracy of different sensor pairs. This often results in limited overlap in SWC information measured using different techniques, and problems with measurement errors within the range of overlap.

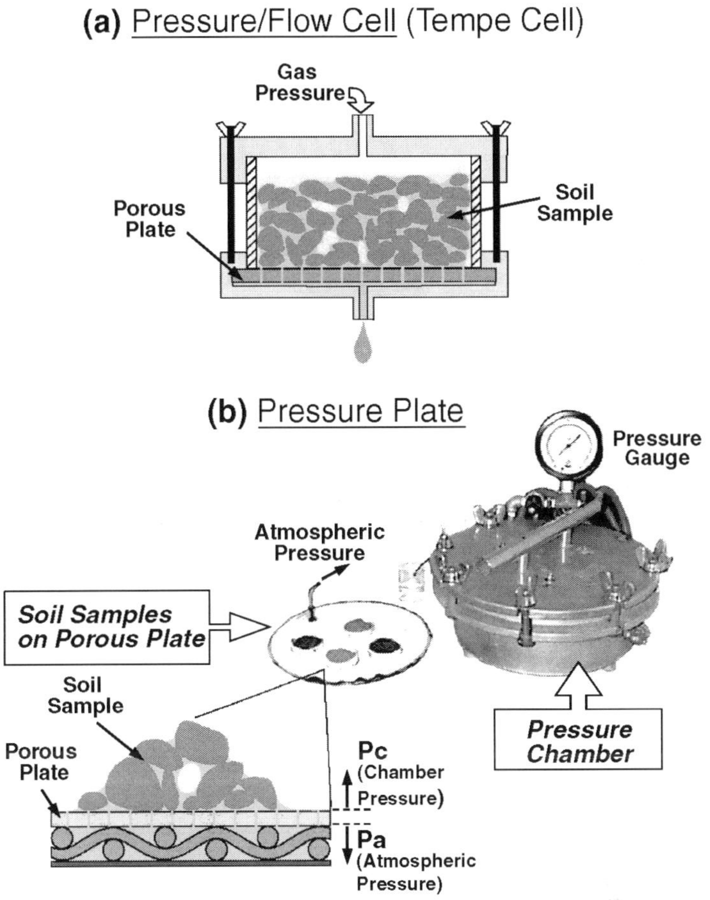

## (a) Pressure/Flow Cell (Tempe Cell)

## (b) Pressure Plate

**FIGURE 3.16** (a) Pressure flow cell (Tempe cell); and (b) pressure plate apparatus used to desaturate soil samples to desired matric potential.

A summary of common methods available for matric potential measurement or inference and their range of application is presented in Figure 3.17. Note that most of the available techniques have a limited range of overlap (or do not overlap at all), and most are laboratory methods not suitable for field applications.

### 3.4.3 FITTING PARAMETRIC SWC EXPRESSIONS TO MEASURED DATA

Measuring a SWC is laborious and time consuming. Measured $\theta - \psi$ pairs are often fragmentary, and usually constitute relatively few measurements over the wetness range of interest. For modeling and analysis purposes and for characterization and comparison between different soils and scenarios, it is therefore beneficial to represent the SWC in a continuous and parametric form. A parametric expression of a SWC model should: (1) contain as few parameters as necessary to simplify its estimation; and (2) describe the behavior of the SWC at the limits (wet and dry ends) while closely fitting the nonlinear shape of $\theta(\psi_m)$ data.

An effective and commonly used parametric model for relating water content to the matric potential was proposed by van Genuchten (1980) and is denoted as VG:

$$\Theta = \frac{\theta - \theta_r}{\theta_s - \theta_r} = \left[ \frac{1}{1 + \left( \alpha \psi_m \right)^n} \right]^m \qquad [3.31]$$

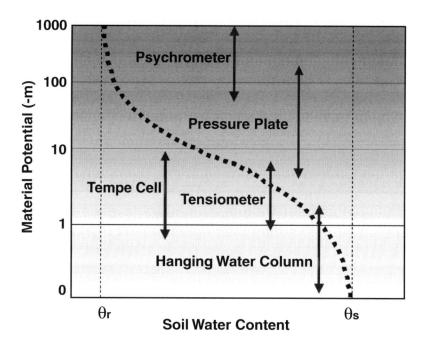

**FIGURE 3.17** Typical ranges of application for some common matric potential measurement or inference methods.

where $\theta_r$ and $\theta_s$ are the residual and saturated water contents, respectively, and $\alpha$, n and m are parameters directly dependent on the shape of the $\theta(\psi)$ curve. A considerable simplification is gained by assuming that m = 1 – 1/n. Thus the parameters required for estimation of the model are $\theta_r$, $\theta_s$, $\alpha$, and n. $\theta_s$ is usually known and is easy to obtain experimentally with good accuracy, leaving only three unknown parameters ($\theta_r$, $\alpha$ and n) to be estimated from the experimental data in many cases. Note that $\theta_r$ may be taken as $\theta_{-1.5\ MPa}$, $\theta_{air\ dry}$, or a similar value, though it is often advantageous to use it as a fitting parameter.

Another well-known parametric model was proposed by Brooks and Corey (1964) and is denoted as BC:

$$\Theta = \frac{\theta - \theta_r}{\theta_s - \theta_r} = \left[ \frac{\psi_b}{\psi_m} \right]^\lambda \qquad \psi_m > \psi_b$$

$$\Theta = 1 \qquad \psi_m \leq \psi_b$$

[3.32]

where $\psi_b$ is a parameter related to the soil matric potential at air entry $\psi_b$ represents bubbling pressure), and $\lambda$ is related to the soil pore size distribution. Matric potentials are expressed as positive quantities in both VG and BC parametric expressions.

Estimation of VG or BC parameters from experimental data requires: (1) sufficient data points (at least 5 to 8 $\theta(\psi_m)$ pairs), and (2) a program for performing nonlinear regression. Recent versions of many computer spreadsheets provide relatively simple and effective mechanisms for performing nonlinear regression. Details of the computational steps required for fitting a SWC to experimental data using commercially available spreadsheet software are given in Wraith and Or (1998). In addition, computer programs for estimation of specific parametric models are also available such as the RETC code (van Genuchten et al., 1991).

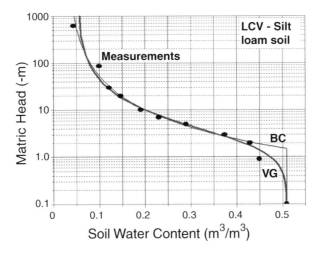

**FIGURE 3.18** van Genuchten (VG) and Brooks-Corey (BC) parametric models fitted to measured water retention data.

Fitted parametric models of van Genuchten (VG) and Brooks and Corey (BC) to silt loam $\theta(\psi)$ data measured by Or et al. (1991) are presented in Figure 3.18. The resulting best fit parameters for the VG model are $\alpha = 0.417$ m$^{-1}$; $n = 1.75$; $\theta_s = 0.513$ m$^3$ m$^{-3}$; and $\theta_r = 0.05$ m$^3$ m$^{-3}$ (with $r^2 = 0.99$). For the BC model the best fit parameters are $\lambda = 0.54$; $\psi_b = 1.48$ m; $\theta_s = 0.513$ m$^3$ m$^{-3}$; and $\theta_r = 0.03$ m$^3$ m$^{-3}$ (with $r^2 = 0.98$). Note that the most striking difference between the VG and the BC models is in the discontinuity at $\psi = \psi_b$ for BC.

Sources of additional experimental and parametric SWC information include the Unsaturated Soil Hydraulic Database (UNSODA) (Leij et al., 1996), which is a computer database compiled by the U.S. Salinity Laboratory which contains an exhaustive collection of retention (SWC) and unsaturated hydraulic conductivity information for soils of different textures from around the world. While the authors/compilers have attempted to provide some indices of data quality or reliability, the user is advised (as always) to use their own experience and discretion in adapting others' data to their own applications. Also, the regression studies by McCuen et al. (1981) and Rawls and Brakensiek (1989) provide a wealth of information on the Brooks-Corey parameter values for many soils including estimation of the hydraulic parameters based on other, often more easily available, soil properties. These estimates may be sufficiently accurate for some applications and could be used to obtain first-order approximations.

### 3.4.4 Hysteresis in the Soil-Water Characteristic Relation

Water content and the potential energy of soil water are not uniquely related because the amount of water present at a given matric potential is dependent on the pore size distribution and the properties of air-water-solid interfaces. A $\theta(\psi)$ relationship may be obtained by: (1) taking an initially saturated sample and applying suction or pressure to desaturate it (desorption), or by (2) gradually wetting an initially dry soil (sorption). These two pathways produce curves that in most cases are not identical; the water content in the drying curve is higher for a given matric potential than that in the wetting branch (Figure 3.19). This is called hysteresis, defined as the phenomenon exhibited by a system in which the reaction of the system to changes is dependent upon its past reactions to change.

The hysteresis in SWC can be related to several phenomena: (1) the ink bottle effect resulting from nonuniformity in shape and sizes of interconnected pores as illustrated in Figure 3.20a; drainage is governed by the smaller pore radius r, whereas wetting is dependent on the larger radius

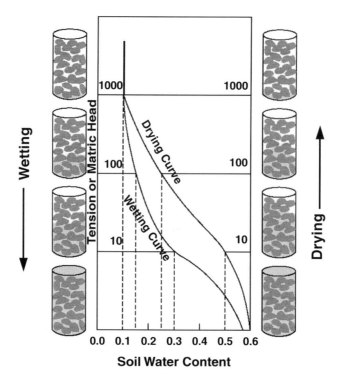

**FIGURE 3.19** Concept of hysteresis in soil-water characteristic relationships.

R; (2) different liquid-solid contact angles for advancing and receding water meniscii (Figure 3.20b); (3) entrapped air in a newly wetted soil (e.g., pore doublet) (Dullien, 1992); and (4) swelling and shrinking of the soil under wetting and drying which may alter the porosity and pore size distribution. Based on early observations of the phenomenon by Haines (1930) as well as present day theories (Mualem, 1984; Kool and Parker, 1987), the role of individual factors remain unclear and subject to ongoing research.

## 3.5  RESOURCES

A number of resources are available for readers interested in additional insight or information concerning soil water content and energy. The following suggestions are by no means inclusive, but should serve to augment and extend the discussions presented here, as well as provide a source of additional references. Readers are advised to consult the references cited in relevant sections of the chapter, as well as various texts, monographs and review chapters. Soil physics and related textbooks which address soil water content and water energy include those of Childs (1969), Hanks (1992), Hillel (1998), Jury et al. (1991), Kirkham and Powers (1972), Kutilek and Nielsen (1994), and Marshall et al. (1996). Several chapters in the monograph edited by Klute (1986), as well as that edited by Carter (1993), contain valuable information. A number of excellent review papers or chapters are also available, including Baker (1990), Campbell and Mulla (1990), and Parker (1986).

Some Internet sites contain valuable information related to soil water content and energy. These include various individual or professional organization pages and discussion groups (e.g., the soil moisture group SOWACS). However, Internet addresses tend to change rather frequently, and new sources may be added. This means it is generally more efficient to conduct keyword searches using one or more of the available search engines, than to rely on potentially outdated Web addresses. Searching the World Wide Web indicates that there are presently few sites having useful or extensive

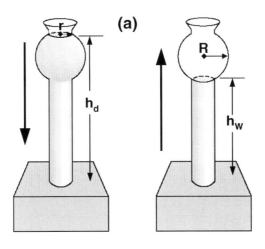

**(a)**

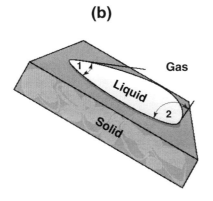

**(b)**

**FIGURE 3.20** The "ink bottle" effect (a), and the contact angle effect (b).

soil water information. However, this may change as the Web becomes more heavily utilized as a repository for soil information and soil property databases. Some government agencies are reportedly moving in this direction.

---

**EXAMPLE 3.2**   Soil Water Potential Conversion

*Problem:*
Convert a soil water head of –0.01 m to kPa and to J/kg.

*Solution:*
1. Conversion from h = –0.01 m to $\psi$ (kPa): the conversion equation is $\psi = \rho_w g h$ where g = 9.8  m/s$^2$ and $\rho_w$ = 1000 kg/m$^3$. Hence, $\psi$ = –0.01*9.8*1000 (m*m*kg)/(s$^2$*m$^3$) = –98 (kg*m/s$^2$)*(1/m$^2$) = –98 Pa or –0.098 kPa (note that Pa = N/m$^2$, and N = kg*m/s$^2$).

2. Conversion from h = –0.01 m to $\mu$ (J/kg): the conversion equation is $\mu$ = gh where g = 9.8 m/s$^2$, hence $\mu$ = –0.01*9.8 (m*m/s$^2$) = –0.098 (m*m/s$^2$); we multiply and divide the expression by kg = –0.098 kg*m*m/(kg*s$^2$) which is –0.098 J/kg (note that a Joule = N*m = (kg*m/s$^2$)*m).

---

**EXAMPLE 3.3**  A bundle of capillaries composed of glass tubes having the following diameters and numbers was dipped (vertically) into a water reservoir. Compute and plot the relative saturation of the capillaries at different elevations above the free water. Remember that $\theta_v$ is equivalent to the water-filled area per total cross-sectional area (Diagram 3.3.1).

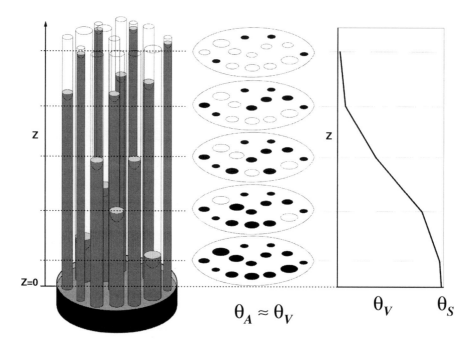

**DIAGRAM 3.3.1**  Schematic representation of the relationship between degree of saturation and matric potential (expressed as height above a reference pool of water), illustrating soil water retention for the bundle of capillaries model.

| # of capillaries | 300 | 400 | 325 | 250 | 150 | 90 | 50 | 10 | 5 | 2 |
|---|---|---|---|---|---|---|---|---|---|---|
| Diameter (mm) | 0.001 | 0.005 | 0.01 | 0.02 | 0.05 | 0.1 | 0.2 | 0.5 | 1 | 2 |

*Solution:* Capillary rise equation to calculate h:

$$h = \frac{2\sigma \cdot \cos\gamma}{\rho_w \cdot g \cdot \dfrac{d}{2}} \quad \frac{\left[\dfrac{kg \cdot m}{s^2 \cdot m}\right]}{\left[\dfrac{kg}{m^3}\right] \cdot \left[\dfrac{m}{s^2}\right] \cdot [m]} \Rightarrow [m]$$

with:
$\rho_s$  surface tension of water (0.0728 N/m at 20°C)
$\rho_w$  density of water (998.2 kg/m³ at 20°C)
$\gamma$   contact angle ($\gamma = 0$)
g   acceleration of gravity (9.81 m/s²)

*Cross-sectional area:*

$$A = \frac{\pi}{4} d^2 * (\# \text{ of capillaries})$$

*Relative Saturation:*

$$\theta_{vrel} = \frac{\text{Cross-sectional area of water-filled pores}}{\text{Total cross-sectional area}}$$

*Results:*

Note how the largest pores, filled only at low matric potentials and thus only near the free water surface, dominate the relative saturation.

| Capillaries [#] | Diameter [m] | Capillary Rise, h [m] | Area [m²] | Saturation [–] |
|---|---|---|---|---|
| 300 | 1.0E-06 | 29.74 | 2.356E-10 | 0.0001 |
| 400 | 5.0E-06 | 5.95 | 7.854E-09 | 0.001 |
| 325 | 1.0E-05 | 2.97 | 2.553E-08 | 0.002 |
| 250 | 2.0E-05 | 1.49 | 7.854E-08 | 0.008 |
| 150 | 5.0E-05 | 0.59 | 2.945E-07 | 0.027 |
| 90 | 1.0E-04 | 0.30 | 7.069E-07 | 0.075 |
| 50 | 2.0E-04 | 0.15 | 1.571E-06 | 0.181 |
| 10 | 5.0E-04 | 0.06 | 1.963E-06 | 0.313 |
| 5 | 1.0E-03 | 0.03 | 3.927E-06 | 0.577 |
| 2 | 2.0E-03 | 0.01 | 6.283E-06 | 1.000 |

**EXAMPLE 3.4** Textural averages of hydraulic parameters based on UNSODA unsaturated soil hydraulic properties database and NRCS soil survey database (Leij et al., 1998).

| Textural Class | Soil Water Retention (1/cm) | | | | | Saturated Hydraulic Conductivity (cm/d) | |
|---|---|---|---|---|---|---|---|
| | N* | $\theta_r$ | $\theta_s$ | $\alpha$ | n | N* | $K_s$ |
| | | | UNSODA | | | | |
| Sand | 126 | 0.058 | 0.37 | 0.035 | 3.19 | 74 | 505.8 |
| Loamy sand | 51 | 0.074 | 0.39 | 0.035 | 2.39 | 31 | 226.5 |
| Sandy loam | 78 | 0.067 | 0.37 | 0.021 | 1.61 | 50 | 41.6 |
| Loam | 61 | 0.083 | 0.46 | 0.025 | 1.31 | 31 | 38.3 |
| Silt | 3 | 0.123 | 0.48 | 0.006 | 1.53 | 2 | 55.7 |
| Silt loam | 101 | 0061 | 0.43 | 0.012 | 1.39 | 62 | 30.5 |
| Sandy clay loam | 37 | 0.086 | 0.40 | 0.033 | 1.49 | 19 | 9.69 |
| Clay loam | 23 | 0.129 | 0.47 | 0.030 | 1.37 | 8 | 1.84 |
| Silty clay loam | 20 | 0.098 | 0.55 | 0.027 | 1.41 | 10 | 7.41 |
| Silty clay | 12 | 0.163 | 0.47 | 0.023 | 1.39 | 6 | 8.40 |
| Clay | 25 | 0.102 | 0.51 | 0.021 | 1.20 | 23 | 26.0 |

| Textural Class | Soil Water Retention (1/cm) | | | | | Saturated Hydraulic Conductivity (cm/d) | |
|---|---|---|---|---|---|---|---|
| | N* | $\theta_r$ | $\theta_s$ | $\alpha$ | n | N* | $K_s$ |
| **Soil Survey** | | | | | | | |
| Sand | 246 | 0.045 | 0.43 | 0.145 | 2.68 | 246 | 712.8 |
| Loamy sand | 315 | 0.057 | 0.41 | 0.124 | 2.28 | 315 | 350.2 |
| Sandy loam | 1183 | 0.065 | 0.41 | 0.075 | 1.89 | 1183 | 106.1 |
| Loam | 735 | 0.078 | 0.43 | 0.036 | 1.56 | 735 | 25.0 |
| Silt | 82 | 0.034 | 0.46 | 0.016 | 1.37 | 88 | 6.00 |
| Silt loam | 1093 | 0.067 | 0.45 | 0.020 | 1.41 | 1093 | 10.8 |
| Sandy clay loam | 214 | 0.100 | 0.39 | 0.059 | 1.48 | 214 | 31.4 |
| Clay loam | 364 | 0.095 | 0.41 | 0.019 | 1.31 | 345 | 6.24 |
| Silty clay loam | 641 | 0.089 | 0.43 | 0.010 | 1.23 | 592 | 1.68 |
| Sandy clay | 46 | 0.100 | 0.38 | 0.027 | 1.23 | 46 | 2.88 |
| Silty clay | 374 | 0.070 | 0.36 | 0.005 | 1.09 | 126 | 0.48 |
| Clay | 400 | 0.068 | 0.38 | 0.008 | 1.09 | 114 | 4.80 |

* Approximate sample size for Soil Survey database.

## ACKNOWLEDGMENTS

Partial funding for this work was provided by the Utah Agricultural Experimental Station (UAES) and the Montana Agricultural Experimental Station (MAES).

## REFERENCES

Babinet, J. 1848. Note sur un atmidoscope, *Compt. Rend.,* 27:529–530.

Babb, A.T.S. 1951. A radio-frequency electronic moisture meter, *Analyst,* 76:428–433.

Baker, J.M. 1990. Measurement of soil water content, *Rem. Sens. Rev.,* 5:263–279.

Baker, J.M. and R.R. Allmaras. 1990. System for automating and multiplexing soil moisture measurement by time-domain reflectometry, *Soil Sci. Soc. Am. J.,* 54:1–6.

Banin A. and A. Amiel. 1970. A correlative study of the chemical and physical properties of a group of natural soils of Israel, *Geoderma,* 3:185–198.

Baumgartner, N., G.W. Parkin, and D.E. Elrick. 1994. Soil water content and potential measured by hollow time domain reflectometry probe, *Soil Sci. Soc. Am. J.,* 58:315–318.

Bolt, G.H. 1976. Soil physics terminology, *Int. Soc, Soil Sci., Bull.,* 49:16–22.

Bouyoucos, G.J. and A.H. Mick. 1940. An electrical resistance method for the continuous measurement of soil moisture under field conditions, *MI Agric. Exp. Sta. Tech. Bull.,* 172:1–38.

Bristow, K.L., G.S. Campbell, and K. Calissendroff. 1993. Test of a heat-pulse probe for measuring changes in soil water content, *Soil Sci. Soc. Am. J.,* 57:930–934.

Brooks, R.H. and A.T. Corey. 1964. Hydraulic properties of porous media, Hydrol. Pap. 3. Colorado State University, Fort Collins.

Brutsaert, W. 1982. Evaporation into the atmosphere. D. Reidel Publishing Company, Dordrecht, Netherlands.

Campbell, G.S. and D.J. Mulla. 1990. Measurement of soil water content and potential, B.A. Stewart and D.R. Nielsen (eds.), Irrigation of agricultural crops, 127–142, *Am. Soc. Agron.,* Madison, WI.

Carter, M.R. (ed.). 1993. Soil sampling and methods of analysis. Lewis Publishers, Boca Raton, FL.

Cassel, D.K. and A. Klute. 1986. Water potential: Tensiometry, *in Methods of Soil Analysis Part 1: Physical and Mineralogical Methods,* A. Klute (ed.), 563–596, Special Publication #9, ASA, Madison, WI.

Childs, E.C. 1969. An introduction to the physical basis of soil water phenomena. John Wiley and Sons Ltd, London.

Corey, A.T. and A. Klute. 1985. Application of the potential concept to soil water equilibrium and transport, *Soil Sci. Soc. Am. J.,* 49:3–11.

Dahiya, I.S., D.J. Dahiya, M.S. Kuhad, and P.S. Karwasra. 1988. Statistical equations for estimating field capacity, wilting point, and available water capacity of soils from their saturation presentage, *J. Agric. Res.,* 110:515–520.

Dalton, F.N., W.N. Herkelrath, D.S. Rawlins, and J.D. Rhoades. 1984. Time-domain reflectometry: simultaneous measurement of soil water content and electrical conductivity with a single probe, *Science,* 224:989–990.

Day, P.R., G.H. Bolt, and D.M. Anderson. 1967. Nature of soil water, R.M. Hagan, H.R. Haise, and T.W. Edminster (eds.), Irrigation of agricultural lands, 193–208, *Am. Soc. Agron.,* Madison, WI.

Dean, T.J., J.P. Bell, and A.J.B. Baty. 1987. Soil moisture measurement by an improved capacitance technique. Part I. Sensor design and performance, *J. Hydrol.,* 93:67–78.

Dobson, M.C., F.T. Ulaby, M.T. Hallikainen, and M.A. El-Rayes. 1985. Microwave dielectric behavior of wet soil, II. Dielectric mixing models, *IEEE Trans. Geosci. Rem. Sens.,* GE-23:35–46.

Dullien, F.A.L. 1992. Porous media: fluid transport and pore structure, 2nd ed., Academic Press, New York.

Evett, S.R. and J.L. Steiner. 1995. Precision of neutron scattering and capacitance type soil water content gauges from field calibration, *Soil Sci. Soc. Am. J.,* 59:961–968.

Gardner, W., O.W. Israelsen, N.E. Edlefsen, and D. Clyde. 1922. The capillary potential function and its relation to irrigation practice (Abstract), *Phys. Rev.,* 20:196.

Gardner, W.H. 1986. Water content, A. Klute (ed.), Methods of soil analysis. Part 1, 2nd ed., 493–544, *Am. Soc. Agron.,* Madison, WI.

Haines, W.B. 1927. Studies in the physical properties of soils: IV. A further contribution on the theory of capillary phenomena in soil, *J. Agric. Res.,* 17:264–290.

Haines, W.B. 1930. Studies in the physical properties of soil. V. The hysteresis effect in capillary properties, and the modes of moisture distribution associated therewith, *J. Agric. Sci.,* 20:97–116.

Hanks, R.J. 1992. Applied soil physics. 2nd ed., Springer Verlag, New York.

Hawkins, L.A. and B.E. Livingston. 1915. The water supplying power of the soil as indicated by osmometers, 49–84, *Carnegie Inst. Washington,* Pub. No. 204.

Hilhorst, M.A. and C. Dirksen. 1994. Dielectric water content sensors: Time domain vs. frequency domain, Proc. for symp. and workshop on time domain reflectometry in environmental, infrastructure, and mining applications, 23–33, Sept. 7–9, 1994. Northwestern University, Evanston, IL, *Bur. Mines Spec. Publ.,* SP 19–94, U.S. Department of the Interior, Washington, DC.

Hillel, D. 1998. Environmental soil physics, Academic Press, San Diego.

Hoekstra, P. and A. Delaney. 1974. Dielectric properties of soils at UHF and microwave frequencies. J. Geophys. Res. 79:1699–1708.

Hubbell, J.M. and J.B. Sisson. 1996. Portable tensiometer for use in deep boreholes, *Soil Sci.,* 161:376–381.

IAEA. 1970. Neutron moisture gauges. Tech. Rep. Ser. No. 112. International Atomic Energy Agency, Vienna, Austria.

Iwata, S., T. Tabuchi, and B.P. Warkentin. 1988. Soil water interactions. M. Dekker, New York, NY.

Jury, W.A., W.R., Gardner, and W.H. Gardner. 1991. Soil physics. John Wiley and Sons, New York, NY.

Kirkham, D. and W.L. Powers. 1972. Advanced soil physics. R.E. Krieger Publishing Company, Malabar, FL.

Klute A. (ed.). 1986. Methods of soil analysis: physical and mineralogical methods, Part 1. 2nd ed., *Soil Sci. Soc. Am.,* Madison, WI.

Kool, J.B. and J.C. Parker. 1987. Development and evaluation of closed-form expressions for hysteretic soil hydraulic properties, *Water Resour. Res.,* 23:105–114.

Kutilek, M. and D.R. Nielsen. 1994. Soil hydrology, *Catena Verlag,* Cremlingen-Destedt, Germany.

Leij, F.J., W.J. Alves, M.Th. van Genuchten, and J.R. Williams. 1996. The UNSODA unsaturated hydraulic database. EPA/600/R-96/095, U.S. Environmental Protection Agency, Cincinnati.

Livingston, B.E. 1908. A method for controlling plant moisture, *Plant World,* 11:39–40.

Livingston, B.E. 1918. Porous clay cones for the auto-irrigation of potted plants, *Plant World,* 21:202–208.

Livingston, B.E. and L.A. Hawkins. 1915. The water relation between plant and soil, 3–48, *Carnegie Inst. Washington,* Pub. No. 204.

Marshall, T.J., J.W. Holmes, and C.W. Rose. 1996. Soil physics, 3rd ed., Cambridge University Press, U.K.

McCuen, R.H., W.J. Rawls, and D.L. Brakensiek. 1981. Statistical analysis of the Brook-Corey and Green-Ampt parameters across soil texture, *Water Resour. Res.,* 17:1005–1013.

Mualem, Y. 1984. A modified dependent domain theory of hysteresis, *Soil Sci.,* 137:283–291.

Nitao, J.J. and J. Bear. 1996. Potentials and their role in transport in porous media, *Water Resour. Res.,* 32:225–250.

Olhoeft, G.R. 1985. Low-frequency electrical properties, *Geophysics,* 50:2492–2503.

Or, D., D.P. Groeneveld, K. Loague, and Y. Rubin. 1991. Evaluation of single and multi-parameter methods for estimating soil-water characteristic curves, Geotechnical Engineering Report No. UCB/GT/91-07, University of California, Berkeley.

Paltineanu, I.C. and J.L. Starr. 1997. Real-time soil water dynamics using multisensor capacitance probes: laboratory calibration, *Soil Sci. Soc. Am. J.,* 61:1576–1585.

Parker, J.C. 1986. Hydrostatics of water in porous media, D.L. Sparks (ed.), *Soil Physical Chemistry,* 209–296, CRC Press, Boca Raton, FL.

Peck, A.J. and R.M. Rabbidge. 1969. Design and performance of an osmotic tensiometer for measuring capillary potential, *Soil Sci. Soc. Am. Proc.,* 33:196–202.

Petersen L.W., P. Moldrup, O.H. Jacobsen, and D.E. Rolston. 1996. Relations between specific surface area and soil physical and chemical properties, *Soil Sci.,* 161:9–21.

Phene, C.J., G.J. Hoffman, and S.L. Rawlins. 1971. Measuring soil matric potential *in situ* by sensing heat dissipation within a porous body: I. Theory and sensor construction, *Soil Sci. Soc. Am. Proc.,* 35:27–33.

Rawls, W.J. and D.L. Brakensiek. 1989. Estimation of soil water retention and hydraulic properties, H.J. Morel-Seytoux (ed.), Unsaturated flow in hydraulic modeling theory and practice. 275–300, NATO ASI Series. Series c: Mathematical and physical science, vol. 275.

Reece, C.F. 1996. Evaluation of line heat dissipation sensor for measuring soil matric potential, *Soil Sci. Soc. Am. J.,* 60:1022–1028.

Richards, L.A. 1928. The usefulness of capillary potential to soil moisture and plant investigators, *J. Agric. Res.,* 37:719–742.

Richards, L.A. (ed.). 1954. Diagnosis and improvement of saline and alkali soils, USDA Handb. 60, U.S. Government Printing Office, Washington, D.C.

Roth, K., R. Schulin, H. Fluhler, and W. Attinger. 1990. Calibration of time domain reflectometry for water content measurement using composite dielectric approach, *Water Resour. Res.,* 26:2267–2273.

Selig, E.T. and S. Mansukhani. 1975. Relationship of soil moisture to the dielectric property, ASCE Geotech. Eng. Div. GT8:755–770.

Slater, P.J. and J.B. Williams. 1965. The influence of texture on the moisture characteristics of soils, 1. A critical comparison of techniques for determining the available water capacity and moisture characteristic curve of a soil, *J. Soil Sci.,* 16:1–15.

Smith-Rose, R.L. 1933. The electrical properties of soils for alternating currents at radio frequencies, *Proc. Roy. Soc.,* 140:359, London.

Spaans, E.J.A. and J.M. Baker. 1992. Calibration of watermark soil moisture sensors for temperature and matric potential, *Plant Soil,* 143:213–217.

Topp, G.C., J.L. Davis, and A.P. Annan. 1980. Electromagnetic determination of soil water content: measurements in coaxial transmission lines, *Water Resour. Res.,* 16:574–582.

Truzelli. 1985. The porosity of soil aggregates from bulk soil and soil adhering to roots, *Plant Soil.*

van Genuchten, M.Th. 1980. A closed-form equation for predicting the hydraulic conductivity of unsaturated soils, *Soil Sci. Soc. Am. J.,* 44:892–898.

van Genuchten, M.Th., F.J. Leij, and S.R. Yates. 1991. The RETC code for quantifying the hydraulic functions of unsaturated soils, EPA/600/2-91/065, U.S. Environmental Protection Agency, Ada, OK.

von Hippel, A.R. 1954. Dielectric materials and applications. Technology Press of M.I.T., New York.

Waters, P. 1980. Comparison of the ceramic and the pressure membrane to determine the 15-bar water content of soils, *J. Soil Sci.,* 31:443–446.

Wraith J.M. and D. Or. 1998. Nonlinear parameter estimation using spreadsheet software, *J. Nat. Resour. Life. Sci. Educ.,* 27:13–19.

# 4  Soil Water Movement

*David E. Radcliffe and Todd C. Rasmussen*

## 4.1  INTRODUCTION

Water movement in soils occurs under both saturated and unsaturated conditions (Figure 4.1). Saturated conditions occur below the water table where water movement is predominately horizontal, with lesser components of flow in the vertical direction. While unsaturated conditions generally predominate above the water table (the vadose zone), localized zones of saturation can exist especially following precipitation or irrigation. As a general rule, water movement in the unsaturated zone is vertical, but can also have large lateral components.

Saturated soils occur when soil pores are entirely filled with water. In this case, the water content ($\theta$) is equal to the total porosity ($\varphi$) and the air filled porosity ($\theta_a$) is zero. While soil pores can be assumed to be fully saturated below the water table, even so-called saturated soils above the water table may retain some residual entrapped air, especially near the soil surface. Here, we will assume that $\theta = \varphi$ and $\theta_a = 0$ under saturated conditions, unless specified otherwise.

Water flow can be either steady or transient. While analytic solutions of the differential equations that govern soil water flow are abundant for saturated conditions, they are less available for steady flow in the unsaturated zone and almost completely lacking for transient, unsaturated flow conditions. Simplifications of the unsaturated flow equations have been developed to describe infiltration because of the importance of this transient flow process. Alternatively, numerical methods are commonly used to solve most transient flow problems in the unsaturated zone, and in saturated media where large variations in material properties make analytic solutions impossible.

The hydraulic gradient, which is the driving force behind water flow, is a vector that describes the slope of the energy distribution within the soil. The principal parameter required to predict saturated flow is the saturated hydraulic conductivity ($K_s$). Predicting unsaturated flow requires the unsaturated hydraulic conductivity [$K(h)$] and water retention [$\theta(h)$] functions. $K_s$, $\theta(h)$ and $K(h)$ are all affected by soil texture and structure. While soil texture is easily measured and not highly variable in space (in many cases), soil structure is difficult to quantify and highly variable in space and time. Although methods have been developed to measure $K_s$ and $K(h)$, each has some disadvantage.

In recent years, soil scientists have expanded their vertical scale of interest beyond the root zone to include all of the vadose zone. At the same time, there is an emphasis on expanding the horizontal scale from the traditional plot scale to the field and watershed scales. At these larger scales, soil scientists need to have a better understanding of shallow groundwater flow which is emphasized in this chapter.

This chapter introduces the conditions and equations that govern water flow, the parameters that quantify these relationships and applications to common problems. The objective is to provide a foundation for understanding the water flow processes in soils, and to direct the reader to other resources that can clarify or extend the introductory material provided here.

## 4.2  FLOW IN SATURATED SOIL

The following sections examine flow under steady and transient conditions. The approach is to first examine the rule, Darcy's Law, that relates the flow rate to the hydraulic gradient and then show how a conservation equation can be coupled to Darcy's Law to establish the equations that govern

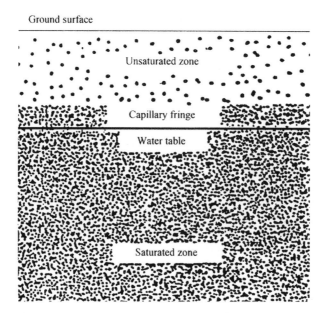

**FIGURE 4.1** Saturated and unsaturated zones (reprinted from Fetter, 1994. Applied hydrogeology. Copyright Macmillan. New York, NY with permission).

both steady and unsteady flow. The resulting set of equations are then applied to problems in one and higher dimensions.

### 4.2.1 SATURATED FLOW EQUATIONS

#### 4.2.1.1 Poiseuille Equation

At the microscopic scale of an individual pore, approximated as a water-filled cylinder of given radius ($r$), the volumetric flow rate ($Q$) is described by Poiseuille's equation:

$$Q = \frac{\pi\, r^4\, \rho\, g\, \Delta H}{8\, \eta\, L} \qquad [4.1]$$

where $\eta$ is the water viscosity, $\rho$ is the water density, $g$ is the acceleration due to gravity and $\Delta H$ is the difference in total head (potential on a weight basis) between two points along the cylinder separated by a distance $L$. Poiseuille's equation has been used to show the disproportionate effect that large-diameter pores (i.e., macropores) have on transmitting water. Flow is proportional to $r^4$ whereas the cross-sectional area of a pore is proportional to $r^2$. Therefore, one large pore with the same cross-sectional area as several smaller pores will transmit considerably more water due to less viscous drag along the pore wall.

#### 4.2.1.2 Darcy's Equation

Because it is usually not feasible to determine the size distribution and interconnectivity of pores, a macroscopic approach is used to describe flow through soils. This approach was first taken by Henri Darcy, an engineer working on sand filters used to purify the water in the city of Dijon, France. Through experimentation, he found that the volumetric flow rate per unit cross-sectional area ($A$) through a sand filter of a given thickness ($L$) was proportional to the total head gradient across the sand ($\Delta H/L$). He called the proportionality constant saturated hydraulic conductivity ($K_s$) (Darcy, 1856):

$$J = \frac{Q}{A} = -K_s \frac{\Delta H}{L} \qquad [4.2]$$

where $J$ is the fluid flux. Using partial derivative notation, this can be written as:

$$J = -K_s \frac{\partial H}{\partial z} \qquad [4.3]$$

Darcy's Law can be extended to two- and three-dimensional flow by noting that the flux can be written in vector notation as:

$$\underline{J} = J_x + J_y + J_z$$

$$= -K_s \frac{\partial H}{\partial x} - K_s \frac{\partial H}{\partial y} - K_s \frac{\partial H}{\partial z} \qquad [4.4]$$

$$= -K_s \left( \frac{\partial H}{\partial x} + \frac{\partial H}{\partial y} + \frac{\partial H}{\partial z} \right)$$

which shows that the total flow can be decomposed into a set of components in the two horizontal $(x,y)$ and one vertical $(z)$ directions. If the vertical component, $J_z$, is zero, then $J$ is just $(J_x, J_y)$. This formulation is suitable for geologic media that have uniform hydraulic conductivity.

Darcy's equation takes the same form as several other important laws in science, among them, Fick's law for chemical diffusion, Ohm's law for current flow, and Fourier's law for heat flow. It clearly shows that for water to flow there must be a difference in water potential, but the rate of water flow will depend on the hydraulic gradient and the hydraulic conductivity of the soil.

---

**EXAMPLE 4.1** A clay liner at the bottom of a swine lagoon is 30-cm thick and has a saturated hydraulic conductivity of $10^{-7}$ cm s$^{-1}$. If the lagoon is filled to an average depth of 3 m and the matric potential head immediately below the clay liner is –50 cm, what can you expect the steady seepage rate through the liner to be?

Set the reference elevation for zero gravitational head at the bottom of the liner:

$$J_w = -K_s \frac{\Delta H}{L}$$

$$= -K_s \frac{H_2 - H_1}{L}$$

$$= -K_s \frac{(p+z)_2 - (p+z)_1}{L}$$

$$= -\left( 10^{-7} \frac{cm}{s} \right) \frac{(300\,cm + 30\,cm)_2 - (-50\,cm + 0\,cm)_1}{30\,cm}$$

$$= -\left( 10^{-7} \frac{cm}{s} \right) \frac{380\,cm}{30\,cm}$$

$$= -1.26 \times 10^{-6} \frac{cm}{s}$$

Although Darcy's equation was developed empirically, it can be derived from first principles using the Navier-Stokes equation if inertial forces are assumed to be negligible compared to viscous forces (Hubbert, 1956). When inertial forces become significant, turbulent flow occurs and the Reynolds number is used to determine when the onset of turbulent flow occurs. For soil pores, turbulent flow may be expected for Reynolds numbers greater than one (Hillel, 1980a). This would occur at 20°C under gravity flow in pores with diameters larger than about 0.15 mm. Under these conditions, Darcy's equation would overestimate flux in larger diameter pores.

In a uniform soil under saturated conditions, Darcy's equation can be used to show that matric potential varies linearly with depth (Jury et al., 1991). Thus, if water is ponded on the surface of a uniform soil column and allowed to drip from the bottom, the matric potential should decrease linearly from the depth of ponding at the surface to zero at the bottom of the column.

---

**EXAMPLE 4.2** Darcy's report on *Les Fontaines Publiques de la Ville de Dijon* was published in 1856 (Hubbert, 1956). The document summarized Darcy's work for the previous several years in modernizing and enlarging the fountains of Dijon, which provided public water for the city. Part of his work involved designing a suitable filter for the system, and he needed to know how large a filter would be required for a given quantity of water per day. Since he could find nothing in the literature to guide him in this design, he set up a series of experiments in one of the city hospitals. His apparatus was a vertical iron cylinder 3.50 m in length and 0.35 m in diameter filled with sand. Water entered the cylinder at the top through a pipe connected to the building water supply and discharged from the bottom of the cylinder through a faucet into a measuring tank. Mercury manometers were connected to the top and bottom of the cylinder to measure pressures. Both the top and bottom manometer readings were recorded in meters of water measured from the bottom of the sand. Hence, Darcy's manometer readings accounted for pressure and gravitational potential and were equivalent to total head. Darcy suffered some of the usual experimental difficulties including fluctuating water pressures in the building water supply system. The similarity in the form of his equation and that of other important laws was probably not coincidental (Narisimhan, 1998). Fourier's work on heat flow was published in 1807, Ohm's work on current flow was published in 1827, and the work of Poiseuille (who was a medical doctor) on blood flow through human and animal veins was published in 1842.

---

### 4.2.1.3 Steady Flow

Coupling a conservation of mass equation, that requires a mass balance at every point within the flow domain, with the Darcy equation under steady flow conditions yields the Laplace equation:

$$\frac{\partial^2 H}{\partial x^2} + \frac{\partial^2 H}{\partial y^2} + \frac{\partial^2 H}{\partial z^2} = 0 \qquad [4.5]$$

which is equivalent to stating that there is no accumulation of water at any point within the flow domain.

#### 4.2.1.4 Unsteady Flow

For conditions of unsteady flow, there will be an accumulation, or release, of water within the flow domain. Under these conditions, the difference between the water flowing into and away from a point within the domain must equal the accumulation, or loss, of water at the point. This is described using the Poisson equation:

$$\frac{\partial}{\partial x}\left(K_s \frac{\partial H}{\partial x}\right) + \frac{\partial}{\partial y}\left(K_s \frac{\partial H}{\partial y}\right) + \frac{\partial}{\partial z}\left(K_s \frac{\partial H}{\partial z}\right) = S_s \frac{\partial H}{\partial t}$$  [4.6]

where $S_s$ is the specific storage of the medium, and $t$ is time.

### 4.2.2 SATURATED FLOW PARAMETERS

#### 4.2.2.1 Saturated Hydraulic Conductivity

For saturated flow, the most important soil parameter is saturated hydraulic conductivity, which is a function of the fluid and soil properties:

$$K_s = \frac{k\,\rho\,g}{\eta}$$  [4.7]

where $k$ is the intrinsic permeability of the soil. It is apparent from this equation that temperature has an effect on $K_s$. It is generally true that liquids other than water will have a different $K_s$ but the same $k$ in a given soil, if the soil structure is not affected.

Soils with low porosity, few large pores and poor interconnectivity between pores have low values of $K_s$. Rawls et al. (1982) compiled values of $K_s$ from 1,323 soils collected over 32 states (Table 4.1). Saturated hydraulic conductivities were highest in coarse-textured soils and declined in fine-textured soil, due to larger pores in the former. This can be corroborated by examining air-entry potentials ($h_a$, the matric potential at which the largest capillary pore empties) which are more negative in the fine-textured soils, indicating that the largest capillary pore was smaller than in coarse-textured soils. Higher $K_s$ occurred in the coarse-textured soils in spite of the generally lower porosity.

The mean values in Table 4.1 may provide unreliable values for $K_s$ and $h_a$ for fine-textured soils, due to the effect of structure. In Figure 4.2, $K_s$ for 395 soil samples from the UNSODA database (Section 4.4.3) are plotted as a function of clay content. There is great variability, except at very low clay contents where there is a cluster of high $K_s$ values. In intact soil, it is apparent that soil structure, as well as soil texture, affects water flow in all but very sandy soils.

In many cases, hydraulic conductivity is not the same in all directions. For the case where a different conductivity is observed in two (or three) different directions, this variability must be accounted for explicitly in the Darcy equation:

$$J_x = -K_x \frac{\partial H}{\partial x} \quad J_y = -K_y \frac{\partial H}{\partial y} \quad J_z = -K_z \frac{\partial H}{\partial z}$$  [4.8]

where the three fluxes are oriented in the three directions of the coordinate system. This can be written in Einstein vector notation as:

$$\underline{J} = -\underline{\underline{K}}_s \left[\frac{\partial H}{\partial x}, \frac{\partial H}{\partial y}, \frac{\partial H}{\partial z}\right]$$  [4.9]

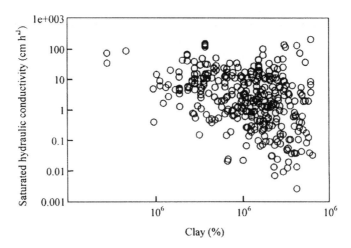

**FIGURE 4.2** Saturated hydraulic conductivity as a function of clay content for 395 soils from the UNSODA database.

**TABLE 4.1**
**Saturated hydraulic conductivity ($K_s$), total porosity ($\phi$), air-entry matric potential ($h_a$), macroscopic capillary length ($\lambda_c$), and microscopic capillary length ($\lambda_m$) of soils of various textures.**

| Texture Class | $K_s$ cm h$^{-1}$ | $\varphi$ cm$^3$ cm$^{-3}$ | $h_a$ cm | $\lambda_c$ cm | $\lambda_m$ cm |
|---|---|---|---|---|---|
| Sand | 21.00 | 0.437 | −16.0 | 2.62 | $2.83 \times 10^{-2}$ |
| Loamy sand | 6.11 | 0.437 | −20.6 | 3.61 | $2.06 \times 10^{-2}$ |
| Sandy loam | 2.59 | 0.453 | −30.2 | 7.48 | $9.92 \times 10^{-3}$ |
| Sandy clay loam | 0.43 | 0.398 | −59.4 | 16.05 | $4.63 \times 10^{-3}$ |
| Loam | 1.32 | 0.463 | −40.1 | 6.72 | $1.11 \times 10^{-2}$ |
| Silt loam | 0.68 | 0.501 | −50.9 | 12.74 | $5.83 \times 10^{-3}$ |
| Clay loam | 0.23 | 0.464 | −56.4 | 16.49 | $4.50 \times 10^{-3}$ |
| Sandy clay | 0.12 | 0.430 | −79.5 | 19.32 | $3.84 \times 10^{-3}$ |
| Silty clay loam | 0.15 | 0.471 | −70.3 | 22.46 | $3.31 \times 10^{-3}$ |
| Silty clay | 0.09 | 0.479 | −76.5 | 24.60 | $3.02 \times 10^{-3}$ |
| Clay | 0.06 | 0.475 | −85.6 | 26.83 | $2.77 \times 10^{-3}$ |

*Source:* From Rawls et al., 1982; Brakensiek and Rawls, 1992.

where the hydraulic conductivity is no longer a constant, but is instead:

$$\underline{\underline{K}}_s = \begin{bmatrix} K_x & 0 & 0 \\ 0 & K_y & 0 \\ 0 & 0 & K_z \end{bmatrix} \qquad [4.10]$$

This formulation implies that the flow direction can be different from the hydraulic gradient. The hydraulic conductivity function ($K_S$) is called a tensor, and represents the magnitude of the hydraulic conductivity as a function of direction. The example above shows $\underline{K}_S$ as having nonzero values along its diagonal, which only occurs when the soil layers are precisely oriented with the $(x,y,z)$ coordinate system (e.g., for perfectly horizontal layers). For conditions when the layers are not coincident with the coordinate system, then the off-diagonal components are no longer zero, and we have:

$$\underline{\underline{K}}_s = \begin{bmatrix} K_{xx} & K_{xy} & K_{xz} \\ K_{yx} & K_{yy} & K_{yz} \\ K_{zx} & K_{zy} & K_{zz} \end{bmatrix} \qquad [4.11]$$

or, equivalently:

$$J_x = -\left[ K_{xx} \frac{\partial H}{\partial x} + K_{xy} \frac{\partial H}{\partial y} + K_{xz} \frac{\partial H}{\partial z} \right]$$

$$J_y = -\left[ K_{yx} \frac{\partial H}{\partial x} + K_{yy} \frac{\partial H}{\partial y} + K_{yz} \frac{\partial H}{\partial z} \right] \qquad [4.12]$$

$$J_z = -\left[ K_{zx} \frac{\partial H}{\partial x} + K_{zy} \frac{\partial H}{\partial y} + K_{zz} \frac{\partial H}{\partial z} \right]$$

The fundamental importance of these expressions lies in the observation that a gradient in one direction can induce flow in a different direction. For example, when soil layers are inclined in the unsaturated zone, a downward gradient caused by gravity often induces lateral drainage. In this case, the downward gradient $(\partial H/\partial z)$ induces a lateral flux $(J_x$ or $J_y)$ due to the nonzero conductivities $(K_{xz}$ or $K_{yz}$, respectively).

### 4.2.2.2 Specific Storage Coefficient

The specific storage coefficient $(S_s)$ is defined as the volume of water released from storage $(V_w)$ per unit volume of soil $(V)$ per unit decline in hydraulic head $(H)$:

$$S_s = -\frac{1}{V} \frac{dV_w}{dH} = -\frac{d\theta}{dH} \qquad [4.13]$$

### 4.2.2.3 Aquifer Transmissivity and Storativity

When flow is uniformly parallel to the soil layers, the flow per unit horizontal distance can be estimated using the transmissivity, $T = b\,K_s$:

$$\frac{Q}{w} = -T \frac{\Delta H}{L} \qquad [4.14]$$

which is identical to Equation [4.2] because $A = b\,w$, where $A$ is the cross-sectional area perpendicular to flow, $b$ is the thickness of the flow region and $w$ is the width region. When the hydraulic conductivity is not uniform with depth (as is generally the case), the transmissivity is defined as the integrated hydraulic conductivity over the layer:

$$T = \int_{z_0}^{z_0 + b} K_s \, dz \quad \text{or} \quad K_s = \frac{dT}{dz} \qquad [4.15]$$

The transmissivity is important because many field situations occur where the thickness of the water conducting unit is unknown, or the hydraulic conductivity is highly variable within the unit. Measured field data can be readily used to estimate the transmissivity, but data may not exist to clearly specify the thickness of the unit. In these cases, the transmissivity serves as a parameter that describes the bulk behavior of the unit.

The aquifer storativity or storage coefficient ($S$) is defined as the volume of water released from storage per unit area of geologic medium, per unit decline in head, or:

$$S = b \; S_s \quad \text{or} \quad S_s = \frac{dS}{dz} \qquad [4.16]$$

Storativity is a function of whether or not the geologic unit is saturated or unsaturated, and if saturated, whether it is confined or unconfined. For unsaturated geologic units, the coefficient is related to the specific water capacity. For unconfined geologic units, the coefficient is related to the specific yield of the unit.

The storage coefficient is especially important for transient flow conditions, where water is released or added to storage. The greater the storage coefficient, the slower an energy response propagates through the system, and vice versa. Confined aquifers show rapid responses to pumping over large areas, while the influence of pumping on unconfined aquifers is more localized. Also, pulses of water moving through the unsaturated zone will be more attenuated when the specific water capacity is small.

### 4.2.2.4  Conductance or Leakance

The conductance, or leakance coefficient, is used for conditions of vertical flow across a low permeability layer, such as below a lake bed, across a confining layer or under a stream. The leakance is another way of writing Darcy's equation for conditions when the thickness of the layer is unknown:

$$J = -C \, \Delta H \qquad [4.17]$$

where $C = K_s/L$ and $L$ is the thickness of the low-permeability layer. In many applications, the head drop ($\Delta H$) across the layer is known, or can be estimated, along with the vertical flux. Because the thickness of the confining layer may not be known, one can use the leakance coefficient as a bulk parameter to describe the behavior of the system.

### 4.2.3  SATURATED FLOW EXAMPLES

### 4.2.3.1  Flow Perpendicular to Layers

Most soils other than Entisols and Inceptisols have well-developed horizons or layers which have different hydrologic characteristics including $K_s$. Above the water table, water usually flows vertically so that flow is perpendicular to these layers. Below the water table, flow can occur parallel to the layers (described in the following section). Steady flow through a layered soil can be described by making an analogy between Darcy's equation and Ohm's equation for current flow (Jury et al., 1991). In this case, the hydraulic resistance is the layer thickness divided by the saturated hydraulic conductivity of the layer. Because flow across $N$ layers is analogous to electrical flow through resistances in series (all of the water must flow through each layer), the effective hydraulic resistance is the sum of the resistances in each layer. The effective saturated hydraulic conductivity is then:

$$K_{eff} = \frac{\sum\limits_{j=1}^{N} L_j}{\sum\limits_{j=1}^{N} \dfrac{L_j}{K_j}} \qquad\qquad [4.18]$$

where $L_j$ and $K_j$ are the thickness and saturated hydraulic conductivity of each layer. Steady vertical flow through the entire soil profile is described with Darcy's equation:

$$J = -K_{eff}\frac{\Delta H}{L_t} = -C\,\Delta H \qquad\qquad [4.19]$$

where potentials are measured at the top and bottom of the profile, $L_t$ is the total thickness of the profile, and $C = K_{eff}/L_t$ is the conductance described above.

Because the flux across all layers is the same under steady flow conditions, Equation [4.19] can be used (once $J$ is known) to determine the matric potential at each layer interface. In this manner, one can show that when water flows down through a high to a low conductivity layer (e.g., a sandy horizon over a more clayey horizon), a positive pressure occurs at the interface that is greater than that present in a uniform soil. This back pressure slows flow through the high permeability layer and speeds flow through the other until there is an equivalent amount of flow through both layers. In the reverse situation (e.g., clay over a sand), the interface matric potential is less than that in a uniform soil, creating a suction which speeds flow through the low permeability layer and slows flow through the other.

In a saturated layered soil under steady flow, the matric potential distribution is linear with depth within a layer, but with a different rate of change in each layer. An abrupt change in slope at boundaries is usually not observed in natural soils because $K_s$ usually varies slowly, leading to a more curvilinear change in matric potential.

For conditions where a large difference in conductance is present, negative pressures can occur at the interface that exceed the air-entry potential of soil in the lower layer. This causes unsaturated conditions in the lower layer which greatly reduce flow through the profile. This often occurs when a surface seal is present. Another way to equalize flow through a layered soil with a high- over a low-permeability layer is for water to flow through only part of the more restrictive layer. This is a form of preferential flow called fingering. Both surface seals and fingering are discussed in Section 4.3.

---

**EXAMPLE 4.3** Determine the effective saturated hydraulic conductivity, steady flux and matric potential at the layer interface in a column consisting of a sand layer 30-cm thick overlying a loam layer 20-cm thick. The saturated hydraulic conductivities of the sand and loam layers are 21.00 and 1.32 cm h$^{-1}$, respectively. Water is ponded at the top of the column to a depth of 5 cm and allowed to drip from the bottom of the column.

Calculate the effective saturated hydraulic conductivity:

$$K_{eff} = \frac{\sum\limits_{j=1}^{N} L_j}{\sum\limits_{j=1}^{N} \dfrac{L_j}{K_j}} = \frac{20\,cm + 30\,cm}{\dfrac{20\,cm}{1.32\,cm\,h^1} + \dfrac{30\,cm}{21.00\,cm\,h^1}} = 3.02\,cm\,h^1$$

Calculate the steady flux using $K_{eff}$ and the conditions at the top and bottom of the column:

$$J_w = -K_{eff} \frac{\Delta H}{L} = -3.02\,cm\,h^1 \frac{(5+50)\,cm\,(0+0)\,cm}{50\,cm} = 3.32\,cm\,h^1$$

Now that the flux is known, set up Darcy's equation using conditions at the bottom of the column and the interface to find the pressure at the interface $p_i$:

$$J_w = -K_{loam} \frac{\Delta H}{L} = -1.32\,cm\,h^{-1} \frac{p_i + 20\,cm\,(0+0)\,cm}{20\,cm} = -3.32\,cm\,h^{-1}$$

$$p_i + 20\,cm = -3.32\,cm\,h^{-1} \frac{20\,cm}{-1.32\,cm\,h^{-1}} = 50.30\,cm$$

$$p_i = 30.30\,cm$$

#### 4.2.3.2 Flow Parallel to Layers

For flow parallel to layers (which usually occurs below the water table), the resistances can be considered in parallel (more water flows through high than low conductivity layers). If the soil layers are of equal thickness, then $K_{eff}$ is the arithmetic average of the individual layer saturated hydraulic conductivities. For layers of different thickness, $K_{eff}$ is a weighted average where the layer thicknesses are the weights. Steady horizontal flow through the profile can be described using:

$$\frac{Q}{w} = K_{eff}\,L_t\,\frac{\partial H}{\partial x} = T\,\frac{\partial H}{\partial x} \qquad [4.20]$$

where $w$ is the unit width of aquifer perpendicular to flow, and $T$ is the aquifer transmissivity.

Flow through heterogeneous aquifers is largely dominated by the highest permeability units. For an aquifer with a log-normally distributed hydraulic conductivity and a unit variance, about half of all flows occurs in only 1% of the thickness, and 90% of the flow occurs in about 16% of the unit.

#### 4.2.3.3 Flow to a Well

Flow to a pumping well occurs when the total head in the well is lower than that in the surrounding aquifer. The hydraulic gradient induces (nearly) horizontal flow into the well casing, with some additional vertical flow due to leakage from above or below the aquifer, and from releases from storage within the aquifer. The Thiem (1906) Equation predicts the steady-state drawdown ($s$) in a confined aquifer with transmissivity ($T$) as a function of distance ($r$) from a well that is pumped at a rate ($Q$):

$$s = \frac{Q}{2\,\pi\,T}\,\ln\left(\frac{r}{r_o}\right) \qquad [4.21]$$

The distance to a constant head boundary ($r_o$) is required to assure that the system reaches steady state. Otherwise, water levels should continue to fall over time, until a source of increased recharge is intercepted by the declining water levels.

Other solutions for flow to a well, including confined aquifer conditions, can be found in Kruseman and deRidder (1990).

### 4.2.3.4  Flow to a Drain

Flow to a horizontal drain is a common concern in regions of high water tables and in arid region irrigation projects where salinity can be a problem. It can be treated as a steady two-dimensional flow problem (Hooghoudt, 1937). Radial ($R_r$) and horizonal ($R_h$) resistances to flow are calculated:

$$R_r = \frac{1}{\pi} \ln\left(\frac{0.7\,D}{r}\right) \quad R_h = \frac{(L - 1.4\,D)^2}{8\,D\,L} \tag{4.22}$$

where $D$ is the distance between the drains and the impermeable layer below and $r$ is the radius of the tile drains. These resistances are used to calculate a factor ($d$) which corrects for radial flow near the drains:

$$d = \frac{L}{8(R_h + R_r)} \tag{4.23}$$

The spacing ($L$) between drains as a function of the desired height ($h_s$) of the water table above the drains (midway between the drains) is given by:

$$L^2 = \frac{8\,K_2\,d\,h_s}{i} + \frac{4\,K_1\,h_s^2}{i} \tag{4.24}$$

where $i$ is the infiltration rate, and $K_1$ and $K_2$ are the saturated hydraulic conductivities above and below the drains, respectively. Since $L$ must be known to calculate $R_h$ and $d$, these equations have to be solved iteratively, but the sensitivity of the final value of $L$ to the initial guess is low. In Figure 4.3, the drain spacing required to maintain the water table at a depth 70 cm below the soil surface is shown as a function of $K_1$. The drains in this example were at a depth of 100 cm with $K_1 = 7.62$ cm h$^{-1}$ and $K_2 = 0.205$ cm h$^{-1}$. As the saturated hydraulic conductivity increased, the spacing between the drains could be increased (thereby decreasing the cost of installation) while maintaining the water table no higher than 70 cm below the surface.

## 4.3  FLOW IN UNSATURATED SOIL

The unsaturated zone, also called the vadose zone, lies between the water table and the soil surface (Figure 4.1). In this region, the water content of the soil is less than saturation ($\theta < \varphi$), many pores are air-filled ($\varphi_a > 0$), and pressure heads are generally negative (h < 0). The water table, which separates the saturated zone from the unsaturated zone, is the surface where the water potential equals the mean atmospheric pressure.

Rising above the water table is the capillary fringe, a zone where water is under tension, but is very near saturation. The capillary fringe is important for water flow because only a slight change in water content or total head can cause a sharp change in the water table position. Also, lateral flow increases in this zone (Liu and Dane, 1996). The thickness of the capillary fringe depends on the water retention curve, and can be approximated by the air-entry pressure head ($h_a$). Sands have thinner capillary fringes than unstructured clays (Table 4.1). Capillary fringes are also found around zones of saturation in the unsaturated zone, such as around injection boreholes, surface seepage basins and losing streams.

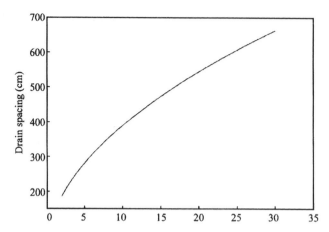

**FIGURE 4.3** Drain spacing as a function of hydraulic conductivity above the drains required to maintain the water table below a depth of 70 cm with $i = 0.25$ cm h$^{-1}$.

In this section, an overview of the governing equations and processes that characterize unsaturated water flow will be presented. While the subsurface should preferably be treated as a uniform porous media, the unsaturated zone possesses several features that prevent this. First, soil layers have a large effect on the behavior of water flow; even small spatial changes in soil properties can induce large variations in the direction and magnitude of water flow, especially under transient conditions. Second, soil properties vary substantially in space within a layer. Third, macropores may allow water to rapidly bypass the bulk of the soil matrix.

### 4.3.1 Unsaturated Flow Equations

#### 4.3.1.1 Buckingham-Darcy Equation

In the unsaturated zone, larger pores drain more readily than smaller ones, as can be noted using the capillary rise equation. Therefore, the hydraulic conductivity is much smaller under unsaturated than saturated conditions due to water moving through small pores or as films along the walls of larger pores. At very low water contents, continuous fluid paths may not exist and water may move in the vapor phase. The unsaturated hydraulic conductivity is therefore represented as a function of pressure head [$K(h)$] or as a function of water content [$K(\theta)$]. Buckingham (1907) modified Darcy's equation for unsaturated flow:

$$J = -K(h)\frac{\partial H}{\partial z} = -K(h)\frac{\partial h}{\partial z} - K(h) = -K(h)\left(\frac{\partial h}{\partial z} + 1\right) \qquad [4.25]$$

where the total head is the sum of the pressure and gravitational heads, $H = h + z$.

#### 4.3.1.2 Richards Equation

For transient flow, the water content and pressure head vary with time ($t$) and the Buckingham-Darcy equation must be extended. This partial differential equation was developed by Richards (1931). For one-dimensional vertical flow, the continuity equation requires that the change in volumetric water stored within a soil element be equal to the net flux into the element and any sources, or sinks within the element:

$$\frac{\partial \theta}{\partial t} = -\frac{\partial J}{\partial z} + s \qquad [4.26]$$

where $s$ is a source/sink term for water per unit time. When the Buckingham-Darcy equation is substituted for $J$, we obtain:

$$\frac{\partial \theta}{\partial t} = \frac{\partial}{\partial z}\left[ K(h)\left(\frac{\partial h}{\partial z} + 1\right)\right] + s \qquad [4.27]$$

which is called the *mixed form* (because it contains two dependent variables [$\theta$ and $h$]) of the one-dimensional Richards equation. It can also be written all in terms of $h$ (the *h form*) by using the chain rule on the left-hand side:

$$C(h)\,\frac{\partial h}{\partial t} = \frac{\partial}{\partial z}\left[ K(h)\left(\frac{\partial h}{\partial z} + 1\right)\right] + s \qquad [4.28]$$

where $C(h)$ is the water capacity function (Section 4.3.2.3). The Richards equation is also a nonlinear partial differential equation. To predict the distribution of $h$ or $\theta$ as a function of depth for a particular time, this equation must be integrated twice with respect to $z$ and once with respect to $t$, with boundary and initial conditions specific to the problem being solved. Except under very limited conditions, these integrals are not known so the Richards equation does not have an analytical solution. Therefore, for transient flow problems, either the equation is simplified in some manner (as happens with infiltration equations, below) or a numerical method is used to solve the equation (Section 4.5).

### 4.3.2 Unsaturated Flow Parameters

The most important parameters in unsaturated flow are the unsaturated hydraulic conductivity and (to a lesser extent) the water capacity functions. Both depend on pore size distribution. Values of $K(h)$ vary by orders of magnitude with pressure head, are sensitive to texture and structure, and are highly variable in space and (in some cases) time.

### 4.3.2.1 Unsaturated Hydraulic Conductivity

A typical $K(h)$ curve for a hypothetical sand and an unstructured clay are shown in Figure 4.4. The saturated hydraulic conductivity [$K(0) = K_s$] is higher in the sand than in the clay, as discussed in Section 4.2.2.1. Conductivity drops rapidly in the sand as pores empty since there is a more narrow distribution of pore sizes compared to the clay. Conductivity declines more gradually in the clay because each decrease in potential empties only a few pores due to the wide distribution in pore size. At low pressure head (water contents), the unsaturated hydraulic conductivity is greater in the clay than in the sand due to more water-filled pores and continuous water films in the clay.

The Gardner (1958) equation is commonly used to describe the unsaturated hydraulic conductivity function:

$$K(h) = K_s \exp(\alpha\, h) \qquad [4.29]$$

where $\alpha$ is a constant. Other commonly used equations are the Brooks and Corey (1964) equation:

$$K(h) = K_s\left(\frac{h}{h_a}\right)^{-2-3\lambda}, \quad h < h_a$$

$$= K_s \qquad\qquad , \quad h \geq h_a \qquad [4.30]$$

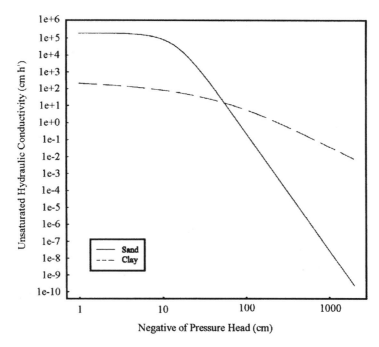

**FIGURE 4.4** Unsaturated hydraulic conductivity as a function of pressure head for sand and clay.

where $h_a$ is the air-entry pressure and $\lambda$ is a constant, the Campbell (1974) equation:

$$K(\theta) = K_s \left( \frac{\theta}{\theta_s} \right)^m \qquad [4.31]$$

where $m$ is a constant, and the Haverkamp equation (Haverkamp et al., 1977):

$$K(h) = \frac{K_s}{1 + \left( \dfrac{h}{a} \right)^N} \qquad [4.32]$$

where $a$ and $N$ are constants. Another useful equation is the van Genuchten (1980) Equation [4.58] described later in Section 4.4.1.2.

### 4.3.2.2  Capillary Length Scales

A useful way to quantify the unsaturated hydraulic conductivity function in a single parameter is in terms of the macroscopic capillary length. This is a $K(h)$-weighted average pressure head over the range of pressure heads of interest, typically the initial pressure head in the soil before water is added ($h_i$) and the soil pressure head in equilibrium with the water source ($h_0$) (White and Sully, 1987):

$$\lambda_c = \frac{\displaystyle\int_{h_i}^{h_0} K(h)dh}{K(h_0) - K(h_i)} \qquad [4.33]$$

**TABLE 4.2**
**Soil texture/structure categories for estimation of macroscopic capillary length ($\lambda_c$).**

| Soil texture/structure category | $\lambda_c$ (cm) |
|---|---|
| Coarse and gravelly sands; may also include some highly structured soils with large cracks and/or macropores | 2.8 |
| Most structured soils from clays through loams; also includes unstructured medium and fine sands | 8.3 |
| Soils which are both fine textured (clayey) and unstructured | 25.0 |
| Compacted, structureless, clayey materials such as landfill caps and liners, lacustrine or marine sediments, etc. | 100.0 |

*Source:* From Elrick and Reynolds (1992).

When the Gardner Equation [4.29] is used to describe $K(h)$ and $K(h_i) << K(h_0)$, $\lambda_c = \alpha^{-1}$. Using this relationship, the effect of the shape of the $K(h)$ function on $\lambda_c$ can be demonstrated. A small value of $\lambda_c$ indicates that $K(h)$ rises sharply at the wet end of the curve, as it might in a structured clay with macropores (Section 4.3.6.1). Typical values for $\lambda_c$ in undisturbed soils are shown in Table 4.2. The effect of structure in fine-textured soils is apparent in that an unstructured clay may have a $\lambda_c$ of 25 cm, whereas a structured clay may have a $\lambda_c$ of 2.8 cm. The macroscopic capillary length is also a measure of the effect of capillarity (the attraction of water to dry soil) as opposed to gravity on water movement. Water flow from a point source into a soil with a large value of $\lambda_c$ (unstructured clay) will have more lateral flow than into a soil with a small value of $\lambda_c$ (sand or well-structured clay).

The macroscopic capillary length can be converted to an average soil pore size called the *microscopic capillary length* ($\lambda_m$) using the capillary rise equation (White and Sully, 1987):

$$\lambda_m = \frac{\sigma}{\rho g \lambda_c} \qquad [4.34]$$

where $\sigma$ is the surface tension.

Brakensiek and Rawls (1992) provided $\lambda_c$ and $\lambda_m$ for soils of different textures (Table 4.1). Fine-textured soils have the largest macroscopic capillary length (capillarity) and the least microscopic capillary length (average pore size).

### 4.3.2.3  Water Capacity Function

The water capacity function $C(h)$ is the slope of the soil water characteristic curve [$\theta(h)$, discussed in more detail in Chapter 3]. It is an important parameter in numerical models of water flow (Section 4.5) and a plot of $C(h)$ versus $h$ is a useful way to display the pore size distribution of a soil.

### 4.3.3  STEADY FLOW

It is useful to consider several cases of steady unsaturated flow because analytical solutions to the Buckingham-Darcy equation have been developed for these special conditions. Although true steady flow may occur only rarely, transient flow usually approaches steady flow at long periods of time and the analytical solutions help to show what factors may be important in nonsteady conditions.

Steady one-dimensional flow in a uniform soil can be described by the Buckingham-Darcy Equation [4.25]. If the water content is uniform with depth, then only gravity causes flow and the flux or infiltration rate is equal to the hydraulic conductivity:

$$J = -K(h) \qquad [4.35]$$

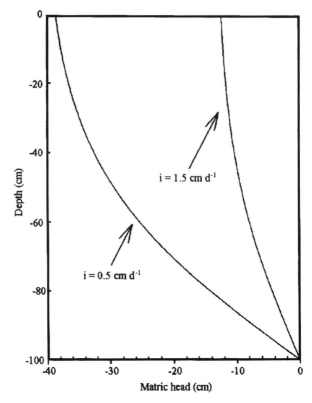

**FIGURE 4.5** Distribution of pressure head as a function of depth for steady infiltration at a low ($i = 0.5$ cm day$^{-1}$) and a high ($i = 1.5$ cm day$^{-1}$) irrigation rate.

In this case the distribution of $h$ with depth is a vertical line (the same at all depths). This is called *gravity flow* or *unit-gradient flow*.

For steady downward flow or infiltration ($J = -i$) in a uniform soil from the surface to a water table at depth $L$ (as might occur in a soil remediation effort designed to flush contaminants from the unsaturated zone down into the capture zone of a well), the Buckingham-Darcy equation can be integrated and solved for $h$ using the Haverkamp equation (Equation [4.32]) for $K(h)$ (Jury et al., 1991):

$$h = a \sqrt{\frac{K_s}{i} - 1} \, \tanh\left[\frac{-\sqrt{\left(1 - \frac{i}{K_s}\right)\frac{i}{K_s}}}{a}(z + L)\right] \tag{4.36}$$

The distribution of $h$ versus $z$ predicted by Equation [4.36] for a Chino clay with $K_s = 1.95$ cm day$^{-1}$, $a = -23.8$ cm, and $N = 2$ (Gardner and Fireman, 1958) is shown for a low ($i = 0.5$ cm day$^{-1}$) and a high ($i = 1.5$ cm day$^{-1}$) irrigation rate in Figure 4.5. The distribution of $h$ is now curvilinear, contrary to a uniform saturated soil (Section 4.2.1.2), especially with the lower irrigation rate which causes a lower $h$ at the surface. At the water table, the pressure head is zero. Near the surface, especially at the high irrigation rate, the curve is nearly vertical indicating that gravity flow is occurring (Equation [4.35]).

Three-dimensional water flow from a point source will approach a steady rate after a period of time. For example, steady infiltration from a shallow ponded ring on the soil surface is usually described using the Wooding (1968) equation:

$$i_s = J \big|_{z=0} = K(h_0)\left(1 + \frac{4\lambda_c}{\pi r}\right)$$ [4.37]

where $i_s$ is the steady infiltration rate, $r$ is the radius of the ring and $K(h_0)$ is the hydraulic conductivity corresponding to the potential of the water supply at the soil surface. Flow into the soil is considered a positive flux, in this case. It is apparent by comparing Equations [4.35] and [4.37] that the second term in Equation [4.37] accounts for the increased infiltration rate from a ring due to lateral (capillary-driven) flow into dry soil. The Wooding equation assumes that the unsaturated hydraulic conductivity function can be described by the Gardner Equation [4.29].

### 4.3.4 INFILTRATION

Infiltration is a key process because it determines how much water from rainfall, irrigation or a contaminant spill enters the soil and how much becomes runoff (or overland flow in hydrology terminology). It is also a key process in erosion in that there can be no erosion without runoff to transport and scour sediment.

During infiltration, a wetting front of higher water content moves down through the soil over time. The abruptness of the wetting front depends on the pore-size distribution and shape of the $K(h)$ function. For coarse-textured soils with a narrow pore-size distribution, the wetting front will be more abrupt and in a fine-textured soil, the wetting front will be more diffuse. The wetting front is a combination of new water added by the rain and old water displaced to lower depths.

Measurements and numerical solutions have shown that the infiltration rate ($i$) in a uniform, initially dry soil when rainfall does not limit infiltration, decreases with time and approaches an asymptotic minimum infiltration rate (Figure 4.6a). Cumulative infiltration ($I$) is the area under the infiltration rate curve (Figure 4.6b). Infiltration rate may be thought of as the Darcy flux at the soil surface with $J_w = -i$ (infiltration is a special case where downward water flow is considered positive). At the soil surface, the water content increases with time so $K(h)$ must also increase with time and cannot account for the decline in $i$. Initially, the wetting front is just below the soil surface and the water potential gradient at the surface is very large. As the wetting front moves deeper in the soil, the gradient at the surface decreases and this has an overriding effect on the increase in $K(h)$ and accounts for the decrease in $i$ with time. Eventually, the distribution of potential with depth near the surface approaches a unit gradient and at that point the infiltration rate asymptotically approaches the field saturated hydraulic conductivity. A high antecedent water content will result in a lower initial infiltration rate due to a diminished potential gradient, but the final infiltration rate will be the same regardless of the antecedent water content.

Experiments have shown that this minimum final infiltration rate is less than the saturated hydraulic conductivity that would be measured on a carefully saturated, intact core in the laboratory (Section 4.4.1.1). This has been attributed to the effect of entrapped air under field conditions that reduces the cross-sectional area available for water flow and the water potential gradient. Entrapped air can be divided into mobile and immobile pockets of air (Faybishenko, 1995). The immobile air resides in the fine and dead-end pores and can only be removed by dissolution. During infiltration, the mobile air is displaced from the finer into larger pores as the fine pores fill by capillarity. Thus entrapped air blocks the larger pores which have a disproportionate effect in reducing flow. It can be a matter of days for all of the mobile air to migrate to the surface (Faybishenko, 1995). Bouwer (1966) recommended using a value for field-saturated hydraulic conductivity ($K_{fs}$) equal to one-half of the true

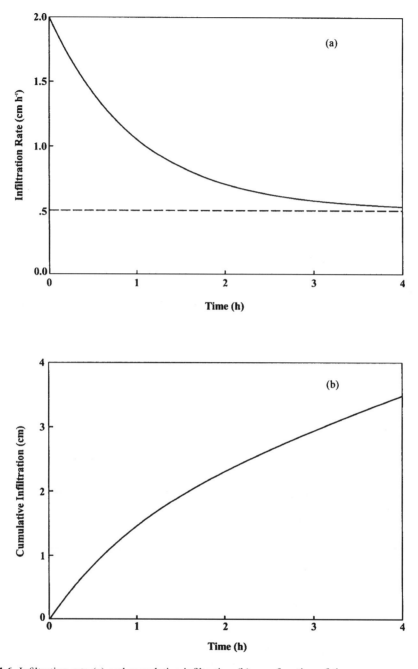

**FIGURE 4.6** Infiltration rate (a) and cumulative infiltration (b) as a function of time.

saturated hydraulic conductivity ($K_s$) for predicting the steady state infiltration rate. In the same way, a field-saturated water content ($\theta_{fs}$) as opposed to $\theta_s$ can be considered.

### 4.3.4.1 Infiltration through Crusts and Layered Soils

Another factor that can cause the infiltration rate to decrease is the formation over time of a surface seal or crust at the soil surface. A surface seal is a very thin layer (1–5 mm) at or just below the soil surface that forms due to the breakdown of soil aggregates and chemical dispersion of clay

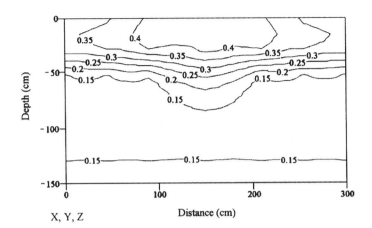

**FIGURE 4.7** Distribution of water after 1.3 hours of infiltration into a clay layer overlying a dry sand layer commencing at a depth of 40 cm. The hydraulic parameters were $K_s = 5$ cm day$^{-1}$, $\theta_s = 0.44$, $\theta_r = 0.15$, $h_a = -50$ cm, Brooks and Corey $\lambda = 0.6$ for the clay layer and $K_s = 100$ cm day$^{-1}$, $\theta_s = 0.40$, $\theta_r = 0.08$, $h_a = -15$ cm, Brooks and Corey $\lambda = 1.0$ for the sand layer. The initial conditions were $h = -100$ cm. The problem was solved with VS2DT (Lapalla et al., 1987; Healy, 1990).

particles under raindrop impact. The clay particles fill the soil pores and create a layer with a saturated hydraulic conductivity several orders of magnitude less than the undisturbed soil (Miller and Radcliffe, 1992). This low conductivity layer can prevent saturation of the soil just beneath the seal due to the suction that occurs at the interface (Section 4.2.3.1), further reducing the infiltration rate. Seals can be prevented by protecting the soil surface from raindrop impact with a mulch or previous crop residue, as is done in conservation tillage, and by preventing chemical dispersion through the use of soluble amendments (Miller and Baharuddin, 1986). Once the seal dries out, it develops a high soil strength due to the increased density of the layer and is called a crust. A crust can reduce seedling emergence, especially in dicots.

Buried clay or dry sand layers near the surface can also reduce infiltration rates. An unstructured buried clay layer will usually have a lower $K_{fs}$ than an overlying coarse-textured layer and reduce $K_{eff}$ (Section 4.2.3.1) and $i$ once the wetting front enters the clay layer. A buried dry sand layer under a fine-textured layer will also impede deeper movement of the wetting front and reduce $i$, but through a different mechanism. The water at the leading edge of the wetting front may be under several thousand cm of tension and cannot enter the smallest pores in the sand layer (which are much larger than the smallest pores in the layer above) until potential at the wetting front increases to the water-entry potential for the sand. This stalls the wetting front until potentials rise to the critical level for entry. Since the pore-size distribution is more narrow in the sand, it is not long after water first enters the soil that the potential is high enough at the wetting front to fill the largest capillary pores in the sand. Once the sand is field-saturated, it no longer impedes flow because $K_{fs}$ is high in the sand compared to the fine-textured layer above. This is shown in Figure 4.7 where a numerical model was used to predict water contents during infiltration from a point source into a soil with a dry sand layer at a depth of 40 cm. After 1.3 hours, the wetting front has spread laterally above the sand layer and just started to penetrate right below the source where the wetting front pressure head is the greatest. Baver et al. (1972) referred to the action of a buried dry layer in temporarily impeding water flow and infiltration as a check valve. This principle is used in the design of golf greens which have a sand surface layer over a coarse gravel layer at about 50 cm. The gravel layer keeps water from frequent light irrigations in the root zone, but if there is a large rainstorm the gravel layer will fill and drain the rootzone so that the green does not become waterlogged.

The effect of soil layers is to introduce bulk anisotropy within the soil profile. The hydraulic conductivity is no longer the same in all directions, hence anisotropic, and is a function of the pressure head, or water content. Such moisture-dependent anisotropy causes distinctly different flow behavior as a function of ambient conditions. When the water content is such that the hydraulic conductivity of both layers is equal, the water moves vertically, but under either wetter or drier conditions there may be substantial lateral flow.

Because of the importance of the infiltration process, simplified solutions to the Richards equation have been developed to predict infiltration.

### 4.3.4.2  Profile Controlled Infiltration

Most of the infiltration equations have been developed for conditions when rainfall does not limit infiltration. In this case, the infiltration rate is less than the rainfall or irrigation rate and runoff occurs. Soil hydraulic properties control the infiltration rate so it is profile controlled (Hillel, 1980b), or ponded, although the depth of ponding may be negligible if surface storage is small.

Green and Ampt (1911) developed a simplified mechanistic equation for infiltration by assuming that the wetting front in a soil was a square wave or sharp front. Although this is approximately true only in coarse-textured soils, there is no error in predicting the infiltration rate as long the amount of water behind the predicted square front is equal to the amount of new water behind the true wetting front. The Green-Ampt equation for cumulative infiltration in a uniform soil, including the effect of gravity, is:

$$I(t) = K(h_0)\, t + \Delta h\, \Delta\theta\, \ln\left(1 + \frac{I(t)}{\Delta h\, \Delta\theta}\right) \qquad [4.38]$$

where $\Delta h = h_0 - h_f$, $\Delta\theta = \theta_0 - \theta_i$, $\theta_i$ is the initial water content, $\theta_0$ is the water content extending from the wetting front to the soil surface, $h_0$ is the pressure head corresponding to $\theta_0$, and $h_f$ is the pressure head at the wetting front. When the rainfall rate is greater than or equal to the infiltration rate, $\theta_0$ and $K(h_0)$ can be approximated by $\theta_{fs}$ and $K_{fs}$, respectively, and $h_0$ can be assumed to be zero. The unknowns in Equation [4.38] are then $I(t)$ and $h_f$. White and Sully (1987) showed that the wetting front pressure head can be approximated using the macroscopic capillary length $\lambda_c$ (Tables 4.1 and 4.2):

$$h_f = -\frac{\lambda_c}{2b} \qquad [4.39]$$

where $b$ is a dimensionless factor that has a theoretical range of ½ to $\pi/4$, but can be assumed to be equal to 0.55 in most cases (Warrick and Broadbridge, 1992). Hence, the only unknown in Equation [4.38] is $I(t)$, but the equation cannot be solved directly for this variable because it appears both inside and outside the natural log function. Therefore it must be solved iteratively. Once the cumulative infiltration curve as a function of time is known, the infiltration rate can be calculated as the instantaneous slope of this curve.

The Green and Ampt Equation [4.38] can also be written as:

$$\frac{K(h_0)\, t}{\Delta h\, \Delta\theta} = \frac{\xi^2}{2} - \frac{\xi^3}{3} - \frac{\xi^4}{4} - \dots \qquad [4.40]$$

because:

$$\ln(1+\xi) = \xi - \frac{\xi^2}{2} + \frac{\xi^3}{3} - \dots \qquad [4.41]$$

where $\xi = I(t)\,\Delta h^{-1}\Delta\theta^{-1}$. Solving for the cumulative infiltration yields an infinite series of the form:

$$I(t) = A_0 t^{\frac{1}{2}} + A_1 t + A_2 t^{\frac{3}{2}} + \dots \qquad [4.42]$$

This is the same form as the Philip (1957a–e) equation for infiltration:

$$I(t) = S_0\, t^{\frac{1}{2}} + A_1\, t + A_2\, t^{\frac{3}{2}} + \dots \qquad [4.43]$$

where $S_0$ is sorptivity (L T$^{-\frac{1}{2}}$). Sorptivity is a measure of the capillary uptake of water and a function of the initial soil water content ($\theta_i$) and the water content at the soil surface ($\theta_0$). It is not surprising that it is related to the macroscopic capillary length ($\lambda_c$) (Table 4.1 and 4.2):

$$S_0 = \frac{\sqrt{\Delta\theta\,[K(h_0) - K(h_i)]\,\lambda_c}}{b} \qquad [4.44]$$

where $b$ is the same dimensionless factor that appears in Equation [4.39].

At early times when capillarity is much more important than gravity, cumulative infiltration can be described by discarding all but the first term in Equation [4.43]:

$$I(t) = S_0\, t^{\frac{1}{2}} \qquad [4.45]$$

Rawls et al. (1990) modified the Green and Ampt Equation [4.38] for soils with a crust by multiplying $K(h_0)$ by a crust factor $CF$:

$$CF = \frac{K_{eff}}{K(h_0)} = \frac{SC}{1 - \dfrac{h_{sc}}{L}} \qquad [4.46]$$

where $K_{eff}$ is the effective hydraulic conductivity of the two-layer system, $h_{sc}$ is the subcrust pressure head (just below the interface between the crust and the underlying soil), $L$ is the depth to the Green-Ampt wetting front [ $= (\theta_0 - \theta_i)/I(t)$ ], and $SC$ is a correction factor accounting for the effect of the subcrust pressure head in causing desaturation, thereby reducing the hydraulic conductivity of the subcrust layer (Section 4.3.4.1). They showed that both $SC$ and $h_{sc}$ varied with texture:

$$SC = 0.736 + 0.0019\, P_s \qquad [4.47]$$

where $P_s$ is the sand content (%) and:

$$h_{sc} = -45.19 + 46.68\, SC \qquad [4.48]$$

Since $L$ appears in Equation [4.46] and depends on $I(t)$, Equations [4.38] and [4.46] must be solved iteratively.

**EXAMPLE 4.4**   How long will it take for 2 cm of rain to infiltrate into a compacted, structureless clay soil with $K_{fs}$ = 0.01 cm h$^{-1}$? The field saturated water content is 0.39 cm$^3$ cm$^{-3}$ and the water content before the event is 0.18 cm$^3$ cm$^{-3}$. Assume that there is negligible ponding at the soil surface throughout the rainfall event.

Calculate $h_f$ using Equation [4.39] and a value of $\lambda_c$ = 100 cm from Table 4.2:

$$h_f = -\frac{\lambda_c}{2b} = -\frac{100\,cm}{2(0.55)} = -90.9\,cm$$

$$h_f = -90.9$$

Calculate $\Delta h$ and $\Delta\theta$:

$$\Delta h = h_0 - h_f = 0\,cm - (-90.9\,cm) = 90.9\,cm$$

$$\Delta\theta = 0.39\,\frac{cm^3}{cm^3} - 0.18\,\frac{cm^3}{cm^3} = 0.21$$

Solve Equation [4.38] for $t$:

$$t = \frac{I - \Delta h \Delta\theta \ln\left[1 + \dfrac{I}{\Delta h\,\Delta\theta}\right]}{K_{fs}}$$

$$= \frac{2\,cm - (90.9\,cm)0.21\ln\left[1 + \dfrac{2\,cm}{(90.9\,cm)0.21}\right]}{0.01\,\dfrac{cm}{h}}$$

$$= 9.79\,h$$

## 4.3.4.3  Supply Controlled Infiltration

During the early stages of a rainfall or irrigation event before ponding occurs, it is likely that the rainfall rate will limit infiltration. In other words, the rainfall rate is less than what the potential infiltration rate would be under ponded conditions (which maximize the potential gradient). In this case, the actual infiltration rate is equal to the rainfall rate (all of the rain infiltrates) and it is supply controlled (Hillel, 1980b). At a later time, called the time to ponding ($t_p$), the potential infiltration rate may drop below the rainfall rate (as the soil wets up and the potential gradient lessens). At that point, ponding (and runoff if surface storage is negligible) commences and rainfall no longer limits the infiltration rate which becomes profile controlled. The crux of the problem for predicting infiltration is to determine the time to ponding and how to correct infiltration rates after ponding for the fact that less water has entered the soil than would have if ponding had occurred from the beginning.

Mein and Larson (1973) used the Green-Ampt equations to solve this problem for a steady rainfall rate ($r$). The cumulative infiltration at the time of ponding is:

$$I_p = \frac{K(h_0)\,\Delta h\,\Delta\theta}{r - K(h_0)}$$ [4.49]

The time of ponding then can be found by dividing $I_p$ by $r$. Then Equation [4.38] for ponded conditions can be used to calculate cumulative infiltration for times later than $t_p$, but if the actual time was used in this equation it would overpredict $I(t)$ because it assumes that infiltration has been proceeding at the maximum (ponded) rate since the beginning of the event. Therefore, the actual time must be corrected by subtracting a time interval that adjusts for the difference between the cumulative rainfall before ponding and $I(t)$ calculated using Equation [4.38]. This corrected time ($t_c$) is calculated as:

$$t_c = t + \frac{I_p - \Delta h\,\Delta\theta \ln\left(1 + \dfrac{I_p}{\Delta h\,\Delta\theta}\right)}{K(h_0)} - t_p$$ [4.50]

Chu (1978) used the Green-Ampt equations to solve the same problem for a nonsteady rainfall rate.

### 4.3.4.4 Curve Number Method

The curve number method is an empirically determined rainfall-runoff relationship that provides an indirect estimate of soil infiltration. The method is based on numerous measurements of runoff for many soil types and considers soil texture, soil drainage class, antecedent moisture conditions and vegetative cover. While not directly estimating infiltration, the method partitions the effective cumulative precipitation ($P_e$) over a watershed into two classes: the depth of water that quickly leaves the watershed as (surface and subsurface) runoff ($Q$) and that water which is abstracted, or retained, on the watershed ($F$) so that $P_e = Q + F$. Abstracted water can emerge later as baseflow, deep groundwater discharge, or as evapotranspiration. The empirical relationship between these variables is:

$$\frac{Q}{P_e} = \frac{F}{S} = \Theta$$ [4.51]

where $S$ is the water holding capacity of the soil (i.e., $F \leq S$), and $\Theta$ is the watershed average soil saturation. This equation implies that the amount of runoff increases as the watershed saturation increases; the runoff is zero when no water is stored ($F \ll S$ implies that $Q = 0$), and all of the rainfall occurs as runoff when the watershed has reached its maximum capacity ($F = S$ implies that $Q = P_e$). The abstraction volume ($F$) can be removed to yield:

$$Q = \frac{P_e^2}{P_e + S}$$ [4.52]

The standard practice is to assume that $P_e$ equals the actual precipitation, minus an initial abstraction equal to $S/5$. The curve number is used as a surrogate for the maximum water holding depth of the watershed:

$$CN = \frac{1000}{S + 10} \quad \text{or} \quad S = \frac{1000}{CN} - 10$$ [4.53]

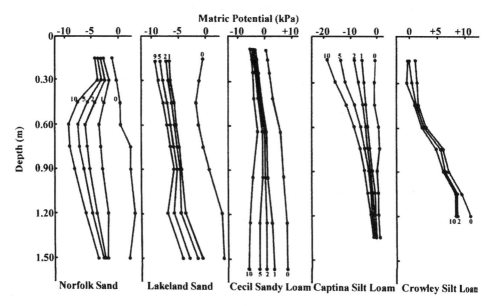

**FIGURE 4.8** Pressure potential as a function of depth for five soils during the first 10 days of drainage following a thorough wetting (Bruce et al., 1980).

Curve numbers are tabulated and range from near zero for a dry, fully vegetated, highly permeable surface to near 100 for an impervious surface (McCuen, 1982).

Several alternative explanations can be used to explain the theoretical basis for the curve number method. A relationship between soil saturation, infiltration rate and overland flow can be readily hypothesized. An alternative explanation rests on the hypothesis that spatially variable soil water holding capacity rather than infiltration rate limits abstraction depths.

### 4.3.5 Redistribution

Water continues to move in a soil for some time after the end of a rainfall or irrigation event. Water will drain from the regions near the soil surface to deeper depths and this process is called redistribution. Changes in profile pressure heads over time are shown in Figure 4.8 for five soils during redistribution (Bruce et al., 1985). For the three coarse-textured soils (two sands and a sandy loam), most of the change in pressure head observed over a 10-day period occurred during the first 24 hours of drainage. This was less true for the two silt loam soils. Redistribution occurs more rapidly in a coarse-textured soil because of the steeper $K(h)$ function, compared to a fine-textured soil (Figure 4.4). In a coarse-textured soil, $K(h)$ is likely to be quite high in the undrained regions near the soil surface where water contents are near saturation so water drains rapidly. Once some of this water drains, however, water contents decrease, $K(h)$ becomes much lower and redistribution effectively stops. In a fine-textured soil, the initial drainage is slower due to the lower $K(h)$ near saturation, but redistribution continues longer because $K(h)$ in the drained region is not negligible.

In the absence of evapotranspiration and any further rainfall or irrigation, water will redistribute until the total potential ($H$) is the same at all depths and pressure head ($h$) increases linearly with depth (at a rate of 1 cm per cm of depth). In a soil with a shallow water table, the magnitude of the pressure head at the surface will be equal to the depth of the water table (Jury et al., 1991).

The water content of the soil near the surface once redistribution becomes negligible has been called field capacity or the drained upper limit. This has been approximated by the water content measured in the field 24–48 h after a thorough wetting of the soil profile or by the water content corresponding to a pressure potential of –33.3 kPa in a clayey soil and –10 kPa in a sandy soil.

Although these approximations are not very satisfying from a theoretical point of view, the field capacity concept is a useful one in irrigation scheduling and crop modeling.

## 4.3.6 PREFERENTIAL FLOW

Preferential flow is used as a general term to describe unusually rapid or deep movement of water or solute through a fraction of the total cross-sectional area available for flow. Preferential flow includes flow through macropores, fingering and funnel flow. Gish and Shirmohammadi (1991) recently reviewed preferential flow.

### 4.3.6.1 Macropore Flow

Macropores are large, continuous voids in soil and include structural, shrink-swell and tillage fractures, old root channels, and soil fauna burrows. Suggested lower limits for macropore diameters and widths are in the 0.03 – 3.00 mm range (Luxmoore, 1981; Beven and Germann, 1982; White, 1985). The lower limit would include some pores that would fill by capillarity and the upper limit would exclude all capillary pores. Macropores are important because they can increase infiltration and may result in bypass flow where water and solutes move rapidly through the profile and do not interact with the soil matrix (Quisenberry and Phillips, 1976).

To a certain extent, the effect of macropores on water flow can be incorporated into conventional flow equations based on Darcy's Law by careful measurement of $K_{fs}$ and the $K(h)$ function. If these parameters are measured in the field on a large enough sample to contain representative macropores, there is little error in using Darcy's equation (in fact the error is that Darcy's equation will overpredict flow in pores with high Reynolds number; under prediction is usually the concern). For example, Jarvis and Messing (1995) used a tension infiltrometer (Section 4.4.2.1) to measure $K(h)$ at values of $h$ between −5 and −150 mm on six Swedish soils of contrasting texture. When the data were plotted ($\ln K$ vs. $h$), the best fit was with two straight lines, the line near saturation being much steeper, especially in the more clayey soils (Figure 4.9). The breakpoint occurred at $h = -30$ to $-50$ mm, which corresponds to a pore diameter near the minimum for macropores ($h = -30$ mm corresponds to a pore diameter of approximately 0.1 mm). The value of $K(h)$ approached at zero pressure (an estimate of $K_{fs}$) was greatest for the four finer-textured soils (in contrast to the data in Table 4.1). This indicated that soil structure and macropores had a greater effect on the $K(h)$ near saturation in the fine- compared to the coarse-textured soils.

Flow in individual water-filled macropores that are cylinders or cracks can be described by a modified Poiseuille equation, but the number and dimensions of macropores in a soil are usually not known (White, 1985). Also, using Poiseuille's law assumes that macropores are open-ended, which is probably not the case. Beven and Germann (1981) developed a kinematic wave equation to describe water flow in macropores that allows flow down the sides of the pores that are not filled with water. It requires an estimate of the size distribution of macropores and the average distance between macropores.

Since macropores are, for the most part, noncapillary pores, it has been assumed that macropore flow cannot occur unless there is free water (incipient ponding conditions) at the soil surface. Experimentally, however, macropore flow has been observed in very dry soils at the onset of a rain when these conditions are unlikely. For example, Shipitalo and Edwards (1996) used intact soil blocks and added water equivalent to a two-year storm with a rainfall simulator. They observed more macropore flow in blocks that were initially dry than in blocks that were at a higher antecedent water content. This may be due to a hydrophobic organic soil surface that develops under dry conditions and causes free water to run across the surface and enter macropores (Miller and Wilkinson, 1979; Edwards et al., 1989).

Seyfried and Rao (1987) used dye patterns to trace water movement through intact columns and found dyed regions of the soil that were not associated with visible macropores. Gupte et al.

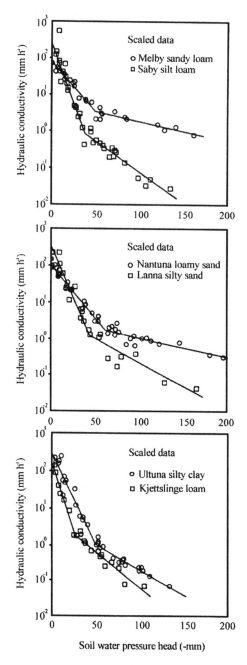

**FIGURE 4.9** Unsaturated hydraulic conductivity as a function of pressure head in six Swedish soils of contrasting textures (Jarvis and Messing, 1995, with permission).

(1996) and Shaw et al. (1997) also used a dye and found regions of dyed cross-sectional area, rather than individual dye-stained macropores. Macropores were visible in some, but not all, dyed areas and there were few if any macropores in the undyed areas. Soil thin sections revealed that the dyed regions had slightly higher total porosity and more large pores than the undyed regions. This may represent spatial variability in soil development. One region of soil may develop more rapidly due to penetration by a root along a plane of weakness, which in turn leads to more intense microbial and faunal activity, more water penetration, and greater differentiation between regions.

This type of flow is analogous to the two-region model of van Genuchten and Weirenga (1976) used to describe solute transport. It assumes that there are immobile regions of soil water in the interior of peds where water movement is much slower than in the mobile interped regions.

### 4.3.6.2  Fingering

When fine-textured soils overlie coarse-textured soils, it has been observed during infiltration that the wetting front does not uniformly penetrate the coarse layer (Hill and Parlange, 1972). Instead, the wetting front becomes unstable and separates into distinct fingers that penetrate only a portion of the coarse-textured layer. Initially, water cannot enter the coarse layer due to the low wetting front pressure head created in the fine-textured layer (Hillel and Baker, 1988). As the wetting front stalls, pressure head increases at the interface. Then when pressure head rises to the value required to enter the smallest pores in the coarse layer, water starts to enter. Pressure head continues to rise until it reaches a value where the flux through the coarse layer is equal to or greater than the flux through the top layer. Baker and Hillel (1990) called this pressure head the "effective water entry" value ($h_e$). If the flux at this potential is greater than the flux through the fine-textured layer, then the only way for flows to be equal is to confine flow through a fraction of the cross-sectional area of the coarse layer (fingering). The effective water entry pressure head corresponded well with the inflection point potential on the water retention curve, which is a measure of the dominant pore size. The fraction of wetted soil in the coarse layer at steady state was equal to the flux through the top layer divided by $K(h_e)$ in the lower layer. Theoretical analyses have shown that fingering can occur under a number of conditions: (1) increasing $K(h)$ with depth (as would occur in a fine-textured layer over a coarse-textured layer), (2) entrapped air that causes substantial air compression at the wetting front, (3) a buried hydrophobic layer, (4) an increase of water content with depth, (5) redistribution in a coarse-textured soil, and (6) continuous nonponding infiltration in a coarse-textured soil (Raats, 1973; Philip, 1975; Parlange and Hill, 1976; Diment et al., 1982; Glass et al., 1989, 1991).

### 4.3.6.3  Funnel Flow

Kung (1990a,b) used a dye to study water flow in sandy Wisconsin soils deposited during glacial outwash. The dye moved down uniformly from the surface until it encountered a lense of coarse sand (common in these soils) which acted as a capillary barrier. Since the lenses were not horizontal, the wetting front did not pause and wetting front pressure heads never increased to the effective water entry value of the coarse sand lenses. The dye moved horizontally along the top of the lense to the downslope end where downward flow continued. Repeated encounters of the dye front with lenses resulted in funneling of the flow into a fraction of the total cross-sectional area of flow.

## 4.4  MEASUREMENT OF HYDRAULIC PARAMETERS

Accurate measurement of hydraulic parameters by both laboratory and field methods is essential to the prediction of water movement in soils. The important laboratory and field methods commonly employed in estimating hydraulic properties are discussed below with additional procedures available in Klute (1986), Smith and Mullins (1991), Topp et al. (1992) and Carter (1993). A multitude of laboratory and field methods have been developed to provide parameter inputs to predictive models because accurate measurement of hydraulic parameters is essential to the credible prediction of water movement in soils. However, uncertainties in parameter estimates arise because of spatial variation, measurement accuracy and scale effects. For example, hydraulic conductivity typically varies over four orders of magnitude over short distances at a site. Small procedural differences in methodology can induce substantial variabilities in estimated hydraulic properties, as well. Reconciling data collected from laboratory versus field techniques often is difficult because bulk parameter estimates resulting from testing at different scales are not commensurate, as discussed in Section 4.2.

### 4.4.1 LABORATORY METHODS

Laboratory methods are used to measure $K_s$ and $K(h)$ on soil cores that are usually 7.6 cm in diameter, while fields, plots or rings are used in field methods. The use of intact rather than packed cores is recommended if the parameters are intended to represent undisturbed soils, especially fine-textured soils.

#### 4.4.1.1  Direct Methods

The constant-head and falling-head methods are common for measuring $K_s$. In the former method, a constant head of water is maintained at the top or bottom of the core and the steady water flux is recorded (Klute and Dirksen, 1986). Using Darcy's equation:

$$K_s = \frac{JL}{H_2 - H_1}$$
[4.54]

where $L$ ($> 0$) is the length of the core, $H_1$ and $H_2$ are total potentials at the top and bottom of the core, respectively. The flux, $J$, is negative in the downward direction so that $K_s$ must always be positive. In the falling-head method, a stand-pipe is attached to the top of the core and the head at the top is allowed to drop over time. The total head at the top of the core is measured using the standpipe at the beginning ($h_1 + L$) and end ($h_2 + L$) of a time interval ($t$). Using Darcy's equation:

$$K_s = \frac{L}{t} \ln\left(\frac{h_1 + L}{h_2 + L}\right)$$
[4.55]

The falling-head method is useful in soils with very low $K_s$ where it is difficult to collect sufficient drainage from a core to measure $J$ accurately within an interval of several hours. With the falling-head method, $h_1$ can be recorded and the core left overnight, for example, before recording $h_2$ and there is no concern about the effect of evaporation on the drainage sample collected.

Values of $K_s$ measured on cores in the laboratory will depend on what measures are used to remove entrapped air from the soil before making the measurement. Flushing the core with $CO_2$, wetting the core slowly from the bottom, and measuring $K_s$ under conditions of an upward flux are commonly used. However, in the field some degree of entrapped air can be expected, especially near the soil surface during infiltration. Hence, one can expect laboratory measurements of saturated hydraulic conductivity ($K_s$), to exceed field measurements ($K_{fs}$).

Unsaturated hydraulic conductivity can be measured on cores in a manner similar to the constant-head method for $K_s$ by fitting the core with porous plates and imposing suctions at the top and bottom of the core (Klute and Dirksen, 1986). The total potential within the core is measured at two heights using miniature tensiometers and the steady flux is recorded. The Buckingham-Darcy equation is solved for $K(h)$ where $h$ in this case is the average pressure head within the core. Steady evaporation from the top of a long core in contact with free water at the bottom can also be used to measure $K(h)$ by placing tensiometers at several heights, recording the steady evaporation rate and calculating the gradient between each pair of tensiometers (Jury et al., 1991).

Steady unsaturated flow can be imposed on a long soil core by supplying water continuously at a rate less than $K_{fs}$. This can be done by applying a crust made of a mixture of gypsum and sand or cement and sand to the soil surface (Bouma et al., 1976, 1983). The steady water flux is measured with a constant positive head above the crust. Because of the low $K_s$ of the crust, flow in the core is unsaturated (Section 4.2.3.1). A tensiometer in the core records $h$ and if a unit-gradient is assumed, then the measured flux is $K(h)$. Successive flux measurements with different crusts, each with a different conductivity, provide measurements of $K(h)$ over the range of pressures achieved. Bouma et al. (1976) used this method in the field, as well.

In all laboratory methods of measuring $K_s$ and $K(h)$, the effect of the chemistry of the added water on dispersion must be considered. The goal is to prevent chemical dispersion so that pore size distribution and connectivity does not change during the measurement, which can be achieved by using a dilute solution with a divalent cation (Chiang et al., 1987) or native water from the site.

### 4.4.1.2 Indirect Method

The unsaturated hydraulic conductivity function of a soil can be estimated from its water characteristic curve and $K_s$ by the indirect method. The characteristic curve provides information on the shape of the $K(\theta)$ function (both depend on the pore-size distribution) and the single value of $K_s$ serves as a matching point to anchor the curve at saturation. The most common approach is based on an equation developed by Mualem (1976):

$$K(\theta) = K_s \Theta^n \left[ \frac{\int_0^\theta \frac{d\theta}{h(\theta)}}{\int_0^{\theta_s} \frac{d\theta}{h(\theta)}} \right]^2 \qquad [4.56]$$

where $\Theta$ is the degree of saturation:

$$\Theta = \frac{\theta - \theta_r}{\theta_s - \theta_r} \qquad [4.57]$$

with $\theta_s$ and $\theta_r$ being the saturation and residual volumetric water contents, respectively. Mualem (1976) developed his model by considering the soil as a bundle of interconnected capillary tubes with Poiseuille's equation describing flow in each tube. The parameter $n$ is related to the tortuosity and connectivity of the tubes and, based on measurements in 45 soils, he recommended a value of $n = \frac{1}{2}$. Some form of the relationship between $\theta$ and $h$ (the water retention function) must be assumed to solve the integrals in Equation [4.56]. Most often the van Genuchten (1980) $\theta(h)$ relationship (Chapter 3) is used and it is assumed that the exponents in this equation are related in the manner $m = 1 - 1/n$. In this case, the integrals in Equation [4.56] can be solved analytically such that:

$$K(\theta) = K_s \Theta^{\frac{1}{2}} \left[ 1 - \left( 1 - \Theta^{\frac{1}{m}} \right)^m \right]^2 \qquad [4.58]$$

The indirect method works well for many coarse- and medium-textured soils, but not as well for fine-textured soils where structure has an important effect (van Genuchten and Leij, 1992). In well-structured soils, the distribution of noncapillary (macro) pores are not reflected in the water characteristic curve. In this case, the indirect method may represent the matrix $K(\theta)$ function accurately, provided a measured value of unsaturated hydraulic conductivity near saturation is used as a matching point in Equation [4.58] instead of $K_s$ (Clothier and Smettem, 1990; Jarvis and Messing, 1995). Durner (1992) developed a bimodal form of the van Genuchten (1980) $\theta(h)$ relationship that could be used to incorporate macropores. In this case, the integrals in Equation [4.56] must be evaluated numerically.

### 4.4.1.3  Laboratory Inverse Method

Inverse methods measure flow under transient conditions and then use a numerical solution of the Richards Equation [4.27] to determine the parameters in an assumed form of the $K(h)$ function. In the one-step outflow inverse method, cumulative outflow is measured from a core during an interval when the pressure at the inlet is increased substantially in a single step. Kool et al. (1985a) used a numerical model with the van Genuchten (1980) equations for water retention (Chapter 3) and $K(\theta)$ (Equation [4.58]) to determine the values of $\alpha$, $\theta_r$ and $m$ that produced the best fit of the model predictions to the observed cumulative outflow. The other parameters in these equations, $K_s$ and $\theta_s$, were measured independently on the core. A program called ONESTEP was used for determining the van Genuchten parameters (Kool et al., 1985b).

### 4.4.2  FIELD METHODS

Several advancements in field methods of measuring $K_{fs}$ and $K(h)$ have occurred in recent years. These include the use of tension infiltrometers and borehole permeameters and the associated solutions to the Richards Equation [4.27], based largely on the equations developed by Philip (1957a-e).

### 4.3.2.1  Infiltrometers

Tension infiltrometers have become a popular method for determining unsaturated hydraulic conductivity and other hydraulic parameters in field soils. They consist of a circular porous plate or membrane which is placed on the soil surface or an excavated soil surface. Water is supplied to the plate under tension using a mariotte bottle arrangement and the rate of water entry into the soil can be measured on a graduated cylinder or with a pressure transducer (Clothier and White, 1981). In some cases, a ring is attached to the tension infiltrometer to allow ponded infiltration measurements at the same location (Perroux and White, 1988). Two approaches are common, one in which the infiltration rate is measured after it attains a steady rate, and the other in which the early non-steady infiltration rate is measured.

Under the steady-rate approach, infiltration is measured at the same location using two or more tensions, starting with the tension farthest from zero. The Wooding Equation [4.37], written in terms of volumetric flow rate ($Q$) instead of infiltration rate ($i$) where $Q = i\pi r^2$, is used to describe the steady-state flow rates ($Q_1$ and $Q_2$) at two negative pressure heads ($h_1 < h_2$). The constant ($\alpha$) in the Gardner $K(h)$ Equation [4.29] can be calculated as:

$$\alpha = \frac{1}{h_1 - h_2} \ln \frac{Q_1}{Q_2} \qquad [4.59]$$

and the field-saturated hydraulic conductivity ($K_{fs}$) can be calculated as:

$$K_{fs} = \frac{\alpha Q_1}{r(4 + \alpha\pi r)\left(\dfrac{Q_1}{Q_2}\right)^P} \qquad [4.60]$$

where $P = h_1/(h_1 - h_2)$ (Reynolds and Elrick, 1991). Knowing $K_{fs}$ and $\alpha$, the unsaturated hydraulic conductivity at any value of $h$ can be predicted using the Gardner Equation [4.29] on the assumption that a constant value of $\alpha$ enables the description of the entire $K(h)$ function (the slope of $\ln K$ versus $h$ is constant), which may not be the case (Jarvis and Messing, 1995). Alternatively, it can be assumed that $\ln K$ versus $h$ is piecewise linear and that $K(h)$ can be described by a piecewise Gardner equation (Reynolds and Elrick, 1991).

The disadvantage of the steady-rate approach is that it may take several hours for flow to become constant (Warrick, 1992). For this reason, non-steady rate approaches have been developed. These methods assume that at early times when capillarity is much more important than gravity, three-dimensional flow from a disk can be described by Philip's equation for one-dimensional infiltration into a dry soil (Equation [4.45]). In this method, $S_0$ is determined from a plot of $I(t)$ versus $t^{1/2}$ during early time measurements with a tension infiltrometer. Two early-time measurements of $S_0$ are made with different water supply potentials ($h_0$) and the unsaturated hydraulic conductivity is calculated at the mean of these two potentials (White and Perroux, 1989). The disadvantage of this method is that two values of $S_0$ are required so that measurements must be made at two locations (soil variability becomes a factor) or the soil must be allowed to dry to the antecedent water content before initiating the second measurement of $S_0$. Also, measurements of the antecedent water content ($\theta_i$) and the water content of the soil in equilibrium with the supply potential of the tension infiltrometer ($\theta_0$) are required. The latter is difficult to measure accurately when it is confined to a narrow depth.

Ring infiltrometers can be used to pond water on the soil surface and measure the infiltration rate. It is usually assumed that the Wooding Equation [4.37] describes the three-dimensional steady infiltration rate. Typical values of $\lambda_c$ (Tables 4.1 and 4.2) can be used to convert steady-flow rates to $K(h_0)$, which in the case of ponded infiltration is a measure of $K_{fs}$. Alternatively, if early-time measurements of the infiltration rate are recorded, $S_0$ can be calculated assuming one-dimensional flow (Equation [4.45]). Assuming that the antecedent water content is low and $K(h_i)$ is negligible, Equation [4.44] may be solved for $\lambda_c$ and substituted into the Wooding equation, which is solved for $K_{fs}$ using $b = 0.55$:

$$K_{fs} = i_s - \frac{2.2\,S_0^2}{(\theta_0 - \theta_i)\,\pi\,r} \qquad [4.61]$$

where $i_s$ is the steady infiltration rate and $r$ is the radius of the ring.

Concentric double rings have been used to create one-dimensional flow in the interior ring. In this case, the steady infiltration rate is an estimate of $K_{fs}$ in a uniform soil. Bouwer (1986) found that for the typical dimensions used in a double ring (20 cm diameter for inner ring and 30 cm diameter for outer ring), flow was not one-dimensional.

### 4.4.2.2  Permeameters

Well permeameters are used to measure saturated hydraulic conductivity below the soil surface in the unsaturated zone. They consist of a mariotte device that maintains water at a constant level in a borehole and they allow measurement of the flow rate into the soil. The most common commercial well permeameters are the Guelph Permeameter and the Compact Constant Head Permeameter (CCHP). Steady flow into a borehole can be described by the equation:

$$i_s = K_{fs}\left(1 + \frac{H\lambda_c}{G\pi r^2} + \frac{H^2}{G\pi r^2}\right) \qquad [4.62]$$

where $H$ is the height of water ponded in the borehole, $r$ is the radius of the hole and $G$ is a dimensionless geometric factor which depends primarily on the ratio of $H/r$ (Elrick and Reynolds, 1992). Bosch and West (1997) fitted a polynomial equation to the data of Elrick and Reynolds (1992) to determine the value of $G$:

$$G = \frac{1}{2\pi}\left[A_1 + A_2\,\frac{H}{r} + A_3\left(\frac{H}{r}\right)^2 + A_4\left(\frac{H}{r}\right)^3\right] \qquad [4.63]$$

---

**TABLE 4.3**
**Coefficients for the polynomial (Equation [4.63]))**
**describing the dimensionless geometric factor G,**
**valid for H/r < 10.**

| Soil texture/structure | $A_1$ | $A_2$ | $A_3$ | $A_4$ |
|---|---|---|---|---|
| Sand | 0.079 | 0.516 | −0.048 | 0.002 |
| Structured loams and clays | 0.083 | 0.514 | −0.053 | 0.002 |
| Unstructured clays | 0.094 | 0.489 | −0.053 | 0.002 |

*Source:* From Bosch and West (1997).

---

The values of the coefficients $A_1...A_4$ in this polynomial depend on texture and structure (Table 4.3).

The similarity between Equation [4.62] and the Wooding Equation [4.37] is apparent. In sequence, the terms inside the brackets in Equation [4.62] account for the effects of gravity, capillarity and hydrostatic pressure in the borehole. There are two unknowns in this equation: $K_{fs}$ and $\lambda_c$. The normal procedure is to measure $i_s$ at two values of $H$ in the same borehole and solve simultaneous equations for $K_{fs}$ and $\lambda_c$. Since changing the level of $H$ in the borehole necessarily changes the region of soil that is being sampled, soil heterogeneity in the form of layering or macropores can result in unrealistic and invalid (i.e., negative) $K_{fs}$ and $\lambda_c$. As many as 30 to 80% of measurements of $K_{fs}$ and $\lambda_c$ in structured soils may be invalid (both negative) according to Elrick and Reynolds (1992). Alternatively, a single measurement of $i_s$ may be used and $\lambda_c$ estimated using Table 4.2. Studies suggest that this method yields values of $K_{fs}$ that are usually accurate to within a factor of 2 (Reynolds et al., 1992).

Another approach that does not require multiple measurements in the same borehole is based on the Glover solution (Zangar, 1953):

$$i_s = K_{fs} \frac{H^2}{G_G \pi r^2} \qquad [4.64]$$

where $G_G$ is the dimensionless geometric factor for the Glover analysis:

$$G_G = \frac{\sinh^{-1}\left(\dfrac{H}{r}\right) - \sqrt{\left(\dfrac{r}{H}\right)^2 + 1} + \dfrac{r}{H}}{2\pi} \qquad [4.65]$$

This approach only considers the effect of hydrostatic pressure in the borehole and ignores the effects of gravity and capillarity (Elrick and Reynolds, 1992). It can overestimate $K_{fs}$ by an order of magnitude or more in dry, fine-textured, structureless soils (i.e., soils where capillarity is most important). On the other hand, the Glover solution can provide good estimates of $K_{fs}$ in wet, coarse-textured or structured soils when the ratio $H/r$ is kept high (> 10, hydrostatic pressure dominates flow).

**EXAMPLE 4.5**  What is the field-saturated hydraulic conductivity for a structured loam soil if the steady state percolation rate measured with a borehole permeameter (CCHP) is 100 cm³ min⁻¹? The radius of the borehole is 3 cm and the depth of ponding is 15 cm.

Calculate the geometric factor using Equation [4.67] and the coefficients $A_1...A_4$ for a structured loam from Table 4.3:

$$G = \frac{1}{2\pi}\left[0.083 + 0.514\left(\frac{15\,cm}{3\,cm}\right) - 0.053\left(\frac{15\,cm}{3\,cm}\right)^2 - 0.002\left(\frac{15\,cm}{3\,cm}\right)^3\right]$$

$$= 0.172$$

Solve Equation [4.66] for $K_{fs}$ using $\lambda_c = 8.3$ cm from Table 4.2:

$$K_{fs} = \frac{i_s}{1 + \dfrac{H\lambda_c}{G\pi r^2} + \dfrac{H^2}{G\pi r^2}}$$

$$= \frac{\dfrac{100\,\dfrac{cm^3}{min}}{\pi(3\,cm)^2}}{1 + \dfrac{(15\,cm)(8.3\,cm)}{0.172\,\pi(3\,cm)^2} + \dfrac{(15\,cm)^2}{0.172\pi(3\,cm)^2}}$$

$$= 0.048\,\frac{cm}{min}$$

### 4.4.2.3  Instantaneous Profile and Field Inverse Methods

The instantaneous profile method is used to measure $K(h)$ in the field during drainage under nonsteady conditions (Green et al., 1985). A soil pedon is thoroughly wetted and then allowed to drain while preventing water fluxes into or out of the soil at the surface. During the ensuing drainage period, profile distributions of water content and total head are measured periodically over the depths of interest. Values of $K(h)$ are calculated at a given depth from an equation developed by integrating the Richards Equation [4.27] with respect to $z$:

$$\frac{\partial}{\partial t}\int_0^{z_1} \theta(z,t)\,dz = K(h)\left.\frac{\partial H(z,t)}{\partial z}\right|_{z_1} \tag{4.66}$$

Water contents may be measured using neutron probes or TDR and gradients are usually measured with tensiometers. The range of pressure head covered by this method is limited (near-saturation to field capacity), but this is the range where significant unsaturated flow occurs. Many measurements of $K(h)$ were made on soils of the southeastern U.S. using this method and are available in state bulletins (Cassel, 1985).

Inverse methods for field data have been developed that do not require the specific set of initial and surface boundary conditions required by the instantaneous profile (saturated profile with no surface flux). Dane and Hruska (1983) used a numerical solution to the Richards Equation [4.27], assuming that the soil profile $\theta(h)$ and $K(h)$ could be described by the van Genuchten (1980) equations. The solution was optimized to find $\alpha$ and $n$ in these equations. The method could be applied to other initial and boundary conditions, such as infiltration, but it assumed the entire profile could be described by a single set of $\theta(h)$ and $K(h)$ functions.

### 4.4.2.4 Borehole Tests

Slug tests are commonly employed to estimate field hydraulic properties below the water table because a minimum of effort is required for conducting the test. A known volume of water is either added or removed from a borehole, and the resulting water level change is recorded. The field saturated hydraulic conductivity is estimated using (Bower and Rice, 1976):

$$K_{fs} = \frac{r^2}{2bt} \ln\left(\frac{R}{r}\right) \ln\left(\frac{h_0}{h}\right) \qquad [4.67]$$

where $r$ is the radius of the borehole, $b$ is the thickness of the permeable unit, $R$ is the radial distance of influence of the test, $h_0$ is the water level at the time of water injection or withdrawal, and $h$ is the water level as a function of time ($t$) in the borehole. This formulation assumes horizontal flow. The radial influence of the test is an unknown, which can be approximated.

Alternatively, a steady pumping (or injection) rate can be established in one borehole, and water level changes can be observed in an observation borehole. Methods described in Section 4.2 can then be used to estimate aquifer hydraulic properties.

### 4.4.3 PARAMETER DATABASES

There are several databases now available that contain hydraulic parameters for a large number of soils. One of these is the Unsaturated Soil Hydraulic Database (UNSODA) (Leij et al., 1996). In 1996, the database contained over 780 soils with data on water retention and saturated and unsaturated hydraulic conductivity, as well as basic soil properties such as particle-size distribution, bulk density, organic matter content, etc. A large portion of the soils are sands and loamy sands. The program can be used to fit various $\theta(h)$ and $K(h)$ functions to the measured data.

## 4.5 NUMERICAL MODELS OF WATER FLOW

Numerical methods of solving differential equations became feasible with the development of high-speed computers, and were first applied to soil water movement in the early 1960s (Ashcroft, 1962; Hanks and Bowers, 1962). For many problems, it is no longer necessary to write computer code in order to apply numerical methods to soil water flow problems. Codes are available from authors or through commercial outlets and some common codes are listed in Table 4.4. Additionally, new mathematical programming languages (e.g., MATLAB, MATHCAD, MATEMATICA) provide alternative techniques for solving of nonlinear partial differential equations. It is important to note that use of numerical codes for modeling unsaturated media requires an appreciation of the difficulty in obtaining even simple solutions.

**TABLE 4.4**
**Examples of current numerical codes and some of their features.**

| Model | Numerical Method | Space Dimensions | Type of Transport | References |
|---|---|---|---|---|
| HYSWASOR | finite element | 1-D | water and solute | Dirksen et al. (1993) |
| LEACHM | finite difference | 1-D | water and solute | Hutson and Wagenet (1992) |
| Dual Porosity | finite element | 1-D | water and solute | Gerke and van Genuchten (1993) |
| HYDRUS-1D | finite element | 1-D | water and solute | Šimůnek et al. (1998) |
| SWIM | finite difference | 1-D | water | Ross (1990) |
| WORM | finite element | 1-D | water and solute | van Genuchten (1987) |
| HYDRUS-2D | finite element | 2-D | water and solute | Šimůnek et al. (1998) |
| SUTRA | finite element | 2-D | water and solute | Voss (1984) |
| SWMS_2D | finite element | 2-D | water and solute | Šimůnek et al. (1994) |
| VS2DT | finite difference | 2-D | water and solute | Lapalla et al. (1987)Healy (1990) |
| VAM2D | finite element | 2-D | water and solute | Huyakorn et al. (1988) |
| VSAFT | finite element | 2-D | water and solute | Yeh et al. (1990) |
| SWMS_3D | finite element | 3-D | water and solute | Šimůnek et al. (1995) |

### 4.5.1 FINITE DIFFERENCE METHOD

Numerical approaches produce a system of simultaneous equations that must be solved in a manner similar to that used to find chemical equilibria. The finite difference and finite element methods are the most commonly used. Despite the similarity in names, the development of the sets of equations is quite different, although the equations themselves may be similar. The finite element method is more suitable for irregularly shaped flow domains, but the accuracy and speed of the two approaches are nearly the same (McCord and Goodrich, 1994). The development of the finite difference equations is much more intuitive and will be discussed further.

The finite difference approach is used to develop a set of algebraic equations from the Richards equation so that it can be solved numerically (since computers cannot integrate or differentiate). The $h$ form (Equation [4.28]) has been used most often to solve problems of transient flow in nonuniform soils, although recent articles indicate that the mixed form (Equation [4.27]) is more accurate (Celia et al., 1990; Ross, 1990). The development of the set of equations for the $h$ form for one-dimensional vertical water movement with no source/sink term and known potentials at the boundaries will be shown.

The first step is to discretize the soil profile into $N$ even depth intervals $\Delta z$ such that $z = i\,\Delta z$ and $i = 0\ldots N$ (Smith, 1985). It is not necessary that the depth intervals be even, but for simplicity we will assume even intervals. The point at which intervals join is called a node so there are $N + 1$ nodes in the profile. Time is also discretized by intervals of $\Delta t$ such that $t = j\,\Delta t$. Then the derivatives in Equation [4.28] are written as discrete differences divided by the appropriate interval:

$$C\left(h_i^{j+\frac{1}{2}}\right)\frac{h_i^{j+1}-h_i^j}{\Delta t} = \frac{\left[K(h)\left(\frac{\partial h}{\partial z}+1\right)\right]\Big|_{i+\frac{1}{2}} - \left[K(h)\left(\frac{\partial h}{\partial z}+1\right)\right]\Big|_{i-\frac{1}{2}}}{\Delta z}$$

$$= \frac{K\left(\frac{h_i+h_{i+1}}{2}\right)\left(\frac{h_{i+1}-h_i}{\Delta z}+1\right) - K\left(\frac{h_{i-1}+h_i}{2}\right)\left(\frac{h_i-h_{i-1}}{\Delta z}+1\right)}{\Delta z} \qquad [4.68]$$

$$= \frac{K\left(\frac{h_i+h_{i+1}}{2}\right)}{(\Delta z)^2}\left(h_{i+1}-h_i+\Delta z\right) - \frac{K\left(\frac{h_{i-1}+h_i}{2}\right)}{(\Delta z)^2}\left(h_i-h_{i-1}+\Delta z\right)$$

Subscripts denote depth nodes and superscripts denote time steps. At this point, the time step for evaluating the terms on the right-hand side of Equation [4.68] is unspecified. Numerical approaches consist of starting with the initial conditions, which give the values of $h$ at all nodes at $t = 0$, then finding $h$ at each node at the next time step ($t = \Delta t$) and repeating the process until the full time period of interest has been covered. Therefore, the crux of the problem is finding the pressure head at the next time step ($h^{j+1}$) given the value at the current time step ($h^j$) at each node.

If all of the terms on the right-hand side of Equation [4.68] are evaluated at the known ($j$) time step and $C(h_i^{j+\frac{1}{2}})$ is approximated by $C(h_i^j)$, then there is only one unknown in the equation ($h_i^{j+1}$) from the term on the left-hand side. The equation can be solved explicitly for $h_i^{j+1}$ at each node ($i = 1 \ldots N\text{-}1$, with $h_0^{j+1}$ and $h_N^{j+1}$ known from the boundary conditions). This is known as the *explicit* finite difference method and has the disadvantage of being unstable for all but very small time steps (small time steps require more computer time). Unstable means that values of $h$ will fluctuate sharply between time intervals and diverge from the true solution.

If all of the terms on the right-hand side of Equation [4.68] are evaluated at the unknown ($j$+1) time step, then there are three unknowns: $h_{i-1}^{j+1}$, $h_i^{j+1}$, and $h_{i+1}^{j+1}$. Collecting coefficients of these terms on the left-hand side gives:

$$a\ h_{i-1}^{j+1} + b\ h_i^{j+1} + c\ h_{1+1}^{j+1} = d \qquad [4.69]$$

and

$$a = -r\ \frac{K_{i-\frac{1}{2}}}{C_i}$$

$$c = -r\ \frac{K_{i+\frac{1}{2}}}{C_i} \qquad [4.70]$$

$$b = 1 + a + c$$

$$d = h_i^j + \Delta z\ (a - c)$$

where $r = \Delta t/(\Delta z)^2$, $K_{i\pm\frac{1}{2}} = K(h_{i\pm\frac{1}{2}}^{j+1})$, $C_i = C(h_i^{j+\frac{1}{2}})$, $h_{i\pm\frac{1}{2}}^{j+1}$ is the arithmetic average of $h_i^{j+1}$ and $h_{i\pm1}^{j+1}$, and $h_i^{j+\frac{1}{2}}$ is the arithmetic average of $h_i^j$ and $h_i^{j+1}$. Equation [4.69] can be written for each node so there are $N$-1 equations with $N$-1 unknowns ($h_i^{j+1}$) to be solved simultaneously. This can be written in matrix notation as:

$$\underline{\underline{A}}\ \underline{u} = \underline{d} \qquad [4.71]$$

where $\underline{\underline{A}}$ is a square tridiagonal matrix with the coefficients $a$, $b$ and $c$ along the sub-diagonal, diagonal and super-diagonal, $\underline{u}$ is a vector of the $N$-1 unknown values of $h$, and $\underline{d}$ is a vector containing the known values of $h$. Computer algorithms are used to invert $\underline{\underline{A}}$ and solve Equation [4.71] for $\underline{u}$ at a given time step. The only remaining difficulty is that the coefficients contain $K_{i\pm\frac{1}{2}}$ and $C_{i+\frac{1}{2}}$, which require pressure heads at the unknown time level in order to be evaluated. An iterative procedure is usually used at each time step wherein $h_i^{j+1}$ is estimated, $K_{i\pm\frac{1}{2}}$ and $C_{i+\frac{1}{2}}$ are evaluated, and the set of equations are solved for an improved estimate of $h_i^{j+1}$. This is repeated until the difference between the two estimates is acceptably small, and then the algorithm goes to the next time step (Paniconi et al., 1991).

Specifying all of the terms on the right-hand side of Equation [4.71] at the $j$+1 time step is known as the *fully implicit* finite difference method. Another approach is used where the terms on the right-hand side of Equation [4.71] are evaluated at the $j$+½ time step (by taking an average or the right-hand side at the $j$ and $j+1$ time step) and this is known as the Crank-Nicolson finite difference method (Smith, 1985). The fully implicit and Crank-Nicolson methods are stable for much larger time steps than the explicit method, but water flow problems that involve large gradients in $h$ such as infiltration still require very small time steps (fast computers).

### 4.5.2 Initial and Boundary Conditions

Numerical solution of the Richards equation requires that the initial distribution of $h$ or $\theta$ in the profile and the conditions at the boundaries be specified, just as they must be with an analytical solution.

#### 4.5.2.1 Initial Conditions

Transient models require the specification of conditions at the beginning of the simulation period because subsequent computations rely on previous states of the system. To avoid numerical difficulties (i.e., unstable solutions), uniform initial conditions are usually preferred. In heterogenous media, however, identifying feasible initial conditions may be difficult, in that the steady flow of water may not result in uniform initial conditions. Instead, a steady flux assumption initially leads to a distribution of potentials that result in $-K(h) = J_z$. Thus, specification of an initial flux yields the steady distribution of potentials as a function of heterogeneous unsaturated hydraulic conductivity. As a practical matter, it is best to allow sufficient time to "warm up" a model so that assumptions regarding initial conditions can be attenuated. Correct specification of initial conditions minimizes initial transients and the likelihood of model instability.

#### 4.5.2.2 Boundary Conditions

Constant head or pressure (first kind or type 1) boundary conditions are used to prescribe known values of the boundary potential. Constant heads are prescribed for circumstances when water levels can be measured or inferred, such as using stage data from a pond, stream or well. Constant pressures are commonly prescribed along seepage faces where water is maintained at atmospheric pressure. For grid systems that do not deform with time, it is relatively easy to convert from pressure to head, and vice versa.

Constant head or gradient (second kind or type 2) boundary conditions are used to prescribe boundary values when the head or pressure varies with time. In these cases, the gradient may remain fixed even when the potential is changing rapidly. For example, a prescribed flux can be specified on the upper surface due to constant infiltration, or a constant gradient can be imposed on the lower surface to represent unit-gradient drainage. Also, a zero flux or gradient can be imposed on the vertical boundaries to represent noflow conditions. Flow to a well can be modeled by specifying a constant flux at the borehole wall. Note that constant flux and constant gradient boundary conditions are not equivalent in the unsaturated zone because the hydraulic conductivity varies with potential.

Mixed potential flux (third kind or type 3) boundary conditions commonly arise for flow across a semipermeable boundary. In this case, the flux and the boundary potential are coupled, such that $J = (H_1 - H_2)/C$, where $J$ and $H_1$ are the unknown boundary flux and head, respectively, $C$ is the hydraulic impedance of the semipermeable membrane, and $H_2$ is a specified head exterior to the domain.

#### 4.5.2.3 Infiltration

During a rainfall or irrigation event, the surface boundary condition must change. Initially, a constant flux into the soil equal to the rainfall rate (which can be variable) is specified (supply controlled infiltration). The pressure head at the surface node is monitored and when sufficient water enters the soil to make this potential zero or positive, the boundary condition is switched to a constant potential of $h$ equal to the depth of ponding, which is often taken as zero or some very small depth to allow for surface storage. This represents the time to ponding ($t_p$) and after this time the infiltration rate will be less than the rainfall rate (profile controlled infiltration) until the rain ends or the rainfall rate decreases. The calculated flux into the soil using the constant potential boundary condition is checked at each time step against the rainfall rate to ensure that the rainfall rate has not decreased below the calculated flux. If this does occur, the boundary condition is switched back to a constant flux equal to the rainfall rate.

### 4.5.2.4 Evapotranspiration

Evaporation also requires a change in the surface boundary condition, much like infiltration except that the flux is reversed. During the first stage of evaporation (immediately after a rainfall or irrigation event), a constant flux from the soil equal to the potential evaporation rate is specified. The pressure head at the soil surface node is again monitored and when sufficient water has left the soil that it drops to some critical level, usually a potential that would correspond to air-dry soil, the surface boundary condition is switched to a constant pressure head equal to the critical level. At this point, evaporation enters the second stage and actual soil evaporation is less than potential. Evaporation rate can vary (as it would diurnally), so the calculated flux is compared to the potential evaporation rate at each time step. When the potential evaporation rate drops below the calculated flux, the boundary condition is switched back to a constant flux equal to the potential rate. Transpiration is usually treated as a distributed sink within the root zone, rather than as a boundary condition at the soil surface.

## 4.6 CONCLUDING REMARKS

Soil water flow is one of the most important processes occurring in soils and is the key to predicting solute transport (Chapter 6). Although important advances have been made in understanding and predicting water flow, many challenges remain. Improvements in the methods for incorporating macropore and structure effects in water movement predictions are required (soil structure is addressed in Chapter 7). Easier, less intrusive methods for measuring saturated and unsaturated hydraulic conductivity are needed. Information on how to scale up these measurements to the field and landscape scale is needed (spatial variability is addressed in Chapter 10). Better integration of soil water and groundwater modeling is also needed.

## REFERENCES

Ashcroft, G., D.D. Marsh, D.D. Evans, and L. Boersma. 1962. Numerical methods for solving the diffusion equation: I. Horizontal flow in semi-infinite media, *Soil Sci. Soc. Proc.*, 26:522–525.

Baker, R.S. and D. Hillel. 1990. Laboratory tests of a theory of fingering during infiltration into layered soils, *Soil Sci. Soc. Am. J.*, 54:20–30.

Baver, L.D., W.H. Gardner, and W.R. Gardner. 1972. Soil physics, John Wiley and Sons, Inc. New York, NY.

Beven, K.J. and P.F. Germann. 1981. Water flow in soil macropores, II. A combined flow model, *Soil Sci.*, 32:15–29.

Beven, K.J. and P.F. Germann. 1982. Macropores and water flow in soils, *Water Resour. Res.*, 5:1311–1325.

Bosch, D.D. and L.T. West. 1997. Hydraulic conductivity variability for two sandy soils, *Soil Sci. Soc. Am. J.*, (in press).

Bouma, J. and J.H. Denning. 1972. Field measurements of unsaturated hydraulic conductivity by infiltration through gypsum crusts, *Soil Sci. Soc. Proc.*, 36:846–847.

Bouma, J., C. Gelmans, L.W. Dekker, and W.J.M. Jeurissen. 1983. Assessing the suitability of soils with macropores for subsurface liquid waste disposal, *J. Environ. Qual.*, 12:305–311.

Bouwer, H. 1969. Infiltration of water into nonuniform soil, *J. Irrig. Drainage Div.*, Proc. ASCE IR4:451–462.

Bouwer, H. 1986. Intake rate: cylinder infiltrometer, *in* A. Klute (ed.) Methods of Soil Analysis: Part 1. Physical and Mineralogical Methods, 2nd ed., *Agronomy*, 9:825–844.

Bouwer, H. 1966. Rapid field measurement of air entry value and hydraulic conductivity as significant parameters in flow system analysis, *Water Resour. Res.*, 1:729–738.

Brakensiek, D.L. and Rawls, W.J. 1992. Comment on "Fractal processes in soil water retention" by Scott W. Tyler and Stephen W. Wheatcraft, *Water Resour. Res.*, 28:601–602.

Brooks, R.H. and A.T. Corey. 1964. Hydraulic properties of porous media. Fort Collins, Colorado State University Hydrology Paper No. 3.

Bruce, R.R., V.L. Quisenberry, H.D. Scott, and W.M. Snyder. 1985. Irrigation practice for crop culture in the southeastern United States, *Advances in Irrigation*, 3:51–106.

Buckingham, E. 1907. Studies on the movement of soil moisture, Bulletin 38. U.S. Department of Agriculture Bureau of Soils, Washington, DC.

Campbell, G.S. 1974. A simple method for determining unsaturated conductivity from moisture retention data, *Soil Sci.*, 117:311–314.

Carter, M.R. 1993. Soil sampling and methods of analysis, Lewis Publishers, Boca Raton, FL.

Cassel, D.K. (ed.). 1985. Physical characteristics of soils of the southern region — Summary of *in situ* unsaturated hydraulic conductivity. Southern Cooperative Series Bulletin 303. North Carolina State University, Raleigh, NC.

Celia, M.A., E.T. Bouloutas, and R.L. Zarba. 1990. A general mass-conservative numerical colution for the unsaturated flow equation, *Water Resour. Res.*, 26:1483–1496.

Chiang, S.C., D.E. Radcliffe, W.P. Miller, and K.D. Newman. 1987. Hydraulic conductivities of three southeastern soils as affected by sodium, electrolyte concentration, and pH, *Soil Sci. Soc. Am. J.*, 51:1293–1299.

Chu, S.T. 1978. Infiltration during an unsteady rain, *Water Resour. Res.*, 14:461–466.

Clothier, B.E. and K.R.J. Smettem. 1990. Combining laboratory and field measurements to define the hydraulic properties of soil, *Soil Sci. Soc. Am. J.*, 54:299–304.

Clothier, B.E. and I. White. 1981. Measurement of sorptivity and soil water diffusivity in the field, *Soil Sci. Soc. Am. J.*, 45:241–245.

Dane, J.H. and S. Hruska. 1983. *In situ* determination of soil hydraulic properties during drainage, *Soil Sci. Soc. Am. J.*, 47:619–624.

Darcy, H. 1856. Les fontaines publiques de la ville de Dijon, Dalmont, Paris.

Diment, G.A., K.K. Watson, and P.J. Blennerhasset. 1982. Stability anlysis of water movement in unsaturated porous materials: 1. Theoretical considerations, *Water Resour. Res.*, 18:1248–1254.

Durner, W. 1992. Predicting the unsaturated hydraulic conductivity using multi-porosity water retention curves, p. 185–202, *in* M. Th. van Genuchten, F.J. Leij, and L.J. Lund, Riverside, CA, University of California, October 11–13, 1989.

Edwards, W.M., M.J. Shipitalo, L.B. Owens, and L.D. Norton. 1989. Water and nitrate movement in earthworm burrows within long-term no-till cornfields, *J. Soil Water Cons.*, 44:240–243.

Elrick, D.E. and W.D. Reynolds. 1992. Infiltration from constant-head well permeameters and infiltrometers, *in* Advances in measurement of soil physical properties: Bringing theory into practice, G.C. Topp, W.D. Reynolds and R.E. Green (eds.), *Soil Sci. Soc. Am.*, Madison, WI, 1–24.

Faybishenko, B.A. 1995. Hydraulic behavior of quasi-saturated soils in the presence of entrapped air: laboratory experiments, *Water Resour. Res.*, 31:2421–2435.

Fetter, C.W. 1994. Applied Hydrogeology, 3rd ed., Macmillan College Publ. Co., New York, NY.

Gardner, W.R. 1958. Some steady-state solutions of the unsaturated moisture flow equation with application to evaporation from a water table, *Soil Sci.*, 85:228–232.

Gardner, W.R. and M. Fireman. 1958. Laboratory studies of evaporation from soil columns in the presence of a water table, *Soil Sci.*, 85:244–249.

Gish, T.J. and A. Shirmohammadi. 1991. Preferential flow, *Am. Soc. Agric. Eng.*, St. Joseph, MI.

Glass, R.J., J.-Y. Parlange, and T.S. Steenhuis. 1991. Immiscible displacement in porous media: Stability analysis of three-dimensional, axisymmetric disturbances with application to gravity-driven wetting front instability, *Water Resour. Res.*, 27:1947–1956.

Glass, R.J., T.S. Steenhuis, and J.-Y. Parlange. 1989. Wetting front instability: 1. Theoretical discusssion and dimensional analysis, *Water Resour. Res.*, 25:1187–1194.

Green, W.H. and G.A. Ampt. 1911. Studies in soil physics: I. The flow of air and water through soils, *J. Agr. Sci.*, 4:1–24.

Green, R.E., L.R. Ahuja, and S.K. Chong, 1985. Hydraulic conductivity, diffusivity, and sorptivity of unsaturated soils: Field methods, *in* A. Klute (ed.), Methods of soil analysis. Part 1. *Agronomy*, 9:771–798.

Gupte, S.M., D.E. Radcliffe, D.H. Franklin, L.T. West, E.W. Tollner, and P.F. Hendrix. 1996. Anion transport in a Piedmont Ultisol: 2. Local-scale parameters, *Soil Sci. Soc. Am. J.*, 60:762–770.

Hanks, R.J. and S.A. Bowers. 1962. Numerical solution of the moisture flow equation for infiltration into layered soils, *Soil Sci. Soc. Proc.*, 26:530–534.

Haverkamp, R., M. Vauclin, J. Tovina, P.J. Wierenga, and G. Vachaud. 1977. A comparison of numerical simulation models for one-dimensional infiltration, *Soil Sci. Soc. Am. Proc.*, 41:285–294.

Healy, R.W. 1990. Simulation of solute transport in variably saturated porous media with supplemental information on modifications to the U.S. Geological Survey's computer program VS2DT, Water-Resources Investigations Report 90-4025, U.S. Geological Survey, Denver, CO.

Hill, R.E. and J.-Y. Parlange. 1972. Wetting front instability in layered soils, *Soil Sci. Soc. Am. Proc.*, 36:697–702.

Hillel, D. 1980a. Fundamentals of soil physics, Academic Press, New York, NY.

Hillel, D. 1980b. Applications of soil physics, Academic Press, New York, NY.

Hillel, D. and R.S. Baker. 1988. A descriptive theory of fingering during infiltration into layered soils, *Soil Sci.*, 146:51–56.

Hooghoudt, S.B. 1937. Bijdregen tot de kennis van eenige natuurkundige grootheden van de grond, *Versl. Landb. Ond.*, 43:461–676.

Hubbert, M.K. 1956. Darcy's law and the field equations of the flow of underground fluids, *Am. Inst. Min. Met. Petl. Eng. Trans.*, 207:222–239.

Huyakorn, P.S., J.B. Kool, and J.B. Robertson. 1989. Documentation and user's guide: VAM2SD — Variably saturated analysis model in two dimensions, NUREG/CR-5352, HGL/89-01, Hydrogeologic, Inc. Herndon, VA.

Jarvis, N.J. and I. Messing. 1995. Near-saturated hydraulic conductivity in soils of contrasting texture measured by tension infiltrometers, *Soil Sci. Soc. Am. J.*, 59:27–34.

Jury, W.A., W.R. Gardner, and W.H. Gardner. 1991. Soil physics, John Wiley and Sons, New York, NY.

Klute, A. 1986. Methods of Soil Analysis: Part 1. Physical and Mineralogical Methods, 2nd ed., Number 9(1), ASA-SSSA, Madison, WI.

Klute, A. and C. Dirksen. 1986. Hydraulic conductivity and diffusivity: Laboratory methods, *in* A. Klute (ed.) Methods of soil analysis. Part 1. *Agronomy*, 9:687–734.

Kool, J.B., J.C. Parker, and M.T. van Genuchten. 1985a. Determining soil hydraulic properties from one-step outflow experiments by parameter estimation: I. Theory and numerical studies, *Soil Sci. Soc. Am. J.*, 49:1348–1354.

Kool, J.B., J.C. Parker, and M.T. van Genuchten. 1985b. ONESTEP, a non-linear parameter estimation program for evaluating soil hydraulic properties from one-step outflow experiments, Virginia Agric. Exp. Stn. Bull. 85-3, Blacksburg, VA.

Kruseman, G.P. and N.A. deRidder. 1990. Analysis and Evaluation of Pumping Test Data, 2nd ed., Publ. 47, International Institute for Land Reclamation and Improvement, P.O. Box 45, 6700 AA Wageningen, The Netherlands.

Kung, K.-J.S. 1990a. Preferential flow in a sandy vadose zone: 1. Field observation, *Geoderma*, 46:51–58.

Kung, K.-J.S. 1990b. Preferential flow in a sandy vadose zone: 2. Mechanism and implications, *Geoderma*, 46:59–71.

Lapalla, E.G., R.W. Healy, and E.P. Weeks. 1987. Documentation of computer program VS2DT to solve the equations of fluid flow in variable saturated porous media, Water-Resources Investigations Report 83-4099, U.S. Geological Survey, Denver, CO.

Leij, F.J., W.J. Alves, and M.Th. van Genuchten. 1996. The UNSODA unsaturated soil hydraulic database, User's manual version 1.0. EPA/600/R-96/095, U.S. EPA. Cincinnati, OH.

Liu, H.H. and J.H. Dane. 1996. Two approaches to modeling unstable flow and mixing of variable density fluids in porous media, *Transport in Porous Media*, 23:219–236.

Luxmoore, R.J. 1981. Micro-, meso-, and macroporosity of soil, *Soil Sci. Soc. Am. J.*, 45:671.

Mualem, Y. 1976. A new model for predicting the hydraulic conductivity of unsaturated porous media, *Water Resour. Res.*, 12:593–622.

McCord, J.T. and M.T. Goodrich. 1994. Benchmark testing and independent verification of the VS2DT computer code, Sandia National Laboratories, Albuquerque, NM.

McCuen, R.H. 1982. A guide to hydrologic analysis using SCS methods, Prentice-Hall, Inc., Englewood Cliffs, NJ.

Mein, R.G. and C.L. Larson. 1973. Modeling infiltration during a steady rain, *Water Resour. Res.,* 9:384–394.

Miller, W.P. and M.K. Baharuddin. 1986. Relationship of soil dispersibility to infiltration and erosion of southeastern soils, *Soil Sci.,* 142:235–240.

Miller, W.P. and D.E. Radcliffe. 1992. Soil Crusting in the Southeastern U.S., *in* M.E. Sumner and B.A. Stewart (eds.) p. 233–266. Soil Crusting: Chemical and Physical Processes, Lewis Publ., Boca Raton, FL.

Miller, R.H. and J.F. Wilkinson. 1979. Nature of the organic coating on sand grains of nonwettable golf greens, *Soil Sci. Soc. Am. Proc.,* 41:1203–1204.

Narasimhan, T.N. 1998. Hydraulic characterization of aquifers, reservoir rocks, and soils: A history of ideas, *Water Resour. Res.,* 34:33–46.

Paniconi, C., A.A. Aldama, and E.F. Wood. 1991. Numerical evaluation of iterative and noniterative methods for the solution of the nonlinear Richards equation, *Water Resour. Res.,* 27:1147–1163.

Parlange, J.-Y. and D.E. Hill. 1976. Theoretical analysis of wetting front instability in soils, *Soil Sci.,* 122:236–239.

Perroux, K.M. and I. White. 1988. Designs for disc permeameters, *Soil Sci. Soc. Am. J.,* 52:1205–1215.

Philip, J.R. 1957a. The theory of infiltration. 1. The infiltration equation and its solution, *Soil Sci.,* 83:345–357.

Philip, J.R. 1957b. The theory of infiltration. 2. The profile at infinity, *Soil Sci.,* 83:435–448.

Philip, J.R. 1957c. The theory of infiltration. 3. Moisture profiles and relation to experiment, *Soil Sci.,* 84:163–178.

Philip, J.R. 1957d. The theory of infiltration. 4. Sorptivity and algebraic infiltration equations, *Soil Sci.,* 84:257–264.

Philip, J.R. 1957e. The theory of infiltration. 5. Influence of initial moisture content, *Soil Sci.,* 84:329–339.

Philip, J.R. 1975. Stability analysis of infiltration, *Soil Sci. Soc. Am. Proc.,* 39:1042–1049.

Quisenberry, V.L. and R.E. Phillips. 1976. Percolation of surface-applied water in the field, *Soil Sci. Soc. Am. J.,* 40:484–489.

Raats, P.A.C. 1973. Unstable wetting fronts in uniform and non-uniform soils, *Soil Sci. Soc. Am. Proc.,* 37:681–685.

Rawls, W.J., D.L. Brakensiek, and K.E. Saxton. 1982. Estimation of soil water properties, Trans. ASAE, 1316–1320,1328.

Reynolds, W.D. and D.E. Elrick. 1991. Determination of hydraulic conductivity using a tension infiltrometer, *Soil Sci. Soc. Am. J.,* 55:633–639.

Reynolds, W.D., S.R. Vieira, and G.C. Topp. 1992. An assessment of the single-head analysis for the constant head well permeameter, *Can J. Soil Sci.,* 72:489–501.

Richards, L.A. 1931. Capillary conduction of liquids in porous mediums, *Physics,* 1:318–333.

Ross, P.J. 1990. Efficient numerical methods for infiltration using Richard's equation, *Water Resour. Res.,* 26:279–290.

Seyfried, M.S. and P.S.C. Rao. 1987. Solute transport in undisturbed columns of an aggregated tropical soil: Preferential flow effects, *Soil Sci. Soc. Am. J.,* 51:1434–1444.

Shaw, J.N., L.T. West, C.C. Truman, and D.E. Radcliffe. 1997. Morphologic and hydraulic properties of soils with restrictive horizons in the Georgia Coastal Plain, *Soil Sci.,* (in press).

Shipitalo, M.J. and W.M. Edwards. 1996. Effects of initial water content on macropore/matrix flow and transport of surface-applied chemicals, *J. Environ. Qual.,* 25:662–670.

Šimůnek, J., K. Huang, and M.Th. van Genuchten. 1998. The HYDRUS code for simulating the one-dimensional movement of water, heat, and multiple solutes in variably-saturated media, Version 6.0, Research Report No. 144, U.S. Salinity Laboratory, USDA, ARS, Riverside, CA, 164 pp.

Šimůnek, J., K. Huang, and M.Th. van Genuchten. 1995. The SWMS_3D code for simulating water flow and solute transport in three-dimensional variably saturated media, Version 1.0. Research Report No. 139, U.S. Salinity Laboratory, USDA, ARS, Riverside, CA.

Šimůnek, J., T. Vogel, and M.Th. van Genuchten. 1994. The SWMS_2D code for simulating water flow and solute transport in two-dimensional variably saturated media, Version 1.21. Research Report No. 132, U.S. Salinity Laboratory, USDA, ARS, Riverside, CA.

Smith, G.D. 1985. Numerical solution of partial differential equations, Clarendon Press, Oxford.

Smith, K.A. and C.M. Mullins. 1991. Soil analysis: Physical methods, Marcel Dekker, New York, NY.

Thiem, G. 1906. Hydrologische Methoden, J.M. Gephardt, Leipzig.

Topp, G.C. 1992. Advances in measurement of soil physical properties: Bringing theory into practice, *Soil Sci. Soc. Am. Spec. Pub.,* 30. Madison, WI.

Tyler, S.W. and S.W. Wheatcraft. 1992. Reply, *Water Resour. Res.,* 28:603–604.

van Genuchten, M.Th. 1980. A closed-form equation for predicting the hydraulic conductivity of unsaturated soils, *Soil Sci. Soc. Am. J.,* 44:892–898.

van Genuchten, M.Th. 1987. A numerical model for solute movement in and below the root zone, Res. Report, U.S. Salinity Laboratory, Riverside, CA.

van Genuchten, M.Th. and F.J. Leij. 1992. On estimating the hydraulic properties of unsaturated soils, p. 1–14, *in* M.Th. van Genuchten, F.J. Leij, and L.J. Lund (eds.), Indirect methods for estimating the hydraulic properties of unsaturated soils, U.S. Salinity Lab, Riverside, CA.

van Genuchten, M.T. and P.J. Wierenga. 1976. Mass transfer studies in sorbing porous media: I. Analytical solutions, *Soil Sci. Soc. Am. J.,* 40:473–479.

Voss, C.I. 1984. A finite-element simulation model for saturated-unsaturated fluid-density-dependent ground-water flow with energy transport or chemically-reactive single species solute transport, U.S. Geological Survey, Denver, CO.

Warrick, A.W. 1992. Models for disc infiltrometers, *Water Resour. Res.,* 28:1319–1327.

Warrick, A.W. and P. Broadbridge. 1992. Sorptivity and macroscopic capillary length relationships, *Water Resour. Res.,* 28:427–431.

White, I. and K.M. Perroux. 1989. Estimation of unsaturated hydraulic conductivity from field sorptivity measurements, *Soil Sci. Soc. Am. J.,* 53:324–329.

White, I. and M.J. Sully. 1987. Macroscopic and microscopic capillary length and time scales from field infiltration, *Water Resour. Res.,* 23:1514–1522.

White, R.E. 1985. The influence of macropores on the transport of dissolved and suspended matter through soil, *in: Advances in Soil Science,* Anonymous, Springer-Verlag, New York, NY.

Wooding, R.A. 1968. Steady infiltration from a shallow circular pond, *Water Resour. Res.,* 4:1259–1273.

Yeh, T.-C.J., R. Srivastava, A. Guzman, and T. Harter. 1993. A numerical model for water flow and chemical transport in variably saturated porous media, *Groundwater,* 31:634–644.

Zangar, C.N. 1953. Flow from a test hole located above groundwater level, p. 69–71, *in* Theory and problems of water percolation, Engineering Monograph No. 8, Bureau of Reclamation, U.S. Dept. of Interior.

# 5 Water and Energy Balances at Soil–Plant–Atmosphere Interfaces*

*Steven R. Evett*

## 5.1 INTRODUCTION

Energy fluxes at soil-atmosphere and plant-atmosphere interfaces can be summed to zero because the surfaces have no capacity for energy storage. The resulting energy balance equations may be written in terms of physical descriptions of these fluxes; and have been the basis for problem casting and solving in diverse fields of environmental and agricultural science such as estimation of evapotranspiration (ET) from plant canopies, estimation of evaporation from bare soil, rate of soil heating in spring (important for timing of seed germination), rate of residue decomposition (dependent on temperature and water content at the soil surface) and many others. The water balances at these surfaces are implicit in the energy balance equations. The soil water balance equation is different from, but linked to, the surface energy balances; a fact that has often been ignored in practical problem solving. In this chapter the energy balance will be discussed first, followed by the water balance in Section 5.3.

Computer simulation has become an important tool for theoretical investigation of energy and water balances at the earth's surface, and for prediction of important results of the mechanisms involved. This chapter will focus more on the underlying principles of energy and water balance processes, and will mention computer models only briefly. More information on computer models that include surface energy and water balance components can be found in Anlauf et al. (1990), ASAE (1988), Campbell (1985), Hanks and Ritchie (1991), Peart and Curry (1998), Pereira et al. (1995) and Richter (1987) to mention only a few.

## 5.2 ENERGY BALANCE EQUATION

The surface energy balance is:

$$0 = Rn + G + LE + H \qquad [5.1]$$

where Rn is net radiation; G is soil heat flux; LE is the latent heat flux (evaporation to the atmosphere) and is the product of the evaporative flux, E, and the latent heat of vaporization, L; and H is sensible heat flux (all terms taken as positive when flux is toward the surface, and in W $m^{-2}$). Each term may be expressed more completely as the sum of subterms that describe specific physical processes, some of which are shown in Figure 5.1. Thus, net radiation includes the absorption and reflection of shortwave radiation (sunlight, Rsi and the reflected portion $\alpha$Rsi), as

---

*\* This chapter was prepared by a USDA employee as part of his official duties and cannot legally be copyrighted. The fact that the private publication in which the chapter appears is itself copyrighted does not affect the material of the U.S. Government, which can be reproduced by the public at will.*

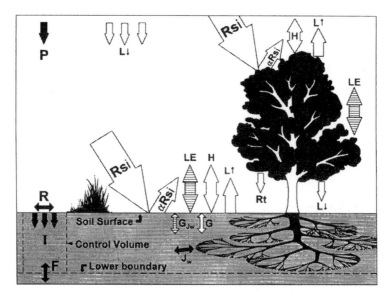

**FIGURE 5.1** Water and energy balance components. Water balance components are in black, energy balance in white. The shared term LE is shaded. Water balance is discussed in Section 5.3.

well as the emission and reception of longwave radiation (L↑ and L↓, respectively, Figure 5.1). Soil heat flux involves not only diffusion of heat, G, as expressed by Fourier's law (see Chapter 9), but also convective heat flux, $G_{Jw}$, as water at temperature T flows at rate $J_w$ into soil at another temperature T′. Both evaporation from the soil and from plants are examples of latent heat flux, but so also is dew formation, whether it wets the soil surface or plant canopy. Finally, sensible heat flux may occur between soil and atmosphere or between plant and atmosphere, and may be short-circuited between soil and plant, for example, when sensible heat flux from the soil warms the plant. In the next few paragraphs, these fluxes and values they may assume will be illustrated with examples from some contrasting surfaces under variable weather conditions.

Values of these energy fluxes change diurnally (Figures 5.2 to 5.4) and seasonally (Figures 5.5 and 5.6). Regional advection is the large scale transport of energy in the atmosphere from place to place on the earth's surface. Regional advection events can change the energy balance greatly as illustrated with measurements taken over irrigated wheat at Bushland, TX (35°11′N Lat; 102°06′W Long) for the 48 h period beginning on day 119, 1992 (Figure 5.2). Total Rsi was 26.1 and 26.7 MJ m$^{-2}$ on days 119 and 120, respectively; close to the expected maximum clear sky value of 28.6 MJ m$^{-2}$ for this latitude and time of year. However, on day 119 strong, dry, adiabatic southwesterly winds (mean 5 m s$^{-1}$, mean dew point 4.1°C, mean $T_{2m}$ 20.1°C) caused H to be strongly positive, providing the extra energy needed to drive total LE to –32.8 MJ m$^{-2}$, even though both Rsi and Rn levels were reduced in the afternoon due to cloudiness. Total LE was much larger in absolute magnitude than Rsi and Rn totals. The next day the total LE was 39% smaller due to the absence of regional advection, even though total Rsi and Rn values were slightly higher. G values were near zero during this period of full canopy cover when leaf area index (LAI) was 7 (leaf area index is defined as the single-sided surface area of leaves per unit land area). Note that net radiation was negative at night. This is indicative of strong radiational cooling of the surface, which radiates heat into the clear, low humidity nighttime skies common to this semi-arid location at 1170 m above mean sea level.

Over alfalfa in late summer, Rsi totals were lower (20.1 and 5.4 MJ m$^{-2}$, respectively, for days 254 and 255, 1997, Figure 5.3). On the very clear day 254, peak Rsi was 798 W m$^{-2}$; and with regional advection occurring, LE flux was high. The 3 h period of negative H just after sunrise was due to the sun-warmed crop canopy being at higher temperature than the air. The arrival of a cool front bringing cloudy skies near midnight causes all fluxes to be much lower on day 255, with Rsi

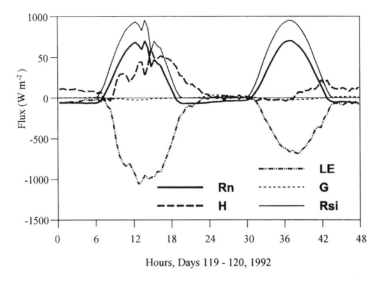

**FIGURE 5.2** Energy balance over irrigated winter wheat at Bushland, TX.

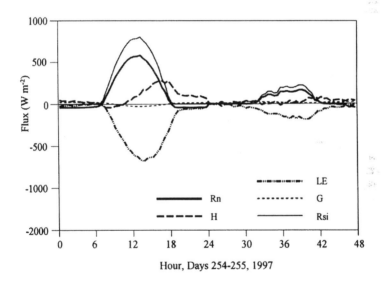

**FIGURE 5.3** Energy balance over irrigated alfalfa at Bushland, TX.

reaching only 220 W m$^{-2}$, and H hovering near zero for much of the day. The arrival of the cloud cover and moist air is signaled near midnight by the abrupt change from negative values of Rn and LE to near zero values. In the case of net radiation, this is due to the increased longwave radiation from the clouds, which were warmer and had higher emissivity than the clear sky that preceded them. Latent heat flux nears zero because the strong vapor pressure gradient from moist crop and soil to dry air is reduced by the arrival of moist air. Note that after sunset, but before midnight, latent heat flux was strong, due to continuing strong sensible heat flux, even though net radiation was negative. Again, due to full crop cover (LAI = 3), G values are low, indicating that very little energy is penetrating the soil surface.

For bare soil, G is often larger, becoming an important part of the energy balance (Figure 5.4). After rain and irrigation totaling 35 mm over the previous two days, the soil was wet on day 193. Latent heat flux totaled –14.4 MJ m$^{-2}$ or 6 mm of evaporation; 77% of Rn. Sensible heat flux was negative for the first few hours after sunrise because the soil was warmer than the air, which had

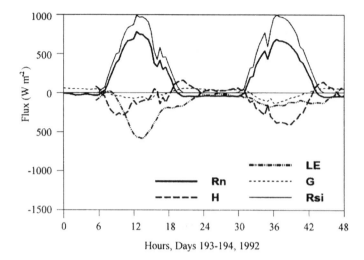

**FIGURE 5.4**  Energy balance for bare Pullman clay loam soil after 35 mm of rain and irrigation at Bushland, TX.

been cooled by a nighttime thunderstorm. Later in the day H and G both approached zero, and near sunset they became positive, supplying the energy consumed in evaporation that continued well into the night hours. Strong radiational cooling occurred on the nights of days 193 and 194 as indicated by negative values of Rn. Evaporation was probably energy-limited on day 193, becoming soil-limited on day 194. Latent heat flux on the second day was reduced to –7.4 MJ m⁻², and peak daytime values were not much larger than those for G. The drying soil became warmer and contributed heat to the atmosphere during almost all daylight hours.

Seasonal variations in daily total energy flux values occur due to changes of sun angle, of distance from the earth to sun (about 3% yearly variation), of seasonal weather and of surface albedo as plant and residue cover changes (Figures 5.5 and 5.6). A curve describing clear sky solar radiation at Bushland, Texas, could be fit to high points of Rsi in Figures 5.5 and 5.6. Net radiation was similar for alfalfa and bare soil except for a rainy period beginning about day 190 when the soil was wet and dark and Rn for the fallow field was markedly larger. The big differences were in LE and H. Latent heat flux from the alfalfa was large, reaching nearly –40 MJ m⁻² (16 mm) on day 136 during a regional advection event that allowed LE to be larger than Rsi. Sensible heat flux was positive during much of the year. Soil heat flux was small during the growing season, becoming larger as the soil cooled during the fall and winter. For the bare soil, LE values were small during the first 150 days, the latter end of a drought. Sensible heat flux was negative during this period, and remained negative after rains began until day 203. Evaporative fluxes were fairly small, rarely reaching 6 mm d⁻¹ even after rains began. In contrast to alfalfa, soil heat flux for bare soil was larger and more variable throughout the year.

Methods of measurement and estimation of the energy fluxes are needed to characterize the energy balance. Examples of the instrumentation* needed to measure components and subcomponents of the energy balance are given in Table 5.1. These will be discussed in the following sections.

### 5.2.1  NET RADIATION

Net radiation is the sum of incoming and outgoing radiation:

$$Rn = Rsi(1-\alpha) - \varepsilon\sigma T^4 + L\downarrow \tag{5.2}$$

---

* The mention of trade or manufacturer names is made for information only and does not imply an endorsement, recommendation or exclusion by USDA-Agricultural Research Service.

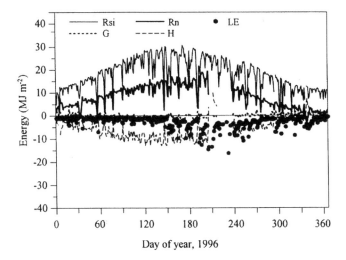

**FIGURE 5.5** Daily totals of energy balance terms for a fallow field (mostly bare Pullman clay loam) at Bushland, TX.

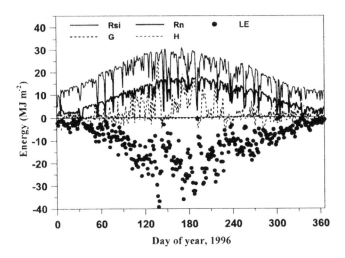

**FIGURE 5.6** Daily totals of energy balance terms for irrigated alfalfa at Bushland, TX.

where Rsi is solar irradiance at the surface, $\alpha$ is the albedo or surface reflectance (0 to 1), $\varepsilon$ is the surface emissivity (0 to 1), $\sigma$ is the Stefan-Boltzmann constant ($5.67 \times 10^{-8}$ W m$^{-2}$ K$^{-4}$), T is surface temperature (K) and L$\downarrow$ is longwave irradiance from the sky. The sun radiates energy like a black body at about 6000 K while the earth radiates at about 285 K. The theoretical maximum emission power spectra for these two bodies overlap very little (Figure 5.7), a fact that leads to description of radiation from the earth (including clouds and the atmosphere) as longwave, and radiation from the sun as shortwave. Note that the radiance of the earth is about 4 million times lower than that of the sun (Figure 5.7). Net radiation may be measured by a net radiometer (Figure 5.8) or its components may be measured separately using pyranometers to measure incoming and reflected short wave radiation, and pyrgeometers to measure incoming and outgoing long wave radiation (first four instruments in Table 5.1). Pyranometers and pyrgeometers are thermopile devices that are equally sensitive across the spectrum.

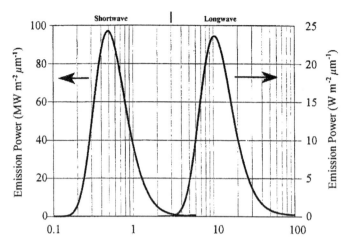

**FIGURE 5.7** Emission power spectra for ideal black bodies at 6000 K (left, shortwave range) and 285 K (right, long-wave range).

**FIGURE 5.8** REBS Q*7 net radiometer.

### 5.2.1.1 Outgoing Long Wave Radiation

The longwave radiance of the earth's surface, $L\uparrow$, is given by the Stefan-Boltzmann law for radiance from a surface at temperature T and with emissivity $\varepsilon$:

$$L\uparrow = \varepsilon\sigma T^4 \tag{5.3}$$

An inverted pyrgeometer (Table 5.1) may be used to measure $L\uparrow$ and, if accompanied by suitable surface temperature measurements, may allow estimation of surface emissivity, $\varepsilon$, by inversion of Equation [5.3]. Surface temperature is often measured by suitably placed and shielded thermocouples, or by infrared thermometer (IRT); though there are problems with either type of measurement (radiational heating of the thermocouples, and uncertainty of the emissivity needed for accurate IRT measurements).

Values of $\alpha$ and $\varepsilon$ for soil and plant surfaces may be estimated from published values relating them to surface properties (see Section 5.1.1.3 and Table 5.4). For soil, the dependence of $\alpha$ on water content is strong, but nearly linear, and amenable to estimation.

### 5.2.1.2 Solar Irradiance

Solar irradiance (Rsi), defined as the radiant energy reaching a horizontal plane at the earth's surface, includes both direct beam and diffuse shortwave. It may be easily measured by pyranometer

## TABLE 5.1
**Instruments and deployment information for bare soil radiation and energy balance experiments at Bushland, TX, 1992 (adapted from Howell et al., 1993). Parameters not shown in Figure 5.1 will be presented later.**

| Parameter | Instrument | Manufacturer[†](Model) | Elevation | Description |
|---|---|---|---|---|
| $R_{si}$ | Pyranometer | Eppley (PSP) | 1 m | Solar irradiance |
| $\alpha R_{si}$ | Pyranometer | Eppley (8-48) | 1 m (I[‡]) | Reflected solar irradiance |
| $L\downarrow$ | Pyrgeometer | Eppley (PIR) | 1 m | Incoming long wave radiation |
| $L\uparrow$ | Pyrgeometer | Eppley (PIR) | 1 m (I) | Outgoing long wave radiation |
| $R_n$ | Net Radiometer | REBS (Q*6) | 1 m | Net radiation |
| $T_s$ | Infrared Thermometer | Everest (4000; 60° fov) | 1 m nadir view angle | Soil surface temperature |
| $T_a$ | thermistor | Rotronics | | Air temperature & relative |
| RH | foil capacitor | (HT225R) | 2 m | humidity |
| $U_2$ | dc generator cups | R.M. Young (12102) | 2 m | Wind speed |
| $U_d$ | potentiometer vane | R.M. Young (12302) | 2 m | Wind direction |
| $T_t$ | Cu-Co thermocouple | Omega (304SS) | –10 mm, –40 mm | Soil temperature(4)[§] |
| $G_{50}$ | plates thermopile | REBS (TH-1) | –50 mm | Soil heat flux (4) |
| $\theta_{v-20}$ | 3-wire | Dynamax | –20 & –40 mm | Soil water content (2) |
| $\theta_{v-40}$ | TDR probe | TR-100/20 cm | horizontal | |
| $E_m$ | lever-scale load cell | Interface SM-50 | Below lysimeter box | Lysimeter mass change |

† Manufacturers and locations are: The Eppley Laboratory, Inc., Newport, RI; Radiation and Energy Balance Systems (REBS), Seattle, WA; Everest Interscience, Inc., Fullerton, CA; Rotronic Instrument Corp., Huntington, NY; R.M. Young Co., Traverse City, RI; Omega Engineering, Inc., Stamford, CT; Dynamax, Inc., Houston, TX; Interface, Inc., Scottsdale, AZ.
‡ I designates instruments that were inverted and facing the ground.
§ Numbers in parentheses indicate replicate sensors.

**FIGURE 5.9** Kipp and Zonen model CM-14 albedometer.

with calibration to international standards (Table 5.1) or by solar cells. Silicon photodetector solar radiation sensors, such as the LI-COR model LI-200SA, are sensitive in only part of the spectrum, but are calibrated to give accurate readings in most outdoor light conditions. Silicon sensors are much cheaper than thermopile pyranometers and have found widespread use in field weather stations. Measurement of both incident (Rsi) and reflected (Rsr) shortwave allows estimation of the albedo from:

$$Rsi(1-\alpha) = Rsi - Rsr \qquad [5.4]$$

This is done using upward and downward facing matched pyranometers (Table 5.1). Specially made albedometers are available for this purpose (e.g., Kipp & Zonen model CM-14) (Figure 5.9).

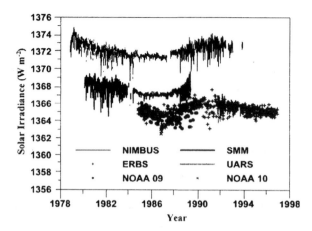

**FIGURE 5.10** Satellite observations of solar irradiance, Ra, outside the earth's atmosphere; corrected for earth-sun distance [Data source: NOAA, 1997].

The solar constant is the flux density of solar radiation on a plane surface perpendicular to the direction of radiation and outside the earth's atmosphere. It is about 1370 W m⁻², with a variation of about ± 3.5%, being largest in January when the sun is closest to the earth, and smallest in July (Jones, 1992). Several satellite observation platforms have recorded the value of solar irradiance over nearly a 20-year span (Figure 5.10) and clearly show the average solar cycle of 11 years. The six sets of data shown range over about 10 W m⁻² or about 0.7% of the mean value. Thus, considering the solar constant to be 1370 W m⁻² will introduce no more than a 1% error in calculations.

Irradiance at the earth's surface is somewhat less, due to absorption and scattering in the atmosphere and due to sun angle effects; not often exceeding 1000 W m⁻². The further the sun is from the zenith, the longer the transmission path through the atmosphere, and the more absorption and scattering occurs. Also, as sun angle above the horizon, $\beta$, decreases (it is highest at solar noon), the radiation density on a horizontal surface decreases, according to Lambert's law:

$$I = I_o \sin \beta \qquad [5.5]$$

where $I_o$ is the flux density on a surface normal to the beam. Sun angle ($\beta$) changes with time of day and year, and can be calculated from:

$$\beta = \sin^{-1}\left[\sin(D)\sin(L) + \cos(D)\cos(L)\cos(H)\right] \qquad [5.6]$$

where L is latitude, D is solar declination and H is solar time angle (all radians). Solar time angle is defined as:

$$H = \frac{(T - T_{SN})2\pi}{24} \qquad [5.7]$$

where T is time (h), and $T_{SN}$ is the time of solar noon, which varies with time of year and longitude according to (recall that 1° longitude = 4 min):

$$T_{SN} = 12 + \frac{4(\text{Longitude} - \text{Local Meridian})}{60} - T_{EQ} \qquad [5.8]$$

**TABLE 5.2**
**Coefficients for calculating the equation of time value from Equation [5.11].**

| | | | | | |
|---|---|---|---|---|---|
| $b_0$ | $4.744 \times 10^{-5}$ | $c_2$ | $9.19 \times 10^{-3}$ | $c_6$ | $-1.29 \times 10^{-3}$ |
| $b_1$ | $-0.157$ | $c_3$ | $-5.78 \times 10^{-4}$ | $c_7$ | $-3.23 \times 10^{-3}$ |
| $b_2$ | $-0.0508$ | $c_4$ | $3.61 \times 10^{-4}$ | $c_8$ | $-2.1 \times 10^{-3}$ |
| $c_1$ | $-0.122$ | $c_5$ | $-5.48 \times 10^{-3}$ | | |

where $T_{EQ}$ is the equation of time value (h), Longitude is in degrees, and the Local Meridian is the longitude (°) for which standard time is calculated for the time zone in question. In the U.S. the meridians for Eastern Standard Time (EST), Central Standard Time (CST), Mountain Standard Time (MST) and Pacific Standard Time (PST) are 75°, 90°, 105° and 120°, respectively. Local or true solar time ($T_{LS}$) for any local standard time ($T_{ST}$) may be calculated with:

$$T_{LS} = T_{ST} - \frac{4(\text{Longitude} - \text{Local Meridian})}{60} + T_{EQ} \qquad [5.9]$$

The declination may be calculated from (Rosenberg et al., 1983):

$$D = 0.4101\cos\left[\frac{2\pi(J-172)}{365}\right] \qquad [5.10]$$

where J is the day of the year.

List (1971) gave equation of time values to the nearest second for the 1st of each month and every 4 days after that for each month (95 values for the year). The following equation reproduces those values with a maximum error of 6 s, and can be used to estimate $T_{EQ}$ in h for any day of the year:

$$T_{EQ} = b_0 + b_1\sin\left(\frac{J}{P_1}\right) + b_2\cos\left(\frac{J}{P_1}\right) + c_1\sin\left(\frac{J}{P_4}\right) + c_2\cos\left(\frac{J}{P_4}\right) + c_3\sin\left(\frac{2J}{P_4}\right)$$
$$+ c_4\cos\left(\frac{2J}{P_4}\right) + c_5\sin\left(\frac{3J}{P_4}\right) + c_6\cos\left(\frac{3J}{P_4}\right) + c_7\sin\left(\frac{4J}{P_4}\right) + c_8\cos\left(\frac{4J}{P_4}\right) \qquad [5.11]$$

where the coefficients $b_i$ and $c_i$ are given in Table 5.2, and $P_1 = 182.5/(2\pi)$ and $P_4 = 365/(2\pi)$.

Jensen et al. (1990) gave a simpler method for $T_{EQ}$:

$$T_{EQ} = 0.1645\sin(2b) - 0.1255\cos(b) - 0.025\sin(b) \qquad [5.12]$$

where $b = 2\pi(J - 81)/364$. The maximum error compared against List's $T_{EQ}$ values is 88 s.

Disregarding air quality, solar irradiance is affected by latitude, time of year and day, and elevation. Latitude and time affect the sun angle, β, and thus affect both the path length of radiation through the atmosphere (and thus absorption and scattering losses), and the flux density at the surface through Equation [5.5]. Elevation affects the path length. Methods for calculating extraterrestrial, Rsa, and clear-sky solar irradiance at the surface, Rso, are given by Campbell (1977, Chapter 5), Jensen et al. (1990, Appendix B), Jones (1992, Appendix 7) and McCullough and Porter (1971). Calculation of Rsa depends on latitude and time of day. Once Rsa is calculated, Rso may be estimated from considerations of adsorption and scattering in the atmosphere, which depend mainly on the path-length through the atmosphere and its density. Thus latitude, time of day and elevation are factors in estimating Rso from Rsa. The value of Rso is an important quantity against which to check measured Rsi; and it can be used in estimates of Rn, either to replace Rsi in Equation [5.2], or using regression relationships of Rn = f(Rso) (see Jensen et al., 1990, Appendix B). Duffie and Beckman (1991) presented the following method of calculating Rsa (MJ m$^{-2}$ h$^{-1}$) for any period, P (h):

$$Rsa = \left[\frac{24(60)}{2\pi}\right] G_{SC} d_r \left\{ \cos(L)\cos(D)\left[\sin(\omega_2) - \sin(\omega_1)\right] + (\omega_1 - \omega_2)\sin(L)\sin(D)\right\} \quad [5.13]$$

where $G_{SC}$ is the solar constant (0.08202 MJ m$^{-2}$ min$^{-1}$), $d_r$ is the relative earth-sun distance, and $\omega_1$ and $\omega_2$ are the solar time angles at the beginning and end of the period, respectively (all angles in radians). The term 24(60)/(2$\pi$) is the inverse angle of rotation per minute. The relative earth–sun distance is given by:

$$d_r = 1 + 0.033\cos\left(\frac{2\pi J}{365}\right) \quad [5.14]$$

where J is the day of year. The factors $\omega_1$ and $\omega_2$ are the solar time angles at the beginning and end of the period in question:

$$\omega_1 = \omega - \frac{\pi P}{24} \quad [5.15]$$

$$\omega_2 = \omega + \frac{\pi P}{24} \quad [5.16]$$

where $\omega$ is the solar time angle at the center of the period (radians), and P is the length of the period in h.

The sunset time angle (angle from noon to sunset) is given by:

$$\omega_s = \cos^{-1}\left[-\tan(L)\tan(D)\right] \quad [5.17]$$

from which it is clear that day length, $T_D$ (h), is:

$$T_D = \frac{24 w_s}{\pi} \quad [5.18]$$

Equation [5.13] can be re-written for total daily Rsa as:

$$Rsa = \left[\frac{24(60)}{2\pi}\right] G_{SC} d_r \left[\cos(L)\cos(D)\sin(\omega_s) + \omega_s \sin(L)\sin(D)\right] \quad [5.19]$$

For example, on day 119 at latitude 35° 11′ N, longitude 102° 6′ W, Rsa calculated using Equation [5.13] on a half-hourly basis was 38.097 MJ m$^{-2}$ compared with 38.100 MJ m$^{-2}$ calculated with Equation [5.19].

Jensen et al. (1990) recommended estimating daily total clear sky solar irradiance as:

$$Rso = 0.75\,Rsa \qquad [5.20]$$

Somewhat in agreement with this, Monteith and Unsworth (1990) stated that direct beam radiation rarely exceeded 1030 W m$^{-2}$, about 75% of the solar constant.

Jones (1992) and Monteith and Unsworth (1990) suggest:

$$Rsi = Rsi_{max}\sin\!\left(\frac{\pi t}{N}\right) \qquad [5.21]$$

for instantaneous values of Rsi on clear days, where $Rsi_{max}$ is the maximum instantaneous irradiance occurring at solar noon, t is time after sunrise (h) and N is daylength (h).

It is more common to know daily total Rsi. Collares-Pereira and Rabl (1979) gave the ratio of hourly irradiance (Rsi,h) to daily irradiance (Rsi,d) as:

$$\frac{Rsi,h}{Rsi,d} = \left(\frac{\pi}{24}\right)\!\left[a + b\cos(\omega)\right]\frac{\cos(\omega) - \cos(\omega_s)}{\sin(\omega_s) - \omega_s\cos(\omega_s)} \qquad [5.22a]$$

where:

$$a = 0.409 + 0.5016\sin\!\left(\omega_s - \frac{\pi}{3}\right) \qquad [5.22b]$$

and:

$$b = 0.6609 - 0.4767\sin\!\left(\omega_s - \frac{\pi}{3}\right) \qquad [5.22c]$$

Equation [5.22] performed well when applied to data from Bushland, Texas (Figure 5.11).

More complex methods of estimating Rso account for attenuation of direct beam radiation using Beer's law, coupled with Lambert's law to calculate irradiance on a horizontal surface, plus an accounting of diffuse irradiance (Jones, 1990; Rosenberg et al., 1983; List, 1971). Beer's law describes the intensity (I) of radiation after passing a distance (x) through a medium in terms of an extinction coefficient (k) and the initial intensity (Ia) as:

$$I = Ia\ e^{kx} \qquad [5.23]$$

For solar radiation the distance is expressed in terms of air mass number (m) as (List, 1971):

$$m = \sec\!\left(\frac{\pi}{2} - \beta\right) \qquad [5.24]$$

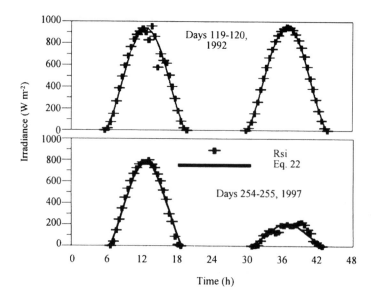

**FIGURE 5.11** Solar irradiance measured at Bushland, TX in 1992 and 1997 on clear and cloudy days; and Equation [5.22] half-hourly predictions.

The air mass is referenced to the length of the path when the sun is directly overhead. For $\beta$ less than 0.175 radians (10°), the measured air mass number is less than that given by Equation [5.24], due to refraction and reflection at these low angles. List (1971) gives corrections, and notes that for pressures (p) less than standard sea-level pressure ($p_0$), that m should be corrected by m = m(p/p₀). Rewriting Equation [5.23] we have:

$$Io = Ia \ e^{k \ \sec(\pi/2-\beta)} \qquad [5.25]$$

where Io is direct beam radiation at the earth's surface. Monteith and Unsworth give a range of values of k for England as 0.07 for very clean air to 0.6 for very polluted air.

Assuming that both direct (Io) and diffuse (Id) radiation are known, the total irradiance at the surface is:

$$Rsi = Io(\sin\beta) + Id \qquad [5.26]$$

Diffuse radiation is quite difficult to estimate because it is so dependent on cloud cover, and aerosol concentration in the air. Yet, summarizing several data sets, Spitters et al. (1986) found that the proportion of Rd to Rsi is a function of the ratio of Rsi to Rsa (Figure 5.12) described for daily total Rsi by:

$$Rd,d = 1 \qquad\qquad\qquad Rsi,d/Rsa,d < 0.07 \qquad [5.27a]$$

$$Rd,d/Rsi,d = 1 - 2.3\left(Rsi,d/Rsa,d - 0.07\right)^2, \qquad 0.07 \le Rsi,d/Rsa,d < 0.35 \qquad [5.27b]$$

$$Rd,d/Rsi,d = 1.33 - 1.46\left(Rsi,d/Rsa,d\right), \qquad 0.35 \le Rsi,d/Rsa,d < 0.75 \qquad [5.27c]$$

$$Rd,d/Rsi,d = 0.23\left(Rsi,d/Rsa,d\right), \qquad 0.75 \le Rsi,d/Rsa,d \qquad [5.27d]$$

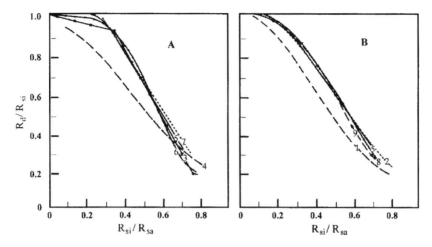

**FIGURE 5.12** Daily (A) and hourly (B) relationships between Rd/Rsi and Rsi/Rsa [Reprinted from Spitters et al., 1986. Agric. Forest Meteorol. 38:217-229 with kind permission of Elsevier Science — NL, Amsterdam, Netherlands].

and for hourly values by:

$$Rd,h = 1, \qquad\qquad Rsi,h/Rsa,h \le 0.22 \qquad [5.28a]$$

$$Rd,h/Rsi,h = 1 - 6.4\big(Rsi,h/Rsa,h - 0.22\big)^2, \qquad 0.22 < Rsi,h/Rsa,h \le 0.35 \qquad [5.28b]$$

$$Rd,h/Rsi,h = 1.47 - 1.66\big(Rsi,h/Rsa,h\big), \qquad 0.35 < Rsi,h/Rsa,h \le K \qquad [5.28c]$$

$$Rd,h/Rsi,h = R, \qquad\qquad K < Rsi,h/Rsa,h \qquad [5.28d]$$

where

$$R = 0.847 - 1.61\sin\beta + 1.04\sin^2\beta \qquad [5.29]$$

and

$$K = \frac{(1.47 - R)}{1.66} \qquad [5.30]$$

### 5.2.1.3 Surface Albedo and Emissivity

Because Rsi provides most of the energy that is partitioned at the earth's surface, albedo plays a major role in the energy balance. The mean albedo of the Earth is $0.36 \pm 0.06$ (Weast, 1982). But albedo varies diurnally (Figure 5.13) with higher albedo corresponding to lower sun angle (see also bare soil data of Monteith and Sziecz, 1961, 1992; Idso et al., 1974; and Aase and Idso, 1975). Soil and plant surfaces are often considered optically rough, but in some cases specular (mirror-like), rather than diffuse, reflection may occur. Some plant leaves are shiny and reflect specularly when the angle of incident radiation is low. Wet soil surfaces may also reflect specularly. These mechanisms lead to higher albedo when sun angle is low. The albedo of plant stands is also lower in midday because more sunlight penetrates deeply within the canopy and is trapped by multiple

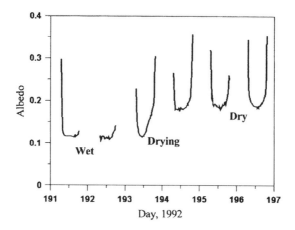

**FIGURE 5.13** Albedo for smooth, bare Pullman clay loam at Bushland, TX when wet and dry.

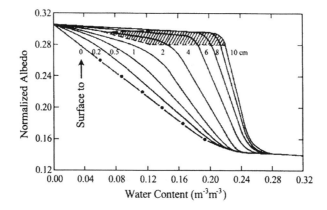

**FIGURE 5.14** Albedo, normalized according to sun angle, vs. soil water content for different surface layer thicknesses of Avondale clay loam at Phoenix, Arizona. Data for shaded area are uncertain. [Reprinted from Idso et al. 1974. Soil Sci. Soc. Am. Proc. 38:831-837 with permission of the Soil Science Society of America].

reflections. Wilting and other physiological changes during the day may also contribute to changes in albedo.

Soil albedo decreases as water content increases. Bowers and Hanks (1965) found the relationship to be curvilinear, as did Skidmore et al. (1975). Idso and Reginato (1974) found that bare soil albedo changed linearly with water content of the surface 2 mm of soil (smooth clay loam) (Figure 5.14). For thicker layers the relationship was curvilinear. The maximum albedo, 0.3, occurred for air dry soil, but the minimum albedo, 0.14, occurred at about 0.23 $m^3$ $m^{-3}$ water content, well before the soil was saturated. This represents field capacity (soil water tension of 30 kPa) for this soil, and Idso and Reginato (1974) postulated that the minimum albedo would occur at field capacity for all soils. Kondo et al. (1992) found a similar relationship for a bare loam with a maximum albedo of 0.24 and minimum of 0.13; and with the minimum attained when soil water content reached about 0.22 $m^3$ $m^{-3}$. Data of Idso et al. (1974, 1975) show that the difference in wet and dry soil albedos was constant despite time of day and day of year. Monteith (1961) measured albedo of clay loam to be 0.18 when dry, decreasing to 0.11 when at field capacity water content of 0.35 $m^3$ $m^{-3}$.

**TABLE 5.3**
**Color (p) and roughness (c) coefficients for Equation [5.31].**

| Surface and Condition | Color Coefficient p | Roughness Coefficient c | Mean $r^2$ |
|---|---|---|---|
| Lakes and ponds, clear water | | | |
| waves, none | 0.13 | 0.29 | 0.82 |
| waves, ripples up to 2.5 cm | 0.16 | 0.70 | 0.74 |
| waves, larger than 2.5 cm with occasional whitecaps | 0.23 | 1.25 | 0.83 |
| waves, frequent whitecaps | 0.30 | 2.00 | 0.85 |
| Lakes and ponds, | | | |
| green water, ripples up to 2.5 cm | 0.22 | 0.70 | 0.90 |
| muddy water, no waves | 0.19 | 0.29 | 0.76 |
| Cotton | | | |
| winds, calm to 4.5 m s$^{-1}$ | 0.27 | 0.27 | 0.80 |
| winds, over 4.5 m s$^{-1}$ | 0.27 | 0.43 | 0.88 |
| Wheat | | | |
| winds, calm to 4.5 m s$^{-1}$ | 0.31 | 0.92 | 0.85 |
| winds, over 4.5 m s$^{-1}$ | 0.37 | 1.30 | 0.85 |

Modified from Dvoracek and Hannabas. 1990. Proc. 3rd Nat. Irrig. Symp., Phoenix, AZ with permission of American Society of Agricultural Engineers.

The interaction of sun angle and soil drying causes complex patterns of soil albedo change over time. Figure 5.13 illustrates low daytime wet soil albedos of 0.11 after irrigation and rain on days 191 and 192, 1992. Rapid soil surface drying on day 193 caused albedo to rise sharply during the day. Additional drying on day 194 completed the change, and diurnal albedo changes on days 195 and 196 reflected only sun angle effects, with a minimum albedo of 0.2 for this smooth soil surface. The same surface in a roughened condition earlier in the year never reached midday albedo values higher than 0.13.

Other than water content, major determinants of soil albedo are color, texture, organic matter content and surface roughness. Dvoracek and Hannabas (1990) presented a model of albedo dependence on sun angle, surface roughness and color:

$$\alpha = p^{(c\sin\beta + 1)} \qquad [5.31]$$

where p was a color coefficient, c was a roughness coefficient and $\beta$ is solar angle. They demonstrated good fits with measured data (Table 5.3). Albedo values modeled using p and c values from Table 5.3 for wheat and cotton (day of year 192, latitude 41°N) appear realistic (Figure 5.15). However, the physical meaning of the p and c coefficients is not well understood.

Daily mean albedos may be calculated as the ratio of daily total reflected shortwave energy to daily total Rsi. Using data from Figures 5.5 and 5.6, daily mean albedos for fallow (soybean residue) and alfalfa differ by about 0.10 when the soil is very dry (Figure 5.16). The gradual decline in fallow albedo in early 1996 may be due to decomposition of the soybean residue. Albedo for the alfalfa field declined at each cutting to nearly that of the fallow field, which was initially rougher than the soil under the alfalfa. But during heavy rains in the latter part of the year, the fallow soil surface was slaked and smoothed and its albedo increased to near that of the alfalfa. Thus, after the 4th cut, the alfalfa field albedo was lower than that of the fallow field for a brief time, probably

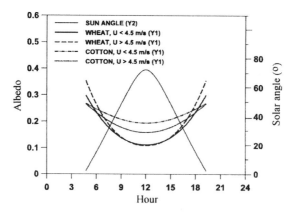

**FIGURE 5.15** Albedo values for wheat and cotton modeled with Equation [5.31].

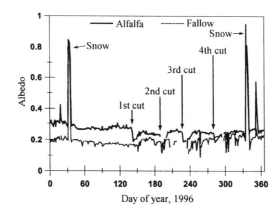

**FIGURE 5.16** Daily means albedos for irrigated alfalfa, and fallow after soybean on Pullman clay loam at Bushland, TX.

because the alfalfa was irrigated and the fallow field had dried out again. Peaks of albedo exceeding 0.8 were due to snow early and late in the year. In contrast to soil, albedo of closed canopies (well-watered) is relatively constant (Table 5.4).

Albedo values for many plant covers may be found in Gates (1980). For surfaces with plants, the amount of radiation reaching the soil surface, Rt (Figure 5.1), depends on the leaf area index (LAI) and the canopy structure. Numerical models have been developed that take into account leaf orientation and distribution in the canopy to calculate absorption of radiation at different levels in the canopy (Goudriaan, 1977; Chen, 1984). Lascano et al. (1987) used Chen's model to calculate polynomials representing the dependence of albedo on LAI, as well as the dependence of the view factor (proportion of sky visible from the soil) on LAI; and incorporated these into their ENergy and WATer BALance model (ENWATBAL). Monteith and Unsworth (1990) present equations describing the albedo of a deep canopy with a spherical distribution of leaves for sun angles higher than 25°. More discussion of these concepts can be found in Russell et al. (1989). For field studies we can either measure albedo, or directly measure the components of net radiation, or use a net radiometer (Table 5.1). The transmitted radiation can be measured below the canopy with tube solarimeters.

**TABLE 5.4**

**Some albedo and emissivity values for various soil and plant surfaces.**

| Surface | Albedo | Emissivity | Source |
|---|---|---|---|
| soils, dark, wet to light, dry | 0.05–0.50 | 0.90–0.98 | Oke, 1978 |
| dry sandy soil | 0.25–0.45 | | Rosenberg et al., 1983 |
| bare dark soil | 0.16–0.17 | | Rosenberg et al., 1983 |
| dry clay soil | 0.20–0.35 | | Rosenberg et al., 1983 |
| quartz sand | 0.35 | | van Wijk and Scholte Ubing, 1963 |
| sand, wet | 0.09 | 0.98 | van Wijk and Scholte Ubing, 1963 |
| sand, dry | 0.18 | 0.95 | van Wijk and Scholte Ubing, 1963 |
| dark clay, wet | 0.02–0.08 | 0.97 | van Wijk and Scholte Ubing, 1963 |
| dark clay, dry | 0.16 | 0.95 | van Wijk and Scholte Ubing, 1963 |
| fields, bare | 0.12–0.25 | | van Wijk and Scholte Ubing, 1963 |
| fields, wet, plowed | 0.05–0.14 | | van Wijk and Scholte Ubing, 1963 |
| dry salt cover | 0.50 | | van Wijk and Scholte Ubing, 1963 |
| snow, fresh | 0.80–0.95 | | Rosenberg et al., 1983 |
| snow, old | 0.42–0.70 | | Rosenberg et al., 1983 |
| snow, fresh | 0.95 | 0.99 | Oke, 1978 |
| snow, old | 0.40 | 0.82 | Oke, 1978 |
| snow, fresh | 0.80–0.85 | | van Wijk and Scholte Ubing, 1963 |
| snow, compressed | 0.70 | | van Wijk and Scholte Ubing, 1963 |
| snow, melting | 0.30–0.65 | | van Wijk and Scholte Ubing, 1963 |
| grass, long (1 m) | 0.16 | 0.90 | Oke, 1978 |
| short (0.02 m) | 0.26 | 0.95 | Oke, 1978 |
| grass, green | 0.16–0.27 | 0.96–0.98 | van Wijk and Scholte Ubing, 1963 |
| grass, dried | 0.16–0.19 | | van Wijk and Scholte Ubing, 1963 |
| prairie, wet | 0.22 | | van Wijk and Scholte Ubing, 1963 |
| prairie, dry | 0.32 | | van Wijk and Scholte Ubing, 1963 |
| stubble fields | 0.15–0.17 | | van Wijk and Scholte Ubing, 1963 |
| grain crops | 0.10–0.25 | | van Wijk and Scholte Ubing, 1963 |
| green field crops full cover, LAI>3 | 0.20–0.25 | | Jensen et al., 1990 |
| leaves of common farm crops | | 0.94–0.98 | Jensen et al., 1990 |
| most field crops | 0.18–0.30 | | Rosenberg et al., 1983 |
| field crops, latitude 22–52° | 0.22–0.26 | 0.94–0.99 | Monteith and Unsworth, 1990 |
| field crops, latitude 7–22° | 0.15–0.21 | 0.94–0.99 | Monteith and Unsworth, 1990 |
| deciduous forest | 0.15–0.20 | 0.96† | Rosenberg et al., 1983 |
| deciduous forest, bare- | 0.15 | 0.97 | Oke, 1978 |
| leaved | 0.20 | 0.98 | Oke, 1978 |
| coniferous forest | 0.10–0.15 | 0.97† | Rosenberg et al., 1983 |
| coniferous forest | 0.05–0.15 | 0.98–0.99 | Oke, 1978 |
| vineyard | 0.18–0.19 | | Rosenberg et al., 1983 |
| mangrove swamp | 0.15 | | Rosenberg et al., 1983 |
| grass | 0.24 | | Jones, 1992 |
| crops | 0.15–0.26 | | Jones, 1992 |
| forest | 0.12–0.18 | | Jones, 1992 |
| water, high sun | 0.03–0.10 | 0.92–0.97 | Oke, 1978 |
| water, low sun | 0.10–1.00 | 0.92–0.97 | Oke, 1978 |
| sea, calm | 0.07–0.08 | | Rosenberg et al., 1983 |
| sea, windy | 0.12–0.14 | | Rosenberg et al., 1983 |
| ice, sea | 0.30–0.45 | 0.92–0.97 | Oke, 1978 |
| ice, glacier | 0.20–0.40 | | |
| ice, lake, clear | 0.10 | | Rosenberg et al., 1983 |
| ice, lake, w/snow | 0.46 | | Rosenberg et al., 1983 |

† van Wijk and Scholte Ubing, 1963

**EXAMPLE 5.1**    Spectral Reflection and Radiation

The discussion of emissivities and albedos of surfaces given here is based on a broadband view of irradiance, reflection and emission that recognizes only short- and long-wave radiation as presented in Figure 5.7. Although these are arguably the most important features from an energy balance perspective, there is much recent work on multi-spectral sensing of radiation reflected and emitted from vegetation and soil surfaces (Robert et al., 1999). This spectral sensing may be done for only a few relatively narrow bands of radiation in the visible and infrared, or may involve hyperspectral scanning that provides sensing of radiation for every nm of the spectrum across a wide range. The advent of fiber optics capable of trans-mitting both visible and infrared light and the development of miniaturized spectrometers (Ocean Optics, Inc. model S2000) has revolutionized the way that researchers view plant and soil surfaces.

An example of multi-spectral sensing is the use of red and near infrared (NIR) reflectance from a cotton canopy (Diagram 5.1.1). The ratio of NIR/Red reflectances is clearly related to leaf area index. However, the relationship is not stable from year to year, and research continues. Other uses include sensing of the onset and progression of plant disease and insect infestation, and sensing of plant water status. Much work remains to be done to make these techniques useful.

### 5.2.1.4  Incoming Long-Wave Radiation

Methods of estimating long-wave irradiance from the sky, $L\downarrow$, usually take the form:

$$L\downarrow = \varepsilon\sigma\left(T_a + 273.16\right)^4$$ [5.32]

where $T_a$ (°C) is air temperature at the reference measurement level (often 2 m), and the emissivity ($\varepsilon$) may be estimated from the vapor pressure of water in air at reference level ($e_a$) (kPa), or using both $e_a$ and $T_a$. The vapor pressure is:

$$e_a = RH\left(e_s\right)$$ [5.33]

where RH is the relative humidity of the air and $e_s$ is the saturation vapor pressure (kPa) at $T_a$ (°C) given by (Murray, 1967):

$$e_s = 0.61078\exp\left(\frac{17.269T_a}{237.3 + T_a}\right)$$ [5.34]

If the dew point temperature, rather than the RH, is known, then:

$$e_a = 0.61078\exp\left(\frac{17.269T_{dew}}{237.3 + T_{dew}}\right)$$ [5.35]

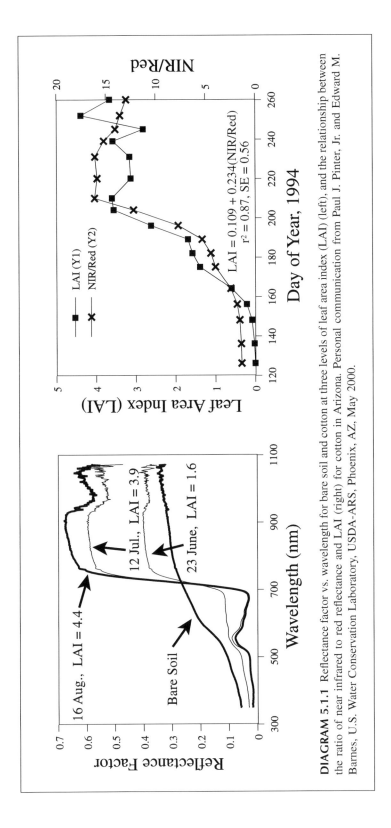

**DIAGRAM 5.1.1** Reflectance factor *vs.* wavelength for bare soil and cotton at three levels of leaf area index (LAI) (left); and the relationship between the ratio of near infrared to red reflectance and LAI (right) for cotton in Arizona. Personal communication from Paul J. Pinter, Jr. and Edward M. Barnes, U.S. Water Conservation Laboratory, USDA-ARS, Phoenix, AZ, May 2000.

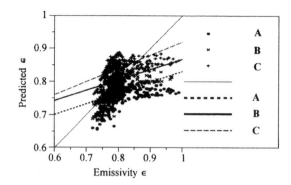

**FIGURE 5.17** Comparison of predictions with measured emissivity for two periods in 1992 at Bushland, TX. Points plotted at extreme right were associated with nighttime and overcast conditions. A = Eq. [5.37], B = Eq. [5.38], C = Eq. [5.36]. Lines are for linear regressions shown in Table 5.5.

Hatfield et al. (1983) compared several methods for estimating $\varepsilon$ and concluded that methods using only air temperature performed less well than those that used vapor pressure or both vapor pressure and air temperature. Among the best methods was Idso's (1981) equation:

$$\varepsilon_a = 0.70 + 5.95 \times 10^{-4} e_a \exp\left(\frac{1500}{T_a + 273.1}\right)$$ [5.36]

where $e_a$ is in kPa. Idso showed fairly conclusively that $\varepsilon$ is a function of both $e_a$ and $T_a$.

Howell et al. (1993) measured L↓ (Table 5.1) and calculated $\varepsilon$ by inverting Equation [5.32]. Applying Equation [5.36] as well as Brunt's (1932) equation:

$$\varepsilon_a = 0.52 + 0.206 e_a^{0.5}$$ [5.37]

and Brutsaert's (1982) equation:

$$\varepsilon_a = 0.767 e_a^{1/7}$$ [5.38]

to their data shows that all three equations gave good predictions for clear sky conditions, but probably underestimated $\varepsilon$ for cloudy and nighttime conditions (Figure 5.17). For regressions of predicted versus measured $\varepsilon$, the Idso equation gave a slightly higher correlation coefficient and a slope closer to unity (Table 5.5). Under heavy clouds, sky emissivity approaches unity, and none of these models predicts this well.

Despite the difficulty of estimating sky emissivity well, uncertainty in the value of L↓ usually causes little difficulty in estimating net radiation, because L↓ is very often a small component of the energy balance.

### 5.2.1.5 Comparison of Net Radiation Estimates with Measured Values

It has become commonplace to have data from field weather stations that includes Rsi, and air temperature ($T_{az}$), wind speed ($U_z$), and relative humdity ($RH_z$), measured at some reference height (z) (often 2 m). Measurement of Rn is still not common, probably due to several factors including additional expense, fragility of the plastic domes used on some models of net radiometer, and problems with calibration. Net radiometer calibration changes with time, and experience shows that even new radiometers may not agree within 10%. If a net radiometer is used, it is prudent, as

**TABLE 5.5**
**Regressions of predicted emissivity ($\epsilon_p$) vs. measured values ($\epsilon$) for data from day 133 through 140, and 192 through 197 (Bushland, TX, 1992).**

| Method | Regression Equation | $r^2$ | SE |
|---|---|---|---|
| Brunt, Equation [5.37] | $\epsilon_p = 0.505 + 0.325\ \epsilon$ | 0.33 | 0.024 |
| Brutsaert, Equation [5.38] | $\epsilon_p = 0.556 + 0.311\ \epsilon$ | 0.32 | 0.024 |
| Idso, Equation [5.36] | $\epsilon_p = 0.522 + 0.398\ \epsilon$ | 0.37 | 0.027 |

with all instruments, to check measured Rn values against estimated ones. Methods presented in previous sections can be used to estimate Rn, but simpler methods exist that are adequate for most cases. Jensen et al. (1990) compared four methods of estimating Rn, including Wright and Jensen (1972), Doorenbos and Pruitt (1977), a combination of Brutsaert (1975) and Weiss (1982), and Wright (1982), against values measured at Copenhagen, Denmark, and Davis, California. The Wright (1982) method was best overall, but underestimated Rn in the peak month at Copenhagen by 9%. The Wright and Jensen (1972) method was almost as good. These methods all assume that surface temperature is not measured, so that only air temperature is used in the calculations.

Jensen et al. (1990) calculated net long-wave radiation, Rnl, as:

$$Rnl = -\left[a\left(\frac{Rsi}{Rso}\right)+b\right]\left(a_1 + b_1 e_d^{0.5}\right)\sigma T_{az}^4 \qquad [5.39]$$

where $e_d$ is the saturation vapor pressure of water in air at dew point temperature (kPa), and the term $(a_1 + b_1 e_d^{0.5})$ is a net emittance ($\epsilon'$) of the surface. The net emittance attempts to compensate for the fact that surface temperature is not measured, the assumption being that $T_{az}$ can substitute reasonably well for both sky and surface temperatures. The coefficients a, b, $a_1$, and $b_1$ are climate specific, a and b being cloudiness factors. Some values are presented by Jensen et al. (1990, Table 3.3).

Many weather stations report only daily totals of solar radiation, and maximum and minimum of air temperature ($T_x$ and $T_n$), respectively (K). If this is the case, the term $\sigma T^4$ can be estimated as:

$$\sigma T^4 \cong \sigma\left(T_x^4 - T_n^4\right)/2 \qquad [5.40]$$

If mean dew point temperature is not available, it may be estimated as equal to $T_n$ in humid areas.

Allen et al. (1994a,b) presented slightly modified versions of the methods presented by Jensen et al. (1990) in a proposed FAO standard for reference evapotranspiration estimation. As an example, estimates of daily total net radiation were made for Bushland, Texas using the following equations from Allen et al. (1990):

$$Rn = (1-\alpha)Rs -\left[a_c\left(\frac{Rsi}{Rso}\right)+b_c\right]\left(a_1 + b_1 e_d^{0.5}\right)\sigma\left(\frac{T_m^4 + T_n^4}{2}\right) \qquad [5.41]$$

where the cloud factors were $a_c = 1.35$ and $b_c = -0.35$, the emissivity factors were $a_1 = 0.35$ and $b_1 = -0.14$, the albedo was $\alpha = 0.23$, Rsi was measured, $e_d$ was calculated from mean dew point temperature, and Rso was calculated from:

$$Rso = (.75 + .00002\ ELEV)Rsa \qquad [5.42]$$

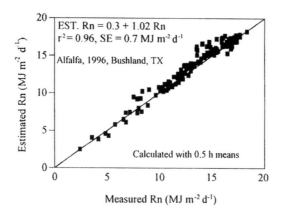

**FIGURE 5.18** Net radiation estimated with methods from Allen et al. (1994a,b) compared to measurements over sprinkler irrigated alfalfa in 1996 at Bushland, TX.

where Rsa is from Equation [5.19], and ELEV is elevation (m) above mean sea level. This is similar to Equation [5.20], but with a correction increasing Rso for higher elevation sites. The mean daily saturated vapor pressure at dew point temperature was estimated from mean daily dew point temperature ($T_d$):

$$e_d = 0.611 \exp\left( \frac{17.27 T_d}{237.3 + T_d} \right) \tag{5.43}$$

Additional estimates were calculated from half-hourly measured values of Rsi, $T_a$, and $T_d$ using equations given by Allen et al. (1994) equivalent to Equations [5.7], [5.8], [5.10], [5.12], [5.13], [5.14], [5.15] and [5.16] to estimate half-hourly Rsa, and Equation [5.41] to estimate half-hourly Rso. Equation [5.42] was applied to half-hourly dew point temperatures to estimate half-hourly $e_d$ values. Equation [5.40] was written for half-hourly values of air temperature ($T_a$) as:

$$Rn = (1-\alpha)Rs - \left[ a_c\left(\frac{Rsi}{Rso}\right) + b_c \right]\left( a_1 + b_1 e_d^{0.5} \right)\sigma T_a^4 \tag{5.44}$$

where the ratio of Rsi to Rso was set to 0.7 for nighttime estimates of Rn.

Comparison of daily Rn estimates, calculated using half-hourly data means, with measurements made with a REBS Q*5 (Seattle, WA) net radiometer over irrigated grass show excellent agreement for alfalfa (Figure 5.18) and grass (Figure 5.19) at Bushland, TX. But there was a consistent bias for Rn estimated from daily means, with underestimation of Rn at high measured values, and overestimation at low measured values (Figure 5.20). The bias evident when daily means and maximum/minimum temperatures were used is probably tied to both poor estimates of vapor pressure from the max/min temperature data and the inadequacy of Equation [5.39].

Estimates of half-hourly net radiation for alfalfa at Bushland, Texas using half-hourly data and these methods also gave good results (Figure 5.21). Allen et al. (1994a, b) give detailed methods for estimating Rn when measurements are missing for Rsi and/or $e_d$.

### 5.2.2 LATENT HEAT FLUX MEASUREMENT

Latent heat flux is the product of the evaporative flux, E (kg s⁻¹ m⁻²), and the latent heat of evaporation, L ($2.44 \times 10^6$ J kg⁻¹ at 25°C). The value of L is temperature dependent, but is well described (in J kg⁻¹ × 10⁶) by:

$$L = 2.501 - 2.370 \times 10^{-3} T \quad \left( r^2 = 0.99995 \right) \tag{5.45}$$

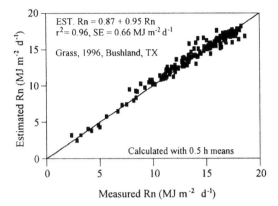

**FIGURE 5.19** Daily net radiation, estimated with methods from Allen et al. (1994a,b) using half-hourly data, compared to measurements with a REBS Q*5 net radiometer over drop irrigated grass in 1996 at Bushland, TX.

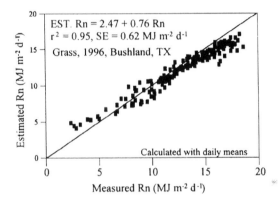

**FIGURE 5.20** Daily net radiation, estimated with methods from Allen et al. (1994a,b) using daily means and temperature maxima and minima, compared to measurements with a REBS Q*5 net radiometer over drip irrigated grass in 1996 at Bushland, TX.

where T is in °C. Methods of measurement of E include weighing lysimeter (including microl-ysimeters), and other mass balance techniques that rely on measurements of change in soil water storage, $\Delta S$, as well as eddy correlation and Bowen ratio measurements. Because $\Delta S$ is a component of the soil water balance, and lysimetry is a key tool for investigations of soil water balance, discussion of lysimetric techniques will be deferred to Section 5.2.

### 5.2.2.1 Boundary Layers

Evaporative fluxes move between plant, or soil, surfaces and the air by both diffusion and convec-tion. Diffusive processes prevail in the laminar sublayer close (millimeters) to these surfaces. In this layer, air movement is parallel to the surface and little mixing occurs. Vapor flux across the laminar sublayer is well described by a Fickian diffusion law relating flux rate to vapor pressure gradient factored by a conductance term. But in the turbulent layer beyond the laminar layer, the flux is mostly convective in nature so that water vapor is moved in parcels of air that are moved and mixed into the atmosphere in turbulent flow. These moving parcels of air are often referred to as eddies, similar to eddies seen in a stream. Usually the eddies are not visible, but in foggy, smoky or dusty air they may be apparent. Certainly anyone who has felt the buffeting of the wind can attest to the force of eddies and the turbulence of the air stream in which they occur. As wind

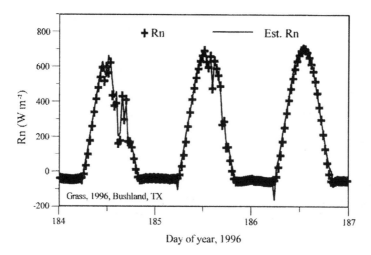

**FIGURE 5.21** Half-hourly measured net radiation compared with values estimated with methods of Allen et al. (1994a,b) using half-hourly data for drip irrigated grass at Bushland, TX.

speeds increase, the depth of the laminar sublayer decreases. Surface roughness enhances this process, resulting in thinner laminar sublayers. Because the resistance to vapor transport across the laminar airstream is much larger than the resistance across a turbulent airstream of similar dimensions, increasing roughness and wind speed both tend to enhance vapor transport. If the air is still, then eddies due to turbulent flow do not exist, but eddies due to free convection may well be present. Free convection occurs when an air parcel is warmer (or colder) than the surrounding air and thus moves upward (or downward) because it is lighter (or heavier). These buoyancy effects can predominate at very low wind speeds when the surface is considerably warmer than the air. As opposed to free convection, transport in eddies due to wind is called forced convection.

A full discussion of the fluid mechanics of laminar and turbulent flow, Fickian diffusion, and forced and free convection is well beyond the scope of this chapter. Discussions relevant to soil and plant surfaces are presented by Monteith and Unsworth (1990), Rosenberg et al. (1983) and Jones (1992). Here we will concentrate on some results and methods of measurement. These methods are valid within the constant flux layer (Figure 5.22), which is a layer of moving air which develops from the point at which the air stream first reaches a surface of given condition, for example the wheat field shown in Figure 5.22. As the air moves over the field it mixes, equilibrating with the new surface condition, and forming a layer of gradually increasing thickness ($\delta$) within which the flux of heat and vapor is constant with height. This is the fully adjusted or equilibrium layer. Within this layer is a layer, extending from the roughness elements (wheat plants in this schema) upward, within which air flow is more turbulent due to the influence of the roughness elements. This is called the roughness sublayer (Monteith and Unsworth, 1990). For any measurement of air temperature, humidity or wind speed, the fetch is the distance upwind from the point of observation to the edge of the new surface. The ratio of the fetch to the value of $\delta$ is dependent on the roughness of the surface, the stability of the air, and the wind speed. For many crop surfaces it may be as small as 20:1 or as large as 200:1. For smooth surfaces such as bare soil the ratio may well be larger than 200:1. Measurements should be made in the constant flux layer but above the roughness sublayer.

### 5.2.2.2 Eddy Correlation Measurements

The observation of turbulent flow and the concept of eddies leads to the eddy correlation method of latent heat flux measurement. The main idea is that if eddies with a vertical velocity component upward are correlated with humidities on average higher than the humidities correlated with

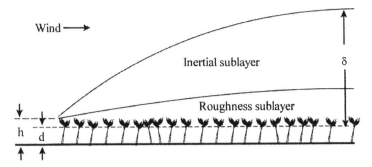

**FIGURE 5.22** Schematic of sublayers of the surface boundary layer over a wheat crop. The height, h, of the crop and the depth, δ, of the constant flux layer are noted. The height d is the zero plane displacement height, which is the height to which a logarithmic wind profile, measured above the crop, would extrapolate to zero wind speed.

downward moving eddies, then the net flux of water vapor is upward. In this method very fast response sensors are used to measure the vertical wind speed and humidity simultaneously at a rate of, for example, 20 Hz. This gives a direct measure of the flux at the measurement height (but see fetch requirements below) according to (Rosenberg et al., 1983):

$$E = \left( \frac{M_w}{M_a P} \right) \rho_a \overline{w' e_a'} \qquad [5.46]$$

where the overscores indicate time averages of vertical wind speed (w′) and vapor pressure (e_a′), the primes indicate instantaneous deviations from the mean, P is atmospheric pressure (Pa), $\rho_a$ is air density, and $M_w$ and $M_a$ are the molecular weights of water and air. The rate of data acquisition must be faster for measurements nearer the surface. Monteith and Unsworth (1992) state that eddy sizes increase with surface roughness and wind speed, and with height above the surface, and they suggest one kHz rates may be needed near a smooth surface, while 10 Hz or slower may be adequate at several meters above a forest. Because the measurements should take place within the fully adjusted boundary layer, simply increasing sensor height will not eliminate the need for fast sensor response. Eddy correlation methods are difficult to carry out, due to the data handling and sensor requirements. Data processing requirements are large, but modern data logging and computing equipment are capable of handling these. Commercial systems including data processing software are now available, although expensive (Campbell Scientific Inc., Logan, Utah; and The Institute of Ecology and Resource Management at the University of Edinburgh, Scotland). The sonic anemometer is the wind sensor of choice for eddy correlation work due to its fast response and sensitivity. At this time a single-axis unit costs about $2,500, and a three-dimensional sonic anemometer costs about $8,000. Suitable vapor pressure sensors include the krypton hygrometer and infrared gas analyzer (IRGA), available at this writing in the $5,000 to $10,000 range.

Eddy correlation measurements may be made for sensible heat flux as well (Section 5.1.4), and if both E and H are measured by eddy correlation, the performance of the system may be checked (Houser et al., 1998) by rearranging Equation [5.1] to:

$$LE + H = -Rn - G \qquad [5.47]$$

and measuring Rn and G (Figure 5.23). Fast response thermocouples for measuring air temperature are used in eddy correlation systems for measuring H. Because these are much less expensive than fast response vapor pressure sensors, it is sometimes sensible to measure Rn and G, and H by eddy correlation, and find LE as the residual:

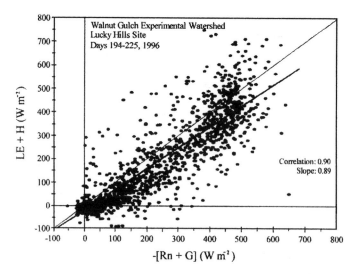

**FIGURE 5.23** Check of eddy correlation system LE and H values against measured Rn and G values [Adapted from Houser, 1998].

$$LE = -Rn - G - H \qquad [5.48]$$

Comparisons of eddy correlation and Bowen ratio systems are found in Houser et al. (1998) and Dugas et al. (1991). Some specifics of eddy correlation system design are given in Unland et al. (1996) and Moncrieff et al. (1997).

### 5.2.2.3  Bowen Ratio Measurement

The Bowen ratio is the ratio of sensible to latent heat flux ($\beta = H/LE$). Introducing this into Equation [5.1] and rearranging gives the Bowen ratio method for estimating LE:

$$LE = \frac{-(Rn + G)}{\beta + 1} \qquad [5.49]$$

In the constant flux layer it is possible to measure temperature and vapor pressure differences at two heights, $z_1$ and $z_2$, and evaluate $\beta$ from a finite difference form:

$$\beta = \frac{\dfrac{K_H \rho_a c_p \left(T_{z_2} - T_{z_1}\right)}{z_2 - z_1}}{\dfrac{K_V \left(\dfrac{\varepsilon_m}{P}\right)\rho_a L \left(e_{z_2} - e_{z_1}\right)}{z_2 - z_1}} \approx \frac{c_p \left(T_{z_2} - T_{z_1}\right)}{\left(\dfrac{\varepsilon_m}{P}\right) L \left(e_{z_2} - e_{z_1}\right)} = \gamma \frac{\left(T_{z_2} - T_{z_1}\right)}{\left(e_{z_2} - e_{z_1}\right)} \qquad [5.50]$$

where the second and third entities assume equivalency of the exchange coefficients for sensible heat flux ($K_H$) and latent heat flux ($K_V$), $\varepsilon_m$ is $M_w/M_a$, and $\gamma = c_p P/(\varepsilon_m L)$ is the psychrometric constant. Commonly, values of T and e are half-hour or hourly means. Because the sensor response time does not have to be very short, Bowen ratio equipment is much less expensive than that for eddy correlation, with complete systems available for under \$10,000. Systems are available from Radiation and Energy Balance Systems (REBS), Seattle, WA; Campbell Scientific, Inc., Logan, UT, and others.

Because slight differences in instrument calibration may lead to large errors, it is advisable to switch instruments between the measurement heights. The moving arm system popularized by REBS is one way to do this. Bowen ratio measurements are usually valid only during daylight hours. At night the sum of Rn and G approaches zero causing Equation [5.49] to become imprecise. For periods just after sunrise, and before sunset the gradients of T and e may become small at the same time that Rn becomes small, leading to instability in Equation [5.49] and imprecision in the estimate of LE. Under advective conditions, Bowen ratio systems tend to underestimate LE when regional sensible heat advection occurs (Todd, 1998b; Blad and Rosenberg, 1974), probably because $K_H/K_V > 1$ under the stable conditions that prevail then (Verma et al., 1978). Four Bowen ratio systems were compared by Dugas et al. (1991), who discuss the merits of different designs. Three eddy correlation systems agreed well with each other, but LE measurements from them were consistently lower than those from the four Bowen ratio systems.

### 5.2.2.4 Fetch Requirements

Both eddy correlation and Bowen ratio methods are sensitive to upwind conditions. The LE and H values from these methods represent an areal mean for a certain upwind area, often called the footprint. Both methods require considerable upwind fetch, often running to hundreds of meters, of surface that is essentially the same as that where the measurement is made, if the measurement is to be representative of that surface. Also, the longer the same-surface fetch is, the deeper is the fully adjusted layer, and the higher the instruments can be placed above the surface. Issues of instrument height and fetch are discussed by Savage et al. (1995 and 1996), who recommended placing the sonic anemometer no closer than 0.5 m above a short grass cover. Because eddies are smaller nearer the surface, placement of the sonic anemometer too near the surface may lead to eddies being smaller than the measurement window of the anemometer. Fetch requirements may be stated as a ratio of fetch distance to instrument height. Heilman et al. (1989) studied fetch requirements for Bowen ratio systems, and concluded that a fetch ratio of 20:1 was adequate for many measurements, down from the 100:1 ratios reported earlier. Fetch requirements increase as measurement height ($z_m$) increases. This poses some additional problems for Bowen ratio systems because these incorporate two sensors and the sensors must be separated enough that the vapor pressure and temperature gradients between them are large enough to be accurately sensed. The rougher the surface the smaller the gradients. For many surfaces, and common instrument resolution, these facts lead to separation distances on the order of a meter. The lower measurement should be above the roughness sublayer, typically at least 0.5 m above a crop (more for a very rough surface such as a forest), so the upper measurement may well be nearly 2 m above the crop surface. This could easily lead to a fetch requirement of 100 m. Analysis of relative flux and cumulative relative flux for an alfalfa field under moderately stable conditions using the methods of Schuepp et al. (1990) leads to rather large fetch requirements (Todd, 1998a) (Figure 5.24). For unstable conditions, mixing is enhanced and the boundary layer becomes adjusted more quickly over a new surface so that fetch requirements are lessened. Fetch requirements are more severe for Bowen ratio than for eddy correlation measurements (Schmid, 1997).

Because of the direct way in which fluxes are measured in eddy correlation schemes, this method is sometimes stated to be the only true measure of latent (or sensible) heat flux. However, consideration of fetch requirements leads to a conclusion that both eddy correlation and Bowen ratio measurements are true only for a constantly changing footprint area upwind of the measurement location. The footprint area and the true flux are poorly defined because the location and size of the footprint change with wind direction and speed. There is strength in this kind of areal averaging, because it reduces noise due to the spatial variability of evaporation. But the measurement cannot be said to be true for any specific location. Indeed, as wind direction changes, the measurement area may change completely. By contrast, the soil water balance methods of estimating E, discussed in Section 5.2, provide measures for specific locations. In the case of weighing lysimeters,

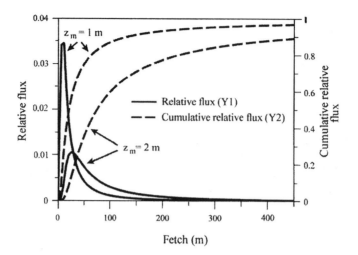

**FIGURE 5.24** Relative and cumulative relative flux of an alfalfa field for measurement heights ($z_m$) of 1 and 2 m, moderately stable thermal conditions, and canopy height of 0.5 m. Cumulative relative flux reaches 0.8 at 65 m for $z_m$ = 1 m, and at 225 m for $z_m$ = 2 m.

these are in fact direct measurements of E, specific to a well-defined location, for all times during which precipitation and runoff are not occurring (neglecting the negligible change in plant mass over short periods).

---

**EXAMPLE 5.2**   Precision Farming Research Tools Needed

Both eddy correlation and Bowen ratio measurements are based on the assumption that, within the fully adjusted layer, the vertical energy fluxes are uniform with height, i.e., that there is no vertical flux divergence. However, we see from footprint analysis that, for non-zero wind speeds, the upwind area contributes to the measured flux. Even in supposedly uniform fields there is spatial variability in soil properties and plant responses, so that there is almost always some horizontal flux divergence, and therefore necessarily some vertical flux divergence. Although both methods have been described as "point" measurements, they are really averages over an area, with closer upwind sub-areas being weighted more heavily, and with wind speed and atmospheric stability causing changes in the relative weighting of sub-areas and the total area involved. Interest in the spatial variability of the energy balance, particularly the LE component, and in precision farming technologies aimed at addressing crop requirements for water and nutrients at scales well below the field size, has led to a need to measure the spatial variability of LE at scales smaller than can reasonably be addressed with Bowen ratio and eddy correlation systems. As will be discussed in the next box, the radiometric surface temperature can be remotely sensed to give the spatial variability of LE based on Equations [5.1] and [5.73]. However, this approach is weakened by problems with quantifying surface and aerodynamic resistances and discriminating between crop and soil contributions to the radiometric temperature. There remains a need for ground truth measurements giving LE for a well-defined area. The water balance methods discussed in Section 5.3.1 are capable in many cases, but costs of deployment are practically insurmountable. Thus, there remains a need for inexpensive, accurate, unattended soil profile water content measurement methods for implementation of the water balance method of LE measurement.

### 5.2.2.5 Penman-Monteith Estimates of Latent Heat Flux

Since Penman (1948) published his famous equation describing evaporation from wet surfaces based on the surface energy balance, there have been developments, additions and refinements of the theory too numerous to mention. Notable examples are the van Bavel (1966) formulation, which includes a surface roughness length term ($z_0$), and the Penman-Monteith (PM) formula (Monteith, 1965), which includes aerodynamic and surface resistances. The van Bavel equation tends to overestimate in windy conditions and is very sensitive to the value of $z_0$ (Rosenberg, 1969). Howell et al. (1994) compared several ET equations for well-watered, full-cover winter wheat and sorghum and found that the PM formula performed best. Because it is widely used in agricultural and environmental research, and because it has been presented by ASCE (Jensen et al., 1990) and FAO (Allen et al., 1994ab) as a method of computing estimates of reference crop water use, we will discuss the Penman-Monteith equation, which is:

$$LE = -\frac{\Delta\left(R_n + G\right) + \rho_a c_p\left(e_s - e_a\right)/r_a}{\Delta + \gamma\left(1 + r_s/r_a\right)} \qquad [5.51]$$

where LE is latent heat flux, $R_n$ is net radiation, and G is soil heat flux (all in MJ m$^{-2}$ s$^{-1}$); $\Delta$ is the slope of the saturation vapor pressure-temperature curve (kPa °C$^{-1}$), $\rho_a$ is air density (kg m$^{-3}$), $c_p$ is the specific heat of air (kJ kg$^{-1}$ °C$^{-1}$), $e_a$ is vapor pressure of the air at reference measurement height z, and $e_s$ is the saturated vapor pressure at a dew point temperature equal to the air temperature at z (kPa), ($e_s - e_a$) is the vapor pressure deficit, $r_a$ is the aerodynamic resistance (s/m), $r_s$ is the surface (canopy) resistance (s/m), and $\gamma$ is the psychrometric constant (kPa °C$^{-1}$). Penman's equation and those derived from it were developed as a means of eliminating canopy temperature from energy balance considerations. Besides measurements of Rn and G, the user must know the vapor pressure of the air ($e_a$) and air temperature (from which $e_s$ may be calculated) at reference measurement height, z (often 2 m). The values of $r_a$ and $r_s$ may be difficult to obtain. The surface or canopy resistance is known for only a few crops and is dependent on plant height, leaf area, irradiance and water status of the plants.

Jensen et al. (1990) and Allen et al. (1994ab) presented methods of calculating E for well-watered, full-cover grass and alfalfa. The following example, drawn from recent studies at Bushland, Texas, employs those methods. Aerodynamic resistance was estimated for neutral atmospheric conditions from:

$$r_a = \frac{\ln\left(\dfrac{z_m - d}{z_{0m}}\right)\ln\left(\dfrac{z_H - d}{z_{0H}}\right)}{k^2 U_z} \qquad [5.52]$$

where $z_m$ (m) is the measurement height for wind speed, $U_z$, (m/s), $z_H$ (m) is measurement height for air temperature and relative humidity, k is 0.41, $z_{0m}$ and $z_{0H}$ are the roughness length parameters for momentum (wind) and sensible heat transport, and d is the zero plane displacement height. The value of $r_a$ calculated from Equation [5.52] will be too high for highly unstable conditions and too low for very stable conditions. Stability corrections should be made to Equation [5.52] for those conditions (Monteith and Unsworth, 1990, p. 234), but were not made for this example.

Surface resistance was calculated from:

$$r_s = \frac{r_1}{0.5 LAI} \qquad [5.53]$$

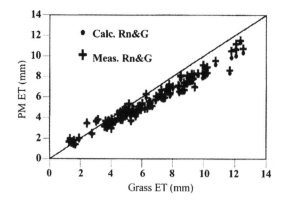

**FIGURE 5.25** Daily Penman-Monteith estimates of ET using both measured and estimated values of Rn and G were not significantly different from each other for well-watered, full-cover mixed fescue grown at Bushland, TX in 1996. Both PM ET values were less than values measured by a weighing lysimeter for values above 4 mm d⁻¹.

where $r_l$ is the stomatal resistance taken as 100 s/m, and the leaf area index (LAI) was taken as:

$$LAI = 5.5 + 1.5\ln(h_C)$$  [5.54]

where the crop height (hc) was taken as 0.12 cm for grass, and 0.5 m for alfalfa.

The zero plane displacement height (d) was calculated as:

$$d = \frac{2}{3}h_c$$  [5.55]

The roughness length for momentum, $z_{0m}$, was calculated as:

$$z_{0m} = 0.123h_C$$  [5.56]

and the roughness length for sensible heat transport was:

$$z_{0H} = 0.1z_{0m}$$  [5.57]

Net radiation was calculated as shown in Section 5.1.1.5. All calculations were on a half-hourly basis. For well-watered mixed fescue grass in 1996, the Penman-Monteith (PM) equation underestimated ET, as measured by a weighing lysimeter, at ET rates exceeding 4 mm per day (Figure 5.25), even though Rn and G were well-estimated. The underestimation of ET was due to systematic error in the surface and/or aerodynamic resistances. For well-watered, full-cover alfalfa in 1996, the PM estimates of ET were close to values measured with a weighing lysimeter (Figure 5.26). Because Rn and G were well-estimated, it is presumed that $r_a$ and $r_s$ were predicted well also. Examination of diurnal dynamics showed that the PM method was capable of closely reproducing those dynamics.

Although important as a research model, the PM method is not much used for direct prediction of LE due to the difficulty of knowing $r_a$ and $r_s$. However, it is commonly used to predict a theoretical reference evapotranspiration, $ET_r$, for use in irrigation scheduling (Allen et al., 1994ab). In this application, crop water use or ET is predicted from daily values of $ET_r$ and a dimensionless crop coefficient ($K_c$), which is dependent on the crop variety and time since planting or growing degree days:

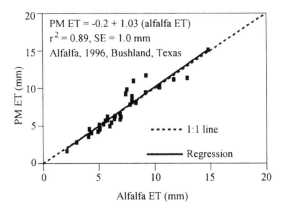

**FIGURE 5.26** Daily Penman-Monteith estimates of ET using estimated Rn and G for well-watered, full-cover (LAI>3) alfalfa at Bushland, TX in 1996.

$$ET = K_c ET_r \qquad [5.58]$$

The crop coefficients are determined from experiments that measure daily crop water use and $ET_r$ and compute:

$$K_c = ET/ET_r \qquad [5.59]$$

Many details on this methodology are found in Jensen et al. (1990).

---

**EXAMPLE 5.3** The Penman Approximation — Then and Now.

At the time that Penman (1948), Monteith (1965) and van Bavel (1966) developed their equations for LE, it was very difficult to measure surface temperature of water or plant canopies. All of these equations are called combination equations because they derive from the combination of the energy balance terms given in Equation [5.1] with heat and mass transfer mechanisms. The transfer mechanisms are usually stated as flux equations in terms of resistance(s) or conductance(s) and a gradient of temperature or vapor pressure between the surface and the atmosphere. Penman (1948) restated Equation [5.1] for a wet surface, and substituted transport mechanisms for LE and H to give a combination equation:

$$LE = f(u)\left(e_z - e_o^*\right) = -\left[Rn + G + \gamma f(u)\left(T_z - T_o\right)\right] \qquad [1]$$

where f(u) is a wind speed dependent conductance or transport coefficient, $\gamma$ is the psychrometric constant ($c_p P/(0.622 L)$), $e_o^*$ is the saturation vapor pressure at the surface temperature, $e_z$ is the air vapor pressure at measurement height z, $T_z$ is the air temperature at measurement height, $T_o$ is the surface temperature, and the transport mechanism for H is analogous to Equation [5.73]. The equation can use $e_o^*$ because it is assumed that the surface is "wet", i.e., a free water surface or the canopy of a crop that is full cover and well supplied with water and thus freely transpiring. Because surface temperature was difficult to measure, Penman (1948) introduced an approximation for $(T_o - T_z)$ that is derived from the slope of the saturation vapor pressure vs. temperature curve, which is (Diagram 5.3.1):

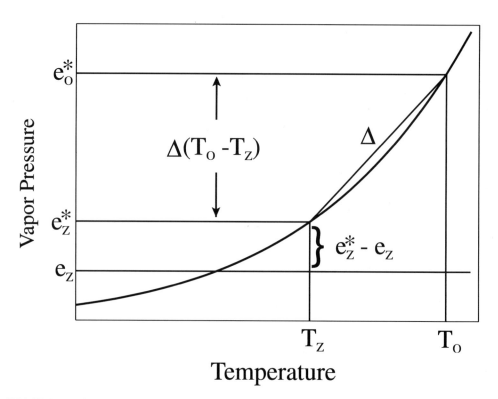

**DIAGRAM 5.3.1**   Quantities used in the derivation of Penman's (1948) combination equation.

$$\Delta = \left(e_o^* - e_z^*\right)\big/\left(T_o - T_z\right) \tag{2}$$

From Equation [2] and Diagram 5.3.1, we see that:

$$e_o^* - e_z = e_z^* - e_z + \Delta\left(T_o - T_z\right) \tag{3}$$

Rearranging to find $(T_z - T_o)$ and substituting into Equation [1] gives:

$$f(u)\left(e_z - e_o^*\right) = -\left[Rn + G + (\gamma/\Delta)f(u)\left(e_z - e_o^* + e_z^* - e_z\right)\right] \tag{4}$$

or

$$f(u)\left(e_z - e_o^*\right) = -\left[Rn + G + (\gamma/\Delta)f(u)\left(e_z - e_o^*\right) + (\gamma/\Delta)f(u)\left(e_z^* - e_z\right)\right] \tag{5}$$

both sides of which hold the identity for $LE = f(u)(e_z - e_o^*)$. Rearranging gives:

$$LE = -\left[\Delta(Rn + G) + \gamma f(u)\left(e_z^* - e_z\right)\right]\big/(\Delta + \gamma) \tag{6}$$

which is the Penman equation and is analogous to Equation [5.51]. Rather than use wind functions of the transport coefficient, Monteith recognized the importance of surface resistance and derived the equation in terms of the aerodynamic and surface resistances shown in Equation [5.51].

Although there is much evidence that the Penman equation and Equation [5.51] are useful for estimating LE from free water and from well-watered full-cover crops, the results have been disappointing for transfer of wind functions and $r_s$ and $r_a$ formulations between different regions and crops. Some facts about the underlying assumptions should lend insight. First, the value of $\Delta$ is usually evaluated at $T_z$. In humid climes, there may be little difference between canopy and air temperatures, particularly if skies are often overcast. But in more arid regions the canopy temperature of a freely transpiring crop may be several degrees cooler than air temperature, causing $\Delta$ to be overestimated — more so at the hottest part of the day than near sunrise and sunset. Second, the net radiation and soil heat fluxes are modified by the quotient $(1 + \gamma/\Delta)$. This is fundamentally incorrect. The value of $\gamma/\Delta$ ranges from 0.80 at 10°C to 0.17 at 40°C; so the modification of Rn and G is not small, and will change with air temperature over the course of a day. Third, the surrogate in the Penman equation for sensible heat flux is parameterized by vapor pressure terms defined at the same height. Thus, unlike real sensible heat flux, the quantity $\gamma f(u)(e_z^* - e_z)$ can never reverse sign (necessarily, $e_z^* \geq e_z$). Also, over the range from 10 to 40°C, the value of $\gamma$ varies from 65.5 to 67.5 Pa K$^{-1}$, while the value of $\Delta$ varies from 85 to 402 Pa K$^{-1}$. To the extent that $\gamma$ is constant, the division of the vapor pressure deficit by $(\Delta + \gamma)$ causes the surrogate term to vary as H would vary. To the extent that the sum $(\Delta + \gamma)$ varies from $\Delta$, and to the extent that $e_z^*$ differs from $e_o^*$, the surrogate differs in value from H.

In Equation [5.51], the sum of Rn and G is further modified by the surface and aerodynamic resistances to mass and heat transfer, even though those resistances have negligible effect on the fundamental mechanisms affecting either quantity. (The effect of $r_s$ and $r_a$ on canopy surface temperature has a negligible effect on Rn.) These facts have much to do with the difficulties encountered in determining appropriate values of $r_a$ and $r_s$, and in transferring these values from one region to another. Also, the fact that wind functions for the Penman equation have been determined to be different for different climates is certainly related to the approximations used in the derivation of the Penman and related equations.

---

**EXAMPLE 5.4**   Exercise Using the Penman-Monteith Equation

Take the heat capacity of air to be 1010 J kg$^{-1}$ K$^{-1}$, and air pressure to be $10^5$ Pa. Use Equation [5.45] to evaluate L for a range of air temperatures from –10 to 45°C, and evaluate $\gamma$ for the same range. Take the derivative of Equation [5.34] with respect to temperature and evaluate $\Delta$ and $\gamma/\Delta$ for the same range. Would you expect crop coefficients, developed using a Penman-Monteith equation for reference ET in a humid area, to be transferrable to an arid area? Is the wind function in Equation [6] equivalent to a transfer coefficient for sensible heat flux or is it for latent heat flux? Are the surface and aerodynamic resistances in Equation [5.51] for latent heat flux or for sensible heat flux, or both? There is much evidence that the transfer coefficients for sensible and latent heat fluxes are not identical, particularly under advective and highly stable conditions. How might this impact the interpretation of data from regions with widely different weather patterns?

At the time that the combination equations were being developed, the instrumentation for measuring net radiation was crude but, compared with that for measuring surface temperature, effective. In the 1960s, the development of infrared radiometers allowed the first radiometric measurements of surface temperature on a large scale, leading to much research on the use of surface temperatures to solve for H using forms of Equation [5.73], and thus to estimate LE from Equation [5.1] (Jackson, 1988). McNaughton (1988) pointed out problems with this method that persist to this day. They include the fact that radiometric surface temperature often differs from aerodynamic surface temperature (the surface temperature that works in Equation [5.73]), the difficulty of evaluating $r_a$ in Equation [5.73] for many surfaces (e.g., partial or mixed canopies), and the spatial heterogeneity of surfaces that leads to spatial heterogeneity of air temperature. However, the continued development of infrared thermometers has led to easy and reliable surface temperature measurement with solid state devices such as the thermocouple infrared thermometer (e.g., the model IRt/c from Exergen, Inc.). Meanwhile, the technology for net radiation measurements has improved, but calibration standards do not exist and it is still common to find differences of 10% or more between competing instruments. Just as the lack of adequate instrumentation for measurement of surface temperature affected the development of theory and practice since 1948, the shift in instrumentation capabilities should affect much of the experimental physics and development of theory and practice for calculation of surface energy and water balances in the next 50 years.

### 5.2.2.6  Bare Soil Evaporation Estimates

Fox (1968) and later Ben-Asher et al. (1983) and Evett et al. (1994) described an LE prediction method based on subtracting the energy balance equations (Equation [5.1]) written for a dry and a drying soil. Because LE is zero for a dry soil, this gives an expression for LE from the drying soil in terms of the other energy balance terms. The method requires a column of dry soil embedded in the field of drying soil; and measurements of the surface temperatures of the dry soil and of the drying field soil. The surface temperature difference between the dry and drying soils explains most of E, but prediction accuracy is only moderately good ($r^2 = 0.82$ for daily predictions, Evett et al., 1994). Evett et al. (1994) showed that the aerodynamic resistance over the dry soil surface was reduced, probably by buoyancy of air heated over the relatively hot surface, and that the resistance was relatively independent of wind speed. They also showed that consideration of the soil albedo change with drying could improve the E estimates. Although the method shows promise it does not provide an estimate of surface soil water content that would be needed to calculate the albedo change.

When the soil is wet, the evaporative flux can be estimated using the Penman or Penman-Monteith equations with surface resistance set to an appropriate low value (Howell et al., 1993). This wet period is the energy-limited stage of evaporation. As the soil dries, E becomes limited by soil properties. Van Bavel and Hillel (1976) addressed this using a finite difference model of soil water and heat flux that later was developed into the CONSERVB model of evaporation from bare soil. This model described one-dimensional soil water movement with Darcy's law, including the dependence of hydraulic conductivity, K (m s$^{-1}$), on soil water potential, h (m), and the soil water retention function, $\theta_v(h)$. The surface energy balance was solved implicitly for surface temperature (T), resulting in calculated values of E, H, Rn and G at each time step. The value of E was used as the upper boundary condition for soil water flux at the next time step. The elements of CONSERVB were included in the ENWATBAL model by Lascano et al. (1987), and the latter model was upgraded to model albedo changes dependent on surface soil water content by Evett and Lascano (1993). Although CONSERVB was not validated against directly measured E, the 1993 version of ENWATBAL was shown to more accurately predict E than either the Penman or Penman-Monteith equations (Howell et al., 1993).

### 5.2.3  Soil Heat Flux

Soil heat flux is discussed in detail in Chapter 9. Briefly, heat conduction in one dimension is described by a diffusion equation:

$$C \frac{\partial T}{\partial t} = \lambda \frac{\partial}{\partial z} \left[ \frac{\partial T}{\partial z} \right]$$

[5.60]

**TABLE 5.6**
**Thermal conductivity (λ) of some soil materials.**

| Soil | Dry $\theta_v$ | λ W m$^{-1}$ K$^{-1}$ | Wet $\theta_v$ | λ W m$^{-1}$ K$^{-1}$ | $\rho_b$ Mg m$^{-3}$ | Source |
|---|---|---|---|---|---|---|
| Fairbanks sand | 0.003 | 0.33 | 0.18 | 2.08 | 1.71 | 1 |
| quartz sand | 0.00 | 0.25 | 0.40 | 2.51 | 1.51 | 1 |
| sand | 0.02 | 0.9 | 0.38 | 2.25 | 1.60 | 2 |
| sand | 0.00 | 0.27 | 0.38 | 1.77 | 1.64 | 3 |
| sand | 0.003 | 0.32 | 0.38 | 2.84 | 1.66 | 4 |
| gravelly coarse sand (pumice) | 0.02 | 0.13 | 0.40 | 0.52 | 0.76 | 5 |
| medium and coarse gravel (pumice) | 0.01 | 0.09 | 0.43 | 0.39 | 0.44 | 5 |
| loamy sand | 0.01 | 0.25 | 0.40 | 1.59 | 1.69 | 6 |
| loam | 0.01 | 0.20 | 0.60 | 1.05 | 1.18 | 6 |
| Avondale loam | 0.08 | 0.46 | 0.23 | 0.88 | 1.35–1.45 | 7 |
| Avondale loam | 0.03 | 0.31 | 0.30 | 1.20 | 1.40 | 9 |
| silt loam | 0.09 | 0.40 | 0.50 | 1.0 | 1.25 | 2 |
| Yolo silt loam | 0.14 | 0.49 | 0.34 | 1.13 | 1.25 | 8 |
| Muir silty clay loam | 0.03 | 0.30 | 0.30 | 0.90 | 1.25 | 9 |
| silty clay loam | 0.01 | 0.20 | 0.59 | 1.09 | 1.16 | 6 |
| Pullman silty clay loam | 0.07 | 0.16 | 0.29 | 0.89 | 1.3 | 10 |
| Healy clay | 0.04 | 0.30 | 0.30 | 0.91 | 1.34 | 1 |
| Fairbanks peat | 0.03 | 0.06 | 0.61 | 0.37 | 0.34 | 1 |
| forest litter | 0.02 | 0.10 | 0.55 | 0.40 | 0.21 | 2 |

1: de Vries, 1963; 2: Riha et al., 1980; 3: Watts et al., 1990; 4: Howell and Tolk, 1990; 5: Cochran et al., 1967; 6: Sepaskhah and Boerma, 1979; 7: Kimball et al., 1976; 8: Wierenga et al., 1969; 9: Asrar and Kanemasu, 1983; 10: Evett, 1994.

where the volumetric heat capacity, C (J m$^{-3}$ K$^{-1}$), and the thermal conductivity, λ (J s$^{-1}$ m$^{-1}$ K$^{-1}$), are assumed constant in space, and vertical distance is denoted by z, time by t, and temperature by T.

The one-dimensional soil heat flux (G) for a homogeneous medium is described by:

$$G = -\lambda \frac{\partial T}{\partial z}$$ [5.61]

The thermal conductivity is a single-valued function of water content and is related to the thermal diffusivity, D$_T$ (m$^2$ s$^{-1}$), by:

$$\lambda = D_T C$$ [5.62]

where the volumetric heat capacity, C (J m$^{-3}$ K$^{-1}$), can be calculated with reasonable accuracy from the volumetric water content, $\theta_v$ (m$^3$ m$^{-3}$), and the soil bulk density, $\rho_b$ (Mg m$^{-3}$), by:

$$C = \frac{2.0 \times 10^6 \rho_b}{2.65} + 4.2 \times 10^6 \theta_v + 2.5 \times 10^6 f_o$$ [5.63]

for a soil with a volume fraction (f$_o$) of organic matter (Hillel, 1980).

Table 5.6 lists thermal conductivities at wet and dry points for several soils. For coarse soils the thermal conductivity vs. water content curve is S-shaped (Campbell et al., 1994), with a rapid rise at water contents corresponding to about 33 kPa soil water tension (about field capacity). For fine soils the relationship is more linear, and the thermal conductivity between dry and wet conditions in Table 5.6 can be linearly interpolated from the values given, with reasonably small errors. But for water contents below the dry value, the thermal conductivity should be taken as the value corresponding to the dry state.

---

**EXAMPLE 5.5**   Estimating Soil Thermal Conductivity — The Sine Wave Approach.

The complicated methods of measuring thermal diffusivity and conductivity mentioned here use computer programs and nonlinear regression fitting of multi-term sine series in order to handle diurnal temperature waves that differ from simple sinusoidal waves, as well as to incorporate measurements of soil water content that vary with depth. However, rough estimates of the thermal parameters may be made from phase differences and amplitude differences observed for temperatures measured at only two depths, with the application of simplifying assumptions of homogeneous water content and soil properties, and a simple sinusoidal diurnal temperature wave describing the temperature T at depth z and time t:

$$T(z,t) = \overline{T} + A_z \sin[\omega t + \phi(z)] \tag{1}$$

where $\overline{T}$ is the mean temperature (i.e., the mean of maximum and minimum for a sine wave), $A_z$ is the amplitude of the wave (i.e., the difference between maximum and minimum temperatures), $\omega$ is $2\pi/\tau$ where $\tau$ is the period (e.g., 24 h), and $\phi(z)$ is the phase angle at depth z (difference in time between the occurrence of the maxima or minima at depth 0 and depth z). Equation [5.60] can be solved using the above equation for T(z, t) yielding:

$$T(z,t) = \overline{T} + A_0 \sin[\omega t - z/d]/e^{z/d} \tag{2}$$

where $A_0$ is the amplitude at depth zero, and d is called the damping depth and is a function of the thermal conductivity and volumetric heat capacity:

$$d = (2\lambda/C\omega)^{1/2} \tag{3}$$

When z equals d, the amplitude is $1/e = 0.37$ of the amplitude at depth zero. Writing Equation [2] for two depths, $z_1$ and $z_2$, and solving for $\lambda/C$ gives:

$$\lambda/C = \frac{\pi(z_2 - z_1)^2}{\tau[2.3\log(A_1/A_2)]^2} \tag{4}$$

Diagram 5.5.1 shows measured diurnal temperature waves for several depths in a Pullman clay loam on two different days. Table 5.5.1 gives the mean water content as measured by time domain reflectometry for each depth. Also shown are the thermal conductivities obtained by fitting a sine series to each depth and evaluating the thermal conductivity of the layer between depths by using the coefficients of the sine series in a sine series solution to T(t) at the next depth and then performing a nonlinear fit of that solution to obtain the thermal conductivity.

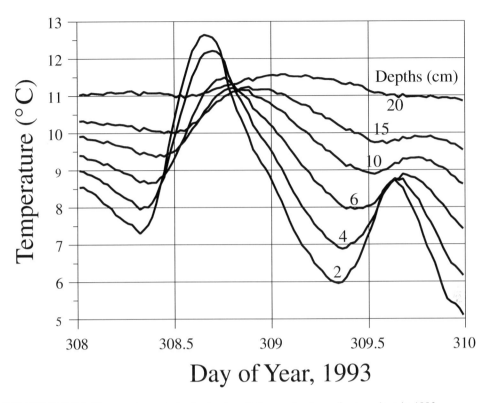

**DIAGRAM 5.5.1** Temperatures at six depths in a Pullman clay loam for two days in 1993

**TABLE 5.5.1**
**Mean soil water contents, θ (m³ m⁻³), and thermal conductivities, λ (J m⁻¹ s⁻¹ K⁻¹), calculated from a nonlinear fit of a sine series solution to Equation [5.60] for a pullman clay loam for two days in 1993.**

| Depth (cm) | $\theta$ (m³ m⁻³) | Amplitude based Equation [4] $\lambda_{\theta = \text{interlayer mean}}$ | Amplitude based Equation [4] $\lambda_{\theta = 0.23}$ | Nonlinear fit sine series soln. $\lambda_{\theta}$ |
|---|---|---|---|---|
| Day 308 | | | | |
| 2 | 0.09 | 0.176 | 0.232 | 0.138 |
| 4 | 0.14 | 0.518 | 0.577 | 0.485 |
| 6 | 0.22 | 0.740 | 0.717 | 0.709 |
| 10 | 0.27 | 1.128 | 1.003 | 1.221 |
| 15 | 0.31 | 1.295 | 1.151 | 1.271 |
| 20 | 0.27 | | | |
| Day 309 | | | | |
| 2 | 0.08 | 0.082 | 0.111 | 0.117 |
| 4 | 0.13 | 2.273 | 2.566 | 0.363 |
| 6 | 0.22 | 3.850 | 3.772 | 0.596 |
| 10 | 0.26 | 3.008 | 2.700 | 0.916 |
| 15 | 0.31 | 2.006 | 1.784 | 1.098 |
| 20 | 0.27 | | | |

**EXAMPLE 5.6**   Exercise for Estimating Soil Thermal Conductivity

Using Equation [4], measure the amplitudes of each vertically adjacent pair of diurnal waves in Diagram 5.5.1 and solve for the thermal conductivity of the layer between. Calculate C from Equation [5.63] assuming zero organic matter, a bulk density of 1.4 Mg m$^{-3}$, and using the mean water content for the layer. Repeat the calculations assuming that only the mean water content for the surface to 30-cm depth (0.23 m$^3$ m$^{-3}$) is known. Why are values for $\lambda$ different than those from the nonlinear fit? How often will the assumption of soil homogeneity that allows Equations [5.60] to be written be true? On these cool fall days, the mean soil temperature increased with depth; but on day 308, there was not much difference between the means at 2 and at 20 cm. However, on day 309, a cool front caused the surface temperature to decline sharply, and the mean temperature was several degrees cooler at 2 cm than at 20 cm. Which set of thermal conductivities in Table 5.5.1 are more likely to be correct, those for day 308 or those for day 309? Explain why the thermal conductivities estimated by nonlinear fit for day 309 are different than those for day 308 even though the water contents differ at most by 0.01 m$^3$ m$^{-3}$. Why is the estimate from Equation [4] for the 15–20 cm layer on day 308 in better agreement with the nonlinear fit estimate than any other?

De Vries (1963) developed a method of estimating soil thermal conductivity from soil texture, bulk density, and water content. The method, while including most important soil properties affecting conductivity, is limited in that it requires knowledge of parameters called shape factors that describe how the soil particles are packed together. The shape factors are specific to a given soil and perhaps pedon and must be measured. They are, in effect, fitting parameters (Kimball et al., 1976). De Vries' method tends to overestimate thermal conductivity at water contents above about 0.15 (Asrar and Kanemasu, 1983; Evett, 1994). Campbell et al. (1994) developed modifications of De Vries' theory that allowed them to match measured values well. They showed that, as temperature increased, the thermal conductivity versus water content curve assumed a pronounced S-shape for the 8 soils in their study, with the curve deviating from monotonicity at temperatures above 50°C.

Horton et al. (1983) developed a measurement method for D$_T$ based on harmonic analysis. The method entailed fitting a Fourier series to the diurnal soil temperature measured at 1-h intervals at 0.01-m depth followed by the prediction of temperatures at a depth, z (0.1 m), based on the Fourier series solution to the one-dimensional heat flux problem using an assumed value of D$_T$. The value of D$_T$ was changed in an iterative fashion until the best fit between predicted and measured temperatures at z was obtained. The best fit was considered to occur when a minimum in the sum of squared differences between predicted and measured temperatures was found (i.e., minimum sum of squared error, SSE). Poor fits with this and earlier methods are often due to the fact that field soils usually exhibit increasing water content with depth and changing water content with time while the method assumes a homogeneous soil. Costello and Braud (1989) used the same Fourier series solution and a nonlinear regression method, with diffusivity as a parameter to be fitted, for fitting the solution to temperatures measured at depths of 0.025, 0.15 and 0.3 m.

Neither Horton et al. (1983) nor Costello and Braud (1989) addressed the dependency of diffusivity on water content or differences in water content between the different depths. Other papers have dealt with thermal diffusivity in nonuniform soils but did not result in functional relationships between thermal properties and water content, probably due to a paucity of depth-dependent soil water content data (Nasser and Horton, 1989, 1990). Soil water content often changes quickly with depth, time and horizontal distance. Moreover, diffusivity is not a single-valued

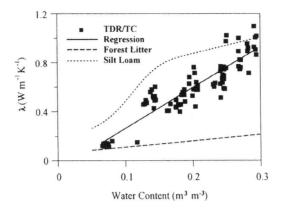

**FIGURE 5.27** Thermal conductivity of Pullman silty clay loam determined from TDR probe and thermocouple arrays compared to functions from Campbell (1985) for forest litter and silt loam.

function of soil water content and so is difficult to directly use in modeling. The ability of time domain reflectometry (TDR) to measure water contents in layers as thin as 0.02 m (Alsanabani, 1991; Baker and Lascano, 1989) provided the basis for design of a system that simultaneously measures water contents and temperatures at several depths. Evett (1994) used measurements of soil temperature at several depths (e.g., 2, 4, 6, 8… cm), coupled with TDR measurements of soil water content at the same depths, to find a relationship between thermal conductivity and water content in a field soil. He used the minimum SSE method of Horton to find the thermal diffusivity for each soil layer between vertically adjacent measurements of water content and temperature. The water content for this layer was used to calculate C and thus $\lambda$ corresponding to that water content. A function of $\lambda$ versus $\theta_v$ was developed by regression analysis on the $\lambda$ and $\theta_v$ data (Figure 5.27). Because both C and $\lambda$ were known for each layer this method also gave the soil heat flux.

Single-probe heat pulse methods have been developed to measure thermal diffusivity; a dual-probe heat pulse method (Campbell et al., 1991) can measure the thermal diffusivity ($D_T$) as well as $\lambda$ and C (Kluitenberg et al., 1995). Noborio et al. (1996) demonstrated a modified trifilar (three-rod) TDR probe that measured $\theta_v$ by TDR, and $\lambda$ by the dual-probe heat pulse method. Their measured $\lambda$ compared well with values calculated from De Vries (1963) theory.

Soil heat flux is commonly measured using heat flux plates (Table 5.1). These are thermopiles that measure the temperature gradient across the plate, and, knowing the conductivity of the plate, allow calculation of the heat flux from Equation [5.61]. Heat flux plates are impermeable and block water movement. Because of this, the plates should be installed a minimum of 5 cm below the soil surface so that the soil above the plate does not dry out or wet up appreciably more than the surrounding soil. Typical installation depths are 5 cm or 10 cm. Even at these shallow depths the heat flux is greatly reduced from its value at the soil surface, and corrections must be applied to compute surface heat flux. The most common correction involves measuring the temperature and water content of the soil at midlayer depths ($z_j$) in N layers (j to N) between the plate and surface, and applying the combination equation over some time period (P) defined by beginning and ending times $t_i$ and $t_{i+1}$:

$$G = G_z + \frac{\sum_{j=1}^{N} \left(T_{zj\_i+1} - T_{zj\_i}\right) \Delta_{zj} C_{zj}}{\left(t_{i+1} - t_i\right)}$$

[5.64]

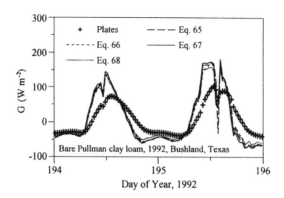

**FIGURE 5.28** Four methods of correcting heat flux, measured with plates at 5-cm depth, to surface heat flux.

where G is the surface heat flux during P, $G_z$ is the flux at depth z, the $T_{zj}$ are temperatures at the N depths ($z_j$) at times $t_i$ and $t_{i+1}$, $\Delta z_j$ is the depth of the layer with midpoint $z_j$, and where the volumetric heat capacities ($C_{zj}$) at depths $z_j$ are calculated from Equation [5.63], rewritten as:

$$C_{zj} = \frac{2.0 \times 10^6 \rho_{bzj}}{2.65} + 4.2 \times 10^6 \theta_{vzj} + 2.5 \times 10^6 f_{ozj} \qquad [5.65]$$

where $\theta_{vzj}$, $\rho_{bzj}$, and $f_{ozj}$ are the water contents, soil bulk densities, and volume fractions of organic matter, respectively, at depths $z_j$. The estimate of G is not much changed by the exact form of the combination equation as shown by data from Bushland, Texas, for four forms of Equation [5.64] (Figure 5.28). For situations where water content and temperature change rapidly with depth, or bulk density or $f_0$ change rapidly with depth, the multiple layer approach will work better.

The four methods of combining temperature and water content data to correct heat flux for bare soil data collected at Bushland, Texas, in 1992 (Figure 5.28) used the following measurements. Temperatures were measured at 2- and 4-cm depths (2 replicates) with thermocouples ($T_2$ and $T_4$), at the surface with a single infrared thermometer ($T_0$), and as a mean temperature of the surface to 5-cm depth soil layer using thermocouples wired in parallel and buried (4 replicates) at 1- and 4-cm depths ($T_{1\_4}$). Water contents were measured by TDR probes (2 replicates) inserted horizontally at 2- and 4-cm depths ($\theta_2$ and $\theta_4$). Soil heat flux at 5-cm depth ($G_5$), was measured with heat flux plates (4 replicates). For all methods the product of soil bulk density and heat capacity of soil solids was set to 1.125 MJ m$^{-3}$. For the first method the surface heat flux ($G_0$) was:

$$G_0 = G_5 + \frac{(1.125 + 4.2\theta_w) \times 10^6 (0.05)(T_{w+1} - T_w)}{1800} \qquad [5.66a]$$

where 1800 was the period in s, the weighted water content for the surface to 5-cm depth layer ($\theta_w$) was:

$$\theta_w = 3\theta_2/5 + 2\theta_4/5 \qquad [5.66b]$$

the weighted temperature for the surface to 5-cm depth layer ($T_w$) was:

$$T_w = 3T_2/5 + 2T_4/5 \qquad [5.66c]$$

and $T_{w+1}$ was the same, but for the previous measurement.

For the second method, $\theta_w$ and the series-wired thermocouple temperature were used:

$$G_0 = G_5 + \frac{\left(1.125 + 4.2\,\theta_W\right) \times 10^6 (0.05)\left(T_{1\times4+1} - T_{1\times4}\right)}{1800} \qquad [5.67]$$

For the third method, $\theta_w$ was used:

$$G_0 = G_5 + \frac{\left(1.125 + 4.2\theta_W\right) \times 10^6 (0.05)\left(T_{024+1} - T_{024}\right)}{1800} \qquad [5.68a]$$

but the depth weighted mean $(T_{024})$ of infrared thermometer temperature and those measured at 2 and 4 cm was used:

$$T_{024} = \frac{\left[\dfrac{\left(T_0 + T_2\right)/2 + T_0}{2}\right]}{5} + \frac{2T_2}{5} + \frac{2T_4}{5} \qquad [5.68b]$$

For the fourth method, a modified layer approach was used:

$$G_0 = G_5 + \frac{\left(1.125 + 4.2\theta_2\right) \times 10^6 (0.05)\left(T_{02+1} - T_{02}\right)}{1800} + \frac{\left(1.125 + 4.2\theta_4\right) \times 10^6 (0.05)\left(T_{4+1} - T_4\right)}{1800} \qquad [5.69a]$$

where the depth weighted mean temperature in the surface to 3-cm deep layer $(T_{02})$ was:

$$T_{02} = \frac{\left[\dfrac{\left(T_0 + T_2\right)/2 + T_0}{2}\right]}{3} + \frac{2T_2}{3} \qquad [5.69b]$$

All of these methods produced similar values of $G_0$, but those using a depth-weighted water content tended to overestimate extreme values, probably because the 2-cm water content was lower than that at 4 cm. The weighted mean approach for both water content and temperature, with surface temperature included (Equation [5.68]), produced generally the largest diurnal swing in $G_0$. Methods that did not include the surface temperature, but used the weighted mean approach for both water content and temperature (Equations [5.66] and [5.67]), produced intermediate results. The layer approach (Equation [5.69]), produced the smallest diurnal swing in $G_0$, despite using the surface temperature, and is probably the most accurate approach. All methods corrected both the amplitude and the phase of the diurnal cycle of $G_0$ appropriately.

Convective heat flux can play an important role in soil heating or cooling. This is the heat transported by moving air or water, the latter denoted in Figure 5.1 by $G_{Jw}$ for heat transported by infiltrating water. Because of the low heat capacity of air, the convective heat flux due to air movement is usually small, but convective heat flux due to infiltration of water can be much larger than that due to diffusion on a diurnal basis. For example, irrigation with 5 cm of water at 15°C on a soil at 25°C with an initial water content of 0.1 m$^3$ m$^{-3}$ and a bulk density of 1.48 would immediately lower the temperature of the 11.6-cm deep-wetted layer to 20°C (assuming negligible heat of wetting, the soil brought to saturation, and a heat capacity of 1.54 MJ m$^{-3}$ K$^{-1}$). The heat of wetting is usually not large enough to be important in heat balance calculations. It can be large for clays with large surface area if they are extremely dry, ranging from 40 J g$^{-1}$ for kaolinitic clays, to 125 J g$^{-1}$ for allophanic clays (Iwata et al., 1988). But it decreases quickly as the initial water content of the soil increases, and is not likely to be important for the normal range of field water contents.

**EXAMPLE 5.7**   Bernoulli, Soil Air, and Convective Heat Flux.

Convective heat flux due to air movement into and out of soil surfaces is commonly ignored in energy balance considerations. Most people do not see any reason for air to move into or out of soil other than changing atmospheric air pressure. However, there are other forces at play that may significantly increase air flow into or out of soil surfaces. Consider Bernoulli's theorem:

$$\frac{v_1^2}{2g} + \frac{p_1}{\rho g} + z_1 = \frac{v_2^2}{2g} + \frac{p_2}{\rho g} + z_2 \qquad [1]$$

which is an equation of conservation of energy where $v_1$ and $v_2$ are fluid velocities at two points, $p_1$ and $p_2$ are the respective fluid pressures, $\rho$ is the density of air and $g$ is the acceleration due to gravity, and $z_1$ and $z_2$ are the elevations of the two points. Taking air as the fluid and placing point 1 in the soil and point 2 in the atmosphere directly above the soil, we see that $v_1 \approx 0$, and $v_2 \geq 0$ so that the equation may be rewritten as:

$$p_1 - p_2 = \frac{\rho v_2^2}{2} + (z_2 - z_1)\rho g \qquad [2]$$

That is, the pressure differential from soil to air is equal only to the elevation difference multiplied by $\rho g$ when wind speed is zero, and increases above that value as the square of wind speed. During sustained winds across flat surfaces, there is a sustained pressure gradient driving air movement from soil to air. For air movement over nonplanar surfaces, the situation is complicated by other aerodynamic effects such as drag, which may increase or decrease the pressure gradient at various places over the surface.

**EXAMPLE 5.8**   Exercise Using Bernoulli's Equation, Soil Air Movement, and Convective Heat Flux.

Now, consider that soil air is usually at nearly 100% relative humidity. For a sustained wind of 10 m/s over a flat surface, calculate the total potential gradient for air movement from the soil to the atmosphere 2 m above the surface, and calculate the convective heat flux due to air movement out of the soil if the soil temperature is 35°C, the air temperature is 15°C, and its relative humidity is 25%. Assume that air is incompressible and that Darcy's law applies (Section 8.1.6.1), and estimate conductivity from Section 8.2.2 and a water saturation of 0.1 (Section 8.4.1). Is this magnitude of heat flux likely to continue for extended periods of time (explain why or why not)? How does it compare with the magnitudes of other energy fluxes described in this chapter? Could this mechanism enhance evaporative loss from the soil? What might happen if the surface were hilly? How would the energy flux change if the surface were covered with a 0.3-m tall wheat stubble? Would depth to the water table have an effect on the outcome if: 1) the depth were 1 m; 2) the depth were 100 m? Why do ice caves form in porous rock formations that are otherwise well above the freezing point of water?

## 5.2.4 Sensible Heat Flux (H)

Sensible heat flux is the transfer of heat away from or to the surface by conduction or convection. Because air is not a very good conductor of heat, most sensible heat flux is by convection (movement) of air. This occurs in eddies of different scales depending on the turbulence of the atmosphere near the surface. Turbulence is influenced by the aerodynamic roughness of the surface, the wind speed and the temperature differential between the surface and the air. Perhaps the most common method of evaluating sensible heat flux is to measure the other terms in Equation [5.1] as accurately as possible and then set H equal to the residual:

$$H = -Rn - G - LE \qquad [5.70]$$

Of course this approach lumps all the errors in the other terms into H. More importantly, it does not allow for a check on the accuracy of the energy balance. By definition, if H is defined by Equation [5.70], then Equation [5.1] will sum to zero. Only an independent measure of H can provide a check sum for Equation [5.1].

As noted in Section 5.2.2.2, eddy correlation is a direct method of measuring H:

$$H = \overline{\rho_a} c_p \overline{w'T'} \qquad [5.71]$$

where the overbars denote short time averages of air density ($\rho_a$), vertical wind speed ($w'$), and air temperature ($T'$), measured at some height within the constant flux layer.

The Bowen ratio method can be applied to sensible heat flux as well as to latent heat flux as outlined in Section 5.2.2.3. For sensible heat flux, the Bowen ratio is (Rosenberg et al., 1983):

$$H = \frac{-(Rn + G)}{\left(1 + \dfrac{1}{\beta}\right)} \qquad [5.72]$$

The considerations of fetch, measurement height, equipment, etc. mentioned in Section 5.1.2 for Bowen ratio and eddy correlation measurements apply as well to sensible heat flux measurements made with these methods.

Though obviously a dynamic and complex process, sensible heat flux, H (W m$^{-2}$), is sometimes estimated using a straightforward resistance equation:

$$H = \frac{\rho c_p (T_z - T_0)}{r_{aH}} \qquad [5.73]$$

where $\rho$ is the density of air ($\rho = 1.291 - 0.00418 T_a$, with less than 0.005 kg m$^{-3}$ error in the $-5$ to 40°C range, $T_a$ in °C), $c_p$ is the heat capacity of air ($1.013 \times 10^3$ J kg$^{-1}$ K$^{-1}$), $T_z$ is the air temperature at measurement height, $r_{aH}$ is the aerodynamic resistance to sensible heat flux (s m$^{-1}$), and $T_0$ is the temperature of the surface. [For vegetation, the surface for aerodynamic resistance is the height at which the logarithmic wind speed profile, established by measurements of wind speed above the surface, extrapolates to zero. This height is d + $z_{om}$ and is often well below the top of the canopy, typically at 2/3 to 3/4 h. Measurements of surface temperature (with, for instance, an infrared thermometer) may not be the mean temperature at the same height as the aerodynamic surface, thus causing some problems with $r_{aH}$ estimation. Also, the roughness length for momentum ($z_{om}$) may be different from that for sensible heat, $z_{oH}$.]

A general form for $r_a$ is:

$$r_a = \frac{1}{k^2 u_z} \left[ \ln \left( \frac{z-d}{z_o} \right) \right]^2 \qquad [5.74]$$

where k is the von Kármán constant = 0.41, $z_o$ is the roughness length (m), z is the reference measurement height (m), $u_z$ is the wind speed (m/s) at that height and d is the zero plane displacement height (m). Equation [5.74] only holds for neutral stability conditions. Unstable conditions occur when the temperature (and thus air density) gradient from the surface upward is such that there is warm air rising through the atmosphere. Stable conditions prevail when the air is much cooler and denser near the surface, thus inhibiting turbulent mixing. Neutral conditions obtain when neither stable nor unstable conditions do.

For bare soil Kreith and Sellers (1975) simplified Equation [5.74] to:

$$r_a = \frac{1}{k^2 u_z} \left[ \ln \left( \frac{z}{z_o} \right) \right]^2 \qquad [5.75]$$

where $u_z$ is the wind speed (m/s) at the reference height (z) (m). They found a value of $z_o = 0.003$ m worked well for smooth bare soil.

For non-neutral conditions, a variety of stability corrections have been proposed (Rosenberg et al., 1983; Monteith and Unsworth, 1990). Because many models of the soil-plant-atmosphere continuum use Equation [5.73] to model H, it is important to note that, while stability corrections can improve model predictions of H and surface temperature, the stability corrections are implicit in terms of H. This leads to a requirement for iterative solution of sensible heat flux at each time step in these models.

Knowledge of appropriate values for d and $z_0$ in the above equations can be hard to come by. Campbell (1977) suggests estimating these from plant height (h) as:

$$d = 0.64 \ h \qquad [5.76]$$

for densely planted agricultural crops; and:

$$z_{0m} = 0.13 \ h \qquad [5.77]$$

for the roughness length for momentum for the same condition. Campbell (1977) gives the roughness length parameters for sensible heat ($z_{0H}$) and vapor transport ($z_{0v}$) as:

$$z_{0H} = z_{0v} = 0.2 z_{om} \qquad [5.78]$$

Note that Equation [5.78] differs from Equation [5.57] where Jensen et al. (1990) used $z_{0H} = 0.1$ $z_{0m}$. For coniferous forest, Jones (1992) gives:

$$d = 0.78 \ h \qquad [5.79]$$

and:

$$z_{0m} = 0.075 \ h \qquad [5.80]$$

for these parameters. As wind speeds increase many plants change form and height, with resulting decrease in h, d and $z_{0m}$. It is unlikely that the relationships given in Equations [5.76-5.80] hold true for high wind speeds.

## 5.3  WATER BALANCE EQUATION

The water balance is written for a control volume of unit surface area, and with a vertical dimension that extends from the soil surface to a lower boundary that is usually assigned a depth at or below the bottom of the root zone (Figure 5.1):

$$0 = \Delta S - P + R - F - E \qquad [5.81]$$

where $\Delta S$ is the change in soil water storage in the profile, P is precipitation or irrigation, R is the sum of runoff and runon, F is flux across the lower boundary of the profile, and E is water lost to the atmosphere through evaporation from the soil or plant or gained by dew formation. The value of P is always positive or zero, but values of $\Delta S$, R, F, and E may have either sign. By convention, R is taken to be positive when there is more runoff than runon, and E is often taken to be positive when flux is out of the control volume. Here, in order to be compatible with the energy balance equation, E is positive toward the surface of the soil. The equation is often rearranged to provide values of E when suitable measurements or estimates of the other terms are available, but it can and has been used to estimate runoff, soil water available for plants, and deep percolation losses (flux downward out of the profile). Here, F is positive when flux is upward across the lower boundary into the control volume. The term F is used rather than P for deep percolation, both to avoid confusion with precipitation, and to avoid the common misconception that flux is only downward when P is used to indicate deep percolation.

Usually, the values of terms in Equation [5.81] are given as equivalent depths of water per unit area (e.g., mm/m²). In the case of E, the units of kg m$^{-2}$ may be conveniently converted to mm by dividing by 1 kg m$^{-2}$ mm$^{-1}$, with little loss of accuracy because the density of water in units of kg L$^{-1}$ is not quite unity (one liter = 1000 cm³ = the volume of a right rectangular prism with sides 1 m, 1 m, and 1 mm). The change in storage ($\Delta S$) is often determined by measuring soil water content changes by methods that give volumetric water content, $\theta_v$ (m³ m$^{-3}$). Multiplying the water content by the depth of the layer gives the depth of water stored. In the United States, the term evapotranspiration (ET) is used to represent the sum of evaporative fluxes from the soil and plant. By convention, ET is positive for fluxes from plant or soil surface to the atmosphere. Thus, ET = $-E/(1$ kg m$^{-2}$ mm$^{-1})$ and the water balance may be rearranged as:

$$\Delta S = P - R + F - ET \qquad [5.82]$$

This provides a use for the ET term for those who prefer to say evaporation rather than evapotranspiration. Examination of Equation [5.82] will satisfy the reader that soil water storage increases with precipitation, decreases if runoff from precipitation occurs, decreases with increasing ET, and increases with flux upward into the control volume.

---

**EXAMPLE 5.9**  Surface Storage — The Human Contribution.

Surface storage is ignored in the discussion of soil water balance here, but it not only has a large influence on the amount of water infiltrated from a precipitation or irrigation event, but it is one of two factors that are the most amenable to human control both on a small and large scale. The other is infiltration rate, a soil property that may be increased by soil tillage over the short term, and decreased by tillage in the long term, and also may be influenced by no tillage or minimum tillage practices, by tree plantings and ground covers designed to protect the soil surface from slaking and crusting, and many other practices. Many farming efforts are aimed at increasing surface storage through practices like plowing to roughen the soil surface, or furrow diking to create infiltration basins and decrease runoff. Soil conservation measures such as terraces, contour bunds, etc. are aimed at decreasing runoff or runoff velocity, and thus increasing the opportunity time for infiltration. On the other hand, water harvesting is a practice of increasing runoff by reducing both surface storage and infiltration rate, with the aim of using the runoff water elsewhere. Among the oldest artifacts of human cultivation are water harvesting systems that included the removal of gravel from desert surfaces to improve runoff, terrace systems for guiding and reducing runoff, and in some cases the combined use of both technologies to concentrate water on an area chosen for cultivation. To the extent that such practices influence infiltration, they will affect the timing, magnitude and spatial patterns of such energy balance terms as sensible and latent heat fluxes, and the convective heat flux accompanying infiltrating water.

### 5.3.1 MEASURING ΔS AND ET

Probably the most accurate method of measuring ΔS is the weighing lysimeter (Wright, 1991). Although large weighing lysimeters involve considerable expense, they can give very precise measurements ($0.05$ mm = $0.05$ kg m$^{-2}$) (Howell et al., 1995). An excellent review of the use of weighing lysimeters is given by Howell et al. (1991). Careful design, installation and operation will overcome any of the serious problems reported with some lysimeters including disturbance of the soil profile (less with monolithic lysimeters), interruption of deep percolation and horizontal flow components, uneven management of lysimeter compared with field soil (Grebet and Cuenca, 1991), and other sources of bias (Ritchie et al., 1996). Other drawbacks include heat flux distortions caused by highly conductive steel walls (Black et al., 1968; Dugas and Bland, 1991) (minimal for large lysimeters) and high cost, e.g., U.S. $ 65,000 (Lourence and Moore, 1991) and U.S. $ 80,000 (Marek et al., 1988).

Schneider et al. (1996) described simplified monolithic weighing lysimeters (Figure 5.29) that were considerably less expensive than, and nearly as accurate as, the monolithic weighing lysimeters described by Marek et al. (1988) (Figure 5.30). These two designs represent contrasts in mode of operation and each presents some advantages and disadvantages. The design in Figure 5.30 allows access to all sides and the bottom of the lysimeter for installation or repair of sensors and weighing or drainage systems. The Campbell Scientific, Inc. CR7 data logger that handles all measurements is installed in the underground chamber and typically is subject to only a small diurnal temperature swing of 1°C, reducing temperature-induced errors in low-level measurements such as load cell transducer bridges and thermocouples. Other equipment installed in the chamber includes a system for time domain reflectometry measurements of soil water content and concurrent measurements of soil temperature, and an automatic vacuum drainage system that continuously monitors drainage rate. The drainage tanks are suspended by load cells from the bottom of the lysimeter tank, allowing measurement of tank mass change without changing the mass of the lysimeter until the tanks are drained (manual but infrequent). The main disadvantages of this design are the shallow soil depth over the ceiling around the periphery of the chamber, and the surface area taken by the entrance hatch. The shallow soil depth can cause uneven plant growth next to the lysimeter, but this problem has been eliminated with the installation of a drip irrigation system to apply additional water to this area. The soil disturbed to install the outer chamber wall appears to have returned to a condition similar to the rest of the field. Access to the lysimeter must be carefully managed to avoid damage to the crop.

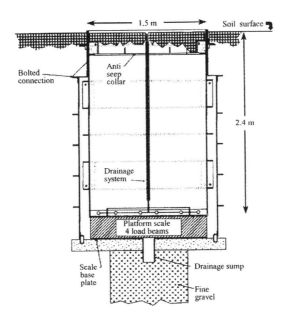

**FIGURE 5.29** Cross-sectional view of the simplified weighing lysimeter installed for grass reference ET measurements at Bushland, TX (Schneider, 1998a).

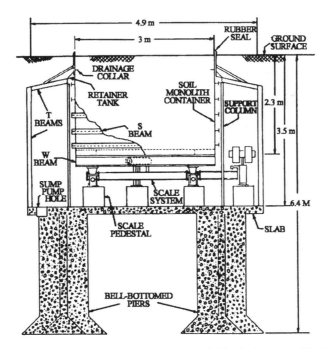

**FIGURE 5.30** Cross-sectional view of one of the four large weighing lysimeters at Bushland, TX (Schneider, 1998b).

Figure 5.29 shows a design that minimizes disturbance to the field during both installation and operation. The monolith was collected a short distance away, and the outer box was installed in a square hole that disturbed only a 15-cm perimeter of soil outside the lysimeter. Because there is no access to the sides or bottom of the lysimeter, there is no reason for personnel to visit the lysimeter area except for crop management and the occasional manual drainage accomplished with

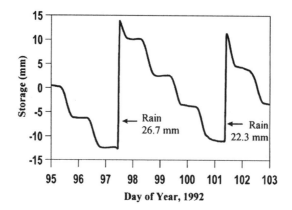

**FIGURE 5.31** Unadjusted weighing lysimeter storage for winter wheat at Bushland, TX.

a vacuum pump and collection bottle. A disadvantage of this design is that continuous drainage rates are not available. The CR7 data logger is located 30 m away in a weather-tight enclosure, and all cables are buried. The location of the CR7 inside the lysimeter chamber in Figure 5.30 allows a four wire bridge to be used for reading the weighing system load cell. The long cable lengths to the external CR7 used with the lysimeter in Figure 5.29, and the large diurnal temperature swing to which cables and CR7 are exposed, both cause a six wire bridge to be needed to eliminate errors due to temperature induced resistance changes when reading the platform scale load cells. Measurement precision with the lever beam scale in Figure 5.30 is 0.05 mm while that with the platform scale in Figure 5.29 is 0.1 mm.

Weighing lysimeters measure mass change over a given time ($\Delta M$). If mass is measured in kg, then dividing the mass change by the surface area in $m^2$ of the lysimeter will give the change in water storage ($\Delta S$), of the lysimeter as an equivalent depth of water in mm, with only slight inaccuracy due to the density of water not being quite equal to 1 kg $L^{-1}$. If only daily ET values are needed, then $\Delta S$ is computed from the 24 h change in lysimeter mass, usually midnight to midnight. Some averaging of readings around midnight may be needed to smooth out noise. By adding any precipitation or drainage, the daily ET is computed. Data from a continuously weighing lysimeter may be presented as a time sequence of mass (or depth of water storage) referenced to an arbitrary zero (Figure 5.31). Often irrigation or precipitation events will show as obvious increases in storage (Figure 5.31), and drainage events will show up as decreases in storage. Adjusting the sequential record of storage amount by adding the rainfall or irrigation depth, or the drainage depth, at the time that these occurred, will remove these changes in storage, and is equivalent to the operations defined by the +P and +F in Equation [5.82], resulting in the monotonically decreasing storage shown in Figure 5.32. Taking the first derivative of the adjusted storage with respect to time gives the adjusted $\Delta S$ rate, and thus ET rate, if R and F are zero (Figure 5.33). In order to compute ET rates on the same time interval as lysimeter mass measurements are made, concurrent measurements of irrigation, precipitation and drainage on the same or a finer recording interval are necessary.

Weighing lysimeters are subject to wind loading, more so when the soil surface is bare, as evidenced by Figure 5.34. In windy regions it may be necessary to smooth the data to remove noise when calculating the ET rate. Gorry (1990), following Savitsky and Golay (1964), described a method for general least squares smoothing that allowed application of different levels of smoothing to both raw data and first derivative. Application of this method to post-processing of data is preferable to real-time smoothing that may eliminate detail in the data. With post-processing we can apply only the amount of smoothing needed to reduce noise to acceptable levels. A computer program to apply Savitsky-Golay smoothing is available (*http://www.cprl.ars.usda.gov/programs/*).

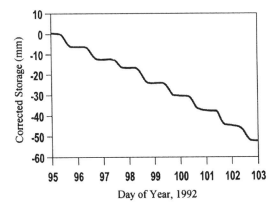

**FIGURE 5.32** Lysimeter storage from Figure 5.31 adjusted by subtracting precipitation amounts.

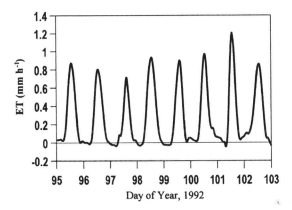

**FIGURE 5.33** Evapotranspiration rate calculated by taking the negative of the first derivative of adjusted storage from Figure 5.32. The negative ET rates shown for some nights are caused by dew formation.

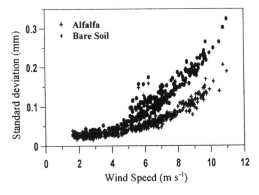

**FIGURE 5.34** Half-hourly standard deviation of lysimeter storage (mm) as affected by wind speed over contrasting surfaces for days 97-105 at Bushland, TX in 1994.

Microlysimeters are small enough to be installed and removed by hand for weighing daily or more often. They can give good precision but are sensitive to spatial variability. Lascano and Hatfield (1992) showed that 182 microlysimeters were required to measure field average E with precision of 0.1 mm d$^{-1}$ at a 90% confidence level when their soil was wet, but only 39 when dry. This was due to the greater variability of E for wet soil. For a precision of 0.5 mm d$^{-1}$ only 7 microlysimeters were required for any soil wetness. To avoid heat conduction to and from the surface, microlysimeter walls should be made of low thermal conductivity materials such as plastic, and to avoid trapping heat at the bottom of the microlysimeter, the bottom end cap should be made of a thermally conductive material such as metal (Evett et al., 1995). Plastic pipe makes good microlysimeters. Typical dimensions are 7.6 or 10 cm in diameter, and from 10 cm to 40 cm high. Beveling the bottom end eases insertion into the soil. Typical practice is to insert the microlysimeter vertically until its top is level with the soil surface, then dig it out, or rotate it to shear the soil at the bottom, and pull it out. After capping the bottom with a watertight seal, it is weighed before reinsertion into the original or a new hole, sometimes lined with a material (e.g., plastic sheet or bag) to prevent sticking of the soil to the microlysimeter outside surface. After a period of time the microlysimeter is reweighed and the difference in initial and final weights is the evaporative loss. Short microlysimeters should be replaced daily, as the water supply is soon used up to the point that the soil inside the lysimeter is no longer at the same water content as the soil outside. In a study of spatial variability of evaporation from bare soil, Evett et al. (1995) used 30-cm high microlysimeters to avoid daily replacement so that the spatial relationship would not be changed. They showed that for their clay loam soil the 30-cm height was adequate for 9 days. If plant roots are present it is recommended to replace microlysimeters daily to lessen errors associated with root water uptake that occurs elsewhere in the field but not in the microlysimeters.

Alternatives to weighing lysimetry include soil water measurement methods for assessing $\Delta S$ for a soil profile of given depth over a given time. In this case the soil volume of interest is unbounded below the surface and F is, strictly speaking, uncontrolled. Measurements of soil water content can give the change in soil water stored in a profile of given depth with good accuracy, and can give good values for E if water flux across the bottom of the profile is known or can be closely estimated. Baker and Spaans (1994) described a microlysimeter with a TDR probe installed vertically to measure the water content. Comparison of E calculated from the change in storage closely matched the E measured by weighing the microlysimeter. Young et al. (1997) showed that a single 800-mm long probe installed vertically from the surface could account for 96% of ET from weighing lysimeters irrigated on a 6-d interval, but standard error for the probe was about 4 times larger than that for the lysimeter (0.46 and 0.07 mm, respectively). In a container study with a sorghum plant, Wraith and Baker (1991) showed that a TDR system could measure ET with high resolution and provide measurements of change in storage on a 15-min interval that compared very well with those measured by an electronic scale.

---

**EXAMPLE 5.10**   Neutron Scattering — Sword to Plowshare.

At the end of World War II, the understanding of nuclear physics had increased tremendously, particularly the interactions of neutrons with atoms, which knowledge was essential to the development of the atomic bomb and fission reactors. There was a concerted effort to turn this knowledge to productive and peaceful uses, most of it funded by the U.S. government. In 1950, the Civil Aeronautics Administration published Technical Report 127, describing a method for measuring soil moisture based on neutron scattering (Belcher et al., 1950). Independently, Wilford Gardner and Don Kirkham developed essentially the same method, which was published in Soil Science (Gardner and Kirkham, 1952). Oddly, the invention was probably aided by a lack of technology; specifically the difficulty of building a detector of fast (~5 million electron volts, MeV) neutrons, and the relative ease of building an efficient detector of thermal neutrons (~0.025 eV). Early detectors were often based on 3He, which has a large cross-section for the (n,p) interaction (Table 5.10.1) in which an alpha particle (+1 charge) is ejected and detected electronically. The method is based on two facts: i) of the elements common in soils, hydrogen is by far the most effective in converting fast neutrons to thermal neutrons through collisions, and ii) of the hydrogen bearing soil constituents, water is usually the most plentiful and the only one that changes rapidly to an important extent. The slowing of neutrons by a particular atomic nucleus is affected both by the mass number (A) of the nucleus and the nuclear cross section ($\sigma_a$). For all scattering angles, Gardner and Kirkham gave the mean ratio ($\varepsilon'$), of the energy of the neutron after a collision to its energy before as:

$$\varepsilon' = 1 - \frac{A-1}{2A} \log_e \frac{A+1}{A-1}$$

Using this equation, calculate values of $\varepsilon'$ for the elements in Table 5.10.1. Note that the nuclear cross section for interaction with a neutron varies in size depending on the speed of the neutron. The larger the cross section, the larger the probability that a neutron will interact with an atom.

**TABLE 5.10.1**
**Values of atomic mass number (A) energy loss ratio ($\varepsilon'$), and neutron scattering cross section ($\sigma_a$) (1 barn = $10^{-24}$ cm$^2$). Adapted from Gardner and Kirkham (1952).**

| | | | $\sigma_a$ (barns) | |
| --- | --- | --- | --- | --- |
| Element | A | $\varepsilon'$ | Fast neutrons 5 MeV | Thermal neutrons 0.025 eV |
| H | 1 | | 2.55 | 47.5 |
| $^3$He | 3 | | 2 (0.37)† | 4 (5400) |
| C | 12 | | 1.60 | 4.6 |
| N | 14 | | 1.0 | 18.0 |
| O | 16 | | 1.5 | 4.2 |
| Cl | 17 | | 2.7 | 40.0 |
| Na | 23 | | 2.6 | 3.6 |
| Mg | 24 | | 2.0 | 3.5 |
| Al | 27 | | 2.5 | 1.6 |
| Si | 28 | | 3.2 | 2.5 |
| P | 31 | | 3.0 | 4.0 |
| S | 32 | | 2.6 | 1.3 |
| K | 39 | | 3.8 | 3.0 |
| Ca | 40 | | 4.9 | 1.5 |
| Mn | 55 | | 3.0 | 12.0 |
| Fe | 56 | | 13.0 | 3.0 |

† Data for $^3$He are from the Evaluated Nuclear Data Files W3 Retrieval System of Brookhaven National Laboratory, Upton, NY 11973-5000 (*http://www.nndc.bnl.gov/nndc/endf/* Verified 15 May 2000). Numbers in parentheses are for the (n,p) interaction in which a neutron is captured and an alpha particle ejected. This is the reaction used in $^3$He neutron detector tubes.

**EXAMPLE 5.11**   Exercise on Neutron Facts of Life.

Given this information, discuss why the neutron scattering method is sensitive to the hydrogen content of soil, but relatively insensitive to other common elements. Discuss why application of large amounts of nitrogenous fertilizer might interfere with neutron probe measurements. Also, discuss why soils with large amounts of chloride salts would require different calibrations for water content vs. count of slow neutrons, and discuss why the use of polyvinylchloride (PVC) plastic for access tubing causes the precision of neutron probe calibrations to decline relative to the precision obtained with aluminum or steel tubing (Dickey et al., 1993). Is it reasonable that soil layers containing large amounts of $CaCO_3$ or $CaSO_4$ would require different calibrations (they do)?

It would be difficult to overestimate the importance of the neutron scattering method in soil science and hydrological research and development over the last fifty years. It was the first useful, nondestructive method of repeatedly sampling the moisture content of soil profiles throughout and below the root zone. It led to the widespread measurement of crop water use values that are essential to irrigation management and the planning of large scale irrigation developments. After fifty years, a panel of scientists, expert in soil water measurement using time domain reflectometry (TDR), capacitance, and neutron scattering methods, convened by the International Atomic Energy Agency, recommended that the neutron scattering method not be replaced in the agency's research and training programs (IAEA, 2000). Three reasons were given: i) the method measures a relatively large volume of soil compared with TDR and capacitance instruments and so integrates across small-scale variability of soil properties and reduces the number of measurements needed, as well as reducing the sensitivity of the method to soil disturbance caused by installation, ii) the method is reliable and easy to use compared with others, and iii) the technology is mature, which brings to bear a large knowledge base of proven solutions to particular problems of use. Also, the large volume of measurement makes field calibration much easier than it is for TDR and capacitance probes.

Evett et al. (1993) showed that change in storage in the upper 35 cm of the profile under winter wheat could be accurately tracked with horizontally placed TDR probes, with an average of 88% of daily $\Delta S$ occurring in the upper 30 cm. But E estimates were incorrect (compared with a weighing lysimeter) when flux across the 30 cm boundary occurred. However, combination of the TDR system with neutron probe measurements of deeper soil moisture allowed measurement of E to within 0.7 mm of lysimeter-measured E over a 16-d period; five times better than the accuracy achieved using only neutron probe measurements.

Figure 5.35 shows the soil water storage (referenced to arbitrary zero) as measured for winter wheat by weighing lysimeter and two TDR arrays. Each TDR array consisted of seven probes inserted horizontally into the side of a pit and the pit backfilled after wheat planting. Probe depths were 2, 4, 6, 10, 15, 20 and 30 cm, and the probes were read every half hour. Rains on days 101, 104 and 106 can be seen as increases in the storage amount. Changes in storage as measured by the two systems were nearly identical in the seven day period shown (Figure 5.36), and ET amounts were closely similar (Figure 5.37).

Water balance measurement intervals commonly range between hours and weeks and are usually no smaller than the required period of LE measurement. Measurement of each variable in Equation [5.81] presents its own unique problems. These include measurement errors in determination of lysimeter mass ($\Delta S$), and errors in P and R measurement. Problems of P and R measurement are essentially identical for either weighing lysimetry or soil profile water content methods, because the surface area of the control volume can be defined for both methods with a watertight border, often consisting of a sheet metal square or rectangle pressed into or partially buried in the

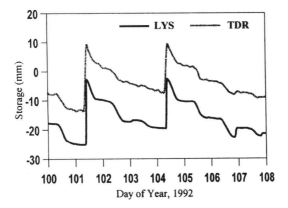

**FIGURE 5.35** Soil water storage in the upper 35 cm of the soil profile as measured by two TDR probe arrays compared to storage in a 2.4-m deep weighing lysimeter. Zero reference is arbitrary. Winter wheat, Bushland, TX, 1992.

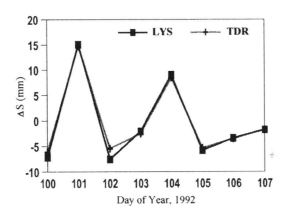

**FIGURE 5.36** Daily change in storage for the 35-cm and 2.4-m profiles from Figure 5.35.

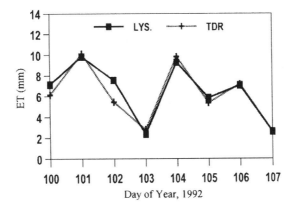

**FIGURE 5.37** Daily evapotranspiration as calculated from the TDR and weighing lysimeter data shown in Figures 5.35 and 5.36.

soil surface. When the soil volume is unbounded below the surface, as in the soil profile water content method, there are additional errors due to uncontrolled horizontal flow components and deep percolation that are difficult to measure or estimate. Nevertheless, the profile water balance technique is applicable in many situations for which lysimetry is inappropriate or impossible and is, in addition, much less expensive. In many cases the horizontal flow components may be assumed to sum to zero, and deep percolation may be nil if the soil profile water content measurements are made to sufficient depth (Wright, 1990).

When neutron scattering alone is used to measure soil water content, the soil water balance method is suitable for periods of several days or more if closure (F = 0) at the bottom of the measured soil profile can be obtained (Wright, 1990). Neutron scattering (NS) is the most common water content measurement technique used (Cuenca, 1988; Wright, 1990) but due to the small changes in water content associated with daily ET and the limited precision of NS near the soil surface this water balance method has usually been restricted to measurement of ET over several day periods (Carrijo and Cuenca, 1992). Evett et al. (1993) showed that time domain reflectometry (TDR) measurements of soil water content near the surface could be coupled with deeper water content measurements by NS to close the water balance considerably more precisely than NS alone, opening up the prospect for daily ET measurements by this method.

## 5.3.2 ESTIMATING FLUX ACROSS THE LOWER BOUNDARY

One of the great advantages of lysimeters is that they control the soil water flux (F) into and out of the control volume. To date, a reliable soil water flux meter has not been developed, so F must be estimated if it is not controlled. If water flux across the lower boundary of the control volume is vertical, it may sometimes be estimated by measurements (preferably multiple) of soil water potential (h) at different depths separated by distance ($\Delta z$) and knowledge of the dependence of hydraulic conductivity, K (m s$^{-1}$), on soil water potential [K(h) curve]. The potential difference ($\Delta h$) coupled with the unit hydraulic gradient for vertical flux, gives the hydraulic head difference ($\Delta H$), driving soil water flux. Averaging the measurements allows estimation of the mean hydraulic conductivity for the soil layer between the measurements from the K(h) curve, and thus estimation of the soil water flux, $J_w$ (m s$^{-1}$), from a finite difference form of Darcy's law:

$$J_w = -K\left(\frac{\Delta H}{\Delta z}\right)$$

[5.83]

---

**EXAMPLE 5.12**   Exercise on Deep Flux Loss or Gain in Soil Water Balance Calculation.

Diagram 5.12.1 shows the profile water content measured for a Pullman clay loam. Assume that the parameters in van Genuchten's equation (Equation [3.31]) are $\theta_s$ = 0.424 m$^3$ m$^{-3}$, $\theta_r$ = 0.01 m$^3$ m$^{-3}$, $\alpha$ = 0.88 m$^{-1}$, n = 1.15, and m = 1 – 1/n. For the 2.10- and 2.30-m depths, use the water contents shown and calculate $\psi_m$ for each depth by inverting Equation [3.31]. Use the same parameter values and water contents in Mualem's equation (Equation [4.62]) and calculate the hydraulic conductivities at each depth. Use Equation [5.83] to calculate the flux between the depths (use the mean hydraulic conductivity). If these conditions persisted for a week during which evaporation averaged 10 mm d$^{-1}$, what would be the absolute error in water content calculated using Equation [5.81] and ignoring flux across the lower boundary? What would the error be as a percentage of the weekly total E? Assume that measurements were taken to only 1.5 m and repeat the calculations using water contents from the 1.3- and 1.5-m depths. What are the errors in E (absolute and percentage), and are the errors larger or smaller? Why? How accurate are the calculations of F? When using the water balance method for E of furrow-irrigated alfalfa, Wright (1990) measured to 3-m depth with the

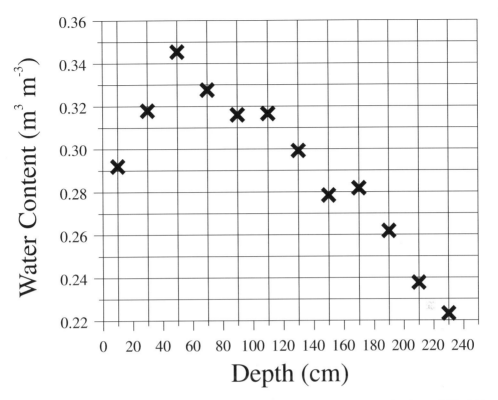

**DIAGRAM 5.12.1**  Soil water content of a Pullman soil measured by neutron scattering in an alfalfa field.

neutron probe, yet he found that flux across the lower boundary or water extraction by plant roots deeper than 3 m caused E estimates to be lower than those from a weighing lysimeter. For sprinkler-irrigated alfalfa, he found that water balance and lysimeter measurements agreed. Why did this happen?

Soil water potential may be measured by tensiometer or other means described in Chapter 3 (van Genuchten et al., 1991). Methods of measuring or estimating the K(h) curve may be found in Chapter 4. For fluxes across boundaries too deep for the installation of tensiometers the soil water content may be measured at two or more depths by neutron scattering (Chapter 3) and the soil water potential inferred by inverting the θ(h) relationship, which may be estimated or measured (Chapter 3 or van Genuchten et al., 1991). Due to the hysteresis of the θ(h) relationship there is more room for error when basing $J_w$ estimates on θ measurements. But for many cases, the soil water potential will be in the range where hysteresis is not a large source of error (drier soils), and hydraulic conductivity is not large either. Thus, both the value of $J_w$ and the error in $J_w$ may be small enough for practical use.

### 5.3.3  PRECIPITATION AND RUNOFF

An in-depth discussion of precipitation and runoff measurement and modeling is beyond the scope of this chapter. A classic and still valuable reference on field hydrologic measurements is the Field Manual for Research in Agricultural Hydrology (Brakensiek et al., 1979). A more up to date and extensive reference is the ASCE Hydrology Handbook, 2nd Edition (ASCE, 1996). Flow measurement in channels is detailed in Flow Measuring and Regulating Flumes (Bos et al., 1983). The monograph Hydrologic Modeling of Small Watersheds (Haan et al., 1982) included useful chapters

**FIGURE 5.38** H-flume and recorder (in white box) for measuring runoff rate from graded bench terrace at Bushland, TX. Note dike diverting flow from uphill flume.

**FIGURE 5.39** Wind shield installed around a heated tipping-bucket rain gauge.

on stochastic modeling, precipitation and snowmelt modeling, runoff modeling, etc., and listed some 75 hydrologic models available at that time. For soil water balance measurements, runoff is often controlled with plot borders or edging driven into the ground or included as the above ground extension of a lysimeter. Steel borders driven into the soil to a depth of 20 cm will suffice in many situations. Sixteen-gauge galvanized steel in rolls 30-cm wide is useful for this, and can be reinforced by rolling over one edge. If runoff must be measured, this can be done with flumes such as the H-flume and recording station shown in Figure 5.38.

Precipitation varies so much from location to location that it is rarely useful to attempt estimating it. Measurement methods include standard U.S. Weather Bureau rain gauges read manually, various tipping-bucket rain gauges, heated gauges to capture snow fall (e.g., Qualimetrics model 6021, Sacramento, CA), snow-depth stations, etc. If possible, a rain gauge should be surrounded by a wind shield to avoid catch loss associated with wind flow over the gauge (Figure 5.39). A standard for the capture area or throat of a rain gauge is that it should be 20 cm in diameter because smaller throats lead to more variability in amount captured. Various designs of tipping-bucket rain gauge have become standard equipment on field weather stations. These are capable of providing precipitation data needed to solve the soil-water balance for short intervals. Two problems are sometimes associated with tipping-bucket type gauges. First, most of these devices count the tips using a Hall effect sensor for detecting the magnetic field of a magnet attached to the tipping bucket, and the sensing system is sometimes susceptible to interference from sources of electromagnetic noise such as vehicle ignition systems. Second, tipping-bucket gauges do not keep up with very high rainfall rates. At Bushland, Texas, tipping-bucket errors have been observed of 10-15% for totals of rainfall from high intensity convective thunderstorms compared with amounts collected in standard rain gauges and sensed by weighing lysimeters. If accuracy is very important, then a tipping-bucket gauge should be supplemented with a standard gauge that captures and stores all the rainfall. For solving the soil-water balance, experience shows that the rain gauge(s) should be placed directly

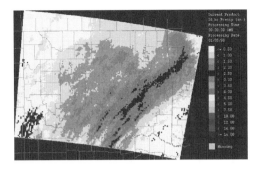

**FIGURE 5.40** Map of 24-h rainfall accumulation over Oklahoma. [From NEXRAD radar data processed by the RPI. Image downloaded from http://ccgwww.ou.edu/rip_images and converted to gray scale, with permission.]

adjacent to the location of $\Delta S$ measurement. Separation of even 100 m can lead to large errors due to the spatial variability of precipitation.

For studies and operations at scales larger than small plot size, there are now precipitation estimates from Doppler radar-based systems that offer calibrated rainfall data on a 24-h basis (Figure 5.40) (Legates et al., 1996; and Vieux and Farajalla, 1996). Although Figure 5.40 shows large grid sizes and only 16 levels of precipitation, grid sizes of 4 km on a side, with 256 levels of rainfall, are available. Data for these maps are generated by the WSR-88D radar system, usually known as the NEXRAD weather radar system, in widespread use in the U.S. The Center for Computational Geosciences at the University of Oklahoma has developed a radar-base precipitation interface (RPI) for the radar data to generate the maps. Radar data are used from two or more stations and calibrated against rain gauge measurements available from, for example, the Oklahoma MESONET system of weather stations.

## ACKNOWLEDGMENTS

Much of the data presented in this chapter was collected by the Evapotranspiration Research Team at the USDA-ARS Conservation & Production Research Laboratory at Bushland, Texas, of which the author is a member. Other past and present team members are (alphabetically) Karen S. Copeland, Donald A. Dusek, Terry A. Howell, Arland D. Schneider, Jean L. Steiner, Rick W. Todd, and Judy A. Tolk. Many technicians supported the data collection efforts including C. Keith Brock, Jim L. Cresap, and Brice B. Ruthardt, among many others. Some of the data shown here have not yet been published and may appear at a later date in a peer-reviewed journal. This work was supported in part by USAID under the subproject title Water Requirements and Management for Maize under Drip and Sprinkler Irrigation, a part of the Agricultural Technology Utilization and Transfer Project, Egypt. "Everything can be taken from a person except the freedom to choose one's attitude in any situation," a favorite quote of Ruby O. Crosby.

## REFERENCES

Aase, J.K. and S.B. Idso. 1975. Solar radiation interactions with mixed prairie rangeland in natural and denuded conditions, Arch. Met. Geoph. Biokl. Ser. B, 23, 255–264.

Allen, R.G., M. Smith, A. Perrier, and L.S. Pereira. 1994a. An update for the definition of reference evapotranspiration, ICID Bulletin 43(2):1–34.

Allen, R.G., M. Smith, A. Perrier, and L.S. Pereira. 1994b. An update for the calculation of reference evapotranspiration, ICID Bulletin 43(2):35–92.

Alsanabani, M.M. 1991. Soil Water Determination by Time Domain Reflectometry: Sampling Domain and Geometry, Ph.D. thesis, Department of Soil and Water Sciences, Univ. of Arizona, Tucson. 209 pp.

Anlauf, R., K. Ch. Kersebaum, L.Y. Ping, A. Nuske-Schüler, J. Richter, G. Springob, K.M. Syring, and J. Utermann. 1990. Models for Processes in the Soil — Programs and Exercises, *Catena Verlag,* Cremlingen, West Germany.

ASAE. 1988. Modeling Agricultural, Forest, and Rangeland Hydrology: Proc. Int. Symp., Dec. 12–13, Chicago, IL, 1988. *Am. Soc. Agric. Eng.,* St. Joseph, MI.

ASCE. 1996. Hydrology Handbook, 2nd ed., *Am. Soc. Civil. Eng.,* New York, NY.

Asrar, G. and E.T. Kanemasu. 1983. Estimating thermal diffusivity near the soil surface using Laplace Transform: uniform initial conditions, *Soil Sci. Soc. Am. J.,* 47:397–401.

Baker, J.M. and R.J. Lascano. 1989. The Spatial Sensitivity of Time-Domain Reflectometry, *Soil Sci.,* 147:378–384.

Baker, J.M. and E.J.A. Spaans. 1994. Measuring water exchange between soil and atmosphere with TDR-microlysimetry, *Soil Sci.,* 158(1):22–30.

Belcher, D.J., T.R. Cuykendall, and H.S. Sack. 1950. Technical Development Report 127, Technical Development and Evaluation Center, Civil Aeronautics Administration, Indianapolis.

Ben-Asher, J., A.D. Matthias, and A.W. Warrick. 1983. Assessment of evaporation from bare soil by infrared thermometry, *Soil Sci. Soc. Am. J.,* 47:185–191.

Black, T.A., G.W. Thurtell, and C.B. Tanner. 1968. Hydraulic load cell lysimeter, construction, calibration, and tests, *Soil Sci. Soc. Am. Proc.,* 32:623–629.

Blad, B.L. and N.J. Rosenberg. 1974. Lysimetric calibration of the Bowen ratio energy balance method for evapotranspiration estimation in the central Great Plains, *J. Appl. Meteorol.,* 13:227–236.

Bos, M.G., J.A. Replogle, and A.J. Clemmens. 1983. Flow Measuring and Regulating Flumes, I.L.R.I., Wageningen, The Netherlands, and U.S. Water Conservation Laboratory, Phoeniz, AZ.

Bowers, S.A. and R.J. Hanks. 1965. Reflection of radiant energy from soils, *Soil Sci.,* 100(2):130–138.

Brakensiek, D.L., H.B. Osborn, and W.J. Rawls. 1979. Field Manual for Research in Agricultural Hydrology. U.S. Dept. Agric., Agricultural Handbook 224. U.S. Gov't. Printing Office.

Brunt, D. 1932. Notes on radiation in the atmosphere, *Quart. J. Roy. Met. Soc.,* 58:389–418.

Brutsaert, W. 1975. The roughness length for water vapor, sensible heat and other scalars, *J. Atm. Sci.,* 32:2028–2031.

Brutsaert, W. 1982. Evaporation Into The Atmosphere, D. Reidel Publ. Co., Boston, MA.

Campbell, G.S. 1977. An Introduction to Environmental Biophysics, Springer-Verlag, New York.

Campbell, G.S. 1985. Soil Physics with BASIC — Transport Models for Soil-Plant Systems, Elsevier, New York.

Campbell, G.S., C. Calissendorff, and J.H. Williams. 1991. Probe for measuring soil specific heat using a heat-pulse method, *Soil Sci. Soc. Am. J.,* 55:291–293.

Campbell, G.S., J.D. Jungbauer, Jr., W.R. Bidlake, and R.D. Hungerford. 1994. Predicting the effect of temperature on soil thermal conductivity, *Soil Sci.,* 158(5):307–313.

Carrijo, O.A. and R.H. Cuenca. 1992. Precision of evapotranspiration measurements using neutron probe, *J. Irrig. Drain. Eng.,* ASCE, 118(6):943–953.

Chen, J. 1984. Mathematical analysis and simulation of crop micrometeorology, Ph.D. diss., Agricultural Univ. Wageningen, The Netherlands.

Cochran, P.H., L. Boersma, and C.T. Youngberg. 1967. Thermal conductivity of a pumice soil, *Soil Sci. Soc. Am. Proc.,* 31:454–459.

Collares-Pereira, M. and A. Rabl. 1979. The average distribution of solar radiation — Correlations between diffuse and hemispherical and between daily and hourly insolation values, *Solar Energy,* 22:155–164.

Costello, T.A. and H.J. Braud, Jr. 1989. Thermal Diffusivity of Soil by Nonlinear Regression Analysis of Soil Temperature Data, Trans. ASAE 32(4):1281–1286.

Cuenca, R.H. 1988. Model for evaptranspiration using neutron probe data, *J. Irrig. Drain. Eng.,* ASCE, 114(4):644–663.

De Vries, D.A. 1963. Chapter 7. Thermal properties of soils, *in* W.R. Van Wijk (ed.), *Physics of Plant Environment,* North-Holland Publ., Amsterdam.

Dickey, G.L., Allen, R.G., Barclay, J.H., Wright, J.L., Stone, J.F., and Draper, B.W. 1993. Neutron gauge calibration comparison of methods, pp. 1136–1144, *in* R.G. Allen and C.M.U. Neale (eds.), Management of Irrigation and Drainage Systems, Integrated Perspectives, *Am. Soc. Civil Eng.,* New York, NY, Proc. Natl Conf. Irrig. Drain. Eng., Park City, UT, July 21–23, 1993.

Doorenbos, J. and W.O. Pruitt. 1977. Guidelines for predicting crop water requirements, FAO Irrigation and Drainage Paper 24, Food and Agriculture Organization of the United Nations, Rome.

Duffie, J.A. and W.A. Beckman. 1991. Solar Engineering of Thermal Processes, 2nd ed., John Wiley & Sons, New York.

Dugas, W.A. and W.L. Bland. 1991. Springtime soil temperatures in lysimeters in Central Texas, *Soil Sci.,* 152(2):87–91.

Dugas, W.A., L.J. Fritchen, L.W. Gay, A.A. Held, A.D. Matthias, D.C. Reicosky, P. Steduto, and J.L. Steiner. 1991. Bowen ratio, eddy correlation, and portable chamber measurements of sensible and latent heat flux over irrigated spring wheat, *Agric. Forest Meteorol.,* 56:1–20.

Dvoracek, M.J. and B. Hannabas. 1990. Prediction of albedo for use in evapotranspiration and irrigation scheduling, *in* Visions of the future, proceedings of the third national irrigation symposium held in conjunction with the 11th annual international irrigation exposition, Oct. 28–Nov. 1, Phoenix, AZ, ASAE, St. Joseph, MI.

Evett, S.R. 1994. TDR-Temperature arrays for analysis of field soil thermal properties, pp. 320–327, *in* Proc. Symp. Time Domain Reflectometry in Environmental, Infrastructure and Mining Applications, Sept. 7–9, 1994, Northwestern University, Evanston, Illinois.

Evett, S.R., T.A. Howell, J.L. Steiner, and J.L. Cresap. 1993. Evapotranspiration by soil water balance using TDR and neutron scattering, pp. 914–921, *in* R.G. Allen and C.M.U. Neale (eds.), Management of Irrigation and Drainage Systems, Integrated Perspectives, *Am. Soc. Civil Eng.,* New York, NY, Proc. Natl. Conf. Irrig. Drainage Eng., Park City, UT, July 21–23, 1993.

Evett, S.R. and R.J. Lascano. 1993. ENWATBAL: A mechanistic evapotranspiration model written in compiled BASIC, *Agron. J.,* 85:763–772.

Evett, S.R., A.D. Matthias, and A.W. Warrick. 1994. Energy balance model of spatially variable evaporation from bare soil, *Soil Sci. Soc. Am. J.,* 58:1604–1611.

Evett, S.R., A.W. Warrick, and A.D. Matthias. 1995. Wall material and capping effects on microlysimeter performance, *Soil Sci. Soc. Am. J.,* 59:329–336.

Fox, M.J. 1968. A technique to determine evaporation from dry stream beds, *J. Appl. Meteorol.,* 7:697–701.

Gardner, W. and D. Kirkham. 1952. Determination of soil moisture by neutron scattering, *Soil Sci.,* 73:391–401.

Gates, D.M. 1980. Biophysical Ecology. Springer-Verlag, New York.

Gorry, P.A. 1990. General least-squares smoothing and differentiation by the convolution (Savitsky-Golay) method, *Anal. Chem.,* 62:570–573.

Goudriaan, J. 1977. Crop micrometeorology: A simulation study, PUDOC, Wageningen, The Netherlands.

Grebet, R. and R.H. Cuenca. 1991. History of lysimeter design and effects of environmental disturbances, pp 10–19, *in* Lysimeters for Evapotranspiration and Environmental Measurements, Proc. Int. Symp. Lysimetry, July 23–25, 1991, Honolulu, Hawaii. ASCE, NY.

Hanks, J. and J.T. Ritchie (eds.). 1991. Modeling Plant and Soil Systems, Agronomy Monograph 31, *Am. Soc. Agron., Crop Sci. Soc. Am., Soil Sci. Soc. Am.,* Madison, WI.

Hann, C.T., H.P. Johnson, and D.L. Brakensiek (eds.). 1982. Hydrologic Modeling of Small Watersheds. ASAE Monograph No. 5, *Am. Soc. Agric. Eng.,* St. Joseph, MI.

Hatfield, J.L., R.J. Reginato, and S.B. Idso. 1983. Comparison of long-wave radiation calculation methods over the United States, *Water Resour. Res.,* 19:285–288.

Heilman, J.L., C.L. Brittin, and C.M.U. Neale. 1989. Fetch requirements for Bowen ratio measurements of latent and sensible heat fluxes, *Agric. and Forest Meteorol.,* 44:261–273.

Hillel, D. 1980. Fundamentals of Soil Physics, Academic Press, San Diego.

Horton, R., P.J. Wierenga, and D.R. Nielsen. 1983. Evaluation of methods for determining the apparent thermal diffusivity of soil near the surface, *Soil Sci. Soc. Am. J.,* 47:25–32.

Houser, P.R. 1998. Personal communication from Paul R. Houser, Hydrological Sciences Branch, Data Assimilation Office, NASA Goddard Space Flight Center.

Houser, P.R., C. Harlow, W.J. Shuttleworth, T.O. Keefer, W.E. Emmerich, and D.C. Goodrich. 1998. Evaluation of multiple flux measurement techniques using water balance information at a semi-arid site, *in* Proc. Special Symp. on Hydrology, Am. Meteorol. Soc. Conf., Phoenix, AZ, pp. 84–87.

Howell, T.A., A.D. Schneider, D.A. Dusek, T.H. Marek, and J.L. Steiner. 1995. Calibration and scale performance of Bushland weighing lysimeters, Trans. ASAE 38(4):1019–1024

Howell, T.A., A.D. Schneider, and M.E. Jensen. 1991. History of lysimeter design and use for evapotranspiration measurements, pp. 1–9, *in* Lysimeters for Evapotranspiration and Environmental Measurements, Proc. Int. Symp. Lysimetry, July 23–25, 1991, Honolulu, Hawaii, ASCE, NY.

Howell, T.A., J.L. Steiner, S.R. Evett, A.D. Schneider, K.S. Copeland, D.A. Dusek, and A. Tunick. 1993. Radiation balance and soil water evaporation of bare Pullman clay loam soil, pp. 922–929, *in* R.G. Allen and C.M.U. Neale (eds.), Management of Irrigation and Drainage Systems, Integrated Perspectives. *Am. Soc. Civil Eng.,* NY, Proc. Natl. Conf. Irrig. Drainage Eng., Park City, UT, July 21–23, 1993.

Howell, T.A., J.L. Steiner, A.D. Schneider, S.R. Evett, and J.A. Tolk. 1994. Evapotranspiration of irrigated winter wheat, sorghum and corn, ASAE paper no. 94-2081, 33 pp.

Howell, T.A. and J.A. Tolk. 1990. Calibration of soil heat flux transducers, *Theor. Appl. Climatol.,* 42:263–272.

IAEA. 2000. Comparison of soil water measurement using the neutron scattering, time domain reflectometry and capacitance methods, Int. Atomic Energy Agency, Vienna, Austria, IAEA-TECDOC-1137. ISSN 1011-4289.

Idso, S.B. 1981. A set of equations for full spectrum and 8- to 14-μm and 10.5- to 12.5-μm thermal radiation from cloudless skies, *Water Resour. Res.,* 17(2)295–304.

Idso, S.B. and R.J. Reginato. 1974. Assessing soil-water status via albedo measurement, *Hydrol. Water Resour.,* Ariz. Southwest 4:41–54.

Idso, S.B., R.D. Jackson, R.J. Reginato, B.A. Kimball, and F.S. Nakayama. 1975. The dependence of bare soil albedo on soil water content, *J. Appl. Meteorol.,* 14(1)109–113.

Idso, S.B., R.J. Reginato, R.D. Jackson, B.A. Kimball, and F.S. Nakayama. 1974. The three stages of drying in a field soil, *Soil Sci. Soc. Am. Proc.,* 38(5),831–837.

Iwata, S., T. Tabuchi, and B.P. Warkentin. 1988. Soil-Water Interactions: Mechanisms and Applications, Marcel Dekker, Inc., NY.

Jackson, R.D. 1988. Surface temperature and the surface energy balance, pp. 133–153, *in* W.L. Steffen and O.T. Denmead (eds.), Flow and Transport in the Natural Environment: Advances and Applications, Springer-Verlag, Berlin, ISBN 0-387-19452-5.

Jensen, M.E., R.D. Burman, and R.G. Allen (ed.). 1990. Evapotranspiration and Irrigation Water Requirements, ASCE Manuals and Reports on Engineering Practices No. 70, ASCE, NY, 332 pp.

Jones, H.G. 1992. Plants and Microclimate: A Quantitative Approach to Environmental Plant Physiology, 2nd ed., The Cambridge University Press, Cambridge, UK.

Kimball, B.A., R.D. Jackson, R.J. Reginato, F.S. Nakayama, and S.B. Idso. 1976. Comparison of field measured and calculated soil heat fluxes, *Soil Sci. Soc. Am. J.,* 40:18–24.

Kluitenberg, G.J., K.L. Bristow, and B.S. Das. 1995. Error analysis of heat pulse method for measuring soil heat capacity, diffusivity, and conductivity, *Soil Sci. Soc. Am. J.,* 59:719–726.

Kondo, J., N. Saigusa, and T. Sato. 1992. A model and experimental study of evaporation from bare-soil surfaces, *J. Appl. Meteor.,* 31:304–312.

Kreith, F. and W.D. Sellers. 1975. General principles of natural evaporation, p. 207–227, *in* D.A. de Vries and N.H. Afgan (eds.), Heat and Mass Transfer in the Biosphere, Part 1, John Wiley & Sons, NY.

Lascano, R.J., C.H.M. van Bavel, J.L. Hatfield, and D.R. Upchurch. 1987. Energy and water balance of a sparse crop: simulated and measured soil and crop evaporation, *Soil Sci. Soc. Am. J.,* 51:1113–1121.

Lascano, R.J. and J.L. Hatfield. 1992. Spatial variability of evaporation along two transects of a bare soil, *Soil Sci. Soc. Am. J.,* 56:341–346.

Legates, D.R., K.R. Nixon, T.D. Stockdale, and G. Quelch. 1996. Soil water management using a water resource decision support system and calibrated WSR-88D precipitation estimates, Proc. AWRA Symp. GIS and Water Resources. Am. Water Resources Assoc., Fort Lauderdale, FL.

List, R.J. 1971. Smithsonian Meteorological Tables, Smithsonian Institution Press, Washington.

Lourence, F. and R. Moore. 1991. Prefabricated weighing lysimeter for remote research stations, pp. 423–439, *in* Lysimeters for Evapotranspiration and Environmental Measurements, Proc. Int. Symp. Lysimetry, July 23–25, 1991, Honolulu, Hawaii, ASCE, NY.

Marek, T.H., A.D. Schneider, T.A. Howell, and L.L. Ebeling. 1988. Design and construction of large weighing monolithic lysimeters, Trans. ASAE 31:477–484.

McCullough, E.C. and W.P. Porter. 1971. Computing clear day solar radiation spectra for the terrestrial ecological environment, *Ecology,* 52:1008–1015.

McNaughton, K.C. 1988. Surface temperature and the surface energy balance: Commentary, pp. 154–159, *in* W.L. Steffen and O.T. Denmead (eds.), Flow and Transport in the Natural Environment: Advances and Applications, Springer-Verlag, Berlin.

Moncrieff, J.B., J.M. Massheder, H.A.R. De Bruin, J. Elbers, T. Friborg, B. Heusinkveld, P. Kabat, S. Scott, H. Soegaard, and A. Verhoef. 1997. A system to measure surface fluxes of momentum, sensible heat, water vapour and carbon dioxide, *J. Hydrol.,* 189:589–611.

Monteith, J.L. 1961. The reflection of short-wave radiation by vegetation, *Q.J. Roy. Meteorol. Soc.,* 85(366):386–392.

Monteith, J.L. 1965. Evaporation and the environment, *in* The State and Movement of Water in Living Organisms, XIXth Symp. Soc. for Exp. Biol., Swansea, Cambridge University Press, pp. 205–234.

Monteith, J.L. and G. Szeicz. 1961. The radiation balance of bare soil and vegetation, *Q.J. Roy. Meteorol. Soc.,* S7 (372):159–170.

Monteith, J.L. and M.H. Unsworth. 1990. Principles of Environmental Physics, 2nd ed., Edward Arnold, London.

Murray, F.W. 1967. On the computation of saturation vapor pressure, *J. Applied Meteorol.,* 6(1):203–204.

Nassar, I.N. and R. Horton. 1989. Determination of The Apparent Thermal Diffusivity of a Nonuniform Soil, *Soil Sci.,* 147(4):238–244.

Nassar, I.N. and R. Horton. 1990. Determination of Soil Apparent Thermal Diffusivity from Multiharmonic Temperature Analysis for Nonuniform Soils, *Soil Sci.,* 149(3):125–130.

NOAA. 1997. Data files UARS96.PLT, ERBS.PLT, NOAA09.PLT, NOAA10.PLT, NIMBUS.PLT, and SMM.PLT. Accessed January 14, 1998.

Noborio, K., K.J. McInnes, and J.L. Heilman. 1996. Measurements of soil water content, heat capacity, and thermal conductivity with a single TDR probe, *Soil Sci.,* 161(1):22–28.

Oke, T.R. 1978. Boundary Layer Climates, Methuen, NY.

Peart, R.M. and R.B. Curry. 1998. Agricultural Systems Modeling and Simulation, Marcel Dekker, Inc., NY.

Penman, H.L. 1948. Natural evapotranspiration from open water, bare soil and grass, Proc. R. Soc. London Ser. A. 193:120–145.

Pereira, L.S., B.J. van den Broek, P. Kabat, and R.G. Allen (eds.). 1995. Crop-Water-Simulation Models in Practice, Wageningen Pers, Wageningen, The Netherlands.

Richter, J. 1987. The Soil as a Reactor: Modeling Processes in the Soil, *Catena Verlag,* Cremlingen, West Germany.

Riha, S.J., K.J. McKinnes, S.W. Childs, and G.S. Campbell. 1980. A finite element calculation for determining thermal conductivity, *Soil Sci. Soc. Am. J.,* 44:1323–1325.

Ritchie, J.T., T.A. Howell, W.S. Meyer, and J.L. Wright. 1996. Sources of biased errors in evaluating evapotranspiration equations, pp. 147–157, *in* C.R. Camp, E.J. Sadler, and R.E. Yoder (eds.), Evapotranspiration and Irrigation Scheduling, Proc. Int. Conf. Nov. 3–6, 1996, San Antonio, TX.

Robert, P.C., R.H. Rust, and W.E. Larson, ed. 1999. Proc. Fourth Int. Conf. Precision Agriculture, July 1998, Minneapolis, MN. Part A (976 pages) and Part B (962 pages); ASA, CSSA, and SSSA.

Rosenberg, N.J. 1969. Seasonal patterns of evapotranspiration by irrigated alfalfa in the Central Great Plains, *Agron. J.,* 61(6):879–886.

Rosenberg, N.J., B.L. Blad, and S.B. Verma. 1983. Microclimate, the biological environment, John Wiley & Sons, NY.

Russell, G., B. Marshall, and P.G. Jarvis (eds.). 1989. Plant Canopies: Their Growth, Form and Function. Cambridge University Press, Cambridge, UK.

Savage, M.J., K.J. McInnes, and J.L. Heilman. 1995. Placement height of eddy correlation sensors above a short turfgrass surface, *Agric. Forest Meteor.,* 74(3–4):195–204.

Savage, M.J., K.J. McInnes, and J.L. Heilman. 1996. The "footprints" of eddy correlation sensible heat flux density, and other micrometeorological measurements, *S. Afr. J. Sci.,* 92:137–142.

Savitsky, A. and M.J.E. Golay. 1964. Smoothing and differentiation of data by simplified least squares, *Anal. Chem.,* 36:1627–1639.

Schmid, H.P. 1997. Experimental design for flux measurements: matching scales of observations and fluxes, *Agric. Forest Meteorol.,* 87:179–200.

Schneider, A.D. 1998a. Personal communication. Drawing of cross-section of simplified weighing lysimeter installed at Bushland, TX.

Schneider, A.D. 1998b. Personal communication. Drawing of cross-section of large weighing lysimeter at Bushland, TX.

Schneider, A.D., T.A. Howell, T.A. Moustafa, S.R. Evett, and W.S. Abou-Zeid. 1996. A simplified weighing lysimeter for developing countries, pp. 289–294, *in* C.R. Camp, E.J. Sadler, and R.E. Yoder (eds.), Proc. Int. Conf. Evapotranspiration and Irrig. Scheduling, Nov. 3–6, 1996, San Antonio, TX, 1166 pp.

Schuepp, P.J., M.Y. LeClerc, J.I. MacPherson, and R.L. Desjardins. 1990. Footprint prediction of scalar fluxes from analytical solutions of the diffusion equation, *Boundary-Layer Meteorol.,* 50:355–373.

Sepaskhah, A.R. and L. Boersma. 1979. Thermal conductivity of soils as a function of temperature and water content, *Soil Sci. Soc. Am. J.,* 43:439–444.

Skidmore, E.L., J.D. Dickerson, and H. Schimmelpfennig. 1975. Evaluating surface-soil water content by measuring reflection, *Soil Sci. Soc. Am. J.,* 39:238–242.

Spitters, C.J.T., H.A.J.M. Toussaint, and J. Goudriaan. 1986. Separating the diffuse and direct component of global radiation and its implications for modeling canopy photosynthesis. Part I. Components of incoming radiation, *Agric. Forest Meteorol.,* 38:217–229.

Todd, R.M. 1998a. Personal communication, analysis of footprint of Bowen ratio system over alfalfa field at Bushland, TX.

Todd, R.M. 1998b. Personal communication, Bowen ratio results for period between 3rd and 4th cuttings of alfalfa at Bushland, TX in 1997.

Unland, H.E., P.R. Houser, W.J. Shuttleworth, and Z-L. Zang. 1996. Surface flux measurement and modeling at a semi-arid Sonoran desert site, *Agric. Forest Meteorol.,* 82:119–153.

van Bavel, C.H.M. 1966. Potential evaporation: The combination concept and its experimental verification, *Water Resour. Res.,* 2:455–467.

van Bavel, C.H.M. and D.I. Hillel. 1976. Calculating potential and actual evaporation from a bare soil surface by simulation of concurrent flow of water and heat, *Agric. Meteorol.,* 17:453–476.

van Genuchten, M. Th., F.J. Leij, and S.R. Yates. 1991. The RETC code for quantifying the hydraulic functions of unsaturated soils. EPA/600/2-91/065. 93 pp. R.S. Kerr Environ. Res. Lab., U.S. Environmental Protection Agency, ADA, OK.

Van Wijk, W.R. and D.W. Scholte Ubing. 1963. Radiation, p. 62–101, *in* W.R. Van Wijk (ed.), *Physics of Plant Environment,* North-Holland Publ. Co., Amsterdam.

Verma, S.B., N.J. Rosenberg, and B.L. Blad. 1978. Turbulent exchange coefficients for sensible heat and water vapor under advective conditions, *J. Appl. Meteorol.,* 17:330–338.

Vieux, B.E. and N.S. Farajalla. 1996. Temporal and spatial aggregation of NEXRAD rainfall estimates on distributed storm runoff simulation, Third Int. Conf. GIS and Environmental Modeling, Jan. 21–25, 1996, Santa Fe, NM.

Watts, D.B., E.T. Kanemasu, and C.B. Tanner. 1990. Modified heat-meter method for determining soil heat flux, *Agric. Forest Meteor.,* 49:311–330.

Weast, R.C. (ed.). 1982. *Handbook of Chemistry and Physics,* CRC Press, FL.

Weiss, A. 1982. An experimental study of net radiation, its components and prediction, *Agron. J.,* 74:871–874.

Wierenga, P.J., D.R. Nielson, and R.J. Hagan. 1969. Thermal properties of a soil, based upon field and laboratory measurements, *Soil Sci. Soc. Am. J.,* 44:354–360.

Wraith, J.M. and J.M. Baker. 1991. High-resolution measurement of root water uptake using automated time-domain reflectometry, *Soil Sci. Soc. Am. J.,* 55:928–932.

Wright, J.L. 1982. New evapotranspiration crop coefficients, *J. Irrig. and Drain. Div.,* ASCE 108(IR1):57–74.

Wright, J.L. 1990. Comparison of ET measured with neutron moisture meters and weighing lysimeters, pp. 202–209, *in* Irrigation and Drainage: Proc. Natl. Conf., Durango, Colorado, July 11–13, 1990. ASCE, N.Y.

Wright, J.L. 1991. Using weighing lysimeters to develop evapotranspiration crop coefficients, *in* R.G. Allen, T.A. Howell, W.O. Pruitt, I.A. Walter, and M.E. Jensen (eds.), Lysimeters for Evapotranspiration and Environmental Measurements, Proc. Int. Symp. Lysimetry, July 23–25, Honolulu, HI, ASCE, NY.

Wright, J.L. and M.E. Jensen. 1972. Peak water requirements of crops in Southern Idaho, *J. Irrig. and Drain. Div.,* ASCE 96(IR1):193–201.

Young, M.H., P.J. Wierenga, and C.F. Mancino. 1997. Monitoring near-surface soil water storage in turfgrass using time domain reflectometry and weighing lysimetry, *Soil Sci. Soc. Am. J.,* 61:1138–1146.

# 6 Solute Transport

*Feike J. Leij and Martinus Th. van Genuchten*

## 6.1 INTRODUCTION

Soil scientists and agricultural engineers have traditionally been interested in the behavior and effectiveness of agricultural chemicals (fertilizers, pesticides) applied to soils for enhancing crop growth, as well as in the effect of salts and other dissolved substances in the soil profile on plant growth. More recently, concern for the quality of the vadose zone and possible contamination of groundwater has provided a major impetus for studying solute transport in soils.

The movement and fate of solutes in the subsurface is affected by a large number of physical, chemical and microbiological processes requiring a broad array of mathematical and physical sciences to study and describe solute transport. A range of experimental and mathematical procedures may be employed to quantify transport in soils. Transport of a dissolved substance (solute) depends on the magnitude and direction of the solvent (water) flux; considerable experimental and numerical effort may be needed to determine the transient flow regime in unsaturated soils. Furthermore, the determination of solute concentrations is not always straightforward, particularly if the solute is involved in partitioning between different phases or subject to transformations.

A vast body of work on solute transport can be found in the soil science literature. An equally vast amount of pertinent studies on solute transport in porous media has been published by civil and environmental engineers, geophysicists and geochemists, physical chemists and others. The scope of this chapter permits only an introductory treatment of the subject. First, the standard transport mechanisms pertaining to the fundamental advection-dispersion equation (ADE) will be introduced in Section 6.2. This equation, also known as the convection-dispersion equation, is most often used to model solute transport in porous media. The movement of a solute that undergoes adsorption by the soil requires modifications of the ADE, particularly if several solute species are present that may participate in a number of different reactions. Section 6.3 is devoted to analytical and numerical methods for quantifying solute concentrations as a function of time and space. The traditional advection-dispersion concept is not always adequate to describe solute transport in field soils. Section 6.4 describes the stream tube model as an example of an alternative transport model that may be better suited to model transport in real world situations.

## 6.2 THE ADVECTION–DISPERSION EQUATION

Consider the transport of a chemical species in a three-phase soil-air-water system, and assume that the chemical species (the solute) is completely miscible with water (the solvent). At the macroscopic level and for one-dimensional flow, the mass balance equation for a solute species subject to arbitrary reactions is given as:

$$\frac{\partial \theta C}{\partial t} = -\frac{\partial J_s}{\partial x} + \theta R_s \qquad [6.1]$$

where $\theta$ is the volumetric water content ($L^3 L^{-3}$), $C$ is the solute concentration expressed as solute mass-per-solvent volume ($M L^{-3}$), $t$ is time (T), $x$ is position (L), $J_s$ is the solute flux expressed in solute mass-per-cross-sectional area of soil-per-unit time (M L$^{-2}$ T$^{-1}$), and $R_s$ denotes arbitrary solute sinks (< 0) or sources (> 0) (M L$^{-3}$ T$^{-1}$). Similar equations may be derived for multidimensional flow and transport. The solute flux is usually distinguished in an advective and a dispersive component according to:

$$J_S = J_w C + J_D \qquad\qquad [6.2]$$

where $J_w$ is a vector quantifying the water flux (LT$^{-1}$), namely, the Darcian velocity expressed as volume of water-per-cross-sectional area of soil-per-unit time, and $J_D$ is a solute flux to quantify transport caused by a gradient in the solute concentration (M L$^{-2}$ T$^{-1}$), also per unit area of soil.

### 6.2.1 TRANSPORT MECHANISMS

The movement of a solute with flowing water, described by the solute flux ($J_w C$), is referred to as advection or convection. Because dissolved substances move in a passive fashion, advective transport can be readily quantified when the solvent flux ($J_w$) is known. The water flux is generally a function of time and position. However, for transport in laboratory soil columns, $J_w$ is often constant while for field studies, approximations may sometimes be made to facilitate a simpler steady-state one-dimensional flow description.

Even if the macroscopic water flux is known or can be measured, the velocity at smaller pore scales is not easily determined. Variations in the microscopic velocity will lead to unequal solute movement in the direction of flow. This movement is quantified by means of the dispersive solute flux. If, during steady water flow, the solute concentration of the solution at the inlet of a water saturated soil column is changed abruptly at time $t = 0$, the observed breakthrough of a solute at the column outlet at times $t > 0$ will not exhibit an equivalent abrupt change (Nielsen and Biggar, 1961). The solute concentration will change more gradually with time as a result of (hydrodynamic) dispersion, which quantifies the effects of both mechanical dispersion and diffusion. Molecular diffusion and mechanical dispersion will be discussed first for free solutions and then for soil solutions.

### 6.2.1.1 Diffusion

Molecular or ionic diffusion is an important mechanism for solute transport in soils in directions where there is little or no water flow. A net transfer of molecules of a solute species usually occurs from regions with higher to lower concentrations as the result of diffusion as described by Fick's first law. For a free or bulk solution, the one-dimensional mass flux [$J_{dif}$(M L$^2$ T$^{-1}$)] due to molecular diffusion is given by:

$$J_{dif} = -D_o \frac{\partial C}{\partial x} \qquad\qquad [6.3]$$

where $D_o$ is the coefficient of molecular diffusion for a free or bulk solution (L$^2$ T$^{-1}$). Many publications exist that provide data on $D_o$ (Kemper, 1986; Lide, 1995).

The experimentally observed proportionality between $J_{dif}$ and the concentration gradient can be described at the molecular or ionic level with a balance of forces. The driving force for particle movement from higher to lower concentrations is the gradient in the chemical potential. For mixtures with ideal behavior, the chemical potential can be expressed in terms of its mole fraction. For solutions with nonideal mixing behavior, the activity coefficient of the solute needs to be determined. Ionic activity coefficients can be estimated with the extended Debye-Hückel equation or the Davies extension for solutions up to 0.5 M (Chapter 10). Activity coefficients for a greater concentration range up to 16 M can be estimated with the Pitzer virial equations (Pitzer, 1979; Harvie and Weare,

1980). Codiffusion or counterdiffusion occurs in systems with multiple ion species. Diffusion rates for individual species as predicted solely by Fick's model would violate the electroneutrality principle. The ionic diffusion flux consists then of a term for ordinary Fickian diffusion and a term accounting for electric transference of ions. The corresponding diffusion coefficient is related to the ionic mobility using the Nernst-Planck equation (Helfferich, 1962).

To characterize diffusion in soils, the diffusivity in a free solution is typically adjusted with terms accounting for a reduced solution phase (a smaller cross-sectional area available for diffusion), and an increased path length. A general treatment of diffusion in soils can be found in Olsen and Kemper (1968) and Nye (1979). The macroscopic diffusive flux per unit area of soil can be written as:

$$J_{dif} = -\theta D_{dif} \frac{\partial C}{\partial x} \qquad [6.4]$$

where $D_{dif}$ is the coefficient of molecular or ionic diffusion for the liquid phase of the soil. The diffusion coefficients for the soil liquid and a free liquid are related by (Epstein, 1989):

$$D_{dif} = \frac{D_o}{\left(L_{dif}/L\right)^2} \equiv \frac{D_o}{\tau^2} \equiv D_o \tau_a \qquad [6.5]$$

where $L_{dif}$ and $L$ are the actual and the shortest path lengths for diffusion (L), $\tau = L_{dif}/L$ is known as the tortuosity, and $\tau^2$ as the tortuosity factor, while $\tau_a = (L/L_{dif})^2$ is often designated as an apparent tortuosity factor. The tortuosity $L_{dif}/L$ appears twice in Equation [6.5] to account for changes in the concentration gradient along the streamline and the travel distance as compared to a higher concentration difference and a shorter travel distance along a straight path with length $L$ in a free solution (Olsen and Kemper, 1968). It should be noted that the terms tortuosity and tortuosity factor have not been used consistently in the literature. Furthermore, some authors include the water content in their definition of tortuosity (Dykhuizen and Casey, 1989) or solute adsorption (retardation) in the expression for either $\tau$ or $D_{dif}$ (Nye, 1979; Robin et al., 1987).

For unsaturated conditions, it is convenient to quantify the dependency of the diffusion coefficient on water content. Assuming that the tortuosity affects diffusion in the liquid phase in the same way as in the gaseous phase, the tortuosity term previously derived for gaseous diffusion in soils by Millington (1959) and Millington and Quirk (1961), can be adapted to describe aqueous diffusion in variably saturated soils. The following expression then results:

$$D_{dif} = \frac{\theta^{7/3}}{\varepsilon^2} D_o \qquad [6.6]$$

where $\varepsilon$ is the soil porosity ($L^3 L^{-3}$). An equivalent of the first expression, using a volumetric air content instead of $\theta$, has been used frequently to describe gaseous diffusion in soils although Jin and Jury (1996) also reported an alternative version.

Diffusion coefficients in soil systems are usually determined by mathematically analyzing solute concentration profiles in the soil as a function of time or position. Van Rees et al. (1991) measured diffusivities by allowing diffusion from (1) a spiked solution into a soil having a zero or low initial concentration, (2) a spiked soil into a solution, and (3) a spiked soil into the soil. In the first two procedures, the concentration of the solution is observed as a function of time. A mathematical solution is then fitted to the observation to determine the diffusion coefficient.

The third procedure of diffusion from a soil with a higher to a lower concentration has been widely applied (Kemper, 1986; Oscarson et al., 1992). Two blocks of soil with different concentrations are brought together at time t = 0. After sufficient time has elapsed for solute diffusion to

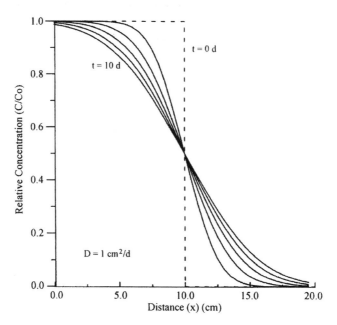

**FIGURE 6.1** Concentration profiles resulting from diffusion ($D_{dif}$ = 1 cm$^2$d$^{-1}$) at different times after two soil columns with $C = 0$ and $C = C_o$ were joined at $t = 0$.

occur from the block with the higher to the lower concentration, the joined soil blocks are sectioned. The solute concentration of each section is determined, for example, by using extraction, centrifugation and chemical analysis of the supernatant liquid. This approach yields a concentration profile versus distance from which the diffusion coefficient may be estimated using an appropriate analytical solution of the governing solute diffusion equation. Consider the diffusion equation:

$$\frac{\partial C}{\partial t} = D_{dif}\frac{\partial^2 C}{\partial x^2}$$

[6.7]

subject to the initial condition:

$$C(x,0) = \begin{cases} C_o & -\infty < x < 0 \\ C_i & 0 < x < \infty \end{cases}$$

[6.8]

and the approximate boundary conditions:

$$C(-\infty, t) = C_o, \quad C(\infty, t) = C_i$$

[6.9]

The solution for this problem is given by (Crank, 1975):

$$C(x,t) = C_i + 0.5(C_o - C_i)\,\text{erfc}\left(x\big/\sqrt{4D_{dif}t}\right)$$

[6.10]

where erfc is the complementary error function (Gautschi, 1964). The distributions of the solute concentration as a function of distance for different times after joining two soil blocks with concentrations $C_i = 0$ and $C_o$, and assuming $D_{dif} = 1$ cm$^2$ d$^{-1}$, are presented in Figure 6.1.

## TABLE 6.1
**Selected diffusion coefficients for aqueous solutions in clay and soil materials.**

| $D$, cm$^2$ day$^{-1}$ | $\tau_a$ - | $\rho_b$ g cm$^{-3}$ | Comments |
|---|---|---|---|
| **van Rees et al. (1991)†** | | | |
| 1.46 | 0.64 | 1.42 | spiked water on top of sediment ($\theta = 0.42$) |
| 1.47 | 0.70 | 1.42 | spiked water on top of sediment ($\theta = 0.42$) |
| 1.66 | 0.79 | 1.35 | spiked water on top of sediment ($\theta = 0.45$) |
| 1.46 | 0.69 | 1.35 | lake water on top of spiked sediment ($\theta = 0.45$) |
| 0.97 | 0.46 | 1.42 | sediment on top of spiked sediment ($\theta = 0.42$) |
| **Robin et al. (1987)‡** | | | |
| 0.19 | 0.11 | 1.63 | 25°C |
| 0.20 | 0.11 | 1.61 | 25°C |
| 0.36 | 0.11 | 1.62 | 60°C |
| 0.40 | 0.12 | 1.61 | 60°C |
| 0.54 | 0.11 | 1.63 | 90°C |
| 0.56 | 0.11 | 1.64 | 90°C |
| **Oscarson et al. (1992)§** | | | |
| 0.33 | 0.19 | 0.90 | |
| 0.27 | 0.15 | 1.12 | |
| 0.17 | 0.10 | 1.31 | |
| 0.08 | 0.05 | 1.50 | |

† $^3$H$_2$O tracer in litoral sediment
‡ $^{36}$Cl tracer in bentonite-sand mixture using a spiked ($C_o \approx 0.27\ M$) and unspiked soil plug for different temperatures
§ $^{125}$I tracer in compacted bentonite using spiked and unspiked clay plugs

Typical values for diffusion coefficients in clays and soils with accompanying $\tau_a$ are provided in Table 6.1. Additional soil diffusivity data are given by, among others, Hamaker (1972) and Nye (1979).

### 6.2.1.2 Dispersion

Local variations in water flow in a porous medium will lead to mechanical dispersion. Several mechanisms that are commonly used to contribute to mechanical dispersion are illustrated with hypothetical tracer particles in Figure 6.2. Dispersion may occur because of (1) the development of a velocity profile within an individual pore such that the highest velocity occurs in the center of the pore, and presumably little or no flow at the pore walls; (2) different mean flow velocities in pores of different sizes; (3) the mean water flow direction in the porous medium being different from the actual streamlines within individual pores, which differ in shape, size and orientation; and (4) solute particles converging to or diverging from the same pore. All of these processes contribute to increased spreading, in which initially steep concentration fronts become smoother during movement along the main flow direction.

The effects of dispersion can be illustrated with a hypothetical laboratory experiment in which water and a dissolved tracer are applied to an initially tracer-free, uniformly packed soil column of length $L$ (Figure 6.3). The column is subjected to steady-state water flow with a uniform water content. As more of the tracer is added, the initially very sharp concentration front near the soil surface becomes spread out because of dispersion. Eventually, a smooth and sigmoidally shaped effluent curve can be monitored at the column exit as shown in Figure 6.3d. In the absence of dispersion, the front of a perfectly inert tracer will travel as a square wave through the column (a

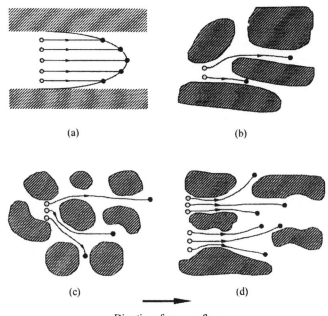

Direction of average flow

**FIGURE 6.2** Schematic concepts contributing to mechanical dispersion.

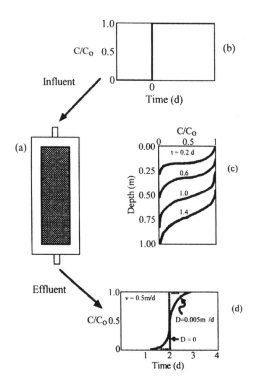

**FIGURE 6.3** Hypothetical laboratory tracer experiment: (a) column of soil, (b) influent curve, (c) concentration distributions inside the column, and (d) breakthrough curves with and without dispersion [Modified after van Genuchten, 1988. p. 360–362. *in* S.P. Parker (ed.) McGraw-Hill Yearbook of Science and Technology, McGraw-Hill, New York].

process often called piston flow) to reach the bottom of the column at time $t = L/v$, where $v$ is the average pore water or interstitial velocity. This velocity is the ratio of the Darcian water flux density ($J_w$), and the volumetric water content ($\theta$). Notice that $v$ is given per unit area of fluid whereas $J_w$ is defined per unit area of (bulk) soil. For piston flow, the tracer reaches the column exit exactly after one pore volume of tracer solution has been injected (or collected at the column exit). Pore volume is defined as the amount of water stored in that column.

The degree of spreading is usually related to the solute travel time, although some constraints do exist on the amount of spreading. Dispersion, as sketched in Figure 6.2a, is limited because of transverse molecular diffusion which causes solutes to move from the center of a tube to areas near the pore walls, or vice versa, in response to local concentration gradients. Such transverse diffusion counteracts spreading caused by variations in the longitudinal flow velocity. Dispersion is also limited since capillaries in a porous medium generally are not independent cylindrical tubes, but branch and join or rejoin each other at distances characteristic of the pore or particle size distribution of the medium. This branching and rejoining promotes lateral mixing of solutes from different pores as sketched in Figure 6.2d.

The macroscopic solute flux due to mechanical dispersion is often conveniently described by Fick's first law of diffusion, despite the conceptual differences between the diffusion and dispersion mechanisms (Scheidegger, 1974). A mathematical foundation for the Fickian description of mechanical dispersion is provided in a classical paper by Taylor (1953). He considered a circular tube, with tube radius $r_o$ (L), filled with water flowing according to a parabolic velocity profile, with $v_o$ being the maximum velocity at the axis (L T$^{-1}$) and a mean velocity $<v> = v_o/2$ over a cross-section of the tube. Taylor (1953) obtained the following expression for the coefficient of mechanical dispersion:

$$D_{dis} = \frac{r_o^2 v_o^2}{192 D_o} = \frac{r_o^2 <v>^2}{48 D_o} \qquad [6.11]$$

which is valid after sufficient time has elapsed. Note that $D_{dis}$ (L$^2$ T$^{-1}$) is inversely proportional to the coefficient of molecular diffusion, $D_o$. The results by Taylor were later derived by Aris (1956) for more general conditions (see examples).

**EXAMPLE 6.1**   Dispersion

Slichter already reported in 1904 on the dispersion of salt during the flow of groundwater in aquifers. The practice of describing dispersion with a single coefficient, referred to as the longitudinal "diffusion" coefficient by Lapidus and Amundson (1952), was established heuristically. However, as is often noted, the justification for describing both diffusion and mechanical dispersion with a single coefficient rests on a convenient similarity in their observed dependency on concentration gradients. A foundation for the Fickian behavior of mechanical dispersion as the result of changing flow paths has been provided using either physical or statistical approaches. The paper by Taylor (1953) provides an analysis of dispersion for a solute injected in a cylindrical tube with laminar flow; this study is often being used as an analog for elucidating dispersion in porous media. Taylor was one of the scientific giants of the twentieth century with seminal contributions on fluid instabilities, turbulence, random walks, rotating flows, shock formation in gases, solute dispersion during blood flow, and electrohydrodynamics (cf. Batchelor, 1996). Taylor's 1953 paper illustrates his typical approach of applying lucid theoretical arguments to complex processes that are supported by experimental results.

For laminar flow in a cylindrical tube filled with water, the parabolic velocity profile is:

$$q(r) = q_o(1 - r^2 / r_o^2) \tag{1}$$

where $q_o$ is the maximum velocity at the center of the tube ($LT^{-1}$), $r$ is the radial distance from the axis (L), and $r_o$ is the tube radius (L). The mean velocity $<q>$ over a cross-section of the tube equals $q_o/2$. If a solute is injected in the tube, the velocity profile tends to stretch its concentration profile while diffusion will mitigate the large (transverse) concentration gradients that would otherwise occur. The solute concentration, $C(x,r,t)$, obeys the following transport equation containing a constant molecular diffusion coefficient $D_o$:

$$\frac{\partial C}{\partial t} = D_o \left( \frac{\partial^2 C}{\partial r^2} + \frac{1}{r} \frac{\partial C}{\partial r} + \frac{\partial^2 C}{\partial x^2} \right) - q_o \left( 1 - \frac{r^2}{r_o^2} \right) \frac{\partial C}{\partial x} \tag{2}$$

In order to quantify the effect of molecular diffusion and a variable fluid velocity on solute spreading, Taylor assumed that diffusion was only significant for radial transport. Some time after the solute has been injected, the concentration profile appears to become steady to an observer moving at the mean flow rate, $<q>$. The effect of the velocity profile is balanced by transversal molecular diffusion. Taylor skillfully simplified the above expression for this scenario. He obtained the following dispersion coefficient from the proportionality between the longitudinal solute flux and the concentration gradient (i.e., Fick's law is applicable):

$$D = \frac{r_o^2 q_o^2}{192 D_o} = \frac{r_o^2 <q>^2}{48 D_o} \tag{3}$$

As mentioned, the dispersion coefficient $D$ is inversely proportional to the coefficient of molecular diffusion, $D_o$. Some caveats are in order when applying these results to dispersion in porous media where the pore geometry is ill-defined and where it is not clear how much time should elapse before solute dispersion becomes Fickian.

Shortly after Taylor's publication appeared, Aris (1956) used his approach to mathematically study ways of improving the accuracy and efficiency of vapor phase chromatography. Lacking Taylor's skill (as Aris, 1991, modestly put it) he relied on moment analysis, which has long since remained customary in chromatography. Aris' expression for the dispersion coefficient is given by:

$$D = (D_o + ka)^2 <q>^2 / D_o \tag{4}$$

where $k$ is a geometry factor and $a$ an effective radius. The analysis of Aris obviates the need to neglect longitudinal diffusion and is applicable to arbitrary geometries.

The Taylor-Aris model for dispersion was adapted early on in soil physics by Nielsen and Biggar (1962). Research to further elucidate dispersion phenomena in porous media has continued along the lines of the original theory (e.g., Gupta and Bhattacharya, 1983).

## REFERENCES

Aris, R. 1956. On the dispersion of a solute in a fluid flowing through a tube, Proc. R. Soc. London, A 235:67–77.

Aris, R. 1991. Longitudinal dispersion due to lateral diffusion, *Current Contents*, 2:8.

Batchelor, G.K. 1996. The Life and Legacy of G.I. Taylor, Cambridge Univ. Press.

Gupta, V.K. and R.N. Bhattacharya. 1983. A New Derivation in the Taylor-Aris Theory of Solute Dispersion in a Capillary, *Water Resour. Res.*, 19:945–951.

Lapidus, L. and N.R. Amundson. 1952. Mathematics of adsorption in beds. VI. The effect of longitudinal diffusion in ion exchange and chromatographic columns, *J. Phys. Chem.,* 56:984–988.

Nielsen, D.R. and J.W. Biggar. 1962. Miscible Displacement: III. Theoretical considerations, *Soil Sci. Soc. Am. Proc.,* 26:216–221.

Slichter, C.S. 1904. The rate of movement of underground waters, Water Supply and Irr. No. 104, U.S. Geol. Survey, Washington, D.C.

Taylor, G.I. 1953. Dispersion of Soluble Matter in Solvent Flowing through a Tube. Proc. R. Soc. London, A 219:186–203.

Because of the complex geometry of the pore space, microscopic flow and transport processes in soils do not easily lend themselves to a relatively simple analysis as is possible for solute transport in a well-defined, water-filled pore. Dispersion in soils can only be approximately described as a Fickian process, particularly at the early stage of solute displacement in which case other models may need to be employed (Jury and Roth, 1990).

The one-dimensional solute flux due to mechanical dispersion in a uniform isotropic soil may be approximated in a similar way as Fick's law:

$$J_{dis} = -\theta D_{dis} \frac{\partial C}{\partial x} \qquad [6.12]$$

where $J_{dis}$ is the dispersive solute flux (M L$^{-2}$ T$^{-1}$).

The above one-dimensional geometry may be too simplistic for many transport problems. Three-dimensional dispersion is quantified with a dispersion tensor. The components of the symmetric dispersion tensor for an isotropic soil are given as (Bear and Verruijt, 1987):

$$D_{ij} = \delta_{ij}\alpha_T|\vec{v}| + (\alpha_L + \alpha_T)v_iv_j/|\vec{v}| \qquad [6.13]$$

where $|\vec{v}|$ denotes the magnitude of the pore water velocity with $v_i$ as the $i$th component (LT$^{-1}$), $\delta_{ij}$ is the Kronecker delta ($\delta_{ij} = 1$, if $i = j$ and $\delta_{ij} = 0$, if $i \neq j$), and $\alpha_L$ and $\alpha_T$ are, respectively, the longitudinal and transverse dispersivity (L). For a one-dimensional system, Equation [6.13] reduces to $D_{dis} = \alpha_L|v|$. Mechanical dispersion, as sketched in an idealized fashion in Figure 6.2a, can be reversed by changing the flow direction to make a smooth front steep again. In soils, however, dispersion is not reversible since mixing erases antecedent concentration distributions, as illustrated in Figure 6.2d. Absolute values are, therefore, used for $v$ in Equation [6.13].

In the case of uniform water flow parallel to the $x$-axis of a Cartesian coordinate system, only the following three main components of Equation [6.13] need to be considered:

$$D_{xx} = \alpha_L v, \quad D_{yy} = \alpha_T v, \quad D_{zz} = \alpha_T v \qquad [6.14]$$

where $D_{xx}$ is the coefficient of longitudinal (mechanical) dispersion, and $D_{yy}$ and $D_{zz}$ are the coefficients of transverse dispersion. This relationship is similar to that derived by Taylor since $<v>$ is inversely proportional to $r_o^2$ in Equation [6.11].

The macroscopic similarity between diffusion and mechanical dispersion has led to the practice of describing both processes with one coefficient of hydrodynamic dispersion ($D = D_{dis} + D_{dif}$). This practice is consistent with results from laboratory and field experiments which do not permit a distinction between mechanical dispersion and molecular diffusion. The hydrodynamic dispersive flux ($J_D$) (Equation [6.2]) consists of contributions from molecular diffusion (Equation [6.4]) and mechanical dispersion (Equation [6.12]):

$$J_D = J_{dif} + J_{dis}$$ [6.15]

Since diffusion is independent of flow, the contribution of diffusion to hydrodynamic dispersion diminishes if the soil water flow rate increases. Hydrodynamic dispersion is often simply referred to as dispersion, as will be done in the remainder of this chapter.

Dispersion coefficients may be determined by fitting a mathematical solution to observed concentrations (Toride et al., 1995). Additional procedures to determine dispersion coefficients are given by Fried and Combarnous (1971) and van Genuchten and Wierenga (1986). Values of the longitudinal dispersivity ($\alpha_L$) for laboratory experiments typically vary between 0.1 and 10 cm with six to 20 times smaller values for $\alpha_T$ (Klotz et al., 1980). Dispersivities for field soils are generally much higher. Gelhar et al. (1992) reviewed published field-scale dispersivities determined in aquifer materials that are typically one or two orders of magnitudes greater, even more so for relatively large experimental scales.

### 6.2.2 ADVECTION-DISPERSION EQUATION

The expressions for the advective and dispersive solute fluxes can be substituted in mass balance Equation [6.1]. The one-dimensional advection-dispersion equation for solute transport in a homogeneous soil becomes:

$$\frac{\partial \theta C}{\partial t} = -\frac{\partial}{\partial x}\left(J_w C - \theta D \frac{\partial C}{\partial x}\right) + \theta R_s$$ [6.16]

In the case where the water content is invariant with time and space, the ADE may be simplified to ($v = J_w/\theta$):

$$\frac{\partial C}{\partial t} = D \frac{\partial^2 C}{\partial x^2} - v \frac{\partial C}{\partial x} + R_s$$ [6.17]

This is a second-order linear partial differential equation. Like the diffusion equation, the ADE is classified as a parabolic differential equation. To complete the mathematical formulation of the transport problem, several concentration types and mathematical conditions will be reviewed in Section 6.3.1.

A variety of solute source or sink terms may be substituted for $R_s$. The most common source/sink term is due to adsorption/desorption and ion exchange stemming from chemical and physical interactions between the solute and the soil solid phase. Many other processes such as radioactive decay, aerobic and anaerobic transformations, volatilization, photolysis, precipitation/dissolution, reduction/oxidation, and complexation may also affect the solute concentration. A further refinement of the transport model is necessary in the case of nonuniform interactions between the solute and the soil, or if there is adsorption on moving particles and colloids. In the following, only interactions at the solid-liquid interface will be considered.

### 6.2.3 ADSORPTION

Dissolved substances in the liquid phase can interact with several soil constituents such as primary minerals, oxides, and inorganic or organic colloids. Dissolved ions in the soil solution counterbalance the surface charge of soil particles caused by isomorphous substitution of one element for another in the crystal lattice of clay minerals, by the presence of hydronium or hydroxyl ions at the solid surface, or other mechanisms. The net surface charge of an assemblage of soil particles produces an electric field that affects the distribution of cations and anions within water films surrounding the soil particles. The mechanisms and characteristics of reactions in solid-solution-solute systems are further discussed in Chapter 3.

Adsorption of solute (adsorbate) by the soil (adsorbent) is an important phenomenon affecting the fate and movement of solutes. The ADE for one dimensional transport of an adsorbed solute may be written as:

$$\frac{\partial C}{\partial t} + \frac{\rho_b}{\theta}\frac{\partial S}{\partial t} = D\frac{\partial^2 C}{\partial x^2} - v\frac{\partial C}{\partial x} \qquad [6.18]$$

where $S$ is the adsorbed concentration, defined as mass of solute per mass of dry soil (M M$^{-1}$). The above equation can be expressed in terms of one dependent variable by assuming a suitable relationship between the adsorbed and liquid concentrations. This is typically done with a simple adsorption isotherm to quantify the adsorbed concentration as a function of the liquid concentration at a constant temperature. In addition to temperature, the adsorption isotherm is generally also affected by the solution composition, total concentration, the pH of the bulk solution, and sometimes the method used for measuring the isotherm. A mathematically pertinent distinction is often made between linear and nonlinear adsorption. Although most adsorption isotherms are nonlinear, the adsorption process may often be assumed linear for low solute concentrations or narrow concentration ranges.

### 6.2.3.1  Linear Adsorption

Consider the general case of nonequilibrium adsorption, where a change in $C$ is accompanied by a delayed change in $S$. The adsorption rate can be described assuming first order kinetics:

$$\frac{\partial S}{\partial t} = kh(C,S) \qquad [6.19]$$

where $k$ is a rate parameter (T$^{-1}$) and $h$ is a function to quantify how far the adsorption or desorption process is removed from equilibrium. A single-valued isotherm for equilibrium adsorption $\Gamma(C)$ as in Equation [6.24], is used to define $h(C,S)$ according to:

$$h(C,S) = \Gamma(C) - S \qquad [6.20]$$

For equilibrium adsorption $k \to \infty$, and hence $h(C,S) \to 0$, which implies that $S = \Gamma(C)$. For a linear adsorption isotherm, the relation between $\Gamma$ and $C$ can simply be given as:

$$\Gamma = K_d C \qquad [6.21]$$

where $K_d$ is a partition coefficient, often referred to as the distribution coefficient, expressed in volume of solvent per mass of soil (L$^3$ M$^{-1}$). For $\Gamma = S$, substitution of Equation [6.21] into Equation [6.18] leads to the following ADE commonly used to describe transport of a solute that undergoes linear equilibrium exchange:

$$R\frac{\partial C}{\partial t} = D\frac{\partial^2 C}{\partial x^2} - v\frac{\partial C}{\partial x} \qquad [6.22]$$

in which the retardation factor $R$ is given by:

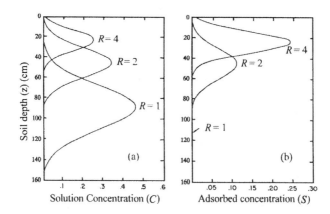

**FIGURE 6.4** Effect of adsorption, as accounted for by the retardation factor $R$, on solution ($C$) and adsorbed ($S$) concentration distributions in a homogeneous soil profile.

$$R = 1 + \frac{\rho_b}{\theta} K_d \qquad [6.23]$$

with $\rho_b$ as soil bulk density. The advective and dispersive fluxes are reduced by a factor $R$ as a result of adsorption. The movement of the solute is said to be retarded with respect to the average solvent movement. If there is no interaction between the solute and the soil ($K_d = 0$), the value for $R$ is equal to unity. The value for R can be readily calculated from $K_d$ as obtained from chemical analyses of the solution and adsorbed phases. Alternatively, $R$ can be estimated from solute displacement studies on laboratory soil columns (Figure 6.3). A mathematical solution may then be used to estimate $R$ from observed concentrations with nonlinear optimization programs. The change in the amount of solute in the soil column should be equal to the net solute flux into the column; the following mass balance can hence be formulated to estimate $R$:

$$v\big[g(t) - C_e\big] = R \int_0^L \big[C(x,t) - f(x)\big] dx \qquad [6.24]$$

where $C_e$ is the effluent concentration and $f(x)$ and $g(t)$ are the initial and influent concentrations.

The effects of linear adsorption on solute transport in a homogeneous soil profile are shown in Figure 6.4. Analytically predicted solution and adsorbed concentrations are plotted four days after the start of a one-day application of influent with a unit solute concentration (units may be selected arbitrarily) to an initially solute-free soil profile subject to steady saturated water flow. Other parameters for this example are $J_w = 10$ cm d$^{-1}$, $\theta = 0.40$ cm$^3$ cm$^{-3}$, and $D = 62.5$ cm$^2$ d$^{-1}$. The pore water velocity ($v = J_w/\theta$) is hence 25 cm d$^{-1}$ and $\alpha_L = 2.5$ cm. Solute distributions are plotted (Figure 6.4) for three values of the retardation factor, $R$. When $R$ is increased from 1.0 to 2.0, the apparent solute velocity ($v/R$) is reduced by one-half (Figure 6.4a), causing a penetration of the solute pulse into the profile. At the same time, the area under the curve in Figure 6.4a is also reduced by one-half. When $R = 4$, the apparent solute velocity and the area under the curve are again reduced by half. Distributions for the adsorbed concentration ($S$) which increases from zero (no adsorption) when $R = 1$ to a maximum when $R = 4$ are similar (Figure 6.4b). Assuming a soil bulk density ($\rho_b$) of 1.25 g cm$^{-3}$ and the same water content as before ($\theta = 0.40$ cm$^3$ cm$^{-3}$), one may calculate, using Equation [6.23], that the distribution coefficient $K_d = 0$, 0.32 and 0.96 cm$^3$ g$^{-1}$ for $R = 1$, 2 and 4, respectively.

**EXAMPLE 6.2**  Estimating Transport Parameters from Solute Displacement Experiments

Many methods have been used to estimate ADE solute transport parameters (e.g., $D$ and $R$) from miscible displacement experiments, including methods of moments, graphical procedures, and least-squares parameter optimization (inverse) methods. Methods of moments were initially adopted from the chemical and petroleum engineering literature (Aris, 1958), and have remained popular to this day. Graphical procedures for estimating D and R were very popular in the soil physics literature in the past. They were generally based on an approximate analytical analysis of the slope and location of a breakthrough effluent curve plotted as a function of the number of pore volumes, or from a logarithmic-normal plot of the breakthrough curve (van Genuchten and Wierenga, 1986).

Least-squares and related parameter optimization methods became popular by the early 1980s, mostly because of the availability of increasingly faster desktop computers. Especially popular was the CXTFIT parameter estimation code of Parker and van Genuchten (1984) for estimating solute transport parameters from a variety of laboratory and field tracer experiments. This code also included a stochastic transport model that considered the effects of areal variations in hydraulic fluxes on field-scale solute transport. An update of that code, called CXTFIT2, was published by Toride et al. (1995) to enable more flexible initial and boundary conditions, as well as additional zero-order production and first-order degradation scenarios. A convenient windows-based graphical interface for CXTFIT2 has recently become available (Šimůnek et al., 2000).

Examples of results obtained from optimizations with CXTFIT are given in Figures 6.5 and 6.12. A more complicated case is shown in Diagram 6.2.1 for iodide transport through a large approximately 6-m deep lysimeter (Caisson B) at Los Alamos National Laboratory packed with crushed Bandelier Tuff (van Genuchten et al., 1987). Solute transport was described with Equation [6.22] subject to a uniform initial concentration of zero and a pulse-type boundary condition of duration $t_o$ at the soil surface. Observed concentrations obtained with hollow fiber suction samplers were assumed to represent flux-averaged concentrations, while $R$ was taken to be unity (thus assuming no sorption).

Diagram 6.2.1 shows the observed iodide breakthrough curves for six depths resulting from the application of a solute pulse beginning at December 6, 1984, under approximately steady flow conditions. The breakthrough curves that were calculated using the fitted parameters are also shown in Diagram 6.2.1. Because of mass balance problems, including uncertainty in the amount of mass that was applied to the lysimeter, the pulse duration $t_o$ was considered unknown rather than using the imposed experimental target value of 6 days. The parameters $v$, $D$ and $t_o$ were fitted simultaneously to the concentration data for all depths, except those for the 2.64-m depth. Notice the poor description of the 2.64-m depth data. Spatial variations in the pore water velocity due to differences in compaction and the presence of a dual-porosity flow field were used as possible explanations for this behavior. Fitted values for $v$, $D$, and $t_o$ were 11.74 ($\pm$ 0.14) cm/day, 11.29 ($\pm$ 3.09) cm$^2$/day, and 5.11 ($\pm$ 0.22) days, respectively. The parameter estimation procedure greatly facilitated calibration of the ADE transport model for this application. The results indicated that even for the relatively homogeneous, artificially constructed medium used in this study, significant variations in water contents and/or hydraulic fluxes can occur within the caisson. More details are given by van Genuchten et al. (1987).

## REFERENCES

Aris, R. 1958. On the dispersion of kinematic waves, *Proc. Royal Soc. London*, Ser. A 245:268–277.
Parker, J.C. and M. Th. van Genuchten. 1984. Determining Transport Parameters from Laboratory and Field Tracer Experiments, *Bull. 84-3*, Virginia Agric. Exp. Sta., Blacksburg, VA, 91 p.

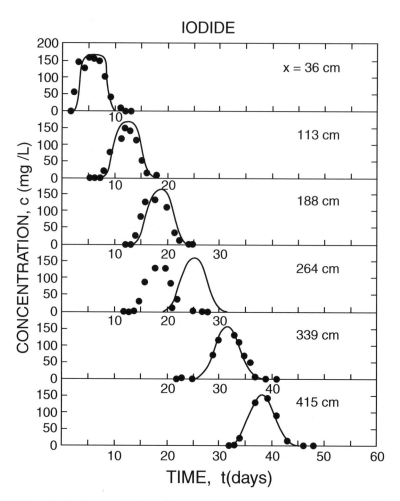

**DIAGRAM 6.2.1** Iodide breakthrough curves measured at various depths, and fitted curves obtained by least-squares parameter optimization using all data except those for the 2.64-m depth location.

van Genuchten, M. Th., J.C. Parker, and J.B. Kool. 1987. Analysis and prediction of water and solute transport in a large lysimeter, *in*: E.P. Springer and H.R. Fuentes (eds.), Modeling Study of Solute Transport in the Unsaturated Zone, *NUREG/CR-4615, LA-10730-MS*, U.S. Nuclear Regulatory Commission, Washington, DC, vol. 2, p. 4–31.

Šimůnek, J., M. Th. van Genuchten, M. Šejna, N. Toride, and F.J. Leij. 2000. STANMOD Studio of analytical models for solving the convection-dispersion equation. Version 1.0. International Ground Water Modeling Center, Colorado School of Mines, Golden, CO.

Toride, N., F.J. Leij, and M. Th. van Genuchten. 1995. The CXTFIT code for estimating transport parameters from laboratory or field tracer experiments, *U.S. Salinity Laboratory Research Report 137*, Riverside CA. 121 pp.

Anion exclusion occurs when negatively charged surfaces of clays and ionizable organic matter are present; anions are repelled from such surfaces and accumulate in the center of pores. Because water flow velocities are zero at pore walls and maximum in the center of pores (Figure 6.2a), the average anion movement will be faster than the average water movement. Many displacement experiments also suggest faster anion than water movement simply because the apparent displacement

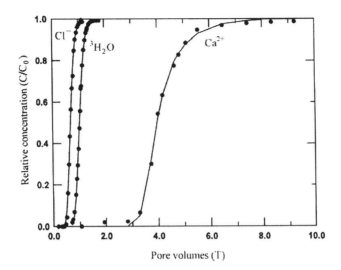

**FIGURE 6.5** Observed and fitted ADE breakthrough curves for three tracers typifying the transport of anions ($Cl^-$), a (nearly) nonreactive solute ($^3H_2O$), and an adsorbing solute ($Ca^{2+}$).

volume is smaller for anions than water. The quantity (1-R) is the relative anion exclusion volume. The exclusion volume-per-unit mass of soil can also be estimated as:

$$V_{ex} \int \left(1 - c/C_o\right) dV \qquad [6.25]$$

where $V_{ex}$ is the exclusion volume ($L^3\ M^{-1}$), $c$ is the local concentration of the anion ($M\ L^{-3}$) and $C_o$ its bulk concentration ($M\ L^{-3}$), and $V$ is the entire volume encompassing the liquid phase. Instead of using $R < 1$, anion transport may be modeled with a model, with $R = 1$, which restricts the accessible liquid volume (Krupp et al., 1972).

Anions are also adsorbed by the soil through surface complexation and adsorption onto positively charged areas of the solid matrix. If the effect of adsorption exceeds exclusion, the anion will be retarded. The retardation factor should be viewed as an effective parameter since it quantifies a variety of adsorption and exclusion processes to which the solute (anion) is subjected.

Breakthrough curves typical for the transport of an excluded anion ($Cl^{-1}$), a nonreactive solute (tritiated water, $^3H_2O$), and an adsorbed cation ($Ca^{2+}$) are presented in Figure 6.5. The first two tracers pertain to transport through 30-cm columns containing disturbed Glendale clay loam soil (P. J. Wierenga, personal communication; van Genuchten and Cleary, 1982), while the $Ca^{2+}$ data are for transport through a 30-cm long column containing a Troup loam and a Savannah fine loam (Leij and Dane, 1989). Analysis of the three breakthrough curves in terms of the ADE, using inverse procedures (Parker and van Genuchten, 1984b), yielded $R$ values of 0.681, 1.027, and 4.120 for $Cl^-$, $^3H_2O$, and $Ca^{2+}$. Hence, the $Cl^-$ curve was strongly affected by anion exclusion, while $^3H_2O$ transport was subject to very minor adsorption/exchange.

### 6.2.3.2 Nonlinear Adsorption

In many cases adsorption, and hence, the retardation factor, cannot be described using a simple $K_d$ approach. For nonlinear equilibrium adsorption, $R$ is given as:

$$R(C) = 1 + \frac{\rho_b}{\theta} \frac{\partial \Gamma}{\partial C} \qquad [6.26]$$

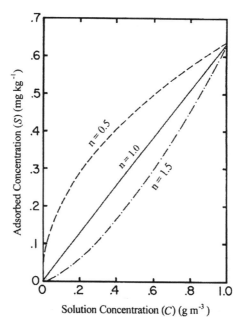

**FIGURE 6.6** Freundlich equilibrium plots for $k_3 = 0.64$ and three values of the exponent $n$.

Two common nonlinear adsorption isotherms are the Langmuir and Freundlich equation:

$$\Gamma = \frac{k_1 C}{1 + k_2 C} \qquad \text{Langmuir} \qquad [6.27]$$

$$\Gamma = k_3 C^n \qquad \text{Freundlich} \qquad [6.28]$$

where $k_1$, $k_2$, $k_3$, and $n$ are empirical constants. Many other equations for adsorption exist, including some that account for differences between adsorption and desorption isotherms (van Genuchten and Sudicky, 1999).

The Freundlich isotherm will be used in the following to illustrate the effects of nonlinear equilibrium adsorption on solute transport. In order to keep the calculations simple, the value of $k_3$ in Equation [6.28] is taken to be 0.64. Three different values of the exponent $n$ are used, viz., 0.5, 1.0, and 1.5, to demonstrate favorable, linear, and unfavorable adsorption isotherms (Figure 6.6). Calculated distributions of the solution ($C$) and adsorbed ($S$) concentrations versus soil depth ($z$) eight days after application of a 4-day-long solute pulse to the soil surface are shown in Figure 6.7. The same pore water velocity is used as for the example illustrated in Figure 6.5, but with a smaller dispersion coefficient of D = 25 cm$^2$ d$^{-1}$ ($\alpha_L$ = 1 cm). Notice that, as in Figure 6.5, the solution concentration distribution for $n = 1$ (linear adsorption) has a nearly symmetrical shape versus depth. The other two $n$ values yield nonsymmetric profiles.

If $n = 0.5$, a very sharp concentration front develops, while the curve near the soil surface becomes more dispersed. The sharp front can be explained by considering the retardation factor ($R$) for nonlinear adsorption (Equation [6.26]), which for $n = 0.5$, $\rho_b = 1.25$ g cm$^{-3}$, $\theta = 0.40$, and $k_3 = 0.64$ leads to $R = 1 + 1/C$. This shows that $R$ increases rapidly when $C$ decreases with the extreme $R \rightarrow \infty$, if $C = 0$. Consequently, the apparent solute velocity $v_a = v/R$ is very small at the lower liquid concentrations, but increases at higher values. Of course, higher concentrations cannot move faster than lower concentrations; front sharpening will lead to a steep solute front. This front never becomes a step function because the large concentration gradient across the front will create a large

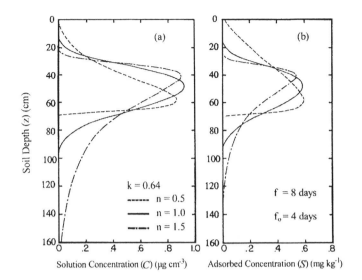

**FIGURE 6.7** Effect of nonlinear adsorption on solution ($C$) and adsorbed ($S$) concentrations in a deep homogeneous soil profile during steady-state flow. The distributions were obtained for the three isotherms shown in Figure 6.6.

diffusion/dispersion flux. When $n \ll 1$, an estimate of the front can be obtained from the average slope of the isotherm between the initial and the maximum concentration. Because in the present example these are zero and approximately one, the average slope of the isotherm is exactly the same as the linear distribution coefficient ($d\Gamma/dC = 0.64$). Substituting this value into Equation [6.26] yields $R = 3$. Hence, the apparent solute velocity ($v_a$) equals 25/3 or 8.33 cm d$^{-1}$, and the solute front after 8 days is located at a depth of about 67 cm (Figure 6.7). Transport of favorably adsorbed solutes is frequently modeled with traveling wave solutions (van der Zee, 1990; Simon et al., 1997).

A reverse scenario occurs if $n > 1$ (unfavorable exchange). Adsorption at the lower concentrations is now relatively small and, as displayed in Figure 6.7, the toe of the front moves through the profile at a velocity nearly equal to that of an inert solute. Adsorption at the higher concentrations, on the other hand, is much more extensive, resulting in a lower apparent solute velocity in the higher range of concentrations. As a result, the concentration front becomes increasingly dispersed over time. Ignoring dispersion, the velocity of the solute front ($v_a = v/R$) at any given value of $C$ is given by:

$$v_a = \frac{v}{1+3\sqrt{C}} \qquad [6.29]$$

while the depth of the solute front can be approximated by:

$$z(C,t) = z(C,0) + v_a t \qquad [6.30]$$

where $z(C,0)$ is the location of a solute concentration with value $C$ at $t = 0$.

The above discussion pertained to adsorption of a single ion species. Cation exchange processes in transport studies involve at least two species. The simplest case arises when two cations of the same valence and total concentration such as $Ca^{2+}$ and $Mg^{2+}$ are considered. The resulting exchange process is then approximately linear for relatively small changes in the composition of the soil solution. Exchange between $Na^+$ and $Ca^{2+}$, on the other hand, is considerably more nonlinear. Equations that quantify the exchange reaction have been proposed by Gapon, Kerr, Vanselow, Eriksson and others.

**EXAMPLE 6.3**   Microbial Transport

Microbial pathogens in the subsurface environment may pose a serious health hazard to humans. This issue has motivated research on microbial transport in soils. Microbial transport thus far has been mostly described according to the advection-dispersion equation augmented with terms for production and decay using zero- and/or first-order terms. However, pathogenic organisms may behave very differently than substances that are traditionally described with the ADE such as salts, dissolved organic compounds and heat. An important distinction is the ability of microorganisms to adapt to their environment. Another complication is that there are vast differences in size, surface properties and survival rates between organisms. The three major groups are protozoa, bacteria and viruses with approximate sizes of 1-12 μm, 0.1-1 μm and 0.02-0.1 μm, respectively. Viruses are of greatest concern because they are not easily trapped in soil while they tend to be more persistent than bacteria (Jin et al., 2000).

Several efforts have been undertaken to model microbial transport (Yates and Yates, 1988). Corapcioglu and Haridas (1984) proposed a general model based on the ADE, which for one-dimensional transport is given by:

$$R_a + \frac{\partial \theta C}{\partial t} = -\frac{\partial J}{\partial x} + R_{df} + R_{gf} \qquad [1]$$

where $C$ is the concentration of suspended particles (viruses or bacteria), $\theta$ is the volume fraction of water contributing to microbial transport, $R_{df}$ and $R_{gf}$ denote the degradation and growth of microbial organisms in the liquid phase, respectively, and $x$ and $t$ are position and time. The rate of deposition onto soil particles is given by $R_a$, which is affected by attachment and detachment from the soil as well as possible microbial decay and growth in the solid phase ($R_{ds}$ and $R_{gs}$). Bacteria may be removed from the solution due to straining by small soil pores, adsorption and sedimentation. Viruses are mainly removed by adsorption while they cannot reproduce outside host cells. Although adsorption can be investigated with batch or column experiments commonly used for kinetic and equilibrium exchange studies in soil chemistry, the interaction between solid and aqueous phases is obviously more complex. Harvey and Garabedian (1991) employed both sorption and filtration kinetics to describe the effect of the solid phase on bacterial transport. Finally, $J$ in [1] represents the microbial mass flux density. This term includes well-known advective and dispersive transport processes, but is also used to describe coordinated movement (chemotaxis) to a substrate source and related random movement (tumbling). Substrate concentrations need to be modeled with a separate mass balance equation.

The message here is that the simulation of microbial transport poses far greater challenges than that of other types of solutes. Process-based description of the relevant processes is much more difficult, while experimental studies are far more cumbersome and expensive. Also, many more model parameters exist that cannot be easily quantified independently, whereas small changes in water content or temperature can have considerable effects.

**REFERENCES**

Corapcioglu, M.Y. and A. Haridas. 1984. Transport and fate of microorganisms in porous media: A theoretical investigation, *J. Hydrol.*, 72:149–169.

Harvey, R.W. and S.P. Garabedian. 1991. Use of colloid filtration theory in modeling movement of bacteria through a contaminated sandy aquifer, *Environ. Sci. Technol.*, 25:178–185.

Jin, Y., M.V. Yates, and S.R. Yates. 2000. Microbial transport, *in* Methods of Soil Analysis, *Soil Sci. Soc. Am.,* Madison, WI.

Yates, M.V. and S.R. Yates. 1988. Modeling microbial fate in the subsurface enviroment, CRC Critical Rev. Env. Cont. 17:307–344.

### 6.2.4 NONEQUILIBRIUM TRANSPORT

Solute breakthrough curves for aggregated soils will exhibit asymmetrical distributions or nonsigmoidal concentration fronts. The concept behind physical nonequilibrium models is that differences between regions of the liquid phase lead to mostly lateral gradients in the solute concentration, resulting in a diffusive type of solute transfer process. Depending upon the exact pore structure of the medium, asymmetry is sometimes enhanced by desaturation when the relative fraction of water residing in the marginally continuous immobile region increases.

Since most of the sorption sites are only accessible after diffusion through the immobile region of the liquid phase, a corresponding delay in adsorption will occur. The delayed adsorption can also be explained with a kinetic description of the adsorption process. Both cases may be described with chemical nonequilibrium models, which distinguish between sites with equilibrium and kinetic sorption.

Bi-continuum or dual-porosity nonequilibrium models are the most widely used. Only two concentrations need to be considered and the equilibrium ADE (Equation [6.18]) can be readily modified for this purpose. The same dimensionless mathematical formulation can be used for physical and chemical nonequilibrium models. If necessary, the ADE can be modified to incorporate additional nonequilibrium processes and continua.

#### 6.2.4.1 Physical Nonequilibrium

Consider one-dimensional solute movement in an isotropic soil with uniform flow and transport properties during steady flow, and assume that the solute is subject to linear retardation, that is, equilibrium sorption can be described with a linear exchange isotherm. The physical nonequilibrium approach is based on a partitioning of the liquid phase into a mobile or flowing region and an immobile or stagnant region. Solute movement in the mobile region occurs by both advection and dispersion, whereas solute exchange between the two regions occurs by first order diffusion (Coats and Smith, 1964). Following van Genuchten and Wierenga (1976), the governing equations for the two region model are:

$$\left(\theta_m + f\rho_b K_d\right)\frac{\partial C_m}{\partial t} = \theta_m D_m \frac{\partial^2 C_m}{\partial x^2} - \theta_m v_m \frac{\partial C_m}{\partial x} - \alpha\left(C_m - C_{im}\right) \qquad [6.31]$$

$$\left[\theta_m + (1-f)\rho_b K_d\right]\frac{\partial C_{im}}{\partial t} = \alpha\left(C_m - C_{im}\right) \qquad [6.32]$$

where $f$ represents the fraction of sorption sites in equilibrium with the fluid of the mobile region, $\alpha$ is a first order mass transfer coefficient ($T^{-1}$), and the subscripts m and im, respectively, refer to the mobile and immobile liquid regions (with $\theta = \theta_m + \theta_{im}$), while $\rho_b$ and $K_d$ are the soil bulk density and distribution coefficient for linear sorption. Transport Equation [6.31] follows directly from addition of a source/sink term ($R_s$) to Equation [6.18].

Anion exclusion can be viewed as a particular example of physical nonequilibrium; the exclusion volume roughly corresponds to the immobile region (Krupp et al., 1972). The physical nonequilibrium concept may, therefore, be adapted to describe transport of excluded anions (van Genuchten, 1981) instead of using a retardation factor of less than one.

---

**EXAMPLE 6.4**   Solute Transport in Structured Media

Field soils generally exhibit a variety of structural features, such as inter-aggregate pores, earthworm or gopher holes, decayed root channels, or drying cracks in fine-textured soils. Water and dissolved chemicals in such soils may move along preferred pathways at rates much faster than what normally can be predicted with the ADE. Mathematical descriptions of preferential flow in structured media are often based on dual-porosity, two-region, or bi-continuum models. Models of this type assume that the medium consists of two interacting pore regions, one associated with the macropore or fracture network, and one with the micropores inside soil aggregates or rock matrix blocks. Different formulations arise depending upon how water and solute movement in the micropore region are modeled, and how the micropore and macropore regions are coupled. The simplest physical nonequilibrium or two-region model is given by Equations [6.31] and [6.32]. This model conveniently assumes that there is no flow in the micropore region and that an averaged value may be used for the concentration in the micropore region.

A process-based description of transport in structured soils is possible when the medium is assumed to contain geometrically well-defined cylindrical, rectangular or other types of macropores or fractures. Models may be formulated by assuming that the chemical is transported by advection, and possibly by diffusion and dispersion, through the macropores, while diffusion-type equations are used to describe the transfer of solutes from the larger pores into the micropores of the soil matrix. As an example, the governing equations for transport through media containing parallel rectangular voids (Figure 1) are given by (van Genuchten and Dalton, 1986):

$$\theta_m R_m \frac{\partial C_m}{\partial t} + \theta_{im} R_{im} \frac{\partial C_{im}}{\partial t} = \theta_m D_m \frac{\partial^2 C_m}{\partial z^2} - \theta_m v_m \frac{\partial C_m}{\partial z} \qquad [1]$$

$$R_{im} \frac{\partial C_a}{\partial t} = D_a \frac{\partial^2 C_a}{\partial x^2} \qquad (-a \leq x \leq a) \qquad [2]$$

$$C_{im}(z,t) = \frac{1}{a} \int_0^a C_a(z,x,t)dx \qquad [3]$$

where the subscripts $m$ and $im$ refer to the mobile (inter-aggregate) and immobile (intra-aggregate) pore regions, respectively, $C_a(z,x,t)$ is the local concentration in the aggregate, $z$ is depth, $x$ is the horizontal coordinate, and $D_a$ is the effective soil or rock matrix diffusion coefficient. Equation [1] describes vertical advective-dispersive transport through the fractures, while [2] accounts for linear diffusion in a slab of width $2a$. Equation [3] represents the average concentration of the immobile soil matrix liquid phase Equations [1] and [2] are coupled using the assumption of concentration continuity across the fracture/matrix interface:

$$C_a(z,a,t) = C_m(z,t) \qquad [4]$$

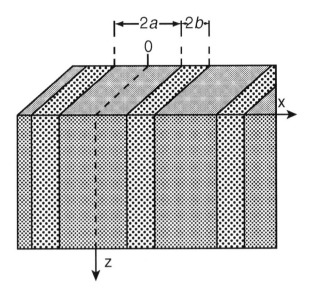

**DIAGRAM 6.4.1** Schematic of rectangular porous matrix blocks of width $2a$, separated by fractures of width $2b$.

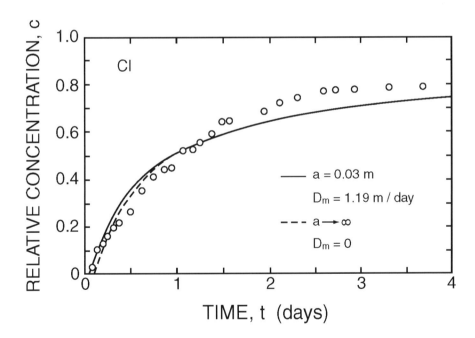

**DIAGRAM 6.4.2** Measured breakthrough curve for Cl transport through fractured clayey till (open circles); data from Grisak et al. (1980). The solid line was obtained with the exact solution of Equations [1] through [4]. The dashed line was obtained by ignoring dispersion in the fractures and allowing the fracture spacing to go to infinity.

Sometimes the water contents are given in terms of the bulk soil volume, especially for dual-permeability approaches that consider (unsaturated) flow in both the inter- and intra-aggregate regions (Gerke and van Genuchten, 1993).

Similar models as above may be formulated for other aggregate or soil matrix geometries (van Genuchten, 1985). Geometry-based transport models have been successfully applied to laboratory-scale experiments as well as to selected field studies involving mostly saturated conditions. As an example, Diagram 6.4.2 shows calculated and observed Cl effluent curves from a 76-cm long undisturbed column of fractured clayey till. The solid line was obtained with the exact solution (van Genuchten, 1985) of Equations [1] through [4] using parameter values given by Grisak et al. (1980). The dashed line was obtained by ignoring dispersion in the macropores ($D_m = 0$) and allowing the fracture spacing to go to infinite ($a \rightarrow \infty$). The extremely skewed (nonsigmoidal) shape of the effluent curve is a direct result of water and dissolved chemicals moving mostly through the fractures and bypassing the soil matrix, but with diffusion taking place between the fractures and the matrix.

While geometry-based models are conceptually attractive, they may be too complicated for routine applications since structured field soils usually contain a mixture of aggregates of various sizes and shapes. The problem of macropore and fluid flow continuity is also not easily addressed with geometry-based flow models. For this reason, two-region models such as Equations [6.31] and [6.32] that are based on a first-order exchange process are far more popular because they can captivate many of the preferential flow features without a need to quantify the intra-aggregate diffusion process. The first-order transfer coefficient, $\alpha$, that characterizes diffusional exchange of solutes between the mobile and immobile liquid phases can still be related to the intra-aggregate diffusion coefficient using a geometry-dependent shape factor and a characteristic length of the aggregate (van Genuchten and Dalton, 1986).

## REFERENCES

Gerke, H.H. and M. Th. van Genuchten. 1993. A dual-porosity model for simulating the preferential movement of water and solutes in structured porous media, *Water Resour. Res.*, 29(2):305–319.

Grisak, G.E., J.F. Pickens, and J.A. Cherry. 1980. Solute transport through fractured media, 2. Column study of fractured till, *Water Resour. Res.*, 16:731–739.

van Genuchten, M. Th. 1985. A general approach for modeling solute transport in structured soils, Proc. 17th. Int. Congress, Hydrogeology of Rocks of Low Permeability, Jan. 7–11, Tucson, AZ, *Memoires Int. Assoc. Hydrogeol.*, 17(2), 513–526.

van Genuchten, M. Th., and F.N. Dalton. 1986. Models for simulating salt movement in aggregated field soils, *Geoderma*, 38, 165–183, 1986.

### 6.2.4.2 CHEMICAL NONEQUILIBRIUM

Sorption of solute, especially for organic chemicals, has often been described with a combined equilibrium and kinetic sorption expression so as to better simulate transport in soils with a wide variety of soil constituents (clay minerals, organic matter and oxides). The lack of an instantaneous equilibrium for the sorption process is sometimes referred to as chemical nonequilibrium. This terminology is somewhat misleading since the rate of adsorption or exchange is usually determined mostly by physical phenomena such as diffusion through the liquid film around soil particles and inside the aggregates (Boyd et al., 1947; Sparks, 1989).

The simplest and by far most popular approach distinguishes between type-1 sites, with instantaneous adsorption, and type-2 sites, where adsorption obeys a kinetic rate law (Selim et al., 1976). In the case of first-order kinetics, the general adsorption rates can be given with a model similar to Equations [6.19] and [6.20] as:

$$\frac{\partial S_1}{\partial t} = \alpha_1 \left[ \Gamma_1(C) - S_1 \right] \qquad [6.33]$$

$$\frac{\partial S_2}{\partial t} = \alpha_2 \left[ \Gamma_2(C) - S_2 \right] \tag{6.34}$$

where $\alpha$ is again a rate constant $(T^{-1})$, $S$ is the actual adsorbed concentration $(M\ M^{-1})$, $\Gamma$ is the final adsorbed concentration at equilibrium as prescribed by the adsorption isotherm, the subscripts 1 and 2 refer to the type of adsorption site, and $\Gamma_1 + \Gamma_2 = \Gamma$. Because type-1 sites are always at equilibrium, $S_1 = \Gamma_1$ and Equation [6.33] can further be ignored. The transport equation becomes:

$$\frac{\partial C}{\partial t} + \frac{\rho_b}{\theta} \frac{\partial \Gamma}{\partial t} + \frac{\alpha_2 \rho_b}{\theta} \left( \Gamma_2 - S_2 \right) = D \frac{\partial^2 C}{\partial x^2} - v \frac{\partial C}{\partial x} \tag{6.35}$$

If the fraction of exchange sites that is at equilibrium (type-1) equals f, and if equilibrium adsorption is governed by the same linear isotherm for both types 1 and 2 ($\Gamma_1 = \Gamma_2$) then:

$$\Gamma_1 + \Gamma_2 = f K_d C + (1 - f) K_d C \tag{6.36}$$

Of course, nonlinear equilibrium isotherms may also be used in nonequilibrium transport models. The complete transport problem can now be written as:

$$\left( 1 + \frac{\rho_b f K_d}{\theta} \right) \frac{\partial C}{\partial t} = D \frac{\partial^2 C}{\partial x^2} - v \frac{\partial C}{\partial x} - \frac{\alpha \rho_b}{\theta} \left[ (1 - f) K_d C - S_2 \right] \tag{6.37}$$

$$\frac{\partial S_2}{\partial t} = \alpha \left[ (1 - f) K_d C - S_2 \right] \tag{6.38}$$

where the subscript for $\alpha$ has been dropped. This two-site chemical nonequilibrium model reduces to a one-site kinetic nonequilibrium model by setting $f = 0$. The two-site chemical nonequilibrium model was applied successfully to describe solute breakthrough curves by Selim et al. (1976), van Genuchten (1981) and Nkedi-Kizza et al. (1983), among others.

### 6.2.4.3   General Nonequilibrium Formulation

The two-site and the two-region nonequilibrium models can be cast in the same (dimensionless) model according to Nkedi-Kizza et al. (1984):

$$\beta R \frac{\partial C_1}{\partial t} = \frac{1}{P} \frac{\partial^2 C_1}{\partial X^2} - \frac{\partial C_1}{\partial X} + \omega \left( C_2 - C_1 \right) \tag{6.39}$$

$$(1 - \beta) R \frac{\partial C_2}{\partial T} = \omega \left( C_1 - C_2 \right) \tag{6.40}$$

where $\beta$ is a partition coefficient, $R$ is a retardation factor, $C_1$ and $C_2$ are dimensionless equilibrium and nonequilibrium concentrations, $T$ is time, $X$ is distance, $P$ is the Peclet number, $\omega$ is a mass transfer coefficient, and the subscripts 1 and 2 refer to the equilibrium and nonequilibrium phases, respectively. The common dimensionless parameters are defined using an arbitrary characteristic concentration $(C_o)$ and length $(L)$:

$$T = vt/L, \quad X = x/L, \quad P = vL/D, \quad R = 1 + \rho_b K_d/\theta \qquad [6.41]$$

For the physical nonequilibrium model, the remaining dimensionless parameters are:

$$\beta = \frac{\theta_m + f\rho_b K_d}{\theta + \rho_b K_d}, \quad \omega = \frac{\alpha L}{\theta v}, \quad C_1 = \frac{C_m}{C_o}, \quad C_2 = \frac{C_{im}}{C_o} \qquad [6.42]$$

whereas for the chemical nonequilibrium model:

$$\beta = \frac{\theta_m + f\rho_b K_d}{\theta + \rho K_d}, \quad \omega = \frac{\alpha(1-\beta)RL}{v}, \quad C_1 = \frac{C}{C_o}, \quad C_2 = \frac{S_2}{(1-f)K_d C_o} \qquad [6.43]$$

In the chemical engineering literature, $\alpha\omega v$ is known as the Damköhler number; it quantifies the rate of the reaction or exchange relative to advective transport.

---

**EXAMPLE 6.5** Hierarchical Approach for Modeling Flow and Transport in Structured Media

There are several approaches to model preferential flow and/or transport in structured media. Diagram 6.5.1 shows a schematic of increasingly complex models that can be used. In the simplest case, the same flow and transport equations are used as for a homogeneous medium but the soil heterogeneity is implicitly accounted for by using effective flow and transport properties such as permeability (Diagram 6.5.1a). This approach can be refined by using composite flow and transport properties while still applying the flow and transport model for a homogeneous medium (Diagram 6.5.1b).

For dual-porosity models, the liquid phase is partitioned into a mobile (fracture or inter-aggregate) and an immobile (matrix or intra-aggregate) region (Diagram 6.5.1c). There is no flow in the immobile region, but the model allows for solute exchange between the two-liquid regions. The physical nonequilibrium model given by [6.31] and [6.32] represents the simplest case with first-order exchange and steady-state flow conditions. As mentioned earlier, the exchange process can be described in greater detail for well-defined aggregates. The model could be extended to more general flow conditions by permitting variably saturated flow in the mobile region and changing immobile water contents due to exchange with the mobile (fracture) region. The latter type of model was described by Zurmühl et al. (1998).

More complex dual-permeability models also allow for vertical flow in the matrix (the immobile or intra-aggregate region). A separate set of flow and transport properties are required for both regions (cf. Diagram 6.5.1d). The challenging issue is to quantify the exchange of water and solute across the interface of the two regions. Several different formulations have been presented (Gerke and van Genuchten, 1993); some models consider more than two domains (Wilson et al., 1999).

Even further refinements are possible by considering transient variably-saturated flow and/or transport in discrete well-defined fractures, either without (Diagram 6.5.1e) or with (Diagram 6.5.1f) interactions between the intra- and inter-aggregate domains. The numerical simulation of transport and flow in porous media with such fracture systems typically involves the simultaneous solution of flow and transport for the matrix and fracture system in a coupled fashion (Shikaze et al., 1994).

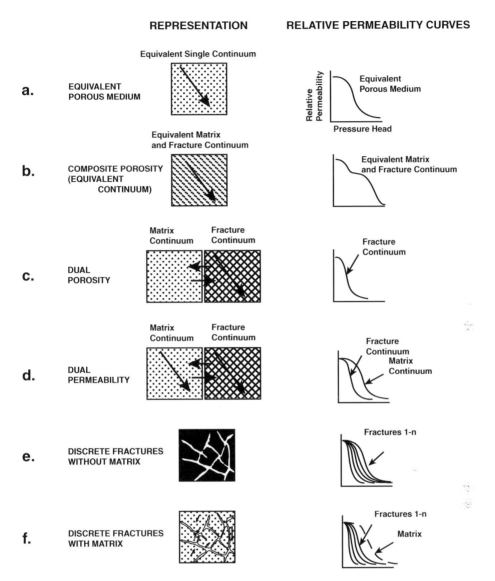

**DIAGRAM 6.5.1** Alternative conceptual models for flow through variably-saturated structured media (after Altman et al., 1996).

## REFERENCES

Altman, S.J., B.W. Arnold, R.W. Barnard, G.E. Barr, C.K. Ho, S.A. McKenna, and R.R. Eaton. 1996. Flow calculations for Yucca Mountain groundwater travel time (GWTT-95), SAND96-0819, Sandia National Laboratories, Albuquerque, NM.

Gerke, H.H. and M. Th. van Genuchten. 1993. A dual-porosity model for simulating the preferential movement of water and solutes in structured porous media, *Water Resour. Res.,* 29(2):305–319.

Shikaze, S.G., E.A. Sudicky and C.A. Mendoza. 1994. Simulation of dense vapor migration in discretely fractured geologic media, *Water Resour. Res.,* 30(7):1993–2009.

Wilson, G.V., P.M. Jardine, and J.P. Gwo. 1992. Modeling the hydraulic properties of a multiregion soil, *Soil Sci. Soc. Am. J.,* 56:1731–1737.

Zurmühl, T. and W. Durner. 1996. Modeling transient water and solute transport in a biporous soil, *Water Resour. Res.,* 32:819–829.

## 6.3  SOLUTIONS OF THE ADVECTION–DISPERSION EQUATION

The research and management of solute behavior in soils almost invariably require that the temporal and spatial solute distribution be known. Solute distributions as a function of time and/or space can be estimated with a variety of analytical and numerical solutions of the ADE, some of which will be briefly reviewed in the following.

### 6.3.1  BASIC CONCEPTS

A complete mathematical formulation of the transport problem requires that the pertinent dependent variable or concentration type is used and that the proper auxiliary conditions are specified.

#### 6.3.1.1  Concentration Types

Concentration is conventionally defined as the amount of solute-per-unit volume of the liquid. Since microscopic concentrations are based on a relatively small scale, a concentration at a larger scale needs to be introduced to allow use of the ADE, which is based on larger macroscopic variables and parameters. For this purpose, a macroscopic resident or volume-averaged concentration ($C_R$) is defined as:

$$C_R = \frac{1}{\Delta V} \iiint c\, dV \qquad\qquad [6.44]$$

where $c$ is the variable local-scale (microscopic) concentration (M L$^{-3}$) in a volume element ($V$) and $V$ is some representative elementary volume (Bear and Verruijt, 1987).

A different concentration type may be encountered at soil boundaries. In many solute displacement experiments, the concentration is determined from effluent samples as the ratio of the solute flux ($J_s$) and water flux ($J_w$) densities:

$$C_F = J_s / J_w \qquad\qquad [6.45]$$

where $C_F$ is the flux averaged concentration. This concentration represents the mass of solute-per-unit volume of fluid passing through a soil cross-section during an elementary time interval (Kreft and Zuber, 1978). For a one-dimensional solute flux consisting of an advective and a dispersive component, the flux-averaged concentration can be derived from the resident concentration according to the transformation:

$$C_F = C_R - \frac{D}{v}\frac{\partial C_R}{\partial x} \qquad\qquad [6.46]$$

The resident concentration may be determined from the flux averaged concentration using (van Genuchten et al., 1984):

$$C_R(x,t) = \frac{v}{D}\exp\left(\frac{vx}{D}\right)\int_x^{\infty}\exp\left(-\frac{v\xi}{D}\right)C_F(\xi,t)\,d\xi \qquad\qquad [6.47]$$

Additional transformations between flux and resident type concentrations are given by Parker and van Genuchten (1984a).

The difference between $C_R$ and $C_F$ is usually small, except when the second term on the right-hand side of Equation [6.46] is relatively large. It should be noted that a distinction between flux and resident type can be made for both the application and the detection of solutes (Kreft and Zuber, 1978, 1986). In soil science, a flux type application mode is often implicitly assumed (Parker and van Genuchten, 1984a). Flux-averaged concentrations are typically used when it is not possible to determine or specify a reliable value for the (resident) concentration. Resident concentrations are used for solute detection with, for example, time domain reflectometry, and to specify most initial conditions. Flux-averaged concentrations, on the other hand, are used for effluent samples, and to specify the influent concentration in most boundary value problems. Unless stated otherwise, it is assumed that solute concentrations are of the resident type.

Averaged concentrations can also be defined in terms of the observation scale, the latter exceeding the macroscopic scale associated with using the ADE. A time-averaged concentration $(C_T)$, is obtained by averaging over a time interval $(\Delta t)$ about a discrete time $(t_o)$ (Fischer et al., 1979):

$$C_T\left(x,t_o\right) = \frac{1}{\Delta t} \int_{t_o-\Delta t/2}^{t_o+\Delta t/2} C(x,t)dt \qquad [6.48]$$

where $C$ is a continuous solution of the solute concentration, which can be obtained by solving the ADE. This type of concentration occurs if solute breakthrough curves are measured using, for example, fraction collectors or gamma ray attenuation. Similarly, a one-dimensional spatial average can be defined as:

$$C_L\left(x_o,t\right) = \frac{1}{\Delta x} \int_{x_o-\Delta x/2}^{x_o+\Delta x/2} C(x,t)dx \qquad [6.49]$$

Such a concentration may be used to describe experimental results obtained for samples with centroid $(x_o)$ and length $(\Delta x)$. This situation occurs, for example, when the measured concentration of a large core sample is to be modeled as a point value (Leij and Toride, 1995).

### 6.3.1.2  Boundary and Initial Conditions

Initial and boundary conditions need to be specified in order to obtain a meaningful solution of the ADE. For a finite or semi-infinite soil, the initial condition can be formulated as:

$$C(x,0) = f(x) \quad x \geq 0 \qquad [6.50]$$

where $f(x)$ is an arbitrary function. Initial concentrations are almost invariably of the resident type.

The selection of the most appropriate boundary conditions for a transport problem is a somewhat esoteric topic that has received considerable attention in the literature. This is partly due to a lack of detailed experimental information for evaluating and applying boundary conditions, and inherent shortcomings of the transport equation itself at boundaries.

Many transport problems involve the application to the soil of a solute, whose influent concentration may be described by a function $g(t)$. The application method may be pumping, ponding or sprinkling. Two different types of inlet conditions are used, which assume either continuity in solute concentration or solute flux density. Simultaneous use of both conditions is seldom possible. It is generally more desirable to ensure mass conservation in the whole system than a continuous concentration at the inlet. The solute fluxes at the inlet boundary are, therefore, equated to obtain the following third or flux type inlet condition:

$$\left( vC - D\frac{\partial D}{\partial x} \right)_{x=0^+} = vg(t) \qquad [6.51]$$

where $0^+$ indicates a position just inside the soil. It is assumed that there is no dispersion outside the soil. The alternative condition requires the concentration to be continuous across the interface at all times. At smaller scales, such continuity will likely exist. However, at the scale of the ADE, it appears difficult to maintain a constant concentration at the interface, particularly during the initial stages of solute displacement for low influent fluxes and high dispersive fluxes in the soil. Mathematically, the first or concentration type condition is expressed as:

$$C(0,t) = g(t) \quad t > 0 \qquad [6.52]$$

The outlet condition can be defined as a zero gradient at a finite or infinite distance from the inlet. The infinite outlet condition:

$$\frac{\partial C}{\partial x}(\infty,t) = 0 \qquad [6.53]$$

is more convenient for mathematical solutions than the finite condition:

$$\frac{\partial C}{\partial x}(L^-,t) = 0 \qquad [6.54]$$

The use of Equation [6.53] implies that there is a semi-infinite fictitious soil layer beyond $x = L$, with identical properties as the actual soil. Such a layer does not affect the movement of the solute in the actual soil upstream of the exit boundary. Since the transport at the outlet cannot be precisely described, the intuitive contradiction of an infinite mathematical condition to describe a finite physical system is often more acceptable than using Equation [6.54], which precludes dispersion inside the soil near the outlet.

The formulation of the boundary and inlet conditions should account for the injection and detection modes in order to arrive at a mathematically consistent formulation of the problem with the same concentration type as independent variable. Only differences in detection mode for finite and semi-infinite systems will be explored here. The ADE in terms of the (usual) resident concentration is given as:

$$\frac{\partial C_R}{\partial t} = D\frac{\partial^2 C_R}{\partial x^2} - v\frac{\partial C_R}{\partial x} \qquad [6.55]$$

subject to a uniform initial condition, a third-type inlet condition, and a finite or infinite outlet condition:

$$C_R(x,0) = C_i \qquad [6.56]$$

$$\left( vC_R - D\frac{\partial C_R}{\partial x} \right)_{x=0^+} = vg(t) \qquad [6.57]$$

$$\frac{\partial C_R}{\partial x}(\infty,t) = 0 \quad \text{or} \quad \frac{\partial C_R}{\partial x}(L,t) = 0 \tag{6.58}$$

This problem can be written in terms of a flux-averaged concentration using Equation [6.48] according to (Parker and van Genuchten, 1984b):

$$\frac{\partial C_F}{\partial t} = D\frac{\partial^2 C_F}{\partial x^2} - v\frac{\partial C_F}{\partial x} \tag{6.59}$$

subject to:

$$C_F(x,0) = C_i \tag{6.60}$$

$$C_F(0,t) = g(t) \tag{6.61}$$

$$\frac{\partial C_F}{\partial x}(\infty,t) = 0 \quad \text{or} \quad \frac{\partial C_F}{\partial x}(L,t) = -\frac{D}{v}\frac{\partial^2 C_R}{\partial x^2}(L,t) \tag{6.62}$$

Notice that the mathematical problem for the flux mode involves a simpler first-type inlet condition with mass being conserved, unlike the use of a first type condition for a resident concentration. The solution for $C_R$ for a semi-infinite system involving a first-type inlet condition is the same as the solution for $C_F$ that conserves mass. As shown by Toride et al. (1994), the transformation is less convenient for nonuniform initial conditions or finite systems. Solutions for $C_F$ are then more easily obtained by transforming $C_R$ according to Equation [6.46].

Differences between the preferred third-type solution for $C_R$ and its first-type solution are usually small except for low values of the dimensionless time [$\xi = v^2 t/(RD)$] (van Genuchten and Parker, 1984). The relative mass balance error in a semi-infinite soil profile if a first-type rather than a third-type inlet condition is used, is given in Figure 6.8 (van Genuchten and Parker, 1984). The error pertains to the transport of a tracer solution of concentration ($C_o$) into an initially solute-free semi-infinite soil profile. Especially for small $\xi$, a substantial error may occur. Unless stated otherwise, a resident concentration is used in conjunction with a third-type inlet condition.

### 6.3.2 ANALYTICAL SOLUTIONS

Analytical solutions can formally be obtained only for linear transport problems. It would appear that analytical solutions are not very useful for transport in field soils where there is (1) spatial and temporal variability of flow and transport parameters, (2) transient flow, especially for unsaturated soils, and (3) nonuniformity in the boundary and initial conditions. However, analytical solutions can still be quite valuable. A nonlinear transport problem may be linearized through a suitable transformation to obtain a problem for which an analytical solution is available. Also, analytical solutions provide quick estimates of solute behavior over large temporal and spatial scales while they may offer insight into the underlying transport processes. Moreover, there is usually a lack of input parameters for field problems, which diminishes the advantage of numerical over analytical model results. Analytical solutions are also routinely used to evaluate the performance of numerical schemes. Finally, the mathematical and physical conditions tend to be well-defined for laboratory settings and an analytical solution can often be used, especially to estimate transport parameters by fitting analytical solutions to experimental data (Parker and van Genuchten, 1984; van Genuchten and Parker, 1987).

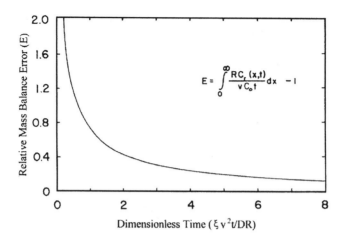

**FIGURE 6.8** Plot of the relative mass balance error versus dimensionless time for a semi-infinite profile when a first-type rather than a third-type inlet condition is used [after van Genuchten and Parker, 1984].

### 6.3.2.1 Variable Transformation

One straightforward way to obtain an analytical solution is to transform the ADE to an equation for which a solution already exists. As an example, consider transport in an infinite system given by:

$$R\frac{\partial C}{\partial t} = D\frac{\partial^2 C}{\partial x^2} - v\frac{\partial C}{\partial x} \tag{6.63}$$

$$C(x,0) = \begin{cases} C_o & x < 0 \\ 0 & x > 0 \end{cases} \tag{6.64}$$

The new coordinates:

$$\xi = x - vt$$
$$\tau = t \tag{6.65}$$

transform the ADE into a heat or solute diffusion problem given by:

$$R\frac{\partial C}{\partial \tau} = D\frac{\partial^2 C}{\partial \xi^2} \tag{6.66}$$

$$C(\xi,0) = \begin{cases} C_o & \xi < 0 \\ 0 & \xi > 0 \end{cases} \tag{6.67}$$

The solution for this problem can be readily found in the literature on diffusion problems (Carslaw and Jaeger, 1959; Crank, 1975):

$$C(x,t) = \frac{C_o}{2}\operatorname{erfc}\left(\frac{Rx - vt}{\sqrt{4RDt}}\right) \tag{6.68}$$

Other transformations to the diffusion problem have been employed as well (Brenner, 1962; Selim and Mansell, 1976; Zwillinger, 1989). Transformation of time to a time-integrated flow variable sometimes allows one to derive an analytical solution of the nonlinear ADE for transient flow (Wierenga, 1977; Parker and van Genuchten, 1984b; Huang and van Genuchten, 1995).

### 6.3.2.2  Laplace Transformation

The ADE is commonly solved directly with the method of Laplace transforms. The solution procedure will be illustrated here for an initially solute-free semi-infinite soil with a constant solute flux ($vC_o$) or concentration ($C_o$) at the inlet boundary. The mathematical problem consists of solving the ADE given by Equation [6.63] subject to:

$$C - \delta \frac{D}{v} \frac{\partial D}{\partial x} = C_o \quad \delta = \begin{cases} 0 & \text{first type} \\ 1 & \text{third type} \end{cases} \qquad [6.69]$$

$$\frac{\partial C}{\partial x}(\infty, t) = 0 \qquad [6.70]$$

with $\delta$ as a coefficient depending on the type of inlet condition. The Laplace transform ($\mathscr{L}$) of the solute concentration with respect to time is defined as (Spiegel, 1965):

$$\overline{C}(x,s) = \mathscr{L}\left[C(x,t)\right] = \int_0^\infty C(x,t) \exp(-st) dt \qquad [6.71]$$

where $s$ is the (complex) transformation variable ($\text{T}^{-1}$). This transformation changes the transport equation from a partial to an ordinary differential equation:

$$\frac{d^2 \overline{C}}{dx^2} - \frac{v}{D} \frac{d\overline{C}}{dx} - \frac{sR}{D} \overline{C} = 0 \qquad [6.72]$$

subject to:

$$\overline{C} - \delta \frac{D}{v} \frac{d\overline{C}}{dx} = \frac{C_o}{s} \qquad [6.73]$$

$$\frac{d\overline{C}}{dx}(\infty, s) = 0 \qquad [6.74]$$

where the bar denotes a transformed variable. The following solution for the concentration in the Laplace domain is obtained with help of the inlet condition:

$$\overline{C}(x,s) = \frac{v}{v - \delta \lambda^- D} \frac{C_o}{s} \exp(\lambda^- x), \quad \lambda^- = \frac{v}{2D} - \left[\left(\frac{v}{2D}\right)^2 + \frac{sR}{D}\right]^{1/2} \qquad [6.75]$$

Inversion of this solution may be done with a table of Laplace transforms, by applying the inversion theorem, or by using a numerical inversion program. It should be noted that the solution for a finite outlet condition is also possible with the Laplace transform, although a bit more cumbersome (Brenner, 1962; Leij and van Genuchten, 1995).

**TABLE 6.2**

**Analytical solutions of the ADE for different boundary conditions after van Genuchten and Alves (1982).**

| Case | Inlet Condition | Exit Condition | Analytical Solution $C(x,t)$ |
|------|-----------------|----------------|------------------------------|
| A1 | $C(0,t) = C_o$ | $\dfrac{\partial C}{\partial x}(\infty,t) = 0$ | $\dfrac{1}{2} erfc\left(\dfrac{Rx - vt}{(4RDt)^{1/2}}\right) + \dfrac{1}{2}\exp\left(\dfrac{vx}{D}\right)erfc\left(\dfrac{Rx + vt}{(4RDt)^{1/2}}\right)$ |
| A2 | $\left(vC - D\dfrac{\partial C}{\partial x}\right)_{x=0} = vC_o$ | $\dfrac{\partial C}{\partial x}(\infty,t) = 0$ | $\dfrac{1}{2} erfc\left(\dfrac{Rx - vt}{(4RDt)^{1/2}}\right) - \dfrac{1}{2}\left(1 + \dfrac{vx}{D} + \dfrac{v^2 t}{DR}\right)\exp\left(\dfrac{vx}{D}\right)erfc\left(\dfrac{Rx + vt}{(4RDt)^{1/2}}\right)$ $+ \left(\dfrac{v^2 t}{\pi RD}\right)^{1/2}\exp\left(-\dfrac{(Rx - vt)^2}{4RDt}\right)$ |
| A3 | $C(0,t) = C_o$ | $\dfrac{\partial C}{\partial x}(L,t) = 0$ | $1 - \displaystyle\sum_{m=1}^{\infty} \dfrac{2\beta_m \sin\left(\dfrac{\beta_m x}{L}\right)\exp\left(\dfrac{vx}{2D} - \dfrac{v^2 t}{4DR} - \dfrac{\beta_m^2 Dt}{L^2 R}\right)}{\beta_m^2 + \left(\dfrac{vL}{2D}\right)^2 + \dfrac{vL}{2D}}$ $\beta_m \cot(\beta_m) + \dfrac{vL}{2D} = 0$ |
| A4 | $\left(vC - D\dfrac{\partial C}{\partial x}\right)_{x=0} = vC_o$ | $\dfrac{\partial C}{\partial x}(L,t) = 0$ | $1 - \displaystyle\sum_{m=1}^{\infty} \dfrac{\dfrac{2vL}{D}\beta_m\left[\beta_m \cos\left(\dfrac{\beta_m x}{L}\right) + \dfrac{vL}{2D}\sin\left(\dfrac{\beta_m x}{L}\right)\right]\exp\left(\dfrac{xv}{2D} - \dfrac{v^2 t}{4DR} - \dfrac{\beta_m^2 Dt}{L^2 R}\right)}{\left[\beta_m^2 + \left(\dfrac{vL}{2D}\right)^2 + \dfrac{vL}{2D}\right]\left[\beta_m^2 + \left(\dfrac{vL}{2D}\right)^2\right]}$ $\beta_m \cot(\beta_m) - \dfrac{\beta_m^2 D}{vL} + \dfrac{vL}{4D} = 0$ |

### 6.3.2.3 Equilibrium Transport

Van Genuchten and Alves (1982) provided a compendium of available analytical one-dimensional solutions for a variety of mathematical conditions and physical processes. Four common analytical solutions for a zero initial condition involving a first- or third-type inlet condition and an infinite or finite outlet condition are listed in Table 6.2. The solutions may be expressed in terms of the dimensionless variables $(P)$, $(T)$ and $(X)$ (Equation [6.41]). Typically, $L$ is equal to the position of the outlet (the column length) for a finite system, whereas for a semi-infinite soil system, $L$ can be assigned to any arbitrary length.

Figure 6.9 contains solute profiles, $[C/C_o(X)]$ according to the solutions listed in Table 6.2 using $R = 1$ and two different values for $P$ and $T$ for a first- (A1) or a third-type (A2) inlet condition assuming an infinite outlet condition, or a first- (A3) or third-type (A4) inlet condition in case of finite outlet condition. The predicted profile for a first-type condition (A1, A3) for the lower Peclet number $(P = 5)$ lies considerably above the line predicted for a third-type condition (A2, A4). The effect of the outlet condition is initially minor, but when the solute front reaches the outlet $(L)$, a

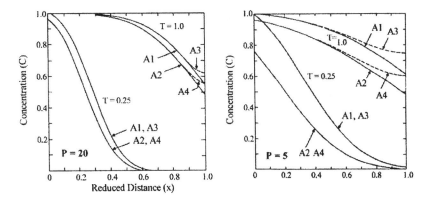

**FIGURE 6.9** Plot of the concentration ($C/C_o$) as a function of distance ($X$) calculated for four different combinations of boundary conditions according to the solutions in Table 6.2 at two different times for a Peclet number of 5 and 20 [after van Genuchten and Alves, 1982].

clear difference between a finite and an infinite outlet condition can be observed for both a first (A1, A3) and a third-type (A2, A4) inlet condition. The simpler solution for a semi-infinite system can, in many cases, be used to approximate the solution for a finite condition; van Genuchten and Alves (1982) formulated the empirical restriction:

$$X < 0.9 - 8/P \qquad\qquad [6.76]$$

on the position for which such an approximation is reasonable. For smaller times ($T \ll 1$), when the solute has not reached the outlet, the finite and infinite outlet condition obviously leads to a similar solution.

For a third-type inlet condition, the concentration at $X = 0$ just inside the soil is not equal to the influent concentration, even at time $T = 1$. Although the jump in concentration is physically odd, mass conservation is ensured. For the higher Peclet number (P = 20), deviations between a first- and a third-type inlet condition are significantly reduced. This is in accordance with Figure 6.8, which shows a smaller error for increased $v^2t/RD$.

Large differences in the predicted concentration may occur if the solute front approaches the outlet at $X = 1$ or $x = L$. Calculated concentrations according to solutions A1, A2, A3 and A4 versus the Peclet number at the outlet for $T = 1$ are illustrated in Figure 6.10. The greatest difference occurs for small Peclet numbers, namely, when hydrodynamic dispersion is relatively important. The nature of hydrodynamic dispersion suggests that $C/C_o$ should be approximately 0.5 for $X = T = R = 1$, the average of the zero initial concentration and the influent concentration ($C_o$). Because of the effect of the boundary conditions, this only happens when the Peclet number exceeds 10 or more, depending on the type of solution. For the first-type inlet condition (A1, A3), $C/C_o$ exceeds 0.5 at low Peclet numbers since a considerable amount of solute is forced to diffuse into the column to establish a constant inlet concentration.

Differences between calculated solute breakthrough curves because of boundary conditions are further depicted in Figure 6.11 for three different Peclet numbers. Notice that for $P = 60$, the curves are almost indistinguishable, considering the margin of error of most solute displacement experiments. The results in Figure 6.11 show that the choice of inlet and outlet conditions for determining parameters from breakthrough experiments becomes less important when $P$ exceeds about 30.

Finally, for displacement experiments involving finite columns, it may be of interest to quantify the amount of solute that can be stored in the liquid and sorbed phases of the soil (Equation [6.24]). When, beginning at $t = 0$, a solution with concentration ($C_o$) is applied to a soil column, holdup can be defined as (Nauman and Buffham, 1983):

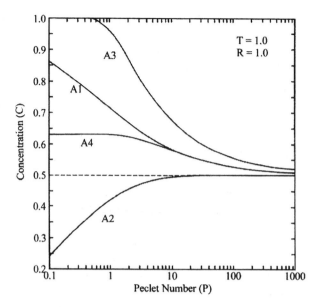

**FIGURE 6.10** Solute concentration predicted with solutions A1, A2, A3, and A4 as a function of the Peclet number at $T = 1$ and $X = 1$ [after van Genuchten and Alves, 1982].

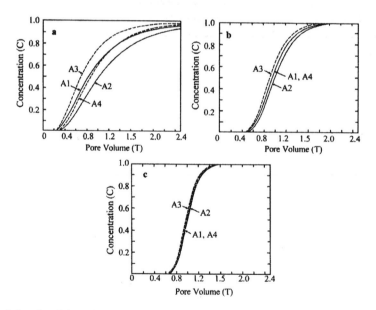

**FIGURE 6.11** Solute breakthrough curves predicted with the four analytical solutions of Table 6.2 for three different Peclet numbers [after van Genuchten and Alves, 1982].

$$H = \frac{v}{L} \int_0^\infty \frac{C_o - C(L,t)}{C_o} \, dt \qquad\qquad [6.77]$$

This amounts to the integration of the complementary solute concentration versus dimensionless time, namely, the area above the breakthrough curves in Figure 6.11. Van Genuchten and Parker (1984) showed that $H = R$ for solutions A1 and A4, $H = R[1+(1/P)]$ for solution A2 and $H = R\{1 - (1/P)[1 - \exp(-P)]\}$ for solution A3. In case of anion exclusion, the relative exclusion volume equals $1 - R$ and the column holdup will be less than one.

---

**EXAMPLE 6.6**  Multidimensional Transport

The use of one-dimensional transport models is mostly restricted to transport in laboratory soil columns or special cases of field transport stemming from a diffuse source. Multidimensional versions of the ADE need to be employed to describe other cases of solute transport. The formulation and solution of the problem is very similar to that of the one-dimensional ADE.

A limited number of multidimensional problems are amenable to analytical solution. As an example, consider steady water flow in the longitudinal direction ($x$) and hydrodynamic dispersion in the longitudinal and two transverse directions ($y$ and $z$). The governing transport equation for a homogeneous soil is of the form:

$$R\frac{\partial C}{\partial t} = D_x \frac{\partial^2 C}{\partial x^2} - v\frac{\partial C}{\partial x} + D_y \frac{\partial^2 C}{\partial y^2} + D_z \frac{\partial^2 C}{\partial z^2} \tag{1}$$

A straightforward method of solving the multidimensional transport equation is the use of the product rule; multiplying the solutions of (simpler) one-dimensional problems may directly give the desired three-dimensional solution. Consider the three one-dimensional problems:

$$R\frac{\partial C_x}{\partial t} = D_x \frac{\partial^2 C_x}{\partial x^2} - v\frac{\partial C_x}{\partial x} \tag{2}$$

$$R\frac{\partial C_y}{\partial t} = D_y \frac{\partial^2 C_y}{\partial y^2} \tag{3}$$

$$R\frac{\partial C_z}{\partial t} = D_z \frac{\partial^2 C_z}{\partial z^2} \tag{4}$$

The reader may verify that the product of the solutions of these individual equations, i.e., $C_x$ $C_y$ $C_z$ satisfies Equation [1]. This technique can be used to determine solutions of transport problems for homogeneous boundary conditions where the product rule can be applied to the initial concentration. Solutions for other problems may be obtained with integral transforms (Leij et al., 1991) or Green's functions (Leij et al., 2000).

Some of the analytical solutions are conveniently evaluated by using software packages. Consider example 2 of the 3DADE program in the Stanmod package of Šimůnek et al. (2000). Solute is applied at a concentration $C_o$, assuming a flux- or first-type condition, to a rectangular area of the soil surface (15x15 cm). The following transport parameters are specified: $v = 10$ cm/d, $D_x = 100$ cm$^2$/d, $D_y = D_z = 10$ cm$^2$/d, and $R = 1$. Diagram 6.6.1 shows the solute distribution in the $xy$-plane ($0 \le x \le 30$, $-20 \le y \le 20$) at $t = 1$ d and $z = 0$ as displayed by Stanmod. These types of distributions are useful for rapid approximate analyses of solute contamination problems and to obtain a good grasp of the effects of different transport parameters.

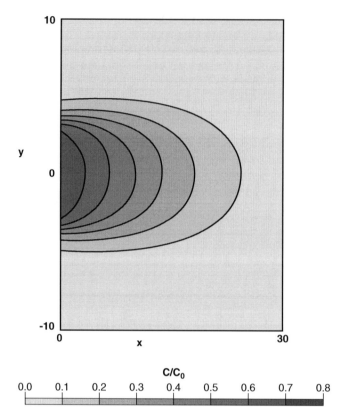

**DIAGRAM 6.6.1** Solute distribution in $xy$-plane for continuous solute application at $x = 0$ and $-7.5 < y < 7.5$.

## REFERENCES

Leij, F.J., E. Priesack, and M.G. Schaap. 2000. Three-dimensional transport from persistent solute sources modeled with Green's functions, *J. Contam. Hydrol.*, 41:155–173.

Leij, F.J., T.H. Skaggs, and M. Th. van Genuchten. 1991. Analytical solutions for solute transport in three-dimensional semi-infinite porous media. *Water Resources Res.*, 27(10):2719–2733.

Šimůnek, J., M. Th. van Genuchten, M. Šejna, N. Toride, and F.J. Leij. 2000. STANMOD Studio of analytical models for solving the convection-dispersion equation. Version 1.0, International Ground Water Modeling Center, Colorado School of Mines, Golden, CO.

### 6.3.2.4 Nonequilibrium Transport

Analytical solutions for one-dimensional bimodal nonequilibrium transport have been presented by, among others, Lindstrom and Narasimhan (1973), Lindstrom and Stone (1974), Lassey (1988) and Toride et al. (1993). The boundary value problem involving solute application with a constant concentration may be specified by Equations [6.39], [6.40] and [6.53] subject to the following conditions:

$$C_1(X,0) = C_2(X,0) = 0 \qquad\qquad [6.78]$$

$$\left( C_1 - \frac{1}{P}\frac{\partial C_1}{\partial X} \right)_{X=0^+} = 1 \qquad [6.79]$$

$$\frac{\partial C_1}{\partial X}(\infty, T) = 0 \qquad [6.80]$$

Solutions for the equilibrium and nonequilibrium concentrations are:

$$C_1 = \int_0^T J(a,b)G(X,\tau)d\tau \qquad [6.81]$$

$$C_2 = \int_0^T [1 - J(b,a)]G(X,\tau)d\tau \qquad [6.82]$$

where the auxiliary (equilibrium) function $G(X,\tau)$ for resident concentrations is defined as:

$$G(X,\tau) = \sqrt{\frac{P}{\pi\beta R\tau}}\exp\left(-\frac{(\beta RX-\tau)^2}{4\beta R\tau/P}\right) - \frac{P}{2\beta R}\operatorname{erfc}\left[\frac{\beta RX+\tau}{\sqrt{4\beta R\tau/P}}\right] \qquad [6.83]$$

and for flux-averaged concentrations by:

$$G(X,\tau) = \sqrt{\frac{\beta RPX^2}{4\pi\tau^3}}\exp\left(-\frac{(\beta RX-\tau)^2}{4\beta R\tau/P}\right) \qquad [6.84]$$

Furthermore, $J$ denotes Goldstein's $J$ function (Goldstein, 1953), which is defined as:

$$J(a,b) = 1 - \exp(-b)\int_0^a \exp(-x)I_0\left(2\sqrt{ab}\right)d\xi \qquad [6.85]$$

with $I_0$ as the zero order modified Bessel function. The variables $a$ and $b$ are given by:

$$a = \frac{\omega\tau}{\beta R}, \quad b = \frac{\omega(T-\tau)}{(1-\beta)R} \qquad [6.86]$$

The above solution for a flux-averaged concentration was used to describe breakthrough data for the pesticide 2,4,5-T (2,4,5-trichlorophenoxyacetic acid) as observed from a 30-cm long soil column containing aggregated (< 6 mm) Glendale clay loam (van Genuchten and Parker, 1987). The nonequilibrium model with three adjustable parameters ($P$, $\beta$, $\omega$) provided an excellent description of the data (Figure 6.12b). The retardation factor ($R$) was estimated independently, using the distribution coefficient obtained from batch experiments, according to Equation [6.23]. A one-parameter ADE fit (using $P$ as adjustable parameter) did not yield a good description of the data (Figure 6.12a) while a similar two-parameter ($P$, $R$) fit, which is not shown, gave results that were only marginally different from those shown in Figure 6.12a.

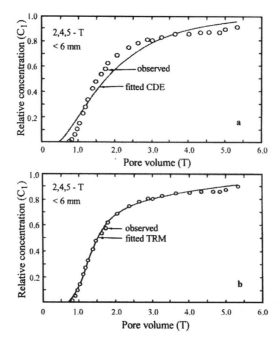

**FIGURE 6.12** Observed and fitted effluent curves for 2,4,5-T movement through Glendale clay loam. The fitted curves were based on: (a) the equilibrium ADE and (b) the nonequilibrium ADE.

As pointed out by van Genuchten and Dalton (1986), the main disadvantage of the first order physical nonequilibrium approach is the obscure dependency of the transfer coefficient ($\alpha$ or $\omega$) on the actual diffusion process in the aggregate, particularly the value for the diffusion coefficient and the aggregate geometry. For well-defined structured or aggregated porous media (media for which the size and geometry of all aggregates are known), the diffusion process inside the aggregate can be modeled, which allows a more detailed description of the concentration inside the aggregate. Analytical solutions are available for several aggregate geometries. The simplified immobile concentration, which is used in the general nonequilibrium formulation, can always be obtained by averaging the more detailed solution over the aggregate volume.

### 6.3.2.5 Time Moments

Moments are frequently used to characterize statistical distributions such as those of solute particles (concentrations) versus time or positions. Analytical expressions for lower order moments are sometimes derived for deterministic transport models, especially when a direct analytical solution may be difficult to obtain. Moment analysis is more widely employed in stochastic than deterministic transport models, since the uncertainty in both model parameters and predicted results is conveniently quantified with moments. Only time moments will be considered here.

The $p$th (time) moment of the breakthrough curve, as obtained from effluent samples collected from a soil column (with length $x = L$) to which a solute pulse is applied during steady water flow, is defined as:

$$m_p(L) = \int_0^\infty t^p C(L,t)dt \quad (p = 0,1,2,\ldots)$$
[6.87]

The zero moment is proportional to the total solute mass, the first moment quantifies the mean displacement, the second moment is indicative of the variance (dispersion), whereas the third moment quantifies the asymmetry or skewness of the breakthrough curve. Normalized moments ($\mu_p$) are obtained as follows:

$$\mu_p = \frac{m_p}{m_0} \tag{6.88}$$

The mean breakthrough time is given by $\mu_1$. Central moments are defined with respect to this mean according to:

$$\mu'_p(L) = \frac{1}{m_0} \int_0^\infty (t - \mu_1)^p C(L,t)\, dt \tag{6.89}$$

The variance of a breakthrough curve, which can be used to assess solute dispersion, is given by the second central moment ($\mu_2'$). The degree of asymmetry of the breakthrough curve is indicated by its skewness ($\mu_3'/(\mu_2')^{3/2}$).

The previous definitions are employed to obtain numerical values for moments from experimental results. Substitution of an analytical solution for the solute concentration into the definitions allows the derivation of algebraic expressions for time moments. Values for transport parameters can be obtained by equating numerical and algebraic moments (Leij and Dane, 1992; Jacobsen et al., 1992). This procedure is not reliable if experimental moments of higher order (p > 2) are needed since even small deviations, at larger times, between experimental and modeled concentrations will greatly bias such moments.

Algebraic moments are normally obtained by using the solution for the concentration in the Laplace domain. The following equality can be established from properties of the Laplace transform (Spiegel, 1965):

$$m_p(x) = (-1)^p \lim_{s \to 0} \frac{d^p \overline{C}(x,s)}{ds^p} \tag{6.90}$$

where, as before, $\overline{C}(x,s)$ is the concentration in the Laplace domain, and $s$ is the (complex) transformation variable. Expressions for moments can hence be obtained by differentiating the solution in the Laplace domain and letting the Laplace variable go to zero (Aris, 1958; van der Laan, 1958). This task is conveniently handled by mathematical software.

Time moments will now be considered for three different transport problems. First, the mathematical problem for physical nonequilibrium transport can be written as:

$$\theta_m R \frac{\partial C_m}{\partial t} + \theta_{im} R \frac{\partial C_{im}}{\partial t} = \theta_m D_m \frac{\partial^2 C_m}{\partial x^2} - \theta v \frac{\partial C_m}{\partial x} \tag{6.91}$$

$$\theta_{im} R \frac{\partial C_{im}}{\partial t} = \alpha \left( C_m - C_{im} \right) \tag{6.92}$$

The conditions for instantaneous solute application to a soil are:

$$C(x,0) = 0 \qquad 0 \le x < \infty \tag{6.93}$$

**TABLE 6.3**
**Mean breakthrough time ($\mu_1$) and variance ($\mu_2'$) for the equilibrium and nonequilibrium solution of the ADE at a distance $x$ from the inlet as a result of a Dirac delta input described with a first-type inlet condition (i.e., flux-averaged concentration).**

| ADE Model | Mean Breakthrough Time, $\mu_1$ | Variance, $\mu_2'$ |
|---|---|---|
| Equilibrium | $\dfrac{Rx}{v}$ | $\dfrac{2DR^2x}{v^3}$ |
| Nonequilibrium | | |
| Two-Region | $\dfrac{Rx}{v}$ | $\dfrac{2\theta_m D_m R^2 x}{\theta v^3} + \dfrac{2\theta(1-\beta)^2 R^2 x}{\alpha v}$ |
| Two-Site | $\dfrac{Rx}{v}$ | $\dfrac{2DR^2x}{v^3} + \dfrac{2(1-\beta)Rx}{\alpha v}$ |

$$C(0,t) = \frac{m_0}{v}\delta(t) \qquad\qquad [6.94]$$

$$\frac{\partial C}{\partial x}(\infty,t) = 0 \qquad\qquad [6.95]$$

where $m_0$ is the solute mass that is applied per unit area of soil solution at $t = 0$, and $\delta(t)$ is the Dirac delta function ($T^{-1}$). The first-type inlet condition is used to describe flux-averaged concentrations such as effluent samples from column displacement experiments. Second, the equilibrium problem is defined by the same set of equations by setting $\theta_m = \theta$, $\theta_{im} = 0$, and $\alpha \to 0$. Third, the chemical nonequilibrium transport equations are as follows:

$$\frac{\partial C}{\partial t} + \frac{\rho_b}{\theta}\frac{\partial S}{\partial t} = D\frac{\partial^2 C}{\partial x^2} - v\frac{\partial C}{\partial x} \qquad\qquad [6.96]$$

$$\frac{\partial S}{\partial t} = \alpha\left(K_d C - S\right) \qquad\qquad [6.97]$$

These equations are also subject to boundary and initial conditions in Equations [6.93] through [6.95].

Formulas for the mean breakthrough time ($\mu_1$) and the variance ($\mu_2'$) of the breakthrough curve predicted for these three models are presented in Table 6.3. The expressions suggest that nonequilibrium conditions do not affect the mean travel time but they do increase solute spreading. Since only the solution in the Laplace domain is needed, moment analysis is particularly useful for more complex transport problems to study the general behavior of the breakthrough curve.

**EXAMPLE 6.7**   Moments

Moment analyses can be conveniently carried out with mathematical software packages. As an example, consider time moments of a breakthrough curve for equilibrium transport in a soil column resulting from the application of a nonreactive solute pulse. The inlet condition is given by:

$$C(0,t) = \begin{cases} C_o & 0 \le t \le t_o \\ 0 & t \ge t_o \end{cases} \qquad [1]$$

The solution of the equilibrium ADE in the Laplace domain follows readily from [6.75]:

$$\overline{C}(x,s) = \frac{C_o}{s}[1 - \exp(-st_o)]\exp(ak - k\sqrt{s+a^2}) \quad , \quad a = v/\sqrt{4D} \quad , \quad k = x/\sqrt{D} \quad [2]$$

The zero-, first- and second-order moments may be determined according to [6.90] using Mathematica (Wolfram, 1999) as well as any other appropriate software. The normalized moments ($\mu_1$ and $\mu_2$) can be determined with [6.88] while the second central moment (i.e., the variance) is given by $\mu_2.\mu_1^2$. From the program output, shown below, it can be readily established that the mean breakthrough time, $\mu_1$, at a position $x$ from the column inlet equals $t_o/2+x/v$, while the variance equals $t_o^2/12+2xD/v^3$. Similar results can be found in Table 6.3 for a Dirac delta input. The use of mathematical software such as Mathematica is particularly convenient for higher-order moments and for more complicated problems such as for non-equilibrium transport for which moment results are also included in Table 6.3.

**f[s_] : = Co (1 – Exp [–s to]) Exp [a k – k Sqrt [s + a ^ 2]] / s**

$$Co\, e^{\left(a-\sqrt{a^2}\right)k}\, to$$

**m0 = Limit [f[s], s → 0]**

**m1 = –Limit [f′[s], s → 0]**

$$\frac{Co\, e^{\left(a-\sqrt{a^2}\right)k}\, to\left(k+\sqrt{a^2}\, to\right)}{2\sqrt{a^2}}$$

**m2 = Limit [f″[s], s → 0]**

**M1 = Simplify [m1/m0]**

$$\frac{Co\, e^{\left(a-\sqrt{a^2}\right)k}\, to\left(3\sqrt{a^2}\, k^2 + 4\left(a^2\right)^{3/2} to^2 + k\left(3+6a^2\, to\right)\right)}{12\left(a^2\right)^{3/2}}$$

**M2 = m2/m0**

**Var = Simplify [M2 – (M1) ^ 2]**

$$\frac{1}{2}\left(\frac{\sqrt{a^2}\, k}{a^2}+to\right)$$

$$\frac{3\sqrt{a^2}\, k^2 + 4\left(a^2\right)^{3/2} to^2 + k\left(3+6a^2\, to\right)}{12\left(a^2\right)^{3/2}}$$

$$\frac{1}{12}\left(\frac{3\sqrt{a^2}\, k}{a^4}+to^2\right)$$

**DIAGRAM 6.7.1**  Input and output of Mathematica for moment analysis.

# REFERENCE

Wolfram, S. 1999. The Mathematica book, 4th ed., Wolfram Media/Cambridge University Press, Champaign, Il.

### 6.3.3 Numerical Solutions

The solution of many practical transport problems requires the use of numerical methods because of changes in water saturation (as the result of irrigation, evaporation and drainage), spatial and temporal variability of soil properties, or complicated boundary and initial conditions. Numerical methods are based on a discretization of the spatial and temporal solution domain, and subsequent calculation of the concentration at discrete nodes in the domain. This approach is in contrast with analytical methods, which offer a continuous description of the concentration. In some cases, a combination of analytical and numerical techniques may be employed (Sudicky, 1989; Moridis and Reddell, 1991; Li et al., 1992).

#### 6.3.3.1 Introduction

The accuracy of the numerical results depends on the input parameters, the approximation of the governing partial differential equation, the discretization, and implementation of the numerical solution in a computer code solving the simulated problem. Numerous texts exist on the numerical modeling of flow and transport in porous media (Pinder and Gray, 1977; Huyakorn and Pinder, 1983; Campbell, 1985; van der Heijde et al., 1985; Istok, 1989).

The many numerical methods for solving the ADE can be classified into three groups (Neuman, 1984): (1) Eulerian, (2) Lagrangian, and (3) mixed Lagrangian-Eulerian. In the Eulerian approach, the transport equation is discretized by the method of finite differences or finite elements using a fixed mesh. For the Lagrangian approach, the mesh deforms along with the flow while it is stable in a moving coordinate system. A two-step procedure is followed for a mixed approach. First, advective transport is solved using a Lagrangian approach and concentrations are obtained from particle trajectories. Subsequently, all other processes including sinks and sources are modeled with an Eulerian approach using finite elements, finite differences, etc.

The method of finite differences (Bresler and Hanks, 1969; Bresler, 1973) and the Galerkin method of finite elements (Gray and Pinder, 1976; van Genuchten, 1978) belong to the first group as do the previously mentioned combination of analytical and numerical techniques. The finite difference and finite element methods were the first numerical methods used for solute transport problems and, in spite of their problems discussed below, are still the most often utilized methods. Numerical experiments have shown that both methods give very good results if significant dispersion exists as quantified with the Peclet number. If advection is dominant, however, numerical oscillations may occur for both methods and small spatial increments should be used. It may not always be possible to decrease the spatial step size due to the associated increase in computations and a variety of approaches have been developed to overcome the oscillations (Chaudhari, 1971; van Genuchten and Gray, 1978; Donea, 1991).

Lagrangian solution methods will result in very few numerical oscillations (Varoglu and Finn, 1982). However, Lagrangian methods, which are based on the method of characteristics, suffer from inherent diffusion and do not conserve mass. They are difficult to implement for two- and three-dimensional problems. Instabilities resulting from inadequate spatial discretization may occur during longer simulations due to deformation of the stream function, especially if the solute is subject to sorption, precipitation and other reactions.

The mixed approach has been applied by several authors (Konikov and Bredehoeft, 1978; Molz et al., 1986; Yeh, 1990). In view of the different behavior of the diffusive (parabolic) and advective (hyperbolic) terms of the ADE, the problem is decomposed into an advection and a diffusion problem. Advective transport is solved with the Lagrangian approach while all other terms are solved with the Eulerian approach. The trajectories of flowing particles are obtained using continuous forward particle tracking (to follow a set of particles as they move through the flow domain), single-step reverse particle tracking (the initial position of particles arriving at nodal points was calculated for each time step), and a combination of both approaches. Ahlstrom et al. (1977), among others, used the attractive random walk model to describe the movement of individual solute particles by viewing the travel distance for a particular time step as the sum of a deterministic and a stochastic velocity component.

In the following, only a brief introduction to the finite difference and finite element methods to solve transport problems is provided. Both methods encompass a wide variety of more specific numerical approaches. As a rule of thumb, the finite difference method is attractive because of its simplicity and the availability, at least early on, of handbooks and computer programs simulating flow and transport in porous media. On the other hand, the finite element method has proven to be more suitable for problems involving irregular geometries of the flow and transport problem, such as situations involving flow to drainage pipes and along sloping soil surfaces.

---

**EXAMPLE 6.8** Numerical Solutions for Variably-Saturated Flow and Solute Transport

Ever more powerful computers allow us now to find solutions to flow and transport problems that were far beyond the means of previous generations of scientists and engineers. Of course, many problems remain, such as conceptualizing flow and transport processes and quantifying relevant model parameters. This section is devoted to the development and implementation of advanced numerical solutions of the Richards equation (Equation [4.30]) for variably-saturated flow, and the advection-dispersion equation (Equation [6.16]) for solute transport. Further details and additional references can be found in van Genuchten and Sudicky (1999).

*Numerical Solution of the Flow Equation*

Early numerical variably saturated flow models generally used classical finite difference methods. Integrated finite differences (Narasimhan and Witherspoon [1976]), control-volume finite element techniques (Therrien and Sudicky, 1996) and Galerkin finite element methods (Huyakorn et al., 1986) became increasingly popular in the mid-seventies. Time and space discretization of the Richards equation using any of these methods leads to a nonlinear system of algebraic equations. These equations are most often linearized and solved using the Newton-Raphson or Picard iteration methods. Picard iteration is widely used because of its ease of implementation, and because this method preserves symmetry of the final system of matrix equations. The Newton-Raphson iteration procedure is more complex and results in nonsymmetric matrices, but often achieves a faster rate of convergence and may be more robust than Picard iteration for highly nonlinear problems. The basic approach for discretizing and solving the Richards equation depends upon the independent variable that is being used in the flow formulation, i.e., the $h$-based, the $\theta$-based, or the mixed formulation (Celia et al., 1990).

*Numerical Solution of the Transport Equation*

As was mentioned in the main text, there are three different approaches to solve the ADE: (1) Eulerian, (2) Lagrangian, and (3) mixed Lagrangian-Eulerian methods. Standard finite difference and Galerkin or control-volume type finite element methods belong to the first group of Eulerian methods. These have been found very reliable and accurate when applied to quasi-symmetric problems when diffusion dominates the transport process. Advection-dominated transport introduces nonsymmetry into the governing solute transport equation and, as a result, Eulerian methods become less reliable for advection-dominated problems. By selecting an appropriate combination of relatively small space and time steps, it is still possible to virtually eliminate most or all oscillations. A well-known guideline recommends an upper limit of 2 for the product of the local Peclet and Courant numbers. Alternatively, the spatial grid system may be refined using a zoomable hidden fine-mesh approach (Yeh, 1990), or by implementing local adaptive grid refinement (Wolfsberg and Freyburg, 1994).

However, an additional computational cost exists with this approach and the handling of natural grid irregularities due to material heterogeneity or other domain features can be problematic. Monoticity conditions and numerical smearing are also influenced by the type of temporal discretization being used. For example, while fully implicit time-weighting schemes are monotone (i.e., concentrations always fall within the physical range), they are more prone to numerical dispersion than central-in-time (Crank-Nicolson type) weighting which is second-order correct. By comparison, monoticity cannot be guaranteed with central weighting unless the grid Peclet and Courant criteria are appropriately satisfied. Upstream weighting methods virtually eliminate numerical oscillations, even for purely advective transport, but a disadvantage is that they may create unacceptable numerical dispersion.

While Lagrangian methods may substantially reduce or even eliminate problems with numerical oscillations, they often have critical deficiencies regarding conservation of mass, difficulty of implementation, and handling of solute reactions.

Mixed Eulerian-Lagrangian approaches may also be successfully used to alleviate numerical dispersion by decomposing the transport problem into purely advective and a remaining (diffusion-dominated) subproblem that are solved with the Lagrangian and Eulerian methods, respectively. Trajectories of flowing particles were obtained by, for example, Neuman (1984) using single-step reverse particle tracking in which the initial position of particles arriving at the end of a time step at fixed nodal points is calculated for each time step. The use of continuous forward particle tracking has similar disadvantages as the Lagrangian approach since complex geometric regions are again difficult to handle. To obtain good results it may be necessary to follow a large number of particles, thereby quickly leading to excessive computer time and memory.

*Matrix Equation Solvers*

Discretization of the governing partial differential equations for water flow and solute transport generally leads to a system of linear matrix equations:

$$[A]\{x\} = \{b\} \tag{1}$$

in which $\{x\}$ is an unknown solution vector, $\{b\}$ is the known right-hand side vector of the matrix equation and $[A]$ is a sparse-banded matrix which is symmetric for water flow if the modified-Picard procedure is used, but asymmetric for water flow if the Newton-Raphson method is used. Matrix $[A]$ is generally asymmetric for solute transport, unless advection is not considered in the formulation. Technological breakthroughs in computer hardware and increased incentives to simulate complex, coupled flow and transport problems in large three-dimensional systems, has spurred the development and use of highly efficient and robust iterative matrix solvers. A brief synopsis of indirect matrix equation solvers, such as preconditioned conjugate gradient methods for symmetric matrices and the generalized conjugate gradient residual method (ORTHOMIN and modifications) for nonsymmetrical matrices, is given by van Genuchten and Sudicky (1999).

## REFERENCES

Celia, M.A., E.T. Bouloutas, and R.L. Zarba. 1990. A general mass-conservative numerical solution for the unsaturated flow equation, *Water Resour. Res.*, 26(7):1483–1496.

Huyakorn, P.S., E.P. Springer, V. Guvanasen, and T.D. Wadsworth. 1986. A three-dimensional finite-element model for simulating water flow in variably saturated porous media, *Water Resour. Res.*, 22:1790–1808.

Narasimhan, T.N. and P.A. Witherspoon. 1976. An integrated finite difference method for analyzing fluid flow in porous media, *Water Resour. Res.*, 12(1):57–64, 1976.

Neuman, S.P. 1984. Eulerian-Lagrangian finite element method of advection-dispersion, *Int. J. Numer. Methods Eng.*, 20:321–337.

Therrien, R. and E.A. Sudicky. 1996. Three-dimensional analysis of variably-saturated flow and solute transport in discretely-fractured porous media, *J. Contam. Hydrol.*, 23:1–44.

van Genuchten, M. Th. and E.A. Sudicky. 1999. Recent advances in vadose zone flow and transport modeling, *in*, M.B. Parlange and J.W. Hopmans (eds.), *Vadose Zone Hydrology: Cutting Across Disciplines*, pp. 155–193, Oxford University Press, NY.

Wolfsberg, A.V. and D.L. Freyberg. 1994. Efficient simulation of single species and multispecies transport in groundwater with local adaptive grid refinement, *Water Resour. Res.*, 30, 2979–2991.

Yeh, G.T. 1990. A Lagrangian-Eulerian method with zoomable hidden fine-mesh approach to solving advection-dispersion equations, *Water Resour. Res.*, 26(6), 1133–1144.

### 6.3.3.2  Finite Difference Methods

For one-dimensional transient solute transport, the dependent variable $[C(x,t)]$ can be discretized according to:

$$C(x,t) = C(i\Delta x, j\Delta t) = C_{ij} \qquad (i = 0,1,2,\ldots,n; j = 0,1,2,\ldots,m) \qquad [6.98]$$

Consider the simulation of the one-dimensional ADE for steady flow of a conservative tracer as given by Equation [6.63] with $R = 1$. Temporal and spatial derivatives are approximated with Taylor series expansions. Assume that the concentrations are known at the current time ($t = j\Delta t$) and that the objective is to calculate the concentration distribution $\{C[i\Delta x,(j+1)t]\}$ at the next time. A forward-in-time finite difference scheme where the unknown concentration is given explicitly in terms of known concentrations, is written as:

$$\frac{C_{i,j+1} - C_{i,j}}{\Delta t} \approx D\frac{C_{i+1,j} - 2C_{i,j} + C_{i-1,j}}{(\Delta x)^2} - v\frac{C_{i+1,j} - C_{i-1,j}}{2\Delta x} \qquad [6.99]$$

On the other hand, a backward-in-time or implicit scheme is given by:

$$\frac{C_{i,j+1} - C_{i,j}}{\Delta t} \approx D\frac{C_{i+1,j+1} - 2C_{i,j+1} + C_{i-1,j+1}}{(\Delta x)^2} - v\frac{C_{i+1,j+1} - C_{i-1,j+1}}{2\Delta x} \qquad [6.100]$$

which contains several concentrations at the next time level. The problem is solved implicitly after combining the difference schemes of all nodes, and subsequently using a matrix equation solver. Finally, a weighted scheme can be defined as:

$$\frac{C_{i,j+1} - C_{i,j}}{\Delta t} \approx D\frac{\omega\left(C_{i+1,j+1} - 2C_{i,j+1} + C_{i-1,j+1}\right) + (1-\omega)\left(C_{i+1,j} - C_{i,j} + C_{i-1,j}\right)}{(\Delta x)^2}$$

$$- v\frac{\omega\left(C_{i+1,j+1} - C_{i-1,j+1}\right) + (1-\omega)\left(C_{i+1,j} - C_{i-1,j}\right)}{2\Delta x} \qquad [6.101]$$

in which $\omega$ is a weighting constant between 0 and 1. The scheme is said to be fully explicit for $\omega = 0$ and fully implicit for $\omega = 1$, while a Crank-Nicolson central-in-time scheme is derived if $\omega = 0.5$. For transient flow, the scheme becomes more complicated; the velocity ($v$) and water content (θ) are usually obtained by solving the Richards equation prior to solving the transport problem.

Errors associated with the discretization and the solution procedure can be evaluated by comparison to analytical results, provided that the problem can be sufficiently idealized (linearized) to permit the use of such solutions. This is true for other numerical solutions as well. For a convergent scheme, the difference between the numerical and analytical solutions should decrease if smaller space and time steps are used in the numerical solution; the difference should become zero if $\Delta x \rightarrow 0$ and $\Delta t \rightarrow 0$, barring round off and computational errors.

An even more important question relates to the stability of the finite difference approximation, namely, the degree to which the numerical solution is affected by errors that occur during the simulation. Such errors can usually not be eliminated completely; they depend on the implemented discretization, the values of the input parameters, and the type of numerical scheme used for approximation of the governing transport equation. Errors are damped during the course of a simulation when a stable scheme is used, while unstable schemes allow such errors to grow unboundedly.

Implicit systems are unconditionally stable, but their results may not be as amenable to changes in grid sizes as is the case with explicit systems. The Crank-Nicolson method offers an attractive compromise of being unconditionally stable and having a truncation error of order $O[(\Delta x)^2 + (\Delta t)^2]$. These properties imply, among other things, that a variable time step can be used independently of the spatial step to effectively balance the needs of an accurate approximation and a limited number of computations. However, oscillations near the concentration front may develop even for unconditionally stable methods due to the hyperbolic (convection) term in the solute transport equation. To avoid these oscillations, stability criteria in terms of the grid Peclet ($P$ or $Pe$) and Courant ($Cr$) numbers are frequently formulated. Huyakorn and Pinder (1983) provided the following conservative guidelines for one-dimensional transport of a nonreactive solute:

$$Pe = v\Delta x/D < 2 \qquad [6.102]$$

$$Cr = v\Delta t/\Delta x < 1 \qquad [6.103]$$

For transport of a reactive solute, the retardation factor R must be included in the denominator of the Courant number.

### 6.3.3.3 Finite Element Methods

The application of the finite element methods for modeling solute transport in soils involves several steps, which will be briefly reviewed in a qualitative manner. The specifics of the method can be found in, among others, Huyakorn and Pinder (1983) and Istok (1989). The (spatial) solution domain should first be discretized. The finite element mesh, consisting of nodal points marking the elements, is usually tailored to the problem at hand. The selection of mesh size is a somewhat subjective process that considers the required degree of accuracy, the geometry of the problem, the ease of mesh generation, and the mathematical complexity associated with the use of a particular mesh. The shape of the finite elements is determined by the dimensionality and the geometry of the transport problem. One-dimensional elements consist of lines between nodal points along the coordinate where (one-dimensional) transport occurs. Examples of two-dimensional elements are triangles, rectangles and parallelograms, while many different three-dimensional elements can be constructed.

A second step in implementing the finite element method is the description of the transport problem for each element. Generally, the method of weighted residuals is used to arrive at an integral formulation for the governing transport equation. An approximate or trial solution for the concentration in each element is formulated as the weighted sum of the unknown concentration at the nodes of that element. A wide variety of so-called basis functions can be selected to assign weights to the nodes; the functions depend on the type of element being selected. The approximation for the element concentration will not be exact; substitution of the element concentration in the ADE yields an expression for this error (residual). The objective is to minimize the residual over the entire domain with some kind of weighting procedure for the elements. The method of weighted residuals forces the weighted average of the residuals at the nodes to be zero and gives an expression for the residual of each element. After applying Green's theorem for each element, the residual can be written in matrix form.

The third step is the assembly of all element matrices into a global matrix system that contains the nodal concentrations and its temporal derivative as unknown variables.

The last step involves the solution of this global matrix equation. Since there are, in general, no obvious advantages for also using a finite element approximation for the temporal derivative, the latter is normally dealt with using the (simpler) finite difference method. The time-step may be constant or variable. As noted earlier, the explicit or forward scheme ($\omega = 0$) is conditionally stable whereas the implicit or backward ($\omega = 1$) and the centered or Crank-Nicolson ($\omega = 0.5$) schemes are unconditionally stable. The nodal concentration can then be solved using standard numerical procedures.

### 6.3.3.4 Application

Numerical oscillations and dispersion are illustrated for several numerical schemes using results by Huang et al. (1997). The numerical solution of the ADE is particularly difficult for relatively sharp concentration fronts with advection-dominated transport characterized by small dispersivities. As mentioned earlier, undesired oscillations can often be prevented with judicious space and time discretizations. The Peclet number increases when advection dominates dispersion; the potentially adverse effect on the numerical solution can be compensated by selecting a smaller grid size. Numerical oscillations can be virtually eliminated if the local Peclet number is always less than 2. However, acceptable results may still be obtained for local Peclet numbers as high as 10 (Huyakorn and Pinder, 1983). The time discretization is based on a second dimensionless number, the Courant number (Equation [6.103]).

Consider one-dimensional, steady-state flow in a soil column with $v = 4$ cm d$^{-1}$. The column was initially free of any solute; a nonreactive solute ($C_o = 1$) was subsequently applied for 20 days. For negligible molecular diffusion and a longitudinal dispersivity ($\alpha_L = 0.02$ cm), the grid Peclet for a spatial discretization ($\Delta x = 2$ cm) is given by:

$$Pe = \frac{v\Delta x}{D} = \frac{\Delta x}{\alpha_L} = 100 \qquad [6.104]$$

This grid Peclet number of 100 indicates solute transport that is dominated by advection. In all calculations, the Courant number was less than 1, which is the stability condition for Eulerian methods.

The concentration predicted with four different finite element methods and with the analytical solution is shown in Figure 6.13. The first numerical method is based on the central Crank-Nicolson scheme for time with weighting constant ($\omega = 0.5$). Second, the analytical solution is shown as a solid line. Third, an implicit scheme ($\omega = 1.0$) is used. The last two numerical methods implement upstream weighting (Huyakorn and Pinder, 1983) in the central and the implicit schemes, respectively. The results obtained with the Crank-Nicolson method have significant numerical oscillations (both overshoot and undershoot) (Figure 6.13). This is not surprising given the large value for *Pe*. The oscillations are reduced by using an implicit scheme. Upstream weighting leads to very similar results for the central and implicit schemes. The oscillations are virtually eliminated but the solute

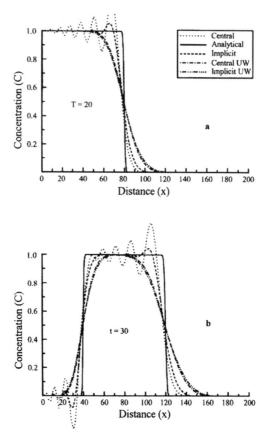

**FIGURE 6.13** Concentration profiles predicted with different numerical schemes for (a) solute infiltration ($t = 30$ d) and (b) solute leaching ($t = 30$ d).

profile exhibits more numerical dispersion; there is a greater discrepancy with the steep analytical profile than for the regular implicit scheme.

---

**EXAMPLE 6.9**   Estimation of Transport Parameters from Transient Flow Experiments

Parameter estimation techniques are now increasingly being used to determine either soil hydraulic properties or solute transport parameters. The latter have typically been determined from solute displacement experiments in laboratory soil columns during steady-state flow using analytical solutions of the ADE. Unsaturated hydraulic properties may be determined inversely by repeated numerical solution of the flow problem describing some transient flow experiment in the laboratory or the field. Hydraulic parameters are then estimated by minimizing an objective function containing the sums of squared deviations between observed and predicted flow variables. This type of inverse problem can also be used to estimate both flow and transport parameters. Such an approach allows one to determine a large number of flow and transport parameters from a broad range of experiments involving nonlinear exchange, transient flow, and nonequilibrium flow and/or transport (Šimůnek et al., 2000).

Fortin et al. (1997) conducted column displacement experiments with flow interruption using bromide and simazine to investigate nonequilibrium sorption models. The data were also analyzed by Šimůnek et al. (2000) using a two-site adsorption model. Equilibrium sorption occurs at type-1 sites according to a linear isotherm while adsorption at type-2 sites

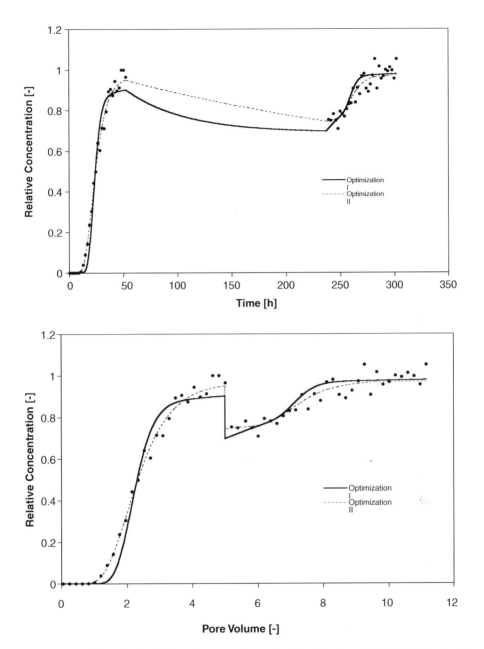

**DIAGRAM 6.9.1** Simazine breakthrough curves determined by Fortin et al. (1997) with the flow interruption technique and optimized using: I. dispersivity estimated from the bromide effluent data, and II. dispersivity estimated from simazine effluent data. Simazine concentration is given as a function of: (a) elapsed time and (b) pore volume.

is described as a first-order kinetic process with rate coefficient $\alpha$ (cf. Toride et al., 1995). The simazine breakthrough curve was optimized by using the dispersivity, $\lambda$, of 0.341 cm from the bromide results (optimization I), or as an optimization parameter along with the nonequilibrium rate coefficient, $\alpha$, the distribution factor, $K_d$, and the fraction of sorption sites with equilibrium sorption, $f$ (optimization II). Optimization I yielded $\alpha = 0.0121$ h$^{-1}$, $K_d = 0.814$ cm$^3$ g$^{-1}$, and $f = 0.533$ whereas the results for optimization II were $\lambda = 1.08$ cm, $\alpha = 0.0128$ h$^{-1}$, $K_d = 1.49$ cm$^3$ g$^{-1}$, and $f = 0.328$. The experimental and optimization results are shown in Diagram 6.9.1 as a function of pore volumes leached through the soil column and as a function of elapsed time. It appears that the second approach, with the increased flexibility to optimize dispersion, provides a better description of the breakthrough curve.

## REFERENCES

Fortin, J., M. Flury, W.A. Jury, and T. Streck. 1997. Rate-limited sorption of simazine in saturated soil columns, *J. Contam. Hydrol.*, 25:229–234.

Šimůnek, J., D. Jacques, J.W. Hopmans, M. Inoue, M. Flury, and M. Th. van Genuchten. 2000. Solute transport during variably-saturated flow: Inverse methods, *in* Methods of Soil Analysis, *Soil Sci. Soc. Am.,* Madison, WI.

Toride, N., F.J. Leij, and M. Th. van Genuchten. 1995. The CXTFIT code for estimating transport parameters from laboratory or field tracer experiments. Version 2.0. Research report No. 137, U.S. Salinity Laboratory, USDA, ARS, Riverside, CA.

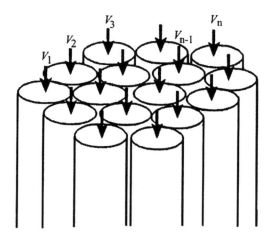

**FIGURE 6.14** Schematic of the stream tube model.

## 6.4  STREAM TUBE MODELS

Considerable errors may be made when applying deterministic methods to the field since model parameters are actually stochastic due to spatial and temporal variability, measurement errors and different averaging scales. Usually, it is not possible or important to obtain discrete values for parameters in deterministic models of field-scale transport. Instead, transport properties and model results are described with statistical functions. Stochastic modeling is no substitute for data collection or model development, but merely a method to deal with uncertainty of model parameters and complexity of flow and transport processes. The scale at which solute movement is observed or modeled is important. The averaging process associated with larger field-scale descriptions tends to filter out the variability at smaller scales.

The stochastic modeling of actual field problems is seldom possible. The stream tube model provides a possible exception (Dagan and Bresler, 1979; Amoozegar-Fard et al., 1982; Rubin and Or, 1993). The field is conceptualized as a system of parallel tubes illustrated in Figure 6.14. A process-based model is used to describe a one-dimensional, autonomous transport in each tube as a function of time and depth. Transport parameters are either deterministic or stochastic. The problem may be solved analytically at the scale of the tube and the field. The stream tube model is suitable for inverse procedures to estimate transport and statistical properties (Toride et al., 1995).

**EXAMPLE 6.10**  Alternatives for the Advection Dispersion Equation

The advection-dispersion equation (ADE) remains by far the most popular model for describing subsurface solute transport, in part because of its convenience to mathematically model solute concentrations. The ADE may also serve as a starting or reference point for more complicated descriptions of solute transport. This chapter is therefore mostly devoted to the ADE, but some alternative approaches are mentioned in the following.

Transfer functions facilitate the use of different transport models (Jury and Roth, 1990; Sardin et al., 1991). In this approach, the breakthrough curve at a location $x = L$ is defined as the convolution between the input concentration at $x = 0$ and the solute residence time distribution according to:

$$C(L,t) = \int_0^t C(0, t - \tau) f(L, \tau) d\tau \qquad [1]$$

The residence time distribution $[f(L, t)]$ whose Laplace transform is the transfer function, may be derived from experimental results or from a transport model. In the latter case when the ADE is involved, the distribution is given by:

$$f(L,t) = \frac{L}{\sqrt{4\pi D t^3}} \exp\left(-\frac{(L - vt)^2}{4Dt}\right) \qquad [2]$$

Note that [2] is a Gaussian distribution. The ADE has typically been validated only with solute displacement experiments in packed laboratory columns for one depth (the column outlet). This may not be very meaningful because a good description of a breakthrough curve can usually be accomplished through manipulation of the model parameters. The ADE is not unique in this regard, as Jury and Roth (1990) demonstrated in a widely cited example. An almost identical description of the breakthrough curve was obtained with a Fickian, a lognormal or a gamma distribution for $f(L, t)$, notwithstanding their obviously different mathematical forms. Khan and Jury (1990) demonstrated that $D$ increased with position for solute transport in undisturbed soil columns, casting doubt on the applicability of the ADE. Clearly, for natural porous media more complicated models are required to properly account for dispersion (e.g., Dagan, 1989).

Recently, fractional advection dispersion equations (FADEs) have been proposed for solute transport in soils (Pachepsky et al., 2000). The simplest form is as follows (Meerschaert et al., 1999):

$$\frac{\partial C}{\partial t} = \frac{D}{2}\left(\frac{\partial^\alpha C}{\partial x^\alpha} + \frac{\partial^\alpha C}{\partial (-x)^\alpha}\right) - v \frac{\partial C}{\partial x} \qquad [3]$$

where $D$ is now a fractional dispersion coefficient and $\alpha$ denotes the order of fractional differentiation ($0 < \alpha \leq 2$). This transport model follows from the use of Lévy statistics to allow for spatial memory, i.e., solute particles moving substantially above or below the mean velocity are more likely to continue doing so in the near future than other particles. Lévy distributions have heavier tails than Gaussian distributions to captivate this feature. The FADE is a special case of more general continuous time random walk (CTRW) models that have been proposed in the physics literature to describe solute transport using particle movement over a random distance during a random time (Montroll and Weiss, 1965).

## REFERENCES

Dagan, G. 1989. Flow and transport in porous formations, Springer Verlag, Berlin.

Jury, W.A. and K. Roth. 1990. Transfer functions and solute movement through soil. Theory and applications, Birkhäuser Verlag, Basel.

Khan, A.U.-H. and W.A. Jury. 1990. A laboratory study of the dispersion scale effect in column outflow experiments, *J. Contam. Hydrol.,* 5:119–131.

Meerschaert, M.M., D.A. Benson, and B. Bäumer. 1999. Multidimensional advection and fractional dispersion, *Phys. Rev.,* E 59:5026–5028.

Montroll, E.W. and G.H. Weiss. 1965. Random walks on lattices. II, *J. Math. Phys.,* 6:167–183.

Pachepsky, Y., D. Benson, and W. Rawls. 2000. Simulating scale-dependent solute transport in soils with the fractional advective-dispersive equation, *Soil Sci. Soc. Am. J.,* (64:1234–1243).

Sardin, M., D. Schweich, F.J. Leij, and M. Th. van Genuchten. 1991. Modeling the nonequilibrium transport of linearly interacting solutes in porous media: A review, *Water Resour. Res.,* 27:2287–2307.

### 6.4.1 MODEL FORMULATION

For one-dimensional transport, the solute concentration at the outlet of a stream tube of length ($L$) may be written with the transfer function approach as (Jury and Roth, 1990):

$$C(L,t) = \int_0^t C(0, t-\tau) f(L, \tau) d\tau \qquad [6.105]$$

The outlet concentration is a convolution of the input signal and the residence time distribution $f(L,\tau)$. The latter is, in effect, a probability density function (pdf) of the time a solute particle resides in the soil between $x = 0$ and $L$. The pdf, which has dimension of inverse time, can be determined from experimental results or it may be a theoretical expression derived from a process-based model such as the ADE. The pdf depends mostly on the transport properties of the soil; it depends to some degree on the mode of solute application and detection, but not on the input and initial concentrations. For the equilibrium problem involving instantaneous solute application to a solute-free soil, the flux-averaged concentration and hence the pdf is given by:

$$f(x,t) = \frac{x}{\sqrt{4\pi D t^3}} \exp\left(-\frac{(x-vt)^2}{4Dt}\right) \qquad [6.106]$$

Note that this is a Gaussian pdf. Expressions for $f(x,t)$ derived at a particular depth can be readily used for predicting concentrations at other depths if the soil is homogeneous.

The stochastic approach is implemented by considering the (constant) column parameters as realizations of a random distribution. The (horizontal) field-averaged solute concentration is considered the ensemble average over the probability distribution. It is assumed that each random parameter obeys a distribution function that is independent of location and that the ensemble of possible concentrations may be estimated from a sufficient number of samples taken at different locations. The average concentration across the field is then identical to the ensemble average:

$$\langle C(x,t) \rangle = \frac{1}{A} \int_A C(x,t) dA = \lim_{n \to \infty} \frac{1}{n} \sum_{i=1}^{n} C_i(x,t) \qquad [6.107]$$

where $A$ denotes the area of the field, $n$ is the number of samples, and $< >$ indicates an ensemble average.

## 6.4.2 APPLICATION

Solute transport in a local-scale stream tube may be described with the one-dimensional ADE. Following Toride and Leij (1996a), the effects of heterogeneity are studied using pairs of random parameters ($v$ and $K_d$). Note that the water content ($\theta$) and bulk density ($\rho_b$) are the same for all stream tubes. The field-scale mean concentration (denoted by ^) is equal to the ensemble average:

$$\hat{C}(x,t) = \langle C(x,t) \rangle = \int_0^\infty \int_0^\infty C(x,t;v,K_d) f(v,K_d) dv \ dK_d \qquad [6.108]$$

where the bivariate lognormal joint probability density function $f(v,K_d)$ is given by:

$$f(v,K_d) = \frac{1}{2\pi\sigma_v \sigma_{Kd} v K_d \sqrt{1-\rho_{vKd}^2}} \exp\left(-\frac{Y_v^2 - 2\rho_{vKd}Y_v Y_{Kd} + Y_{Kd}^2}{2(1-\rho_{vKd}^2)}\right) \qquad [6.109]$$

with

$$\rho_{vKd} = \langle Y_v Y_{Kd} \rangle = \int_0^\infty \int_0^\infty Y_v Y_{Kd} f(v,K_d) dv \ dK_d \qquad [6.110]$$

$$Y_v = \frac{\ln(v) - \mu_v}{\sigma_v}, \quad Y_{Kd} = \frac{\ln(K_d) - \mu_{Kd}}{\sigma_{Kd}} \qquad [6.111]$$

where $\mu$ and $\sigma$ are the mean and standard deviation of the log transformed variable, and $\rho_{vKd}$ is the correlation coefficient between $Y_v$ and $Y_{Kd}$ ($K_d$ tends to increase with $v$ for positive $\rho_{vKd}$ while $K_d$ decreases with $v$ for negative $\rho_{vKd}$). Ensemble averages of $v$ and $K_d$ are given by (Aitcheson and Brown, 1963):

$$\langle v \rangle = \exp\left(\mu_v + \frac{1}{2}\sigma_v^2\right), \quad \langle K_d \rangle = \exp\left(\mu_{Kd} + \frac{1}{2}\sigma_{Kd}^2\right) \qquad [6.112]$$

with coefficients of variation (CV):

$$CV(v) = \sqrt{\exp(\sigma_v^2) - 1}, \quad CV(K_d) = \sqrt{\exp(\sigma_{Kd}^2) - 1} \qquad [6.113]$$

Based upon the detection mode of the local-scale concentration, three types of field-scale concentrations can be defined: (1) the ensemble average of the flux-averaged concentration ($<C_F>$); (2) the field-scale resident concentration ($\hat{C}_R$), which is equal to the ensemble average of the resident concentration ($<C_R>$); and (3) the field-scale flux-averaged concentration ($\hat{C}_F$) which is defined as $<vC_F>/<v>$. The second type of concentration is obtained from averaging values of the resident concentration at a particular depth across the field. The third type is defined as the ratio of ensemble solute and water fluxes in a similar manner as for deterministic transport. It should be noted that $<vC_F>/<v> \neq <C_F>$ because $v$ is a stochastic variable.

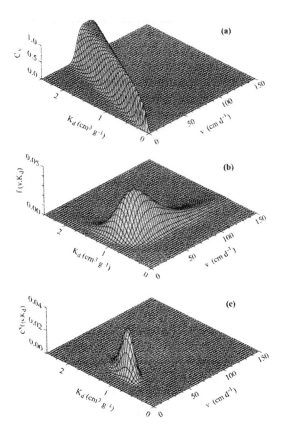

**FIGURE 6.15** Illustration of the stream tube model for field-scale transport for a 2-d solute application with $D = 20$ cm$^2$ d$^{-1}$ and $\rho_b/\theta = 4$ g cm$^{-3}$: (a) the local-scale concentration, $C_R$, as a function of $v$ and $K_d$ at $x = 100$ cm and $t = 5$ days; (b) a bivariate lognormal pdf for $\rho_v K_d = 0.5$, $<v> = 50$ cm d$^{-1}$, $\sigma_v = 0.2$, $<K_d> = 1$ cm$^3$ g$^{-1}$, and $\sigma_{Kd} = 0.2$; and (c) the expected concentration, $C^R$, at $x = 100$ cm and $t = 5$ days.

The use of the stream tube model for field-scale transport is illustrated in Figure 6.15. The local-scale concentration only depends on the particular realizations of the two stochastic parameters ($v$ and $K_d$) after the independent variables ($t$ and $x$) have been specified. The solution for the ADE at $x = 100$ cm and $t = 5$ days as a function of $v$ and $K_d$ is shown in Figure 6.15a and the bivariate lognormal pdf for $v$ and $K_d$ in Figure 6.15b; the distribution is skewed with respect to $v$ since $\sigma_v$ is fairly high; the smaller $\sigma_{Kd}$ results in a more symmetric distribution for $K_d$. The negative $\rho_v K_d$ results in an increasing $v$ with a decreasing $K_d$, and vice versa. The expected concentration is shown in Figure 6.15c, which is obtained by weighting the local concentration (Figure 6.15a) by multiplying it with the joint pdf (Figure 6.16b). The peak in Figure 6.15c suggests that stream tubes with approximately $v = 25$ cm d$^{-1}$ and $K_d = 1$ cm$^3$ g$^{-1}$ contribute the most to solute breakthrough when $x = 100$ cm and $t = 5$ days. The volume of the distribution in Figure 6.15c corresponds to the ensemble average ($<C_R>$).

Variations in the local-scale concentration between stream tubes, at a particular depth and time, can be characterized by its variance. The variance across the horizontal plane is given by (Bresler and Dagan, 1981; Toride and Leij, 1996b):

$$Var[C(x,t)] = \int_0^\infty \int_0^\infty \left[C(x,t) - \langle C(x,t) \rangle\right]^2 f\left(v, K_d\right) dv \ dK_d = \langle C^2(x,t) \rangle - \langle C(x,t) \rangle^2 \quad [6.114]$$

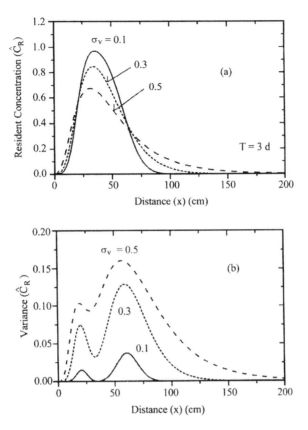

**FIGURE 6.16** The effect of variability in the pore-water velocity, $v$, on (a) the field scale resident concentration profile, and (b) the distribution of the variance for $\hat{C}_R$ in the horizontal plane; for a 2-d application of a nonreactive solute ($R = 1$) assuming $<v> = 20$ cm d$^{-1}$ and $D = 20$ cm$^2$ d$^{-1}$.

For a deterministic distribution coefficient ($K_d$) Equation [6.108] reduces to:

$$\langle C(x,t)\rangle = \int_0^\infty C(x,t;v)f(v)dv \qquad [6.115]$$

where the lognormal pdf for the single stochastic variable (v) is given by:

$$f(v) = \frac{1}{\sqrt{2\pi}\sigma_v v}\exp\left(-\frac{\left[\ln(v)-\mu_v\right]^2}{2\sigma_v^2}\right) \qquad [6.116]$$

The mean (Figure 6.16a) ($\hat{C}_R = <C_R>$) and the variance (Figure 6.16b) are shown according to Equation [6.114], as a function of depth for three values of $\sigma_v$ at $t = 3$ days for a nonreactive solute. More solute spreading occurs in the $\hat{C}_R$-profile when $\sigma_v$ increases (Figure 6.16a). The variation in the local-scale $C_R$ also increases with $\sigma_v$, indicating a more heterogeneous solute distribution in the horizontal plane (Figure 6.16b). Because flow and transport become more heterogeneous as $\sigma_v$ increases, more observations are needed to estimate $\hat{C}_R$ for $\sigma_v = 0.5$ than for $\sigma_v = 0.1$. The double peak in the variance profiles of Figure 6.16b is due to the relative minimum for this pulse application at around $x = 30$ cm, where the highest concentration occurs.

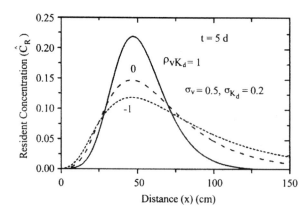

**FIGURE 6.17** The field-scale resident concentration profile, $\hat{C}_R$ $(x)$, for three different correlations between $v$ and $K_d$ with $<v>$ = 50 cm d$^{-1}$, $D$ = 20 cm$^2$d$^{-1}$, $<K_d>$ = 1 g$^{-1}$ cm$^3$, $\sigma_{Kd}$ = 0.2, $<R>$ = 5, and $\rho_b$ /$\theta$ = 4 g cm$^{-3}$.

For reactive solutes, the variability in the distribution coefficient ($K_d$) must also be considered. The field-scale resident concentration ($\hat{C}_R$) at $t$ = 5 days is plotted versus depth in Figure 6.17 for perfect and no correlation between $v$ and $K_d$. The figure shows that the negative correlation between $v$ and $K_d$ causes additional spreading. Such a negative correlation seems plausible since coarse-textured soils generally have a relatively high $v$, due to their higher hydraulic conductivity, and a small $K_d$, due to their lower exchange capacity. The opposite may hold for fine-textured soils. Robin et al. (1991) reported a weak negative correlation between $K_d$ and the saturated hydraulic conductivity for a sandy aquifer.

## REFERENCES

Ahlstrom, S.W., H.P. Foote, R.C. Arnett, C.R. Cole, and R.J. Serne. 1977. Multicomponent mass transport model theory and numerical implementation. Battelle Pacific Northwest Laboratory Report BNWL–2127, Richland, WA.

Aitcheson, J. and J.A.C. Brown. 1963. The lognormal distribution, Cambridge University Press, London, UK.

Amoozegar-Fard, A., D.R. Nielsen, and A.W. Warrick. 1982. Soil solute concentration distribution for spatially varying pore water velocities and apparent diffusion coefficients, *Soil Sci. Soc. Am. J.*, 46:3–9.

Aris, R. 1956. On the dispersion of a solute in a fluid flowing through a tube, *Proc. Roy. Soc. London Ser.*, 235:67–77.

Aris, R. 1958. On the dispersion of linear kinematic waves, *Proc. Roy. Soc. London Ser.*, 245:268–277.

Bear, J. and A. Verruijt. 1987. Modeling groundwater flow and pollution. Kluwer Academic Publishers, Norwell, MA.

Boyd, G.E., A.W. Adamson, and L.S. Meyers. 1947. The exchange adsorption of ions from aqueous solutions by organic zeolites. II. Kinetics, *J. Am. Chem. Soc.*, 69:2836–2848.

Brenner, H. 1962. The diffusion model of longitudinal mixing in beds of finite length. Numerical values, *Chem. Eng. Sci.*, 17:229–243.

Bresler, E. 1973. Simultaneous transport of solutes and water under transient unsaturated flow conditions, *Water Resour. Res.*, 9:975–986.

Bresler, E. and G. Dagan. 1981. Convective and pore scale dispersive solute transport in unsaturated heterogeneous fields, *Water Resour. Res.*, 17:1683–1693.

Bresler, E. and R.J. Hanks. 1969. Numerical method for estimating simultaneous flow of water and salt in saturated soils, *Soil Sci. Soc. Am. Proc.*, 33:827–832.

Campbell, G.S. 1985. Soil physics with BASIC: Transport models for soil-plant systems, Elsevier, NY.

Carslaw, H.S. and J.C. Jaeger. 1959. Conduction of heat in solids, Clarendon Press, Oxford, UK.

Chaudhari, N.M. 1971. An improved numerical technique for solving multidimensional miscible displacement, *Soc. Petrol. Eng. J.*, 11:277–284.

Coats, K.H. and B.D. Smith. 1964. Dead end pore volume and dispersion in porous media, *Soc. Petrol. Eng. J.,* 4:73–84.

Crank, J. 1975. The mathematics of diffusion, Clarendon Press, Oxford, UK.

Dagan, G. and E. Bresler. 1979. Solute dispersion in unsaturated heterogeneous soil at field-scale. I. Theory, *Soil Sci. Soc. Am. J.,* 43:461–467.

Donea, J. 1991. Generalized Galerkin methods for convection dominated transport phenomena, *Appl. Mech. Rev.,* 44:205–214.

Dykhuizen, R.C. and W.H. Casey. 1989. An analysis of solute diffusion in rocks, *Geochim. Cosmochim. Acta,* 53:2797–2805.

Epstein, N. 1989. On tortuosity and the tortuosity factor in flow and diffusion through porous media, *Chem. Eng. Sci.,* 44:777–780.

Fischer, H.B., E. List, R.C.Y. Koh, J. Imberger, and N.H. Brooks. 1979. Mixing in inland and coastal waters, Academic Press, NY.

Fried, J.J. and M.A. Combarnous. 1971. Dispersion in porous media, *Adv. Hydrosci.,* 9:169–282.

Gautschi, W. 1964. Error function and Fresnel integrals. p. 295–329, *in* M. Abramowitz and I.A. Stegun handbook of mathematical functions, *Appl. Math. Ser.,* 55, National Bureau of Standards, Washington, D.C.

Gelhar, L.W., C. Welthy, and K.R. Rehfeldt. 1992. A critical review of data on field-scale dispersion in aquifers, *Water Resour. Res.,* 28:1955–1974.

Goldstein, S. 1953. On the mathematics of exchange processes in fixed columns. I. Mathematical solutions and asymptotic expansions, *Proc. Roy. Soc. London Ser.* A, 219:151–185.

Gray, W.G. and G.F. Pinder. 1976. An analysis of the numerical solution of the transport equation, *Water Resour. Res.,* 12:547–555.

Hamaker, J.W. 1972. Diffusion and volatilization, p. 341–397, *in* C.A.I. Goring and J.W. Hamaker (eds.), Organic chemicals in the soil environment, Marcel Dekker, Inc., NY.

Harvie, C.E. and J.H. Weare. 1980. The prediction of mineral solubilities in natural waters: The Na-K-Mg-Ca-Cl-SO$_4$.H$_2$O system from zero to high concentration at 25°C, *Geochim. Cosmochim. Acta,* 44:981-997.

Helfferich, F. 1962. Ion exchange, McGraw-Hill, NY.

Huang, K. and M. Th. van Genuchten. 1995. An analytical solution for predicting solute transport during ponded infiltration, *Soil Sci.,* 159:217–223.

Huang, K., J. Šimůnek, and M. Th. van Genuchten. 1997. A third-order numerical scheme with upwind weighting for modeling solute transport, *Int. J. Numer. Meth. Eng.,* 40:1623–1637.

Huyakorn, P.S. and G.F. Pinder. 1983. Computational methods in subsurface flow, Academic Press, NY.

Istok, J. 1989. Groundwater modeling by the finite element method, Am. Geophys. Union, Washington, D.C.

Jacobsen, O.H., F.J. Leij, and M. Th. van Genuchten. 1992. Parameter determination for chloride and tritium transport in undisturbed lysimeters during steady flow, *Nordic Hydrol.,* 23:89–104.

Jin, Y. and W.A. Jury. 1996. Characterizing the dependence of gas diffusion coefficient on soil properties, *Soil Sci. Soc. Am. J.,* 60:66–71.

Jury, W.A. and K. Roth. 1990. Transfer functions and solute movement through soil. Theory and applications, Birkhäuser Verlag, Basel, Switzerland.

Kemper, W.D. 1986. Solute Diffusivity, p. 1007–1024, *in* A. Klute (ed.), Methods of soil analysis. I. Physical and mineralogical methods, *Soil Sci. Soc. Am.,* Madison, WI.

Klotz, D., K.P. Seiler, H. Moser, and F. Neumaier. 1980. Dispersity and velocity relationship from laboratory and field experiments, *J. Hydrol.,* 45:169–184.

Konikov, L.F.,and J.D. Bredehoeft. 1978. Computer model of two-dimensional solute transport and dispersion in groundwater, *in* Techniques of water resources investigation, Book 7, U.S. Geological Survey, Reston, VA.

Kreft, A. and A. Zuber. 1978. On the physical meaning of the dispersion equations and its solutions for different initial and boundary conditions, *Chem. Eng. Sci.,* 33:1471–1480.

Kreft, A. and A. Zuber. 1986. Comments on "Flux-averaged and volume-averaged concentrations in continuum approaches to solute transport" by J.C. Parker and M. Th. van Genuchten, *Water Resour. Res.,* 22:1157–1158.

Krupp, H.K., J.W. Biggar, and D.R. Nielsen. 1972. Relative flow rates of salt and water in soil, *Soil Sci. Soc. Am. Proc.,* 36:412–417.

Lassey, K.R. 1988. Unidimensional solute transport incorporating equilibrium and rate-limited isotherms with first-order loss. 1. Model conceptualizations and analytic solutions, *Water Resour. Res.*, 3:343–350.

Leij, F.J. and J.H. Dane. 1992. Moment method applied to solute transport with binary and ternary exchange, *Soil Sci. Soc. Am. J.*, 56:667–674.

Leij, F.J. and N. Toride. 1995. Discrete time- and length-averaged solutions of the advection-dispersion equation, *Water Resour. Res.*, 31:1713–1724.

Leij, F.J. and M. Th. van Genuchten. 1995. Approximate analytical solutions for transport in two-layer porous media, *Trans. Porous Media*, 18:65–85.

Li, S-G, F. Ruan, and D. McLaughlin. 1992. A space-time accurate method for solving solute transport problems, *Water Resour. Res.*, 28:2297–2306.

Lide, D.R. 1995. *Handbook of Chemistry and Physics*, CRC Press, Boca Raton, FL.

Lindstrom, F.T. and M.N.L. Narasimhan. 1973. Mathematical theory of a kinetic model for dispersion of previously distributed chemicals in a sorbing porous medium. SIAM, *J. Appl. Math.*, 24:469–510.

Lindstrom, F.T. and W.M. Stone. 1974. On the start up or initial phase of linear mass transport of chemicals in a water saturated sorbing porous medium. SIAM, *J. Appl. Math.*, 26:578–591.

Millington, R.J. 1959. Gas diffusion of porous solids, *Science*, 130:100–102.

Millington, R.J. and J.P. Quirk. 1961. Permeability of porous solids, *Trans. Farad. Soc.*, 57:1200–1206.

Molz, F.J., M.A. Widdowson, and L.D. Benefield. 1986. Simulation of microbial growth dynamics coupled to nutrient and oxygen transport in porous media, *Water Resour. Res.*, 22:1207–1216.

Moridis, G.J. and D.L. Reddell. 1991. The Laplace transform finite difference method for simulation of flow through porous media, *Water Resour. Res.*, 27:1873–1884.

Nauman, E.B. and B.A. Buffham. 1983. Mixing in continuous flow systems, John Wiley and Sons, NY.

Neuman, S.P. 1984. Adaptive Eulerian-Lagrangian finite element method for advection-dispersion, *Int. J. Numer. Meth. Eng.*, 20:321–337.

Nielsen, D.R. and J.W. Biggar. 1961. Miscible displacement in soils. I. Experimental information, *Soil Sci. Soc. Am. Proc.*, 25:1–5.

Nkedi-Kizza, P., J.W. Biggar, M. Th. van Genuchten, P.J. Wierenga, H.M. Selim, J.M. Davidson, and D.R. Nielsen. 1983. Modeling tritium and chloride 36 transport through an aggregated oxisol, *Water Resour. Res.*, 19:691–700.

Nkedi-Kizza, P., J.W. Biggar, H.M. Selim, M. Th. van Genuchten, P.J. Wierenga, J.M. Davidson, and D.R. Nielsen. 1984. On the equivalence of two conceptual models for describing ion exchange during transport through an aggregated Oxisol, *Water Resour. Res.*, 20:1123–1130.

Nye, P.H. 1979. Diffusion of ions and uncharged solutes in soils and soil clays, *Adv. Agron.*, 31:225–272.

Olsen, S.R. and W.D. Kemper. 1968. Movement of nutrients to plant roots, *Adv. Agron.*, 20:91–151.

Oscarson, D.W., H.B. Hume, N.G. Sawatsky, and S.C.H. Cheung. 1992. Diffusion of iodide in compacted bentonite, *Soil Sci. Soc. Am. J.*, 56:1400–1406.

Parker, J.C. and M. Th. van Genuchten. 1984a. Flux-averaged concentrations in continuum approaches to solute transport, *Water Resour. Res.*, 20:866–872.

Parker, J.C. and M. Th. van Genuchten. 1984b. Determining transport parameters from laboratory and field tracer experiments, VA Agric. Exp. Stat. Bull. 84-3.

Pinder, G.F. and W.G. Gray. 1977. Finite elements in surface and subsurface hydrology, Academic Press, NY.

Pitzer, K.S. 1979. Activity coefficients in electrolyte solutions, CRC Press, Boca Raton, FL.

Robin, M.J.L., R.W. Gillham, and D.W. Oscarson. 1987. Diffusion of strontium and chloride in compacted clay-based materials, *Soil Sci. Soc. Am. J.*, 51:1102–1108.

Robin, M.J.L., E.A. Sudicky, R.W. Gillham, and R.G. Kachanoski. 1991. Spatial variability of Strontium distribution coefficients and their correlation with hydraulic conductivity in the Canadian Forces Base Borden aquifer, *Water Resour. Res.*, 27:2619–2632.

Rubin, Y. and D. Or. 1993. Stochastic modeling of unsaturated flow in heterogeneous soils with water uptake by plant roots: The parallel columns model, *Water Resour. Res.*, 29:619–631.

Scheidegger, A.E. 1974. The physics of flow through porous media, University of Toronto Press, Toronto, Canada.

Selim, H.M., J.M. Davidson, and R.S. Mansell. 1976. Evaluation of a two-site adsorption-desorption model for describing solute transport in soil, p. 444–448, *in* Proc. Computer Simulation Conference, American Institute of Chemical Engineering, Washington, DC.

Selim, H.M. and R.S. Mansell. 1976. Analytical solution of the equation for transport of reactive solutes through soils, *Water Resour. Res.*, 12:528–532.

Simon, W., P. Reichert, and C. Hinz. 1997. Properties of exact and approximate traveling wave solutions for transport with nonlinear and nonequilibrium sorption, *Water Resour. Res.*, 33:1139–1147.

Sparks, D.L. 1989. Kinetics of soil chemical processes, Academic Press, San Diego.

Spiegel, M.R. 1965. Theory and problems of Laplace transforms, Schaum's Outline Series, McGraw-Hill, NY.

Sudicky, E.A. 1989. The Laplace transform Galerkin technique: A time-continuous finite element theory and application to mass transport in groundwater, *Water Resour. Res.*, 25:1833–1846.

Taylor, G.I. 1953. Dispersion of soluble matter in solvent flowing through a tube, *Proc. Roy. Soc. London Ser.*, 219:186–203.

Toride, N. and F.J. Leij. 1996a. Convective-dispersive stream tube nodel for field-scale solute transport: I. Moment analysis, *Soil Sci. Soc. Am. J.*, 60:342–352.

Toride, N. and F.J. Leij. 1996b. Convective-dispersive stream tube nodel for field-scale solute transport: II. Examples and calibration, *Soil Sci. Soc. Am. J.*, 60:352–361.

Toride, N., F.J. Leij, and M. Th. van Genuchten. 1993. A comprehensive set of analytical solutions for nonequilibrium solute transport with first-order decay and zero-order production, *Water Resour. Res.*, 29:2167–2182.

Toride, N., F.J. Leij, and M. Th. van Genuchten. 1994. Flux-averaged concentrations for transport in soils having nonuniform initial solute distributions, *Soil Sci. Soc. Am. J.*, 57:1406–1409.

Toride, N., F.J. Leij, and M. Th. van Genuchten. 1995. The CXTFIT code for estimating transport parameters from laboratory or field tracer experiments. Version 2.0. U.S. Salinity Lab. Res. Rep. 137, Riverside, CA.

van der Heijde, P., Y. Bachmat, J. Bredehoeft, B. Andrews, D. Holtz, and S. Sebastian. 1985. Groundwater management: The use of numerical models. 2nd ed., Am. Geophys. Union, Washington, DC.

van der Laan, E.T. 1958. Notes on the diffusion-type model for the longitudinal mixing in flow, *Chem. Eng. Sci.*, 7:187–191.

van der Zee, S.E.A.T.M. 1990. Analytical traveling wave solutions for transport with nonlinear and nonequilibrium adsorption, *Water Resour. Res.*, 26:2563–2578.

van Genuchten, M. Th. 1978. Mass transport in saturated-unsaturated media. One-dimensional solutions, Princeton University Research Rep. 78-WR-11.

van Genuchten, M. Th. 1981. Non-equilibrium transport parameters from miscible displacement experiments, U.S. Salinity Lab. Res. Rep. 119, Riverside, CA.

van Genuchten, M. Th. 1988. Solute transport, p. 360–362, *in* S.P. Parker (ed.), *McGraw-Hill Yearbook Sci. Tech.*, McGraw-Hill Book Co., NY.

van Genuchten, M. Th. and W.J. Alves. 1982. Analytical solutions of the one-dimensional convective-dispersive solute transport equation, USDA Tech. Bull. 1661.

van Genuchten, M. Th. and R.W. Cleary. 1982. Movement of solutes in soils: Computer-simulated and laboratory results, p. 349–386, *in* G.H. Bolt (ed.), *Soil Chem.*, B. Physico-chemical models. Elsevier, Amsterdam, Netherlands.

van Genuchten, M. Th. and F.N. Dalton. 1986. Models for simulating salt movement in aggregated field soils, *Geoderma*, 38:165–183.

van Genuchten, M. Th. and W.G. Gray. 1978. Analysis of some dispersion corrected numerical schemes for solution of the transport equation, *Int. J. Num. Meth. Eng.*, 12:387–404.

van Genuchten, M. Th. and J.C. Parker. 1987. Parameter estimation for various contaminant transport models, p. 273–295, *in* C.A. Brebbia and G.A. Keramidas (eds.), Reliability and robustness of engineering software, Elsevier, NY.

van Genuchten, M. Th. and J.C. Parker. 1984. Boundary conditions for displacement experiments through short laboratory columns, *Soil Sci. Soc. Am. J.*, 48:703–708.

van Genuchten, M. Th. and E.A. Sudicky. 1999. Recent advances in vadose zone flow and transport modeling, p. 155–193, *in* M.B. Parlange and J.W. Hopmans (eds.), Vadose zone hydrology: Cutting across disciplines, Oxford University Press, NY.

van Genuchten, M. Th., D.H. Tang, and R. Guennelon. 1984. Some exact solutions for solute transport through soils containing large cylindrical macropores, *Water Resour. Res.*, 20:335–346.

van Genuchten, M. Th. and P.J. Wierenga. 1976. Mass transfer studies in sorbing porous media. I. Analytical solutions, *Soil Sci. Soc. Am. J.*, 40:473–480.

van Genuchten, M. Th. and P.J. Wierenga. 1986. Solute dispersion coefficients and retardation factors, *in* A. Klute (ed.), *Methods of Soil Analysis.* I. Physical and mineralogical methods, *Soil Sci. Soc. Am.,* Madison, WI.

van Rees, K.C.J., E.A. Sudicky, P.S.C. Rao, and K.R. Reddy. 1991. Evaluation of laboratory techniques for measuring diffusion coefficients in sediments, *Env. Sci. Tech.,* 25:1605–1611.

Varoglu, E. and W.D.L. Finn. 1982. Utilization of the method of characteristics to solve accurately two-dimensional transport problems by finite elements, *Int. J. Num. Fluids,* 2:173–184.

Wierenga, P.J. 1977. Solute distribution profiles computed with steady-state and transient water movement, *Soil Sci. Soc. Am. J.,* 41:1050–1055.

Yeh, G.T. 1990. A Lagrangian-Eulerian method with zoomable hidden fine-mesh approach to solving advection-dispersion equations, *Water Resour. Res.,* 26:1133–1144.

Yeh, G.T. and V.S. Tripathi. 1989. A critical evaluation of recent developments in hydrogeochemical transport models of reactive multichemical components, *Water Resour. Res.,* 25:93–108.

Yeh, G.T. and V.S. Tripathi. 1991. A model for simulating transport of reactive multispecies components: Model development and demonstration, *Water Resour. Res.,* 27:3075–3094.

Zwillinger, D. 1989. *Handbook of Differential Equations,* Academic Press, San Diego.

# 7 Soil Structure

## B. D. Kay and D. A. Angers

Soil structure has a major influence on the ability of soil to support plant growth, cycle C and nutrients, receive, store and transmit water, and to resist soil erosion and the dispersal of chemicals of anthropogenic origin. Particular attention must be paid to soil structure in managed ecosystems where human activities can cause both short- and long-term changes that may have positive or detrimental impacts on the functions that soil fulfills.

The importance of soil structure was recognized by researchers more than 150 years ago and the large volume of research on its nature has been summarized in a number of comprehensive reviews (Harris et al., 1966; Oades, 1984; Dexter, 1988; Kay, 1990; Horn et al., 1994). This paper builds on previous reviews and provides an overview to assist readers in interpreting data on soil structure in relation to land use. Emphasis will be placed on soils containing less than 40 to 60% clay from the temperate regions of the world where the relations between soil structure and land use have been studied most extensively. Although the nature of these relations changes with depth in the profile, attention will be primarily directed to the A horizon where the effects of land-use practices on soil structure are most pronounced. Because research on soil structure has, until recently, been carried out primarily in the context of agriculture, much of the following discussion is necessarily focused on this aspect.

## 7.1 CHARACTERISTICS, SIGNIFICANCE, AND MEASUREMENT OF SOIL STRUCTURE

Field experience and decades of research have given rise to a multitude of descriptions of soil structure. The term tilth is often used to describe the quality of soil structure for plant growth and is a popular term that predates modern agriculture. The term soil tilth embodies an integration of many of the characteristics of soil structure and reflects the practical experience of generations of people who have worked the soil and the understanding that they have of conditions that lead to greatest productivity and ease of management. A soil considered to possess good tilth is one that readily fractures whether the stress arises from tillage, emerging seedlings or growing roots, and provides an optimal environment for the growth of plants and microorganisms. Consequently, Gupta (1986) has attempted to define tilth using an index incorporating several variables. Tilth has also been defined using single characteristics such as those related to compaction (Scott Blair, 1937) and tensile failure (Utomo and Dexter, 1981a). Characteristics of soil structure that influence different functions of soil (including plant growth) will be considered in this chapter and, although tilth will not be considered in detail, several of these characteristics relate to the concept of tilth.

### 7.1.1 FORM

The term structural form applies to a group of characteristics that describe the heterogeneous arrangement of void and solid space that exists in soil at any given time. An assessment of soil structure at a site normally involves a visual assessment of structural form that is complemented

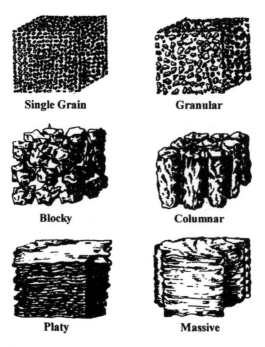

**FIGURE 7.1** Diagrammatical representation of types of structure.

by quantitative analyses of samples collected from the site. Visual analyses include a description of the morphology of soil at the surface and the variation in morphology with depth. Most national soil classification systems include procedures and terminology for describing and classifying soil structure based on morphology (Soil Survey Staff, 1975; Hodgson, 1978; Canada Expert Committee on Soil Survey, 1987). For example, the Canadian system classifies soil structure in terms of grade or distinctness (structureless, weak, moderate, strong), class or size (very fine, fine, medium, coarse, very coarse) and type (single grain, amorphous or massive, angular blocky, subangular blocky, granular, platy, prismatic, columnar). Grades and types of soil structure are defined on the basis of visual evidence of the extent of development of peds and their shape (peds are secondary structural units made up of primary particles that are distinguished from adjacent structures on the basis of failure zones and are formed by natural processes, in contrast with clods which are formed artificially). Grades of soil structure are distinguished as follows by the Canada Expert Committee on Soil Survey (1987): (1) *weak:* peds are barely observable in place; (2) *moderate:* peds are moderately evident; on disturbance, soil will break down into a mixture of many distinct entire peds, some broken peds, and little disaggregated material; and (3) *strong:* peds are quite evident in undisturbed soil; on disturbance, peds retain their identity with some broken peds, and little disaggregated material. The types of structure are diagrammatically illustrated in Figure 7.1 and details on class or size of the different types are given in Table 7.1.

Although morphological descriptions are largely qualitative, an experienced pedologist can make some quantitative predictions from these analyses (McDonald and Julian, 1966; McKeague et al.,1982; Wang et al., 1985). Visual analyses during a site examination may also include an examination of the distribution of roots, soil color, or the infiltration of precipitation and, together with observations of morphology, can provide evidence of the influence of structural form on functions that the soil is fulfilling. Visual analyses are normally complemented by quantitative measurements of soil physical properties. A number of structural characteristics have been used to characterize the soil matrix.

**TABLE 7.1**
**Details on types of structure and their size [Canada Expert Committee on Soil Survey, 1987].**

| Type | Class | | | | |
|---|---|---|---|---|---|
| | v. fine | fine | medium | coarse | v. coarse |
| | | | mm | | |
| Single grain: loose incoherent mass of individual particles as in sands | — | — | — | — | — |
| Amorphous (massive): coherent mass showing no evidence of any distinct arrangement of soil particles | — | — | — | — | — |
| Angular blocky: faces of peds rectangular and flattened, vertices sharply angular | <5 | <10 | 10–20 | 20–50 | >50 |
| Subangular blocky: faces of peds subangular, vertices mostly oblique or subrounded | <5 | <10 | 10–20 | 20–50 | >50 |
| Granular: spheroidal peds characterized by rounded vertices | <1 | <2 | 2–5 | 5–10 | — |
| Platy: horizontal planes more or less developed | <10 | <2 | 2–5 | >5 | — |
| Prismatic: vertical faces of peds well defined and edges sharp | <10 | <20 | 20–50 | 50–100 | >100 |
| Columnar: vertical edges near top of columns are not sharp (columns may be flat or round topped, or irregular) | | <20 | 20–50 | 50–100 | >100 |

#### 7.1.1.1 Pores

*Total Porosity and Bulk Density*

Total porosity is seldom measured directly but is normally calculated from the bulk density and the particle density. Total porosity is not very sensitive to the variation in particle density that is normally encountered in the field, and consequently, discussions on total porosity are often presented using bulk density as the measured variable. Details on the measurement of these characteristics are given in Chapter 1.

Bulk density is strongly influenced by texture and organic C content and, for a given soil, reflects the impact of stresses arising from activities such as traffic and tillage. Interpreting data on bulk density from different soils with respect to the extent of compaction that has occurred is, therefore, complicated by the influence of variation in texture and organic C. Interpreting data on bulk density from different soils with respect to processes related to the growth of plants or water flow, has also proven to be difficult. This difficulty is due, in part, to the importance of pore size distribution on these processes and the fact that differences in management practices, as well as texture and organic C contents, can result in different pore size distributions as well as total porosity.

*Pore Size Distribution and Continuity*

Pores of different size are often arbitrarily grouped into different classes (Table 7.2). Pores > 30 μm include biopores, shrinkage cracks, and other interaggregate pores, and the largest pores within aggregates and peds. This porosity is included in the class referred to as structural porosity (Stengel, 1979; Derdour et al., 1993) or aeration capacity (Thomasson, 1978) and can represent as much as a third of the total porosity of soils. These pores are influenced by texture and organic C content and are very sensitive to management. Macropores and, to a lesser extent mesopores, are the least stable of all pore size classes, and collapse when they experience stresses originating from a range of sources.

The volume fraction of pores > 30 μm and the continuity or connectivity of these pores have a major influence on water and solute flow (White, 1985; Blackwell et al., 1990b; Ahuja et al.,

**TABLE 7.2**
**Pore size classification [After Soil Science Society of America, 1997].**

| Class | Class Limits Equivalent Diam. (μm) |
|---|---|
| Macropores | >75 |
| Mesopores | 30–75 |
| Micropores | 5–30 |
| Ultramicropores | 0.1–5 |
| Cryptopores | <0.1 |

1993), aeration (Thomasson, 1978), a range of soil mechanical characteristics (Carter, 1990b), and on root development (Jakobsen and Dexter, 1988). At water potentials close to zero (most of these pores are water filled), water and solute flow primarily occurs in mesopores and macropores and the rest of the pore system is bypassed. These pores drain quickly and, as air is redistributed in the soil, aeration is enhanced. Macroporosity and mesoporosity represent failure zones of low strength and therefore a reduction in the volume fraction of these pore classes results in an increase in tensile strength and resistance to penetration.

Pores with equivalent diameters of 0.1 to 30 μm are often referred to as storage pores and include the volume fraction of pores defining the water that is potentially available to plants (Veihmeyer and Hendrickson, 1927). These pores also provide a habitat for microorganisms and smaller soil fauna. Micropores and ultramicropores are strongly influenced by texture and organic C content (Ratliff et al.,1983; da Silva and Kay, 1997a), but are not strongly influenced by increases in bulk density that can arise from traffic or other stresses (da Silva and Kay, 1997a).

Pores < 0.1 μm are least influenced by management of any of the pore classes. Although these pores remain water filled for a much larger proportion of time than the other pore classes, little of the water in these pores is available to plants and the rate of water flow through them is very slow. These pores remain largely inactive biologically because they are not penetrated by roots or by most microorganisms and are, therefore, only accessible to molecular sized byproducts of biological activity. The relative inaccessibility of these pores to microorganisms may also give rise to one of their most important functions, namely the physical protection of organic C.

Pore size distribution can be estimated from the total porosity and measurements during the progressive removal of water or the intrusion of Hg (Danielson and Sutherland, 1982; Carter and Ball, 1993). Details of the shape, tortuosity and connectivity of macro- and mesopores can be obtained from an examination of thin sections that often utilizes image analyses techniques (Moran et al. 1989; Crawford et al.,1995) or from measurements of gas movement (Ball and Smith, 1991). The nature of macropores has also been studied using computed tomography (Warner et al., 1989).

### 7.1.1.2  Failure Zones

The application of mechanical stress (such as that associated with tillage) to undisturbed bulk soil results in failure of the soil matrix into peds, broken peds and disaggregated material. Failure occurs through zones made up of elemental volumes in which the bonding between the elementary particles is small and, therefore, failure zones include pores as well as elemental volumes containing less cementing materials than the surrounding matrix. Failure zones, therefore, reflect the characteristics of both the void and the solid space.

Failure zones can be characterized in terms of their average strength (or strength at a given percentile of failure in a frequency distribution) and the distribution of the magnitude of these strengths. Zones with a given strength will fracture when the force that is applied exceeds this value. The spatial distribution of failure zones of different strengths leads to the creation of aggregates with characteristic size distributions. The application of increasing energy causes fragmentation to occur along zones of increasing strength and closer proximity, thereby resulting in smaller aggregates. The increase in the tensile strength of aggregates with decreasing size of aggregates has been used to quantify the friability of soils (Utomo and Dexter, 1981b).

Characterizing the strength of failure zones and their distribution involves the application of stress and measurement of the nature of failure that has occurred. The strength of failure zones has been determined from measurements of the maximum tensile stress required to cause failure of the matrix (Braunack et al., 1979), and measurements of aggregate size distributions after the sample has fallen through a given distance (a drop-shatter test) (Marshall and Quirk, 1950; Ingles, 1965).

### 7.1.1.3  Peds, Aggregates, and Clods

Morphological descriptions of soil structure include reference to the extent of development and shape of peds (Table 7.1). Peds are considered to be naturally formed (in contrast to formation by anthropogenic means) and the extent of ped development relates to the strength of failure zones; peds that are very visible (structure is strongly developed) are separated by failure zones of very low strength. The term aggregate may be used interchangeably with ped where the type of soil structure is granular. A granular structure is often found in topsoils that have been continuously under grass for many years (McKeague et al., 1987). The term aggregate is also used for structural units that result from fragmentation of the soil matrix through the application of mechanical energy. The energy may be applied in the field through tillage or seedbed preparation and in the laboratory through crushing or sieving. Aggregates > 10 cm are commonly referred to as clods. The distinction between aggregates and peds becomes greater as more energy is applied to the bulk soil. Most studies on soil structural units in the laboratory have involved the application of energy and, therefore, the term aggregate will be used in the remainder of this chapter in relation to these structural units. It is important to note, however, that the characteristics of aggregates reflect, in part, the amount of energy that has been applied during the fragmentation process.

Aggregates may be characterized by their size, shape and surface roughness, although aggregate size has the greatest practical importance. Aggregate size is estimated by distributing the sample across sieves with openings of different size and weighing the amount of soil on each sieve (Kemper and Rosenau, 1986; White, 1993). Any soil sample will contain aggregates of different size and the soil can be characterized in terms of the distribution of sizes of aggregates and the properties of aggregates in the different size fractions. The distribution of weights of aggregates among the different size fractions (dry aggregate size distribution), has been described using parameters related to the center of the distribution such as mean weight diameter and to the shape of the distribution. An assessment of different parameters used for this purpose is given by Perfect et al. (1993). Aggregates < 250 μm are often referred to as microaggregates, in contrast to larger aggregates which are referred to as macroaggregates. This distinction is based on the hypotheses that large aggregates are made up of small aggregates (a hierarchical structure), that the building blocks of the large aggregates of most soils are aggregates that are < 250 μm, and that the cementing materials within the microaggregates are different from those that bind the microaggregates into macroaggregates (Tisdall and Oades, 1982). Individual primary soil particles that are released from the matrix when stress is applied represent disaggregated material (Table 7.1) and may include gravel, sand, silt and clay-sized material. Part of the clay, and to a lesser extent the silt-sized material, may be readily suspended in water and are often measured as a dispersible fraction. Material that becomes dispersed simply as a consequence of wetting aggregates to saturation is referred to as spontaneously dispersible material, whereas material that only becomes dispersed after energy is applied to the aggregates is referred to as mechanically dispersible material.

The distribution of aggregate sizes in the seedbed is related to type of tillage, number of passes and soil water content. When energy input is kept constant, dry aggregate size distribution is influenced by soil properties and water content at the time of tillage, and reflects the influence of these factors on the characteristics of the failure zones in the soils. While the effects of tillage or cropping practices on aggregate size distribution can disappear quickly after tillage due to crop growth and climatic factors, these effects can have a significant impact on the germination, emergence and early growth of seedlings. Measurements of aggregate size distributions are most relevant to the germination and early growth of plants on soils that are tilled, structurally stable, and not compacted by traffic. The measurements will have less relevance to later growth, or to early growth on untilled soils, or tilled soils that are unstable or compacted by traffic.

The size distribution of aggregates in the seedbed can also influence the flux of water and solutes. Distributions dominated by large aggregates will have a higher proportion of macropores (Dexter et al., 1982), and consequently under saturated or near saturated conditions, will have higher hydraulic conductivities with larger fractions of water and solutes bypassing the rest of the pore network. These pores will exist between aggregates. Little information on the distribution of pore sizes within aggregates is provided directly by aggregate size distribution (Wu and Vomocil, 1992), although the development of fractal theory to describe the organization of solid and void space in a manner that is scale independent offers the potential for predicting transport characteristics from information on solid space (Crawford, 1994).

## 7.1.2 STABILITY

The term structural stability is used to describe the ability of soil to retain its arrangement of solid and void space when exposed to different stresses. Stability characteristics are specific for a characteristic of structural form and the type of stress applied. Stresses may arise from processes as diverse as tillage, traffic and wetting or drying. The response of both the void and the solid space to the application of these stresses reflects, in part, the strengths of the failure zones and their spatial distribution. Extensive research has been directed to understanding the stability of aggregates when they are exposed to stresses arising from wetting under different conditions and then exposed to mechanical stresses arising from sieving, shaking in water, or exposed to ultrasonic energy. Consequently, the term, "structural stability," is often considered to be synonymous with aggregate stability. There are, however, important characteristics that reflect the ability of a soil to retain its arrangement of solid and void space when other stresses are applied, and conventional measurements of aggregate stability may not provide meaningful measures of these properties. For instance, the compressibility index is a measure of the ability of the total porosity to withstand compressive stresses. This parameter is particularly important where changes in agricultural practices are leading to greater traffic.

### 7.1.2.1 Pores

Pores experience stress arising from tillage, traffic, wetting or drying, and root growth. Compaction of soil by vehicular traffic or other means results in a decrease in total pore space. The resistance of pores to compression can be measured using compression cells of various types (Bradford and Gupta, 1986). Compression indices can be calculated and related to soil properties and management (Larson et al., 1980; Angers et al., 1987). Although macropores and mesopores tend to be lost first during compaction, as discussed earlier, the stability of macropores may also relate to the orientation of the pores in relation to the applied stress. The stability of biopores is greatest if the stress is applied parallel to the axis of the pores (Blackwell et al., 1990a).

Pore space may experience stress during wetting if irrigation or precipitation rates result in rapid wetting (Mitchell et al., 1995). The stresses arise from differential swelling of clay minerals and from air entrapment subsequently causing increased air pressures. The stability of the pore system under these circumstances can be characterized by comparing the permeability of the soil to water to the permeability with a nonreactive fluid such as air (Reeve, 1953) or by determining changes in pore size distribution arising from fast and slow wetting rates (Collis-George and Figuero, 1984).

In many soils the pore space also varies with water content. The loss of porosity upon drying can be characterized by the shrinkage curve. Smaller slopes of the shrinkage curve under no-till versus tilled soils (McGarry and Smith, 1988) or under cropped versus bare soils (Mitchell and van Genuchten, 1992) were attributed to greater pore stability in the case of no-till and cropped soils, respectively.

### 7.1.2.2 Aggregates

Aggregates may experience stresses related to tillage, traffic, wetting, and mechanical abrasion by flowing water. The ability of aggregates to resist stresses from wetting followed by mechanically shaking in water is referred to as wet aggregate stability. Although this parameter was originally used to characterize the erodibility of soil (Yoder, 1936), it has been increasingly used to study the cohesion of aggregates and the dynamics of changes in the nature of bonding between particles. The stability of aggregates is strongly dependent on the rate of wetting. Aggregate stability declines as the rate of wetting increases and, as for pores, the decline has been attributed to increased stresses related to entrapment of air within the aggregates and differential swelling of clays (Panabokke and Quirk, 1957; Concaret, 1967; Emerson, 1977).

Measurements of the loss of stability due to rapid rates of wetting are most relevant to aggregates at the soil surface in environments where the soil surface is very dry and rapid wetting occurs from intense rains or irrigation. In environments where wetting occurs more slowly (surface aggregates that have high water contents or experience less intense rainfall events, or aggregates beneath the surface), measurements of stability in which air entrapment and differential swelling are minimized are more relevant. Soil samples are often not dried if measurements involve slow wetting and, in these cases, stability is very strongly influenced by the water content of the sample prior to saturation (Section 7.2.4.2).

Measurements of stability in water can be restricted to macroaggregates, microaggregates, dispersible material or can include a wide range of size classes. The measurements will be strongly influenced by size class of the aggregates selected for study, their initial water content, conditions under which wetting occurs, and size classes that are measured after application of the stress. Early methods include those of Yoder (1936) and Hénin et al. (1958). Details on current methodologies for measuring wet aggregate stability are summarized in Kemper and Rosenau (1986) and in Angers and Mehuys (1993).

The stability of aggregates to stresses representative of those encountered during tillage have also been measured (Watts et al., 1996a,b), although this characteristic has been explored in much less detail than wet aggregate stability. Watts and coworkers used a falling weight to impose different specific amounts of mechanical energy on aggregates and measured mechanically dispersible clay as the stability parameter.

### 7.1.3 RESILIENCY

The term structural resiliency describes the ability of soil to recover its structural form through natural processes when the stresses are reduced or removed. Recovery relates, in the first instance, to changes in pore characteristics and the distribution of strengths of failure zones. Resiliency arises from a number of processes such as freezing/thawing, wetting/drying and biological activity (e.g., root development and activity of soil fauna). Soils differ in their ability to recover structural form when stresses are reduced or removed. Soils that are most resilient are known as self-mulching soils (Wenke and Grant, 1994). The structural form of such soils are not particularly sensitive to mechanical processes. For instance, if much of the macroporosity of the upper few cm are destroyed through tillage when the soil is wet, a desirable structure can be recreated by wetting and drying.

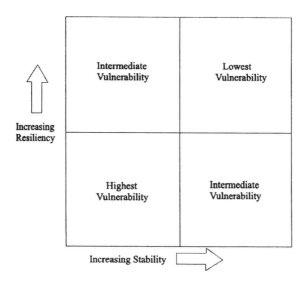

**FIGURE 7.2** Diagrammatic representation of the relation between stability, resiliency and vulnerability [Kay, 1998. Reprinted with permission from Lal et al. 1998. Soil processes and the carbon cycle. CRC Press, Boca Raton, FL].

The creation of failure zones by wetting events has been referred to as tilth mellowing by Utomo and Dexter (1981a). Resiliency is also exhibited by the partial recovery (rebound or relaxation) of porosity after removal of a compressive stress (Guérif, 1979; Stone and Larson, 1980; McBride and Watson, 1990). The maximum change or recovery that is possible by the different mechanisms, and the rate at which this change can occur, are particularly important characteristics of soils where management practices are leading to the application of increasing stress (Kay et al., 1994a).

### 7.1.4 VULNERABILITY

The term structural vulnerability reflects the combined characteristics of stability and resiliency (Figure 7.2). Soils which are least stable under a given stress and which do not recover when the stress is reduced or removed (or which recover very slowly) are most vulnerable to stress, whether the origin of the stress is natural or anthropogenic. Conversely, soils which have a great resistance to stress and are very resilient are the least vulnerable.

### 7.1.5 SPATIAL AND TEMPORAL VARIABILITY AND SIGNIFICANCE OF SCALE

The size of pores, the distance between failure zones of similar strength, and the size of peds or aggregates can range across several orders of magnitude. Consequently, stability and resiliency can also be considered at different scales. At any given scale, these characteristics are spatially variable and Dexter (1988) has argued that this spatial variability or heterogeneity is, itself, a fundamental characteristic of soil structure. Sampling strategies should therefore be designed to provide data on the spatial variability in structural characteristics.

    Structural characteristics also vary temporally and changes can occur at scales ranging from hours to decades. The extent of temporal variation is also dependent on the spatial scale of interest. For instance, macroporosity is very dynamic and can change over short time periods (e.g., as a consequence of tillage) whereas pores < 0.1 µm are much less dynamic but may change over several years or decades (as a consequence of changes in organic C contents arising from management practices). Sampling strategies should therefore also take into account the scale of temporal variation in the structural characteristics being measured. Data presented in Figure 7.3 illustrate the temporal variation in the proportion of water-stable aggregates during a four-year period in a silty clay soil

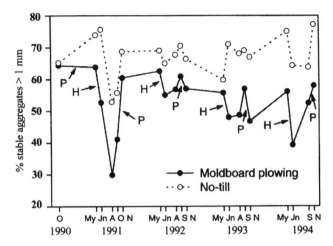

**FIGURE 7.3** Temporal variation in the proportion of water stable aggregates > 1 mm under two contrasting tillage systems in silty clay soil (Humic Gleysol, Humic Haplaquept) under barley production in eastern Canada. P = Plowing event, H = Disk harrowing event [J. Laford, D.A. Angers and D. Pageau, unpublished].

submitted to two contrasting tillage systems. Short-term variations (within a growing season) represent the immediate effects of climatic events and tillage practices whereas long-term trends are a result of cumulative effects of both these factors. For instance, the drop in aggregation observed in months 9 to 13 coincides with conditions of dry soils during which aggregates were subject to greater slaking. Despite these short-range variations, the difference in the level of stable aggregation between the two tillage treatments kept increasing with time reflecting the cumulative effects of less disturbance and greater accumulation of C at the surface of the no-till soil.

## 7.2 SOIL FACTORS INFLUENCING STRUCTURE

The dominant soil characteristics influencing structure are texture and clay mineralogy, organic matter, inorganic noncrystalline materials, composition of the pore fluid and adsorbed or exchangeable solutes, plants and soil organisms, and depth in the profile. Few, if any, of these factors function in isolation. In addition, the magnitude of their influence on soil structure may vary with factors unrelated to soils such as climate, land use or management practices.

### 7.2.1 TEXTURE AND MINERALOGY

Soil texture has a major influence on the form, stability and resiliency of soil structure as well as the response of soil structure to weather, biological factors and management. The nature of the influence of texture on structural characteristics is related to the soil matrix. For instance, sands may be viewed, in the simplest case, as being made up of single grains and all characteristics of structural form are determined by the distribution of grain sizes, and modification introduced by tillage or traffic. Clay or silt sized materials that may be present exist as coatings on the sand grains or fill the interstices between the sand grains. The structure does not shrink or swell and is not very responsive to freezing. Organic materials, together with fine clays and other amorphous and crystalline inorganic materials, provide what little cementation exists between the grains. The resulting aggregates have low stability. The single grain structure is, however, relatively stable under compressive stresses. The structural form possesses limited resiliency and has, therefore, low to intermediate vulnerability (Figure 7.2). As the clay content increases, the characteristics of the soil matrix (including both structural form and stability) are increasingly dominated by the characteristics of the clay (including mineralogy and exchangeable ions), the nature and quantities of cementing materials and the composition of the pore fluid. Rengasamy et al. (1984) has suggested that the physical properties of soils change with increasing clay content until the clay content reaches about 30%; at greater clay contents, the behavior of the soil depends on the type of clay, and cannot be predicted from clay content alone.

Structural resiliency, related to weather, biological processes and management play an important role, and the magnitude of the impacts of these processes are determined by the characteristics of the matrix. For instance, the formation of pores and other failure zones by biological processes may be more important in medium-textured soils or low-activity clay soils that undergo limited shrinking and swelling than is the case for finer-textured soils that are particularly responsive to wetting and drying (Oades, 1993). In the latter soils, biological processes may complement abiotic processes which exert the dominant control over soil structure.

### 7.2.1.1   Form

The effect of texture on pores, failure zones and aggregation can be influenced by mineralogy and depth in the profile and, therefore, it is difficult to identify broad generalizations. For instance, Manrique and Jones (1991) found, using data largely from the United States, that bulk density was positively correlated with clay content in the B and C horizons but was not correlated with clay in the A horizon. Difficulties in relating bulk density to texture in the A horizon are created by the strong influence of tillage, traffic and organic C content on bulk density and variation in their influence with texture. When the data used by Manrique and Jones (1991) were grouped by soil order, the bulk density was positively related to clay content in Aridisols, Entisols, Inceptisols and Mollisols, negatively related in Ultisols and Vertisols, and not related in Oxisols and Spodosols. Differences may relate to mineralogy with clay content having a negative relation with bulk density as the amount of swelling clay minerals increase.

Few studies have related pore size distribution to texture or mineralogy. The contribution of shrinkage cracks to macroporosity would be expected to be positively related to the amounts of swelling clay minerals present, with the magnitude of the contribution relating to exchangeable ions and the composition of the pore fluid. The influence of texture and mineralogy on pores 0.2–30 µm equivalent diameter in nonswelling soils can be estimated from the water release curve. The influence of texture on these curves has been described by regression equations (pedotransfer functions) although the descriptions have not always been consistent (Kay et al., 1997). Inconsistent relations between texture and the water release curve may be due, in part, to variations in particle size distribution not reflected in broad textural characteristics (clay, silt or sand content) which can alter the water release curve (Schjønning, 1992). Inconsistent relations between texture and the water release curve may also arise from variation in mineralogy. Few studies have systematically examined the role of mineralogy on pore size distribution. Williams et al. (1983) found that montmorillonite, Fe oxide, vermiculite and quartz introduced greater variation in the water release curve of Australian soils than illite, kaolinite, halloysite or randomly interstratified materials. The influence of texture and mineralogy on aggregates reflects the influence of these variables on failure zones and their response to stress (stability). Aggregate size distributions that are dominated by large aggregates and clods are more common in finer textured soils.

### 7.2.1.2   Stability

The influence of the clay fraction on stability varies with the nature of the stress applied. Larson et al. (1980) found that the compression index, a measure of the susceptibility to compressive stress, increased approximately linearly with clay content up to about 33% and thereafter remained approximately constant. Soils dominated by kaolinite or Fe oxides were less compactible (lower compression index) than soils with predominantly 2:1 type clays. On the other hand, the stability of aggregates to wetting and mechanical abrasion during sieving usually increases with increasing clay content, but the relation varies with depth in the profile. For instance, measurements of the stability of vapor wetted aggregates from soils across the western United States and Canada indicated that aggregate stability increased curvilinearly as the clay content increased to 90% (Kemper and Koch, 1966). Similar results have been obtained from other regions of the world and over narrower ranges of texture (le Bissonnais, 1988; Elustondo et al., 1990; Rasiah et al., 1992). The influence of mineralogy is most apparent at the scale of dispersible mineral material where the strength of bonds between platelets, quasicrystals and domains are controlled by mineralogy, exchangeable ions and composition of the pore fluid (Rengasamy et al., 1993; Rengasamy and Sumner, 1998) (Section 7.2.4).

### 7.2.1.3   Resiliency and Vulnerability

The resiliency of soils related to weather is more strongly influenced by the clay fraction than that due to biological factors. The degree of reversibility of soil compaction by development of micro-cracks caused by wetting and drying increases with shrink/swell potential (Barzegar et al., 1995). Wenke and Grant (1994) have noted that soils that are self-mulching have clay contents > 35%. Large changes in the strength of failure zones can also be caused by freezing and thawing events. The morphology of the ice lenses are strongly influenced by texture as well as mineralogy (Czeratzki and Frese, 1958). Although many of the pores associated with these lenses are unstable and collapse on melting (Kay et al., 1985), their spatial distribution corresponds with the distribution of failure zones and associated aggregates (Pawluk, 1988). Resiliency arising from rebound or relaxation following compaction also varies with mineralogy, being larger in soils containing expanding than nonexpanding clays (Stone and Larson, 1980).

### 7.2.2   ORGANIC MATTER

The most general characteristic of soil organic matter (SOM) is the total organic C content, and the impact of SOM on soil structure is often described using this parameter. Structural form and stability generally improve with increasing SOM contents; however, exceptions to this generalization exist because SOM is not homogeneous in composition or location. SOM represents materials of plant, animal and microbial origin that are in various stages of decomposition and are associated with the mineral fraction with different degrees of intimacy. In most recent conceptual and predictive models, SOM is defined as being composed of three pools which vary in chemical composition, physical location and decomposition kinetics representing a continuum along the decomposition process. The forms and location of organic C associated with these various pools are first briefly reviewed as an introduction to this section.

### 7.2.2.1   Forms and Location of Organic Matter and Association with Minerals

*Particulate Organic Matter*
Plant and animal materials enter the soil through a range of different processes and the nature of the space that is occupied by these materials and their degree of contact with the soil vary with the material and the process. For instance, plant shoots that have been incorporated into the soil by tillage will be located in macropores and the contact area between the crop residue and the soil will be less than that of plant roots and root hairs which will be located in smaller pores and be in more extensive contact with the soil. Plant and animal materials are attacked by soil flora and fauna at the outset of the decomposition process. Organic C that is relatively free, or has least association with the mineral fraction, is the lightest and coarsest fraction (the density of this fraction may be < 1.6 g cm$^{-3}$ and its size often defined as > 50 μm) and is often referred to as particulate organic matter (POM). Golchin et al. (1994) distinguished between POM that was free and that which was occluded (requiring ultrasonic dispersion for separation). They found the two fractions to be present in roughly similar proportions in the light fraction of five cultivated soils and to make

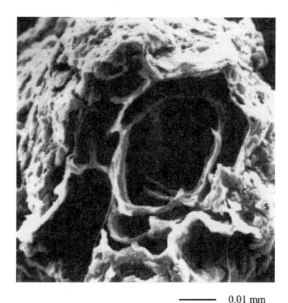

———— 0.01 mm

**FIGURE 7.4** Particulate organic matter (POM) encrusted with mineral soil particles. Remnants of conducting vessels are coated by clay and silt particles. Conventional SEM. White bar is 10 μm, c = clay particles, s = silt particles [P. Puget and C. Chenu, unpublished].

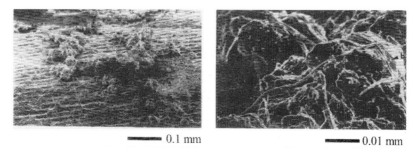

———— 0.1 mm                                              ———— 0.01 mm

**FIGURE 7.5** Aggregate development on decomposing wheat straw residues. Straw was incubated for 3 months in a silty soil. Fungi enmesh the soil aggregates within mycelial strands and the newly stabilized aggregates are also colonized by bacteria. Low temperature SEM. Bar is 100 μm (a) and 10 μm (b), b = bacteria, f = fungi [D.A. Angers, C. Chenu, and S. Recous, unpublished].

up as much as 38% of the total organic C. Although these proportions vary with soil type and cropping history (Cambardella and Elliott, 1993; Besnard et al., 1996; Gregorich et al., 1997), it is clear that a large proportion of the POM is in close contact with the mineral particles and is often occluded within the soil matrix (Figures 7.4 and 7.5).

### Microbial Biomass and Metabolites

The presence of readily mineralizable C will result in growth in the microbial biomass and their predators and the production of extracellular materials including polysaccharides and other compounds. Electron micrographs show that the microorganisms (Figure 7.5) and extracellular material (Foster, 1988) can become intimately associated with the mineral material at this stage. Part of the microbial biomass may grow into pores adjacent to the organic residue or may be squeezed into this space along with the extracellular material. Dense mucilage may impregnate the soil fabric as much as 50 μm from the plant residue (Foster, 1988). The polysaccharides may exist as fibrous or granular materials and become more closely associated with the mineral surface as drying occurs (Foster, 1988). The matrix of decomposing plant residues, microorganisms, extracellular materials and mineral material has been proposed by Golchin et al. (1994) to coincide with material in the 1.8 to 2.0 g cm$^{-3}$ density fraction. Its chemical composition changes during decomposition. As decomposition of plant residues proceeds, less C is quantitatively and qualitatively available to the decomposers and microbial activity decreases. Chemical analyses suggest that after consumption of the more labile portions of the organic debris (proteins and carbohydrates), the more resistant plant structural materials (which have low O-alkyl, high alkyl and aromatic content) are concentrated as occluded particles within soil aggregates. Electron micrographs obtained by Waters and Oades (1991) show an abundance of void spaces within the aggregates that would fall in the range between 0.1 and 30 μm diameter. Some of these pores are associated with cellular debris while others represent voids within aggregates. These pores may be completely closed, have one entrance (bottle-shaped pores) or may be tubes with two or more entrances. As mineralization approaches advanced stages, the site of the decomposing plant material may become a failure zone of low strength.

### Recalcitrant Materials

Chemical and biological transformations of the organic materials ultimately lead to their stabilization in the soil (Chapters 2 and 4). At this stage, the organic material is intimately associated with the mineral phase and would appear in the dense fraction of the soil (> 2.0 g cm$^{-3}$). This fraction accounted for 25 to 50% of the organic C in the soils considered by Golchin et al. (1994). Analyses of this fraction using $^{13}$C CP/MAS NMR spectroscopy indicated that about half of its C was O-alkyl or carbohydrate C. There was no microscopic evidence of particulate material with a cellular structure in that fraction, and spectroscopic analyses indicated that the composition was relatively uniform irrespective of the composition of the original C source (Golchin et al.,1994) suggesting that the material is largely of microbial origin. If so, its recalcitrance must be partially due to physical protection from attack by microorganisms as a consequence of being adsorbed or located in pores that are too small to be accessible. A more complete discussion on the characteristics of the stabilized SOM is presented in Chapter 2.

## 7.2.2.2   Effects of Organic Matter on Pores and their Stability

### Pores

Organic matter influences total porosity and pore size distribution. The total porosity that is measured at any given time in nonswelling soils has generally been found to increase (and bulk density to decrease) with increasing organic C content (Anderson et al., 1990; Lal et al., 1994; Schjønning et al., 1994). The magnitude of the increase varies with the soil order (Manrique and Jones, 1991). The dynamic nature of pores with an equivalent diameter > 30 μm may account for the poor correlations between SOM content and macro- or mesoporosity (Thomasson and Carter, 1992) or inconsistent correlations (Douglas et al., 1986; Kay et al., 1997). The influence of organic C on these pore size classes would be most obvious where there is least variation in texture, mineralogy, pore fluid composition, climate and management conditions, and variability in these factors may also contribute to the difficulty in generalizing across different studies.

The role of organic C on 0.1–30 μm pores has frequently been overlooked or underestimated. Data from site-specific studies from various parts of the world (Karlen et al., 1994; Schjønning et al., 1994; Emerson, 1995) have shown an increase in the volume of 0.2–30 μm pores ranging from 1–10 mL g$^{-1}$ organic C. However, Anderson et al. (1990) found that, although increases in organic C due to the annual additions of manure to a montmorillonitic soil for 100 years increased the volume fraction of pores > 25 μm, the volume fraction of 5–25 μm pores decreased, suggesting that the latter pore class may be less responsive to increases in organic C in soils that are dominated by swelling clay minerals. The increase in the volume fraction of 0.1–30 μm pores with organic

C may vary with texture, although there are inconsistencies in the conclusions that have been drawn by several researchers (Emerson, 1995; Hudson 1994; Kay et al., 1997) and further research is required. The increase in the volume fraction of 0.1–30 μm pores with organic C may be attributed to several mechanisms. Although none have been rigorously assessed, possible mechanisms include: pores in POM, sites of plant or animal residue from which most of the C has been mineralized but which are encased in mineral materials enriched in microbial decomposition products, and pores within a matrix of extracellular polysaccharides (Kay, 1998). The volume fraction of small pores also tends to increase with increasing organic C. Calculations based on data from Emerson (1995), Schjønning et al. (1994) and Karlen et al. (1994) show that the increase in the volume of pores < 0.2 μm ranged from 0–3 mL g$^{-1}$ organic C.

*Stability of Pores*

Where the process of creating the pore does not substantially increase the organic C content of the pore wall, the stability of the pore wall would be expected to be equivalent to the stability of the surrounding matrix which is often inadequate to allow the pore to withstand the effective stress created by drying, rewetting or swelling pressures, or stresses from overburden or traffic. For instance, the organic C content of the pore walls is not increased during tillage and many of the pores created by this process are unstable (Derdour et al., 1993). Gusli et al. (1994) showed that the structural collapse of beds of aggregates on wetting and draining was caused by the development of failure zones on wetting followed by consolidation on draining and that the extent of collapse was greater for soils having lower organic C contents. Macropores created by shrinkage or by earthworms may also be unstable in some situations. For instance, Mitchell et al. (1995) found that these pores failed to provide preferential flow paths within 10 minutes of the onset of flood irrigation of a swelling soil with low organic C content (0.9% C). Pores created by the tap roots of alfalfa (*Medicago sativa* L.) were, however, much more stable, presumably due to increased C content of the pore wall and consequent enhanced stability.

Traffic can cause a loss of porosity and a decrease in pore continuity, the extent of which is determined by the magnitude of the stresses imposed (contact pressure, the number of passes) and the compactibility of soils. Increasing organic C content diminishes the impact of stress on total porosity (Soane, 1975; Angers and Simard, 1986) and on pore continuity (Ball and Robertson, 1994). The influence of organic C on the stability of pores under compaction may also vary with texture and form of the organic C (Soane, 1990). Angers (1990) found increasing organic C only resulted in reduced compactibility of soils at clay content > 35%. Ball et al. (1996) reported a strong negative correlation between maximum bulk density after compaction and NaOH extractable carbohydrates. Fresh and decomposing plant residues may have variable effects on the stability of pores under compaction (Guérif, 1979; Gupta et al., 1987; Rawitz et al., 1994).

### 7.2.2.3 Effects of Organic Matter on Failure Zones and Strength

The influence of organic matter on failure zones is dependent on the nature of the failure zones, the forms of the organic C, and its spatial distribution and care must be taken into account in interpreting these relations. An increase in porosity, particularly macro- and mesoporosity, would be expected to result in a decrease in strength. An increase in organic C in soil with the same water content can, however, also cause a decrease in water potential (because of the effect on the pore size distribution) with a concomitant increase in effective stress and strength. Kay et al. (1997) used pedotransfer functions to assess the influence of increasing organic C contents on the resistance to penetration of soils with a range of clay contents that were used for corn production under two different tillage treatments and found that: (1) an increase in the organic C content caused a decrease in soil resistance when the water potential was constant; and (2) the magnitude of the decrease increased with clay content and with declining potential.

Increases in organic C content that result in increasing levels of polysaccharides would also be expected to increase the cementation between mineral particles (Chenu and Guérif, 1991). However, the impact of increasing polysaccharide content on failure zones may vary with the nature of the soil and the spatial distribution of the organic C. The strength of some soils is strongly influenced by dispersible clay and silt material which functions as cementing material as soils dry. Under these conditions, increasing organic C content (and in particular, the fraction closely associated with the mineral phase) reduces the amounts of spontaneously dispersible mineral material resulting in reduced strength on drying. This mechanism appears to be particularly important in hardsetting soils (Young and Mullins, 1991; Chan, 1995). However, an increase in the strength of bonds between mineral material may also result in an overall increase in strength, if the cementation occurs uniformly throughout failure zones. Measurements on disturbed soils (44% clay, 1.5–2.3% C) packed to bulk densities of 1.25 g cm$^{-3}$ have indicated (Davies, 1985) that shear strength increased with organic C content when measurements were made at similar potentials. Emerson et al. (1994) have inferred a similar trend from measurements of penetration resistance noting that high levels of polysaccharide gels could lead to compacted parts of the mineral matrix becoming so strongly bonded together that, at a water potential of 0.01 MPa, the penetration resistance would be sufficiently high to limit root penetration. Fine roots and fungal hyphae enmesh the soil matrix, and therefore, management practices that increase the amount of this form of organic C can result in an increase in the strength of failure zones. Additional details on the impact of plant roots and fungi on the strength of soils are given in Sections 7.2.5 and 7.2.6.

## 7.2.2.4 Effects of Organic Matter on Aggregates and their Stability

The role of SOM on aggregates reflects the cumulative effects of the method of incorporation, the dynamics of decomposition (including the transformation of different forms of organic C, their spatial location and their degree of association with the mineral materials), and the influence of organic C on pore characteristics and failure zones (and therefore aggregate stability). Several studies have found a close relationship between the level of aggregation and SOM (some are reviewed by Tisdall and Oades, 1982), and numerous others have shown that incorporation of organic residues usually results in the formation of water-stable aggregates (Lynch and Bragg, 1985). Organic matter and its various fractions can contribute to both the formation and the stabilization of soil aggregates but, under some circumstances, specific fractions of organic matter can also destabilize aggregates and increase the dispersibility of clay and silt sized materials.

Increasing total organic C content normally results in an increase in the size and stability of aggregates, irrespective of the origin of the stress and the trends are most obvious in soils with widely varying SOM contents. The response of soils to wetting with mechanical abrasion will be used to illustrate this generalization. In a comprehensive study of soils from western United States and Canada (< 0.6–10% C), Kemper and Koch (1966) found that the stability of vacuum saturated soils increased with log C with different relations for subsurface layers, and surface layers under sod and cultivation, respectively. The stability decreased most rapidly at organic C contents below 1.2–1.5%. Greenland et al. (1975) found a critical level of 2% C below which soils from England and Wales were very liable to structural deterioration, especially in the absence of CaCO$_3$. Albrecht et al. (1992) also suggested that for a vertisol and a ferrisol, the aggregates were stable if they contained a minimum of 2% organic C.

The physical mechanisms by which organic matter affects aggregate stability are complex. The stability of soil aggregates is controlled by two opposing factors: the development of stresses within the aggregate pore space during wetting due to the compression of the entrapped air, and the strength of the interparticle bonds (Yoder, 1936; Hénin, 1938 cited by Concaret, 1967). SOM influences stability by reducing the rate of wetting and increasing the resistance to stresses generated during wetting (Monnier, 1965; Quirk and Murray, 1991; Rasiah and Kay, 1995; Caron et al., 1996b).

As is the case for other properties that are strongly dependent on the characteristics of failure zones, the water stability of aggregates from a soil can exhibit large changes due to cropping treatments before changes in the total organic C content are observed (Baldock and Kay, 1987). Changes in aggregate stability have been, therefore, attributed to changes in the amounts of various organic fractions such as POM (including fine roots and fungal hyphae), polysaccharides and lipids. These materials are considered to be labile and represent only a fraction of the total C content. Their amounts in the soil at any given time are determined by the rates of plant C input and the mineralization of this C and the microbial byproducts. It is the macroaggregate fraction (> 250 μm) which is mostly influenced by the labile SOM fractions. This generalization is consistent with the model which suggests that for most soils, the nature of the SOM responsible for providing stability varies with the size of the aggregates; water-stable macroaggregates are stabilized by transient and relatively undecomposed organic binding agents, while microaggregates would be stabilized by more processed SOM (Tisdall and Oades, 1982). Incubation and isotopic studies have confirmed that the turnover and nature of SOM varies with aggregate size and that macroaggregates are enriched in young and labile SOM (Elliott, 1986; Puget et al., 1995).

The role of POM in stabilizing aggregates can be through several mechanisms. First, as suggested by Tisdall and Oades (1982), fungal hyphae and fine roots can directly bind soil particles (Sections 7.2.5 and 7.2.6). Second, POM serves as substrate for microbial activity which can produce microbial bonding material (Golchin et al., 1994; Cambardella and Elliott, 1993; Jastrow, 1996; Besnard et al., 1996). Tiessen and Stewart (1988) and Oades and Waters (1991) have observed that many aggregates have cores of plant debris, leading to the proposal that encrustation of plant fragments by mineral particles was a mechanism of aggregate formation. POM entering the soil is rapidly colonized by microbes (Figure 7.5), resulting in byproducts that have strong adhesive properties, causing mineral particles to adhere to them. The dynamics of aggregation and decomposition are linked. Plant residues and POM can be assigned different aggregating potentials depending on their intrinsic decomposability and stage of decomposition, namely, on their ability to support microbial growth (Monnier, 1965; Golchin et al., 1994). Finally, it has been proposed that organic matter can enhance aggregate stability by obstructing soil pores, thereby reducing the rate of water entry and the extent of slaking (Caron et al., 1996). It is conceivable that POM could play this role.

Because polysaccharides are strongly adsorbed on mineral materials, they are particularly effective, at the interparticle level, in strengthening failure zones (Chenu, 1989; Chenu and Guérif, 1991), accounting for the frequent significant correlations that have been observed between aggregate stability and polysaccharide or readily extractable carbohydrate content (Chaney and Swift, 1984; Haynes and Swift, 1990; Roberson et al., 1991; Angers et al., 1993). Further supporting evidence comes from observations showing: (1) an improvement in aggregate stability following addition of microbial polysaccharides (Lynch and Bragg, 1985); and (2) disruption of aggregates by periodate treatment (Greenland et al., 1962; Clapp and Emerson, 1965; Cheshire et al., 1983). Using ultrathin sections and specific staining, Foster (1981) observed that polysaccharides were widely distributed in soils and suggested that their locations in very small pores and as clay coatings explained their inaccessibility to microbial degradation and their role in stabilizing clay into aggregates. Because polysaccharides are readily mineralized, their effects on the strength of failure zones will be transient (Tisdall and Oades, 1982) unless the residue source is continually renewed or the polysaccharides are physically protected from attack by microorganisms and extracellular enzymes. Guckert et al. (1975) proposed that following labile organic matter additions, polysaccharides were involved in the early stage of aggregate formation and stabilization and that longer-lasting effects involved more humified substances.

Decomposition of organic materials yields increasing proportions of long-chain aliphatic compounds (lipids) which may contribute to biologically induced changes in aggregate stability (Dinel et al., 1991, 1992). Capriel et al. (1990) and Monreal et al. (1995) found high correlations between the quantity of lipidic compounds and the stability of field soils under various management systems.

Giovannini et al. (1983) extracted a hydrophobic fraction with benzene and found that this treatment reduced the water stability of aggregates from a water repellent soil. The rate of water entry into an aggregate largely determines its resistance to slaking. Lipidic compounds are hydrophobic and would, therefore, reduce the rate of water entry thus increasing water stability (Capriel et al., 1990). The effects of the long-chain fatty acids was also attributed to polyvalent cation bridging (Dinel et al., 1992).

A significant proportion of the SOM is composed of organic molecules with chemical structures which do not allow them to be classified as biomolecules. These so-called humic substances have strong adhesive properties due to their charge characteristics. Correlations have been found between aggregate stability and the content of humic material (Chaney and Swift, 1984), and their addition to the soil can increase aggregate stability (Piccolo and Mbagwu, 1994). Extraction of humic compounds complexed with Fe and Al with either pyrophosphate (Baldock and Kay, 1987) or acetylacetone (Giovannini and Sequi, 1976) has confirmed the role of humic substances in aggregate stability. Monnier (1965) suggested that humic substances, as the ultimate organic matter transformation products, had smaller but longer-lasting effect on aggregate stability than early fermentation products such as polysaccharides. Further, Chaney and Swift (1986a,b) concluded that humic substances provided long-lasting stability to microaggregates whereas microbial polysaccharides were more involved in the short-term binding of these microaggregates into macroaggregates, which is in accordance with the Tisdall and Oades (1982) conceptual model.

Although it is clear that specific fractions of SOM have different roles in aggregate formation and stabilization, it is also doubtless that under field conditions most if not all fractions would be involved but at different degrees and at different scales depending on prevailing soil, climatic and cropping conditions. This may explain why in several field studies, correlation analysis shows that most fractions studied are correlated with one another and with aggregate stability (Capriel et al., 1990; Haynes and Francis, 1993; Angers et al., 1993).

Specific fractions of SOM may also, in some circumstances, destabilize aggregates, enhance dispersion or stabilize materials in suspension that have become dispersed. The effects appear to be most pronounced in kaolinitic soils with low SOM content (Emerson, 1983). The fractions that are responsible are not well-defined although soluble organic anions have been shown to cause dispersion (Shanmuganathan and Oades, 1983; Visser and Caillier, 1988). Shanmuganathan and Oades (1983) found the extent of dispersion by soluble organic anions was determined by the magnitude of the change in net surface charge arising from adsorption of the anions. Organic materials exuded by the growing roots of some plants (particularly maize) have also been reported to exhibit dispersive characteristics (Reid and Goss, 1981), although Pojasok and Kay (1990), using exudates extracted from growing maize plants, were unable to demonstrate this behavior.

### 7.2.2.5  Effects of Organic Matter on Resiliency and Vulnerability

The role, if any, that SOM plays in self-mulching is not well understood although it is probable that it alters the rate of wetting and drying. The rebound of soils after removal of a compressive stress increases with increasing SOM (Guérif, 1979), but the magnitude of recovery is inhibited by high initial bulk densities (McBride and Watson, 1990). Although the strength of failure zones appears to be particularly sensitive to wetting/drying or freezing/thawing events, there is less evidence that pore characteristics are as responsive and very little is known about the role of SOM on changes in either failure zones or pore characteristics. There is a dearth of information on the effects of SOM on soil structural vulnerability. Although the vulnerability of soils probably decreases with increasing C content (since increasing organic C generally increases stability and is unlikely to decrease resiliency), research is needed to confirm this speculation. The degree of vulnerability of soil structure has an important bearing on the long-term impact of land use practices on processes controlling crop productivity and environmental quality. Consequently, the role of organic C on structural vulnerability needs to be much better understood.

### 7.2.3 Inorganic Sesquioxides and Amorphous Materials

#### 7.2.3.1 Inorganic Cementing Materials

Sesquioxides which increase bonding between mineral particles have the greatest influence on structural characteristics by increasing the strength of failure zones. These materials exist as coatings on the surface of mineral particles and are amorphous or poorly crystallized. Examples include Fe and Al oxides, precipitated aluminosilicates or combinations of these materials that may include SOM. Because of their variable charge and pH dependent solubility, the extent of cementation can be modified by pH. In addition, their effectiveness may be related to the mineralogy of the matrix, exchangeable ions, pore fluid composition and water content. The interaction of Al and Fe oxides with clay minerals and their effect on soil physical properties have been reviewed by Goldberg (1989).

#### 7.2.3.2 Effects of Inorganic Materials on Failure Zones, Pores, Aggregates, and their Stability

The role of cementing materials has been studied most extensively in soils of high strength, with less emphasis on their influence on pore and aggregate characteristics. The strength of hardsetting soils and fragipans has been correlated positively with extractable Si and negatively with Al (Chartres and Fitzgerald, 1990; Franzmeier et al., 1996; Norfleet and Karathanasis, 1996). Hardsetting soils have low wet and high dry strength. Franzmeier et al. (1996) have hypothesized that hardsetting horizons are formed when silica bonds with Fe oxide minerals on clay surfaces and the bond is strengthened on drying. The influence of amorphous materials on the properties of soils of lower strength is more ambiguous.

Amorphous inorganic materials have dominantly a variable-charge nature and can alter the surface charge characteristics of clay minerals when adsorbed, or existing as a surface film. Changes in surface charge characteristics result in changes in particle-particle interaction. For instance, addition of Fe(III) polycations to clay suspensions from a red-brown earth resulted in clay flocculation, increased microaggregation and volume of 40–100 $\mu$m pores, and decreased bulk density and modulus of rupture (Shanmuganathan and Oades, 1982). Under field conditions, however, many of the positive charges on the oxides may be balanced by SOM functional groups and the role of the inorganic oxides may be linked to that of SOM. For instance, Kemper and Koch (1966) found that the stability of aggregates from 519 soils from the western United States and Canada was positively correlated with $Fe_2O_3$ but not with $Al_2O_3$. In a review of inorganic oxide-clay mineral interactions, Emerson (1983) concluded that such effects could be mainly due to the strong interaction between Fe and SOM. Bartoli et al. (1988) have attempted to assess the relative importance of SOM and poorly ordered Fe and Al hydroxides, and concluded that no single constituent is likely to adequately describe aggregate stability. In soils with < 6% C and small quantities of metal hydroxides, Bartoli et al. (1988) found evidence that soil polysaccharides were most involved in the aggregation process. The contribution of metal oxides to aggregate stability in soils with low levels of organic C might be expected to increase as the metal oxide content increases. In a study of the aggregation of B horizon of Alfisols and Inceptisols, Barberis et al. (1991) concluded that Fe oxides did enhance aggregation and that oxides in the amorphous form were more effective than those in the crystalline form (Al oxides were not found to have a major effect on stability). In soils with > 6% C and richer in noncrystalline materials, Bartoli et al. (1988) found that the cementing materials were mainly poorly ordered Fe and Al hydrous oxides associated with Ca or Al higher molecular weight organic materials. These analyses suggest that there is a complex relation between noncrystalline inorganic cementing materials and particle-particle interaction, and that the relation is strongly influenced by a number of other characteristics of the soil environment. The sensitivity of failure zones and aggregate stability to variation in surface charge characteristics diminishes with increasing crystallinity of the sesquioxides. For instance, Oxisol aggregates rich in gibbsite are extremely stable and nonelectrostatic bonds are much more important than electrostatic ones (Bartoli et al., 1992). Stability increases with SOM content, but the increases have been attributed to lower porosities and greater hydrophobicity (Bartoli et al., 1992).

### 7.2.4 PORE FLUID

#### 7.2.4.1 Solute Composition, Exchangeable Ions and their Relation to Structural Form and Stability

Pore fluid composition and the exchangeable ion suite influence the interaction between mineral particles, the extent of swelling and, ultimately, the dispersibility of the particles. Changes in the ease of dispersion of particles, particularly at the scale of clay sized particles, result in changes in strength characteristics and the stability of pores and aggregates. The attraction or bonding between mineral particles is sensitive to soil pH, exchangeable Na and/or Mg percentage (ESP, EMgP), and electrical conductivity (EC) of the pore fluid. The influence of one factor, for example, composition of the pore fluid, can be influenced by other factors, such as pH. In addition, the impact of these factors on the dispersibility of clay is strongly influenced by clay mineralogy. The cumulative effect of these factors on clay dispersibility is reflected in the difference in osmotic pressure between the diffuse electrical double layer and the pore fluid, as the difference in the osmotic pressure increases clays become more dispersible (Rengasamy et al., 1993; Rengasamy and Sumner, 1998). The manifestation of differences in osmotic pressure on the dispersibility of clay is also influenced by organic and inorganic cementing materials, which can restrain particles from swelling. Among common clay minerals, only smectites with a high ESP show extensive swelling (Churchman et al., 1993); the extent of swelling decreases as the EC (or osmotic pressure) of the pore fluid increases. Illites with a high ESP often remain dispersed in solutions of high EC, partly because the shape of the particles prevents strong cohesion. Increasing proportions of Ca as exchangeable or dissolved species (and to a lesser degree K [Chen et al., 1983]) reduce dispersion. The dispersion of kaolinites varies with pH since a significant part of their charge is variable. The sensitivity of clay sized materials in soils to dispersion with changes in pH may also be related to the effect of pH on the charge of metal oxides; as pH increases, the oxides become more negatively charged and clay minerals which may be coated by these oxides become more dispersible in the presence of Na. Exchangeable Mg enhances the dispersion arising from exchangeable Na, but apparently to a greater extent in illitic soils than in smectitic soils. Clay dispersion is discussed in detail by Emerson (1983) and Rengasamy and Sumner (1998).

   As the clay becomes more dispersible, aggregate stability declines and the formation of crusts become more common. Pore stability decreases, dispersed clay lodges in pore necks, and the hydraulic conductivity is reduced. In addition, increased dispersible clay leads to dramatic increases in strength of most soils on drying (Figure 7.6). The effects of dispersion on soil physical properties are reviewed in more detail by So and Aylmore (1993).

#### 7.2.4.2 Water Content/Potential and Relation to Structural Form and Stability

Soil water content varies continuously under field conditions. Water loss through drainage or evapotranspiration results in decreasing potentials and can cause shrinkage. The decrease in potential increases the effective stress (Bishop and Blight, 1963; Groenevelt and Kay, 1981), with a concomitant increase in strength. In addition, the effectiveness of bonds between cementing materials and mineral particles may increase with decreasing water content. Changes in the effectiveness of cementing materials with decreasing water content can arise from changes in orientation or position or from adsorption/precipitation. Dispersed clay can be reoriented or accumulate in the menisci between larger mineral particles as soils dry (Brewer and Haldane, 1957; Horn and Dexter, 1989). Adsorption, precipitation or crystallization of solutes occurs when drying concentrates solutes in the remaining pore fluid (Gifford and Thran, 1974; Kemper et al., 1989). Changes in the

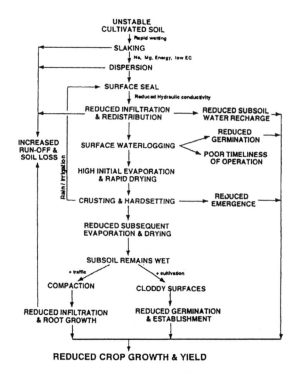

**FIGURE 7.6** Diagrammatic representation of processes arising from increased dispersion of soil clays [Reprinted from So and Aylmore, 1993. Aust. J. Soil Res. 31:761-777, with permission].

effectiveness of organic cementing materials may also occur (Caron et al., 1992b). These effects may be only slowly reversible and increases in water content can result in decreases in the cohesion of aggregates and increases in the dispersibility of clay that may continue for several days (Caron and Kay, 1992; Caron et al., 1992c).

The dependence of shrinkage, effective stress and cementation on soil water content cause many other structural characteristics to vary with water content. Measurements of aggregate stability made under conditions in which air entrapment and differential swelling are minimized have shown that the stability of a soil declines with increasing water content (Reid and Goss, 1982; Gollany et al., 1991). The loss in stability applies at the scale of macroaggregates (> 250 $\mu$m) as well as at the scale of material < 2 $\mu$m (Perfect et al., 1990a; Kay and Dexter, 1990). In cases where air entrapment and differential swelling are present, slaking can occur and water content may have a different effect on stability (Concaret, 1967). In such cases, stability initially increases with water content and then decreases. The relationship can vary with soil type and cropping system (Kemper et al., 1985; le Bissonnais, 1988; Angers, 1992). Threshold water contents can be used to identify the lower limit of the range in water contents in which a structural characteristic is sensitive to changes in water content (Emerson 1983; Watts et al., 1996a,b). For instance, after imposing different levels of mechanical energy on aggregates at different water contents, Watts et al. (1996a) found that the dispersibility of clay in aggregates was a function of both energy input and water content. Below a certain threshold water content, approximating the plastic limit, even relatively large energy inputs had little effect on clay dispersion. As the water content increased above the plastic limit, the soil became increasingly sensitive to mechanical disruption.

### 7.2.5  SOIL MICROORGANISMS

Evidence for the involvement of microorganisms in soil aggregate formation and stabilization comes from early work on the effects of adding organic materials to soil. Organic matter additions have little or no effect unless microorganisms are present (Lynch and Bragg, 1985). The role of plant residues in providing a nucleus for aggregate formation was discussed earlier (Section 7.2.2). There are several mechanisms by which microorganisms are involved in stabilizing soil structure: first, by directly providing mechanical binding between soil particles; second, by producing cementing materials during the decomposition of organic materials; and last, microorganisms also serve as a substrate for further microbial growth.

Fungi are believed to be the most efficient group of soil microorganisms in terms of soil aggregation (Lynch and Bragg, 1985). Soil fungal hyphae ramify throughout the soil and bind soil particles together (Dorioz et al., 1993). The degree of soil aggregation has been correlated with hyphal length and biomass indicators (Tisdall et al., 1997; Chantigny et al., 1997). Fungi are believed to be involved in both the formation and stabilization of aggregates (Degens et al., 1996; Tisdall et al., 1997). In addition to providing physical enmeshment by networking in the soil, thereby contributing to the formation of aggregates, fungi also produce polysaccharides (Chenu, 1989; Beare et al., 1997) and other proteic and lipidic compounds (Lynch and Bragg, 1985; Wright and Upadhyaya, 1996) which promote aggregate stability. Both mycorrhizal and saprophytic fungi were found to be effective. Mycorrhizae are believed to be partially responsible for the aggregation of soil particles in the rhizosphere and the effect of roots on aggregation has often been associated with vesicular arbuscular mycorrhiza (VAM) supported by the root systems of some plant species (Tisdall and Oades, 1979; Thomas et al., 1986; Jastrow, 1987). This role was also clearly demonstrated in studies of the revegetation of unstable maritime sand dunes (Sutton and Sheppard, 1976; Forster, 1990). Saprophytic fungi, as part of the suite of microorganisms developing during residue decomposition, are also involved in soil aggregation. Laboratory incubations of soils with or without organic amendments have shown the role of fungi in the initial increase in soil aggregation and aggregate stabilization (Molope et al., 1987; Metzger et al., 1987). As pointed by Degens (1997), the growth of saprophytic fungi is restricted to the soil surrounding the decaying organic residue and their aggregating efficiency will, therefore, depend on the even distribution and fragmentation of the residues in the soil.

Other filamentous microorganisms were found to be involved in soil aggregation such as streptomycetes and microalgae (Lynch and Bragg, 1985), but their mechanisms of action have been much less studied. Metting (1987) has investigated the potential use of a microalgae as a soil conditioner.

Although bacteria are often associated with mineral particles, because of their size they are not likely involved in the direct binding of soil particles to form aggregates (Dorioz et al., 1993). It is rather the product of their activity which is the active binding mechanism. Several compounds are produced by bacteria and microorganisms in general during decomposition of organic residues. Extracellular polysaccharides having binding capacities can be produced in large quantities by soil bacteria (Lynch and Bragg, 1985), and their role in stabilizing soil aggregates was discussed earlier as well as those of long-chain aliphatics or lipids which are also produced during residue decomposition (Section 7.2.2.4).

## 7.2.6  PLANTS

Root development is strongly influenced by soil structure. In return, plants also influence the form, stability and resiliency of the structure.

### 7.2.6.1  Effects on Structural Form

Growing in existing pores or through the soil matrix, roots create compressive and shear stresses which result in the creation of new pores and the enlargement of existing ones. Root influence on pores varies with the stage of growth and decay. Infiltration rates can be reduced by actively growing roots which fill pores. Transport efficiency in root channels is enhanced as the root decays and tissue remnants and associated microflora remain as pore coatings on channel walls (Barley, 1954). For example, alfalfa (*Medicago sativa* L.) is characterized by a large diameter, long and straight tap root, and has been reported to create macropores that improve water and solute flow properties (Meek et al., 1989 and 1990; Mitchell et al.,1995; Caron et al., 1996a) and reduce subsoil penetration resistance (Radcliffe et al., 1986). While probably less noticeable than the effect on macropores, the effect of plant roots on smaller pores may be very important to other soil structural properties. Radial and axial pressures exerted by the growing roots will affect the small pores (Guidi et al., 1985; Bruand et al., 1996) by compressing the soil in the vicinity of the roots and decreasing pore size in that immediate zone.

Depending on their clay content and type, some soils show potential to shrink and swell with variations in water content. During shrinking, the loss of water is associated with a loss of volume and the development of cracks in planes of weakness. The type of vegetation present has a marked influence on the soil cracking pattern since the plants cause shrinkage of the soil when extracting water for their use. Grevers and de Jong (1990) found differences due to grass species in macropore structure of a swelling clay soil. They attributed the structural differences to differences in water uptake and thus desiccation of the soil. The greater the plant biomass production, the greater were the area and length of macropores. Since plant distribution varies in space because of associated cultural practices, the cracking pattern will also vary in relation to plants. For row crops, water is used first at the row position and the cracking pattern will develop at the interrow position (Fox, 1964; Chan and Hodgson, 1984). Wetting and drying cycles also influence aggregate formation and fragmentation (Horn and Dexter, 1989), and plant growth as well as precipitation or watering will influence the magnitude, frequency and effects of these cycles on aggregation (Caron et al., 1992a; Materechera et al., 1992). Materechera et al. (1992) attributed the production of small aggregates associated with root growth to soil cracking caused by tensile stress induced by heterogeneous water uptake by the plants.

### 7.2.6.2  Effects on Structural Stability

The root systems of many plant species form a dense network in soils leading to soil stabilization and reinforcement as found on streambanks (Kleinfelder et al., 1992). Soil shear resistance was found to be greatly improved by the root systems of various plants (Waldron and Dakessian, 1982). On a smaller scale, plant roots and root hairs can also directly enmesh and stabilize soil aggregates of millimeter size (Tisdall and Oades, 1982). Visual (Figure 7.7) and microscopic observations (Foster and Rovira, 1976; Tisdall and Oades, 1982; Dormaar and Foster, 1991) clearly show that aggregates are formed and stabilized in the immediate vicinity of plant roots. Field and greenhouse studies have demonstrated that growing plants induce the rapid formation and stabilization of soil aggregates (Tisdall and Oades, 1979; Reid and Goss, 1981; Angers and Mehuys, 1988; Stone and Buttery, 1989). Statistical correlations have been found between root length or mass and soil aggregation (Dufey et al., 1986; Miller and Jastrow, 1990). Although fine roots can form a dense network which can probably entangle or enmesh soil particles and form aggregates, indirect effects such as associated microbial activity or the release of binding material have most often been invoked to explain the apparent relationship between fine roots and aggregate stability.

Soil structural stability is influenced by soil water content and its variations with time (Section 7.2.4.2). Water uptake by the plant and the consequent decrease in soil moisture will usually result in increased soil strength (Horn and Dexter, 1989; Lafond et al., 1992). At the aggregate level, soil cohesion is greatly enhanced by decreasing water content and the dispersion of clay size material decreases accordingly (Caron and Kay, 1992). The drying of soil by roots may also act synergistically with the aggregate binding material produced in the rhizosphere and increase soil structural stability. As will be discussed later, organic materials released by the roots and microbial population of the rhizosphere can be efficient in binding soil particles. The drying that occurs in the zone of mucilage production contributes to the efficiency of the binding agents through increased sorption of the binding material onto colloid surfaces (Reid and Goss, 1981;

**FIGURE 7.7** Aggregate formation around timothy roots [Reprinted from Angers and Caron, 1998. *Biogeochem.* 42:55-72, with kind permission from Kluwer Academic Publishers].

Caron et al., 1992b). Plant roots can promote soil aggregation by releasing material, which can directly stabilize soil particles, or by favoring microbial activity in the rhizosphere which, in turn, will affect soil structure. Morel et al. (1991) provided evidence that intact mucilage released by maize root tips significantly increased soil aggregate stability independent of any microbial activity, since it occurred immediately after the incorporation of the exudates in the soil. Pojasok and Kay (1990) found that the release of nutrient ions and C in the exudates of bromegrass and corn roots increased aggregate stability.

The rhizosphere presents a very high level of microbial activity induced by root exudation and mucilage, root sloughing and favorable aeration and water conditions (Bowen and Rovira, 1991). The rhizosphere microbial community is also very diverse. The presence of mycorrhiza in the rhizosphere of many plants is noticeable and their role in soil aggregation has been discussed earlier (Section 7.2.5). Few studies have investigated the contribution of other specific rhizosphere microorganisms to soil aggregation (Gouzou et al., 1993). Microbial exocellular polysaccharides are found in the rhizosphere of plants (Bowen and Rovira, 1991) and could act as cementing materials, but their effect cannot be distinguished easily from that of plant mucilages. Rhizobial polysaccharides have also been shown to be efficient in promoting soil aggregation (Clapp et al., 1962). Much remains to be determined about the mechanisms of aggregate formation and stabilization in the plant rhizosphere, and the respective contributions of roots and specific rhizosphere microorganisms is still unclear. Moreover, both biological and physical processes such as drying contribute to aggregate formation and stabilization in the immediate vicinity of the roots through complex interactions.

Aside from the immediate and short-term effects of roots on soil structure which have already been described, plant roots and litter also have a longer term influence through their contributions to bulk SOM. A large proportion of the C fixed by plants is distributed below ground. Consequently, in many ecosystems, plant roots constitute the most important source of SOM, and so have a predominant effect on biologically induced changes in soil structure. Moreover, in many ecosystems, a large part of the aboveground plant is returned to the soil as litter or crop residue, which also constitutes an important C source potentially affecting soil structure.

### 7.2.7 SOIL FAUNA

Soil is the habitat of a large number of animals whose biomass and work may well exceed those outside it. Through their activity, animals affect soil chemical, biological and physical properties in various ways (Hole, 1981), but their effects on soil structure are mostly through their movement in the soil and their feeding/excreting activities. Because earthworms have been by far the most widely studied, the following discussion will be limited largely to their effects on soil structure.

#### 7.2.7.1  Effects on Porosity

Soil fauna form channels as they move in soil, contributing greatly to the formation and maintenance of porosity. Among the different groups of soil animals, earthworms, termites and ants are believed to contribute to the formation of pore whereas smaller animals like microarthropods are confined to preexisting burrows because of their small size (Lee and Foster, 1991). The pores formed by earthworms are often large (macropores) and cylindrical. When connected to the soil surface, they contribute to water infiltration (Ehlers, 1975) and gas exchange. Water infiltration rate is well correlated with earthworm population (Bouché and Al-Addan, 1997). Pores formed by soil animals also improve root growth (Edwards and Lofty, 1978). Anecic species, which feed at the soil surface, create vertical pores whereas endogenic species feed within the soil and provide more or less horizontal voids, which can be connected with vertical voids. Biopores formed by earthworms are believed to be relatively stable, although this may not be the case in swelling soils in which lateral pressure can be high (Mitchell et al., 1995). Dexter (1978) showed that the soil around burrows was not compacted and concluded that earthworms contributed more to the formation of pores by ingesting soil than by compressing it. Pores are coated with a thin lining called the drilosphere (Bouché, 1975, cited by Lee and Foster, 1991), enriched in various components and oriented clays, which presumably provide some stability to the walls.

Termites and ants live in colonies, and therefore, their effect on soil porosity is concentrated in discrete areas (Lee and Foster, 1991). Although their population may be very high, they affect, at a given time, only a small proportion of the soil volume. More work is required to determine the effects of termites on soil structure.

#### 7.2.7.2  Aggregate Formation and Stabilization

In the case of earthworms, ingestion of organic and fine-textured inorganic materials results in intimate mixing and the excretion of casts which have structural characteristics vastly different from the bulk soil. Aggregates created in the presence of earthworms are generally more stable than those created in their absence (Blanchart, 1992), and the size of the formed aggregates depends on the size of the earthworms (Blanchart et al., 1997). However, earthworm casts can be unstable while they are fresh and wet (Shipitalo and Protz, 1988; Marinissen and Dexter, 1990). Preexisting bonds between soil particles prior to ingestion are probably disrupted in the gut of the animal where large amounts of watery mucus are produced and the soil is mechanically dispersed. At excretion, casts are very moist, slurry-like and relatively unstable, but will gain stability with drying and aging (Shipitalo and Protz, 1988). Microbial activity and, in particular, fungi (Marinissen and Dexter, 1990) and the presence of coarse organic fragments (Shipitalo and Protz, 1988) also stabilize earthworm casts. Marinissen et al. (1996) concluded that orientation of clay particles following passage in the worm also leads to closer contact resulting in the stabilization of the casts.

As emphasized by Lee and Foster (1991), little quantitative information is available on effects of termites or ants on aggregate formation and stabilization. A termite species (*Thoracotermes macrothorax*) was shown by Garnier-Sillam et al. (1988) to have a positive influence on structural stability, which was attributed to the enrichment of the termitary soil with organic matter and cations.

### 7.3  OTHER FACTORS INFLUENCING SOIL STRUCTURE

#### 7.3.1  CLIMATE

Climate controls the temporal variation in the water content and temperature of soils, and has direct effects on a range of physical and biological processes linked to soil structure. Combeau (1965) illustrated how temperature and soil moisture controlled the seasonal variation in structural stability for both tropical and temperate soils. The impact of temporal variation in water content on structural characteristics is strongly influenced by the rate at which the water content changes, and the swelling

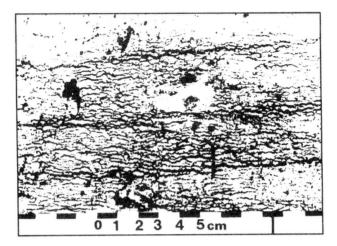

**FIGURE 7.8** Macromorphology of ice enriched frozen silt loam [Reprinted from Kay et al., 1985. *Soil Sci. Soc. Am. J.* 49:973-978, with permission of the Soil Science Society of America].

characteristics of the soils. Slow changes in water content arising from low intensity rainfall events, drainage and evapotranspiration can cause changes in the pore characteristics in both swelling and nonswelling soils. In the case of swelling soils, the changes are a result of swelling or shrinkage and crack formation, whereas in nonswelling soils, changes are restricted to the seedbed, where decreasing water content and a concomitant increase in effective stress causes a loss of interaggregate pore space and a progressive consolidation of the seedbed.

The seasonal variation in structural characteristics that are strongly influenced by failure zones (tensile strength and aggregate stability) is often much larger than changes caused by different cropping practices, and this variation can be related to wetting events preceding the time of sampling (Kay et al., 1994b). The magnitude of the variation in water content and its rate of change decrease rapidly with depth in the profile. It is therefore not surprising that major differences in the strength of failure zones can be found with differences in depth within the A horizon as small as 5 cm (Kay et al., 1994b).

Changes in the liquid water content of surface soils, as a consequence of ground freezing, are common in many soils. The pore water may freeze *in situ,* and where the soil is saturated, the increased volume can create stresses that may be expected to result in failure zones. A much more common feature, particularly on medium- and fine-textured soils, is the migration of water in response to gradients in water potential at the freezing front and the subsequent accumulation and freezing of water in the form of lamella or ice lenses just behind the freezing front (Figure 7.8). The pores that are created under these conditions do not appear to be stable and most are lost as the soil consolidates during the thaw period (Kay et al., 1985). The sites of ice lenses must, however, represent zones of very low strength, and undoubtedly contribute to the loss in stability (Willis, 1955; Bullock et al., 1988) and strength (Utomo and Dexter, 1981b; Voorhees, 1983; Douglas et al., 1986) of large aggregates that are often observed as a consequence of freezing. Zones between ice lenses are desiccated and the rearrangement and flocculation of clay sized particles (Rowell and Dillon, 1972; Richardson, 1976) in these areas may account for the increased stability of small aggregates that is often observed after freezing (Perfect et al., 1990b,c).

## 7.3.2 MANAGEMENT

Land is managed for a range of uses and each use may involve a suite of management practices. Management of soil, water and plants influences soil structure by controlling the form and amount of C entering the soil and its spatial distribution in the soil. These practices also influence the populations of soil macro- and microorganisms and the mineralization rate of C. The nature of soil cover provided by plants or crop residue influences soil structure through its influence on raindrop impact, soil water content, the rate of wetting and the depth of freezing. The nature of the root system of the crops selected will influence the depth of water extraction, and therefore the depth to which shrinkage may occur. In addition, macropores can be created by tillage and destroyed by traffic.

There are an infinite number of combinations of management options used in crop production under various soil and climatic conditions. Consequently, the number of case studies reported in the literature is large. The following discussion is an attempt to illustrate how selected management combinations affect soil structure under selected soil and climatic conditions. Detailed reviews on management effects on soil structure, with particular reference to aggregation and organic matter, are available for various soil and climatic conditions (Haynes and Beare, 1996; Dalal and Bridge, 1996; Feller et al., 1996; Angers and Carter, 1996).

### 7.3.2.1  Effects of Crops

The specific effects of plants on both soil structural form and stability and the mechanisms involved were discussed in Section 7.2.6. This section will review some site-specific studies, mostly from eastern Canada, to illustrate the effects of various crop sequences on soil aggregation and aggregate stability. For a discussion on the role of plants on porosity, the reader is referred to Section 7.2.6.

Beneficial effects of perennial forages on soil aggregation are well-known. Several studies have shown that water-stable aggregation increases rapidly when arable or degraded lands are put into continuous perennial forages. Maximum stability is often achieved after 3–5 years (Low, 1955; Perfect et al., 1990c; Angers, 1992). Choice of perennial forage crop species and cultivar will also influence the extent of aggregate formation and stabilization. Alfalfa was as efficient as bromegrass but slightly less than timothy (*Phleum pratense* L.) and reed canarygrass (*Phalaris arundinacea* L.) in improving the water-stable aggregation of two fine-textured soils (Chantigny et al., 1997). Carter et al. (1994) showed that the potential for increasing the stability of a loamy soil from the eastern seaboard of Canada varied with grass species but also with cultivar. Management of perennial forages should also affect aggregation through factors which influence crop C production. However, Perfect et al. (1990c) did not observe any effect of a cutting regime of red clover (*Trifolium pratense* L.) or N fertilization of bromegrass on aggregation. Experimental results on the effects of short-term rotations and cover crops on soil aggregation have been inconclusive. In some cases, cover crops and rotation with a legume or a grass legume mixture significantly improved soil macroaggregation (Webber, 1965; Raimbault and Vyn, 1991), but in others little or no effects were observed (MacRae and Mehuys, 1987; Carter and Kunelius, 1993). The effects of short-term rotations and cover crops on water-stable aggregation are highly dependent on crop species and, in particular, the amount of residues that each crop of the rotation returns to the soil in the form of either roots or aboveground residues, as well as the tillage and fertilizer practices, and the soil water uptake pattern associated with the plant grown in each phase of the rotation.

### 7.3.2.2  Organic Amendments and Fertilization

There is a multitude of organic materials applied to agricultural soils. The incorporation of organic amendments such as cattle manure, compost, and industrial and municipal organic wastes usually results in a decrease in bulk density and a consequent increase in soil porosity (Hardan and Al-Ani, 1978; Pagliai et al., 1981; Churchman and Tate, 1986; N'Dayegamiye and Angers, 1990). Pagliai et al. (1981) found that the different pore size fractions were all increased by the addition of municipal sludge. The beneficial effects of farmyard manure are also observed on soil aggregation under various soil and climatic conditions (Mazurak et al., 1977; Ketcheson and Beauchamp, 1978; Sommerfeldt and Chang, 1985; Dormaar et al., 1988; N'Dayegamiye and Angers, 1990). The application of other organic materials such as compost, sewage sludge and lignocellulosic materials is also probably beneficial to soil aggregation, although they vary widely in composition and decomposability and their efficiency to promote the aggregation. Their aggregating efficiency has been related to the decomposability of the material (Martin and Waksman, 1940; Monnier, 1965; Tisdall et al., 1978).

Commercially produced organic polymers have been used to stabilize structure, thereby reducing seal formation, increasing infiltration, and diminishing runoff and erosion. Their effectiveness varies with polymer characteristics (type and amount of charge, configuration and molecular weight), soil properties (clay content and mineralogy, EC and pH), methods of application (spraying on soil surface or applying in irrigation water) and amounts applied (Stewart, 1975; Seygold, 1994). The cost of commercially produced polymers and the need for repeated applications continue to constrain their widespread use in agriculture.

Mineral fertilizers, in particular N, can have both positive and negative effects on soil aggregation. In the short term, N fertilization can accelerate mineralization of organic binding agents (Acton et al., 1963), whereas in the long term, N fertilization increases C content by improving crop yield and crop residue inputs to the soil and potentially results in increased aggregation (Campbell et al., 1993). Kay (1990) suggested that the potential improvement in soil aggregation associated with increased production of roots due to fertilization may not be fully realized if fertilization also causes increased rates of mineralization of the binding material.

### 7.3.2.3  Tillage Practices and Controlled Traffic

The impact of tillage on soil structural form will depend on the type of equipment used, initial structural form, soil water content at time of tillage and frequency of tillage. In the short term, tillage usually results in a decrease in bulk density and therefore increases total porosity in the tilled soil layer. Large pores are usually created which favor fluid transmission and root growth in the surface soil (Ehlers, 1977; Carter, 1988; Klute, 1982). However, tillage can disrupt the continuity of the pore system created by root and faunal activity in the surface soil and between this layer and deeper nontilled horizons (Ball, 1981; Goss et al., 1984). It may also create compacted zones at the bottom of the plowed layer (Bowen, 1981).

In the long term (> 5 yr), the effects of the absence of tillage on total porosity compared to plowed soils may be either positive, negative or absent (Voorhees and Lindstrom, 1984; Heard et al., 1988). More importantly, nontilled soils usually show greater macropore continuity, which results in higher hydraulic conductivity (Heard et al., 1988; Edwards et al., 1988). van Lanen et al. (1992) observed a better structure, as assessed by greater porosity and hydraulic conductivity, in a permanent grassland soil than in a young arable soil after 8 yr, due to intensive tillage in the latter.

As discussed earlier, compaction by vehicular traffic or animals can be very detrimental to soil structure as it reduces soil porosity and increases soil strength, and ultimately can reduce crop yield and increase the risk of water erosion. Although damage caused by compaction can, in some cases, be alleviated by subsoiling or by natural processes of wetting/drying, it is generally agreed that prevention by controlling traffic or maintaining SOM levels is more efficient. Controlled traffic involves traversing the same tracks in the field, thus restricting structural damage to zones of traffic (Unger, 1996).

Tillage practices also have a strong influence on soil aggregation and structural stability. Several studies have shown that surface soils (0–10 cm) under no till contain larger and more stable aggregates than their tilled counterparts (Carter, 1992; Beare et al., 1994a), due to the combined effects of crop residue accumulation at the surface and the absence of mechanical disturbance in non-tilled systems. Mechanical stress would result in both a direct breakup of the aggregates and oxidation of SOM which stabilizes the aggregates. Labile C fractions which are involved in stabilizing aggregates such as polysaccharides and fungal hyphae are generally present in lower concentrations in tilled than in nontilled soils (Beare et., 1994b; Angers et al., 1993). The spatial distribution of incorporated crop residue, which may influence C dynamics and subsequently aggregate stability, varies with the type of tillage (Allmaras et al., 1996).

### 7.3.2.4 Managing Water

Management of soil water influences soil structure primarily by altering the resistance of structural form to deformation when stress is applied. Soil water can be considered from the perspective of both quantity and quality and is primarily managed through irrigation and drainage, and to a lesser extent through management of runoff and evaporation. Farmers have supplemented rainfall with irrigation to meet the needs of growing crops for centuries. There is ample historical evidence of the need to prevent the accumulation of Na in irrigated soils and the associated decline in stability, infiltration and drainage, and increased runoff and hardsetting at the soil surface (So and Aylmore, 1993).

Drainage is used to remove excess water and, in rain-fed environments, the improved aeration and timeliness of operations can result in changes in both soil and crop management practices. The effects of drainage on soil structure arise from changes in organic C dynamics (including the quantity and quality of C added as well as the rates of mineralization of soil C and added C), and changes in the probability of degradation by tillage and traffic (related to changes in stability and the magnitude of stress applied). Little research has related changes in soil structure to changes in drainage. Hundal et al. (1976) found that drainage increased alfalfa and timothy yields on a clay soil resulting in greater hydraulic conductivity, lower strength and decreased bulk density after 16 yr. The tillage/traffic and yield response accounted for the improvements in structure with improved drainage. Although yields of corn increased with improved drainage in soils of similar texture (Fausey and Lal, 1989), plant-available water was lower on the drained treatments (Lal and Fausey, 1993). These changes may have been less influenced by the production of corn (for 5 yr) than by the decline in soil organic C content with improved drainage (Fausey and Lal, 1992).

---

**EXAMPLE 7.1**   Discussion of role of organic carbon on the volume fraction of pores with effective diameters ranging from 0.2-30 μm.

Water that is most readily available to plants is held in pores with effective diameters ranging from 0.2 up to 30 μm. The volume fraction of pores in this size range is generally observed to be positively correlated with organic carbon content and this observation provides a compelling reason for land managers to adopt practices that maintain or increase organic carbon contents. Speculate on the mechanisms that may account for the positive correlation.

Issues to be taken into account include:

roots, crop residue and organic amendments make up a large part of the organic carbon entering soil, and are initially located in pores that are larger than 150 μm; organic carbon in the rhizosphere may be located in smaller pores that are adjacent to the macropores

comminution, redistribution and decomposition of organic residues is achieved by soil fauna and the decomposition process is continued by soil flora

pores of different sizes are also created by freezing, drying and tillage; pores are lost or diminished in size through thawing and consolidation, rewetting, and compaction induced by traffic or other mechanisms

the water content of this size class of pores is more dynamic than that of any other pore size class; pores smaller than 0.2 μm are water-filled much of the time and pores larger than 30 μm are air-filled much of the time

## 7.4 INTERPRETING DATA ON SOIL STRUCTURE

Soil structure has a major influence on the functions of soils in ecosystems to support plant growth, cycle carbon and nutrients, receive, store and transmit water, and resist soil erosion and the dispersal of chemicals of anthropogenic origin. A case has been made that particular attention must be paid to soil structure in managed ecosystems where human activities can cause both short- and long-term changes that may have positive or detrimental impacts on these functions. Consequently, emphasis will be placed on interpreting data on soil structure in relation to functions that soils fulfill in managed ecosystems, particularly agroecosystems.

There have been two major practical objectives for doing research on soil structure in agroecosystems during the past 50 yr. The first one emerged from observations that the ability of soil to fulfill different functions was related to its resistance to various types of stresses (initially those related to raindrop impact and rapid wetting, wind, and later those stresses caused by tillage and traffic). This led to a large body of work on various aspects of soil aggregation and structural stability (particularly aggregate stability). The second purpose was to relate its various characteristics to seedling development, plant growth and productivity in an attempt to better understand factors controlling yield. More recently, interest related to the protection of the environment has stimulated research in two additional areas: the role of soil structure in C and nutrient cycling, and in controlling water and solute flow (particularly with respect to macropore or bypass flow). The degree to which data on soil structure can be interpreted with respect to the different functions of soil is therefore related to their evolutionary stage. Emphasis here will be placed on the relation of structure to plant growth, C and nutrient cycling, and soil erosion while the aspects of soil structure dealing with the ability of soil to receive, store and transmit water and to resist the dispersal of chemicals are covered in Chapters 4 and 6.

### 7.4.1 PLANT GROWTH
#### 7.4.1.1 Germination, Emergence and Growth of Seedlings

Structural conditions in the seedbed that promote rapid germination, uniform emergence and rapid growth of seedlings are desirable. Where seedbeds are created by tillage or seeding operations, the principal characteristic of soil structure that can be varied is aggregate size distribution which determines the availability of $O_2$ and water under a given climatic condition and the resistance offered to emerging shoots and roots. Ideal conditions for a seedbed are produced by stable aggregates not < 0.5–1.0 mm and not > 5–6 mm diameter (Russell, 1973, Schneider and Gupta, 1985). Qualitative ratings have been developed for aggregate size distributions by assigning different weighting values to aggregate size fractions (MacRae and Mehuys, 1987; Braunack and Dexter, 1989). Although the relative values of the weightings that have been used by investigators differ, the greatest weighting is applied to 1–2 mm aggregates and lowest to > 5 mm and < 0.2 mm aggregates. While these criteria represent useful broad generalizations, the optimum aggregate size distribution will also vary with both crop and climate. For instance, finer seedbeds may be required when crops with smaller seeds are planted (Hadas and Russo, 1974) or where precipitation is low (Braunack and Dexter, 1989). Measurements of aggregate size distributions are most relevant to the germination and early growth of plants on soils that are tilled, structurally stable, and not compacted by traffic. The measurements will have less relevance to later growth, or to early growth on untilled soils or tilled soils with characteristics that do not persist.

The persistence of the seedbed characteristics during the relatively short period of germination, emergence and early growth is related to the structural stability and the amount of energy that may be applied. During this period, the surface of the seedbed is particularly vulnerable to stresses arising from rainfall that can break up aggregates and disperse clay and fine silt if the structural stability is low. Subsequent drying results in the formation of a crust. A crust reduces infiltration rates and the availability of $O_2$, but has the most profound impact on the emergence of seedlings. When the mechanical strength of the crust approaches or exceeds the force that can be exerted by the growing seedling, emergence becomes nonuniform or even prevented. Although there are few generalized measures of soil structure to interpret the susceptibility of soils to form crusts of high strength, the mechanical strength of the crust that arises from a storm of a given intensity must be related to the susceptibility to disaggregate and disperse (le Bissonnais, 1996), and to the susceptibility of the resulting matrix to achieve high strength on drying.

## 7.4.1.2  Plant Growth, Development and Yield

Subsequent to emergence, the growth and development of plants and final crop yield continue to be influenced by structural characteristics that control aeration, the availability of water and the resistance offered to penetration of the soil by the growing roots. The structure will reflect the coalescence of the seedbed, traffic-related compaction and soil and environmental characteristics. The extent to which soil structure influences water and $O_2$ supply, and the mechanical impedance offered to root development strongly depends on water content and evaporative demand. Under rain-fed conditions, this means that the impact of a given structure on plant growth varies with climate.

After tillage, the coalescence of the structure of the seedbed will be controlled in part by the pattern of rainfall and subsequent evapotranspiration (and associated stresses) and structural stability. Macropores will be progressively lost (Kwaad and Mucher, 1994), relative compaction will increase (Carter, 1990), and hydraulic conductivity (particularly near saturation) will decline (Mapa et al., 1986). The rates of change in these characteristics may vary with the initial conditions of the seedbed (including those related to position relative to the row [Cassel, 1983]) and are not readily predictable, since the stresses related to wetting and drying will vary from year to year depending on climatic conditions (Carter, 1990a). In addition, there have been few attempts to define quantitative relations between stability characteristics, stress and the rate of coalescence of the seedbed.

Traffic and associated compaction also lead to changes in the structure of the seedbed. The processes and mechanics of soil compaction are described in Chapter 2. As mentioned earlier, attempts to relate soil aggregate characteristics to susceptibility to compaction, which is related to the preconsolidation state of the soil and its water content, have been relatively unsuccessful (Angers et al., 1987, Angers, 1990). Soil structural characteristics more likely to be related to compactibility are, therefore, initial bulk density and soil water content which determine soil strength. The degree of compaction of soils, whether arising from traffic or from coalescence, is best described by the relative compaction (RC), namely, the observed bulk density divided by the maximum bulk density under a standard compaction treatment (Häkansson, 1990; Carter, 1990a), when soils with different inherent characteristics (texture and organic C contents) are compared. The RC tends to normalize bulk density with respect to variation in inherent characteristics, and differences in RC primarily reflect changes in macro- and mesoporosity (Carter, 1990a). Studies of the relation between RC and relative yields (observed yield divided by the maximum yield) have shown similar relations in Scandinavia (Häkansson, 1990) and eastern Canada (Carter, 1990a); relative yields of cereals vary curvilinearly with RC and are maximum at RC values between 0.77 and 0.84. The relation between RC and relative yield (including the optimum values of RC) must reflect an integration of functional relations between (1) compaction and the structural characteristics controlling the availability of $O_2$ and water to the plant as well as resistance of the soil to penetration by roots, (2) compaction and the temporal variation in water content, and (3) plant response to stresses arising from detrimental soil structure/water content conditions. The degree to which these relations may vary with soils, climate and crop species (or perhaps even varieties) is not well understood. Although RC appears to be a particularly useful characteristic to quantify the extent of compaction of different soils arising from wetting and drying or traffic, interpretations with respect to the yield of different crops under different climates should be made with caution.

The availability of $O_2$ to plant roots is determined by $O_2$ diffusion rates, and decreases as an increasing proportion of the pore space becomes water filled, and approaches zero when there are no continuous air-filled pores remaining. This limiting condition is strongly influenced by the volume fraction and continuity of macro- and mesopores. It is therefore not surprising that aeration is spatially variable between these pores and within aggregates (Zausig et al., 1993). While direct measurements of $O_2$ diffusion rates are the preferred way to assess the influence of soil structure on aeration since the measurements reflect the influence of both pore size distribution and continuity (Hodgson and MacLeod, 1989a), such measurements often cannot be made and aeration must be inferred from measurements of other structural characteristics, most often those related to pore size distribution. One such characteristic, the volume fraction of air-filled pores, is about 0.10 $m^3$ $m^{-3}$ (Xu et al., 1992), although limiting values of 0.05 and 0.145 $m^3$ $m^{-3}$ have been reported (Ehlers et al., 1995; Hodgson and MacLeod, 1989b) when the diffusion rates approach zero. Other inves-tigators have found no relation between diffusion rates and air-filled porosity (Zausig et al., 1993). Inconsistent interpretations of the influence of air-filled porosity on $O_2$ diffusion may reflect variation in the contribution of pore continuity to the measurements. The air-filled porosity, as a measure of aeration, has also been linked to cessation of rapid drainage (Thomasson, 1978). Under the soil conditions in England and Wales, Thomasson (1978) found that the water potential of mineral topsoils approached a value of –0.005 MPa as rapid drainage ceased (this potential corre-sponds to pores with an equivalent diameter of > 60 µm being air-filled) and he referred to the air-filled porosity at this potential as aeration capacity. Soils with an aeration capacity > 0.15, 0.10–0.15, 0.05–0.10 and < 0.05 $m^3$ $m^{-3}$ were considered to have aeration characteristics that were very good, good, moderate and poor, respectively. He also noted that the aeration capacity requirement was dependent on climate and should be increased in wetter parts of the country. Aeration conditions that are limiting for plant growth will also vary with plant species and with the demands for $O_2$ by the soil microbial population.

Pore characteristics also determine the amount of water that is potentially available to plants and the rate at which water is transported to the growing roots. Thomasson (1978) defined potentially available water as that held between potentials of –0.005 and –1.5 MPa (corresponding to water held in pores with equivalent diameters of 60 to 0.2 µm) and considered values of 0.20, 0.15–0.20, 0.10–0.15 and < 0.10 $m^3$ $m^{-3}$ to be very good, good, moderate and poor, respectively. Once again, it was recognized that the qualitative ratings were climate dependent and the water contents should be adjusted downward in wetter regions.

Different sized pores dominate the influence of pores on aeration and on potentially available water and, therefore, there is value in trying to represent the effects of pore size distribution on both aeration and potentially available water. Thomasson (1978) applied qualitative assessment to both aeration and available water (Figure 7.9) to develop a classification of the structural quality of topsoils. Although the numerical values in such a representation may be climate dependent, there is obvious value in combining the two structural characteristics for the purpose of interpreting soil structure information in relation to plant growth. Perhaps the greatest limitation in this approach is that there is no direct provision for the influence of structure on the resistance of soil to penetration by roots. Root growth can occur in macropores which have dimensions similar to those of the root, but most growth involves deformation of the existing pore structure to accommodate the root. Root growth decreases as the resistance offered to penetration increases. Soil resistance to penetration has been measured with metal probes (penetrometers) with diameters ranging from 150 µm to about 1 cm. Neither the size, shape nor friction characteristics of these probes simulate roots very well, but the penetration resistance that is measured is well correlated with root length and elongation rate in soils with few macropores. However, in soils with an abundance of macropores, measurements of penetration resistance may be less relevant (Stypa et al., 1987). Although the functional dependence of root growth on penetration resistance depends on a number of factors including species (Materechera et al., 1991), root growth is often found to approach minimum values when the soil resistance is about 2 MPa (Taylor et al., 1966). Resistance to penetration increases with bulk density and decreasing water content.

Aeration, available water and soil strength have been incorporated into a single characteristic referred to as the least (or non) limiting water range (LLWR) (Letey, 1985). The LLWR is defined

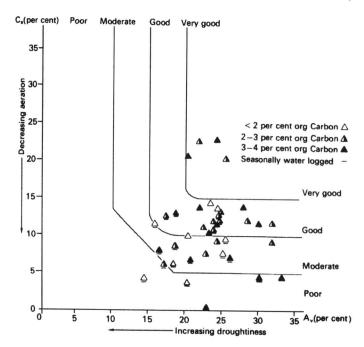

**FIGURE 7.9** A classification of structural quality of topsoils based on aeration ($C_a$) and available water ($A_v$) [Reprinted from Thomasson, 1978. *J. Soil Sci.* 29:38-46, with permission from Blackwell Science Ltd].

by water contents at which aeration, water potential and mechanical impedance reach values that are critical or limiting to plant growth. The upper limit of this range is defined by the water content at field capacity or the water content at which aeration becomes limiting, whichever is smaller. The lower limit is defined by the water content at the permanent wilting point or the water content at which penetration resistance becomes limiting, whichever is higher. The LLWR integrates many of the characteristics of pores into a single parameter and does so in a way that is directly related to plant growth. Using limiting values for aeration and penetration resistance of 0.10 $m^3$ $m^{-3}$ and 2.0 MPa, respectively, and the potential at field capacity and permanent wilting point of 0.01 and 1.5 MPa, respectively, da Silva and Kay (1997b) calculated the LLWR for different soils under humid temperate conditions in southwestern Ontario, Canada, and found that the growth of corn plants decreased linearly with increasing frequency of the soil water content falling outside of the LLWR. Under similar climatic conditions, the frequency of this occurrence would be expected to increase as the LLWR gets smaller in soils that drain freely, and was found to follow a logistic type function (da Silva and Kay, 1997c). These data could be used to define the quality of soil structure for plant growth using classes of the LLWR based on classes for available water given by Thomasson (1978) (Table 7.3). Studies in Australian orchards (Emerson et al., 1994), using the same limiting values for aeration, available water and resistance to penetration as used by da Silva and Kay (1987b), showed that small values of the LLWR coincided with a paucity of peach roots.

The value of using the LLWR to interpret the effects of soil structure on plant growth rather than using one of aeration capacity, potentially available water or soil resistance to penetration is illustrated in a survey of eight Canadian soils (Topp et al., 1994). Over 90% of the horizons tested developed a penetrometer resistance > 2 MPa at water potentials higher (less negative) than –1.5 MPa and nearly 50% of the horizons had aeration limitations at field capacity. Preliminary studies (McKenzie et al., 1988) on a self-mulching Vertisol under cotton production in New South Wales, Australia, suggested, however, that standard techniques to measure aeration and strength on soils with vertical macropores may lead to anomalous interpretations of the LLWR. Additional complications of the use of LLWR to interpret the effects of soil structure on plant growth arise when there is little or no correlation between the frequency that the water content falls outside of the

**TABLE 7.3**
**Illustration of the application of classes of available water to the least limiting water range (LLWR) and normalized growth rates (Growth rate at a give value of LLWR divided by the maximum growth rate) Calculated from da Silva and Kay (1997b).**

| Structural Quality | Least Limiting Water Range (m³ m⁻³) | Normalized Growth Rate |
|---|---|---|
| very good | >0.20 | >0.97 |
| good | 0.15–0.20 | 0.93–0.97 |
| moderate | 0.10–0.15 | 0.82–0.93 |
| poor | <0.10 | <0.83 |

LLWR and the magnitude of the LLWR. This condition would be expected to exist when the water content is largely controlled by water flow characteristics at other depths in the profile (Gardner et al., 1984). On the basis of these studies, the LLWR merits further evaluation as a measure of the structural quality of soils for crop production.

---

**EXAMPLE 7.2** Discussion of choice of bulk densities in designing laboratory studies to assess the influence of texture and organic carbon contents on biological processes.

Researchers who wish to assess the influence of texture and organic carbon content on specific biological processes in controlled environments often use soil that has been sieved. Under such circumstances, the researcher must make a decision about the densities to which the soil must be packed. Options include packing all soils to the same bulk density, packing each soil to the density observed in the field at the time of sampling and packing all soils to the same relative compaction. Discuss the merits of each option. Assume that the studies will be done with all soils held at the same water potential.

Issues to be taken into account include:

bulk density varies with texture, organic carbon, soil and crop management practices, and if the soil has any potential to shrink and swell, the bulk density also varies with water content (or water potential)

relative compaction is much less dependent on texture and organic carbon than is bulk density

bulk densities of different soils calculated from a constant relative compaction will reflect the variation in bulk densities arising from management practices that give rise to similar compaction

---

## 7.4.2 Carbon and Nutrient Cycling

Soil structure is a dominant control of microbially mediated processes in soils (van Veen and Kuikman, 1990; Juma, 1993; Ladd et al., 1996; Golchin et al., 1998). Soil structure controls C and nutrient cycling through its influence on habitable pore space and soil water content (Killham et al., 1993). Soil structure is characterized by its high spatial heterogeneity even on a very small scale, which contributes to heterogeneity in microbial activities. As pointed out by van Veen and Kuikman (1990), large amounts of readily decomposable compounds can be found in the vicinity of starving microbial populations, suggesting that organic matter may be rendered inaccessible to decomposers through chemical and physical mechanisms. The latter is often referred to as physical protection.

In many soils, a large proportion of the pores is too small to be accessible to microbes (Elliott et al., 1980; van Veen and Kuikman, 1990), and the SOM located therein may be physically protected from decomposition. Assessment of critical pore neck sizes necessary to constrain specific biological activities is difficult considering the wide diversity of organisms involved in the processes of C and nutrient turnover in soils (Ladd et al., 1996). A few attempts have been made to measure the influence of soil pore characteristics on specific microbial activities. Killham et al. (1993) observed that glucose turned over faster when incorporated into larger pores (6–30 μm) than into smaller pores (<6 μm). Compaction, which reduces total pore volume and increases the percentage of small pores less accessible to decomposers, may retard the decomposition of labile C compounds (van der Linden et al., 1989 cited by van Veen and Kuikman, 1990) and may reduce N mineralization from plant residues (Breland and Hansen, 1996).

The biological formation and stabilization of soil aggregates also provide stability to the organic compounds located inside aggregates. Indirect evidence for this feedback mechanism is found in experiments showing slower turnover of organic molecules in aggregated than in nonaggregated soils (Adu and Oades, 1978). Encrustation of plant residues with minerals provides protection from decomposition (Golchin et al., 1994). Within aggregates, POM had a slower apparent decomposition rate than free POM (Besnard et al., 1996; Gregorich et al., 1997) due to physical protection by the aggregates. The formation of stable aggregates was held partly responsible for the accrual of stable C observed in restored prairie soils or under conservation management (Jastrow, 1996; Beare et al., 1997; Angers and Chenu, 1998). Disruption of soil aggregates usually leads to a flush in C and N mineralization, which has been attributed to the release or increased accessibility of physically protected SOM (Rovira and Greacen, 1957; Hassink, 1992; Beare et al., 1994a). Further discussion on mechanisms providing recalcitrancy to SOM in soils can be found in the reviews previously cited. Although soil structure undoubtedly influences carbon and nutrient cycling, the heterogeneous nature of soil structure and the complexity of the relationships between organic compounds and the mineral material do not yet allow definitive interpretations and generalizations in the form of models.

---

**EXAMPLE 7.3**   Discussion of role of structure on decomposition of organic residues in the soil.

Organic matter is supplied to the soil either in the form of crop residues and roots or as organic amendments. When incorporated into the soil, organic materials undergo decomposition by the soil microfauna and microflora. Decomposition involves the biochemical transformation of the residues into more stable organic matter and the mineralization of the carbon as $CO_2$. Decomposition of organic residues in the soil is controlled by various physical, biological and chemical factors. Discuss the possible role of soil structure (both form and stability) in controlling the decomposition of organic residues.

Issues to be taken into consideration include:

the location of fresh organic residues can vary and has an impact on their decomposition

colonization of the residues by soil microorganisms depends on the degree and extent of contact between soil particles and the residues

decomposition is most often aerobic and an adequate oxygen supply is therefore necessary

the diffusion coefficient of oxygen is approximately $10^4$ greater in air than in water

sealing of the soil surface can reduce the supply of oxygen

the decomposing activity of soil microorganisms is greatly influenced by moisture, temperature and the supply of nutrients

when fresh organic residues are incorporated in the soil, their decomposition starts quickly and the residues get incrusted with mineral particles to form aggregates

## 7.4.3 EROSION

Very early on erosion by both wind and water was recognized to be largely dependent on the degree and stability of soil aggregation. Erosion by water is determined by a number of factors related to soil and climatic conditions, and land management. Soil erodibility, which defines the intrinsic susceptibility of soils to erosion, is largely a function of soil aggregate stability, which is in turn influenced by intrinsic soil properties. As mentioned earlier, aggregate stability was originally used to characterize erodibility (Yoder, 1936). Increasing the aggregate stability reduces the detachability of the particles and reduces the susceptibility to surface crusting, which are two factors favoring erosion. Due to the difficulty of making field erodibility measurements on a large scale, predictive tools based on easily measured soil properties are necessary. However, assessment of aggregate stability to predict water erodibility has not always been successful. This is due to the complexity of the processes involved and the numerous interactions among factors. Among those, the nature and rate of wetting of aggregates which should simulate natural processes are probably the most significant. As discussed earlier, methods involving rapid wetting are most relevant to simulate conditions of environments involving intense rains or quick irrigation, whereas slow wetting is more relevant to subsurface soil or in areas with less intense rainfall. le Bissonnais and Arrouays (1997) illustrated the close relationship that exists between the proportion of water-stable aggregates, soil organic matter and the susceptibility to crust formation and erodibility of French loamy soils. Their results emphasized the importance of wetting procedures in assessing erodibility from aggregate stability measurements. Their best relationship was with stability measurements involving slow wetting. They proposed that rapid wetting would be more appropriate in discriminating between soils of different stability.

Wind erodibility is directly influenced by the size, shape and density of the structural units, and by their mechanical stability (Chepil and Woodruff, 1963). Predictions of wind erodibility from soil structural properties have been attempted. The Wind Erosion Equation relates proportion of aggregates $> 0.84$ µm and wind erodibility (Woodruff and Siddoway, 1965). Black and Chanasyk (1989) showed that wind erosion was difficult to predict from intrinsic soil properties because of the great variability in the amount of dry aggregates $> 0.84$ µm due to management and in particular tillage. Using a different dry soil aggregate stability test for wind erodibility based on the resistance of clods to mechanical stress, Skidmore and Layton (1992) could predict stability from clay or water content.

Although there is no doubt that water and wind erodibility are directly related to soil structural form and stability, our ability to predict and model erodibility from soil structural information is still limited.

## REFERENCES

Acton, C.J., D.A. Rennie, and E.A. Paul. 1963. Dynamics of soil aggregation, *Can. J. Soil Sci.,* 43:201–209.

Adu, J.R. and J.M. Oades. 1978. Utilization of organic materials in soil aggregates by bacteria and fungi, *Soil Biol. Biochem.,* 10:117–122.

Ahuja, L.R., D.G. Decoursey, B.B. Barnes, and K. Rojas. 1993. Characteristics of macropore transport studied with the ARS Root Zone Water Quality Model, Trans. ASAE 36:369–380.

Albrecht, A., L. Rangon, and P. Barret. 1992. Effets de la matière organique sur la stabilité structurale et la détachabilité d'un vertisol et d'un ferrisol (Martinique). Cah. ORSTOM, sér. Pédol. 27:121–133.

Allmaras, R.R., S.M. Copeland, P.J. Copeland, and M. Oussible. 1996. Spatial relations between oat residue and ceramic spheres when incorporated sequentially by tillage, *Soil Sci. Soc. Am. J.,* 60:1209–1216.

Anderson, S.H., C.J. Gantzer, and J.R. Brown. 1990. Soil physical properties after 100 years of continuous cultivation, *J. Soil Water Cons.,* 45:117–121.

Angers, D.A. 1990. Compression of agricultural soils from Québec, *Soil Till. Res.,* 18:357–365.

Angers, D.A. 1992. Changes in soil aggregation and organic C under corn and alfalfa, *Soil Sci. Soc. Am. J.,* 56:1244–1249.

Angers, D.A. and R.R. Simard. 1986. Relation entre la teneur en matière organique et la masse volumique apparente du sol, *Can. J. Soil Sci.,* 66:743–746.

Angers, D.A. and G.R. Mehuys. 1988. Effects of cropping on macroaggregation of a marine clay soil, *Can. J. Soil Sci.,* 68:723–732.

Angers, D.A. and G.R. Mehuys. 1993. Aggregate stability to water, p. 651–657, *in* M.R. Carter (ed.), Manual on soil sampling and methods of analysis, CRC Press, Boca Raton, FL.

Angers, D.A. and M.R. Carter. 1996. Aggregation and organic matter storage in cool, humid agricultural soils. p. 193–211, *in* M.R. Carter and B.A. Stewart (eds.), Structure and organic matter storage in agricultural soils, CRC Press, Boca Raton, FL.

Angers, D.A. and C. Chenu. 1998. Dynamics of soil aggregation and C sequestration, p. 199–206, *in* R. Lal et al. (ed.), Soil processes and the carbon cycle. CRC Press, Boca Raton, FL.

Angers, D.A. and J. Caron. 1998. Plant-induced changes in soil structure: processes and feedbacks, *Biogeochem.,* 42:55–72.

Angers, D.A., B.D. Kay, and P.H. Groenevelt. 1987. Compaction characteristics of a soil cropped to corn and bromegrass, *Soil Sci. Soc. Am. J.,* 51:779–783.

Angers, D.A., N. Samson, and A. Légère. 1993. Early changes in water-stable aggregation induced by rotation and tillage in a soil under barley production, *Can. J. Soil Sci.,* 73:51–59.

Baldock, J.A. and B.D. Kay. 1987. Influence of cropping history and chemical treatments on the water-stable aggregation of a silt loam soil, *Can. J. Soil Sci.,* 67:501–511.

Ball, B.C. 1981. Pore characteristics of soils from two cultivation experiments as shown by gas diffusivities and permeabilities and air-filled porosities, *J. Soil Sci.,* 32:483–498.

Ball, B.C. and E.A.G. Robertson. 1994. Effects of uniaxial compaction on aeration and structure of ploughed or direct drilled soils, *Soil Till. Res.,* 31:135–149.

Ball, B.C. and K.A. Smith. 1991. Gas movement, p. 511–549, *in* K.E. Smith and C.E. Mullins (eds.), *Soil Analysis,* Marcel Dekker, Inc., NY.

Ball, B.C., M.V. Cheshire, E.A.G. Robertson, and E.A. Hunter. 1996. Carbohydrate composition in relation to structural stability, compactibility and plasticity of two soils in a long-term experiment, *Soil Till. Res.,* 39:143–160.

Barberis, E., F.A. Marsan, V. Boero, and E. Arduino. 1991. Aggregation of soil particles by iron oxides in various size fractions of soil B horizons, *J. Soil Sci.,* 42:535–42.

Barley, K.P. 1954. Effects of root growth and decay on the permeability of a synthetic sandy loam, *Soil Sci.,* 78:205–211.

Bartoli, F., R. Philippy, and G. Burtin. 1988. Aggregation in soils with small amounts of swelling clays: Aggregate stability, *J. Soil Sci.,* 39:593–616.

Bartoli, F., G. Burtin, and J. Guérif. 1992. Influence of organic matter on aggregation in Oxisols rich in gibbsite or in geothite. II. Clay dispersion, aggregate strength and water stability, *Geoderma,* 54:259–274.

Barzegar, A.R., P. Rengasamy, and J.M. Oades. 1995. Effects of clay type and rate of wetting on the mellowing of compacted soils, *Geoderma,* 68:39–49.

Beare, M.H., P.F. Hendrix, and D.C. Coleman. 1994a. Water-stable aggregates and organic matter fractions in conventional and no-tillage soils, *Soil Sci. Soc. Am. J.,* 58:777–786.

Beare, M.H., M.L. Cabrera, P.F. Hendrix, and D.C. Coleman. 1994b. Aggregate-protected and unprotected organic matter pools in conventional and no-tillage soils, *Soil Sci. Soc. Am. J.,* 58:787–795.

Beare, M.H., S. Hu, D.C. Coleman, and P.F. Hendrix. 1997. Influences of mycelial fungi on soil aggregation and organic matter storage in conventional and no-tillage soils, *Appl. Soil Ecol.,* 5:211–219.

Besnard, E., C. Chenu, J. Balesdent, P. Puget, and D. Arrouays. 1996. Fate of particulate organic matter in soil aggregates during cultivation, *Eur. J. Soil Sci.,* 47:495–503.

Bishop, A. and G.E. Blight. 1963. Some aspects of effective stress in saturated and partly saturated soils, *Geotech.,* 13:177–197.

Black, J.M.W. and D.S. Chanasyk. 1989. The wind erodibility of some Alberta soils after seeding: Aggregation in relation to field parameters, *Can. J. Soil Sci.,* 69:835–847.

Blackwell, P.S., T.W. Green, and W.K. Mason. 1990a. Responses of biopore channels from roots to compression by vertical stresses, *Soil Sci. Soc. Am. J.,* 54:1088–91.

Blackwell, P.S., A.J. Ringrose-Voase, N.S. Jayawardane, K.A. Olsson, D.C. McKenzie, and W.K. Mason. 1990b. The use of air-filled porosity and intrinsic permeability to air to characterize structure of macropore space and saturated hydraulic conductivity of clay soils, *J. Soil Sci.,* 41:215–228.

Blanchart, E. 1992. Restoration by earthworms (Megascolecidae) of the macroaggregate structure of a destructed savanna soil under field conditions, *Soil Biol. Biochem.*, 24:1587–1594.

Blanchart, E., P. Lavelle, E. Braudeau, Y. le Bissonnais, and C. Valentin. 1997. Regulation of soil structure by geophagous earthworm activities in humid savannas of Côte d'Ivoire, *Soil Biol. Biochem.*, 29:431–439.

Bouché, M.B. and F. Al-Addan. 1997. Earthworms, water infiltration and soil stability: some new assessments, *Soil Biol. Biochem.*, 29:441–452.

Bowen, G.D. and A.D. Rovira. 1991. The rhizosphere, p. 641–669, *in* Y. Waisel et al. (ed.), Plant roots. The hidden half, Marcel Dekker, Inc., NY.

Bowen, H.D. 1981. Alleviating mechanical impedance, p. 21–53, *in* G.F. Arkin and H.M. Taylor (eds.), Modifying the root environment to reduce stress, *Am. Soc. Agr. Eng.*, St. Joseph, MI.

Bradford, J.M. and S.C. Gupta. 1986. Compressibility, p. 479–492, *in* A.L. Page (ed.), Methods of soil analysis. Part 1. Physical and mineralogical methods, *Soil Sci. Soc. Am.*, Madison, WI.

Braunack, M.V. and A.R. Dexter. 1989. Soil aggregation in the seedbed: a review. 2. Effect of aggregate sizes on plant growth, *Soil Till. Res.*, 14:281–298.

Braunack, M.V., J.S. Hewitt, and A.R. Dexter. 1979. Brittle fracture of soil aggregates and the compaction of aggregate beds, *J. Soil Sci.*, 30:653–67.

Breland, T.A. and S. Hansen. 1996. Nitrogen mineralization and microbial biomass as affected by soil compaction, *Soil Biol. Biochem.*, 28:655–663.

Brewer, R. and A.D. Haldane. 1957. Preliminary experiments in the development of clay orientation in soils, *Soil Sci.*, 84:301–9.

Bruand, A., I. Cousin, B. Nicoullaud, O. Duval, and J.C. Bégon. 1996. Backscatter electron scanning images of soil porosity for analyzing soil compaction around roots, *Soil Sci. Soc. Am. J.*, 60:895–901.

Bullock, M.S., W.D. Kemper, and S.D. Nelson. 1988. Soil cohesion as affected by freezing, water content, time and tillage, *Soil Sci. Soc. Am. J.*, 52:770–776.

Cambardella, C.A. and E.T. Elliott. 1993. Carbon and nitrogen in aggregates from cultivated and native grassland soils, *Soil Sci. Soc. Am. J.*, 57:1071–1076.

Campbell, C.A., D. Curtin, S. Brandt, and R.P. Zentner. 1993. Soil aggregation as influenced by cultural practices in Saskatchewan: II. Brown and dark brown Chernozemic soils, *Can. J. Soil Sci.*, 73:597–612.

Canada Expert Committee on Soil Survey. 1987. The Canadian system of soil classification. Agriculture Canada Publ. 1646. Canadian Government Publication Centre, Ottawa, Canada.

Capriel, P., T. Beck, H. Borchert, and P. Harter. 1990. Relationship between soil aliphatic fraction extracted with supercritical hexane, soil microbial biomass, and soil aggregate stability, *Soil Sci. Soc. Am. J.*, 54:415–420.

Caron, J. and B.D. Kay. 1992. Rate of response of structural stability to a change in water content: Influence of cropping history, *Soil Till. Res.*, 25:167–185.

Caron, J., B.D. Kay, and E. Perfect. 1992a. Short term decrease in soil structural stability following bromegrass establishment on a clay loam soil, *Plant Soil*, 145:121–130.

Caron, J., B.D. Kay, and J.A. Stone. 1992b. Improvement of structural stability of a clay loam with drying, *Soil Sci. Soc. Am. J.*, 56:1583–1590.

Caron, J., B.D. Kay, J.A. Stone, and R.G. Kachanoski. 1992c. Modeling temporal changes in structural stability of a clay loam soil, *Soil Sci. Soc. Am. J.*, 56:1597–1604.

Caron, J., O. Banton, D.A. Angers, and J.P. Villeneneuve. 1996a. Preferential bromide transport through a clay loam under alfalfa and corn, *Geoderma*, 69:175–191.

Caron, J., C.R. Espindola, and D.A. Angers. 1996b. Soil structural stability during rapid wetting: Influence of land use on some aggregate properties, *Soil Sci. Soc. Am. J.*, 60:901–908.

Carter, M.R. 1988. Temporal variability of soil macroporosity in a fine sandy loam under molboard ploughing and direct drilling, *Soil Till. Res.*, 12:37–51.

Carter, M.R. 1990a. Relative measures of soil bulk density to characterize compaction in tillage studies on fine sandy loams, *Can. J. Soil Sci.*, 70:425–433.

Carter, M.R. 1990b. Relationship of strength properties to bulk density and macroporosity in cultivated loamy sand to loam soils, *Soil Till. Res.*, 15:257–268.

Carter, M.R. 1992. Influence of reduced tillage systems on organic matter, microbial biomass, macro-aggregate distribution and structural stability of the surface soil in a humid climate, *Soil Till. Res.*, 23:361–372.

Carter, M.R. and B.C. Ball. 1993. Soil porosity, p. 581–588, *in* M.R. Carter (ed.), Manual on soil sampling and methods of analysis, CRC Press, Boca Raton, FL.

Carter, M.R. and H.T. Kunelius. 1993. Effect of undersowing barley with annual ryegrasses or red clover on soil structure in a barley-soybean rotation, *Agric. Ecosys. Environ.,* 43:245–254.

Carter, M.R., D.A. Angers, and H.T. Kunelius. 1994. Soil structural form and stability, and organic matter under cool-season perennial grasses, *Soil Sci. Soc. Am. J.,* 58:1194–1199.

Cassel, D.K. 1983. Spatial and temporal variability of soil physical properties following tillage of Norfolk loamy sand, *Soil Sci. Soc. Am. J.,* 47:196–201.

Chan, K.Y. 1995. Strength characteristics of a potentially hardsetting soil under pasture and conventional tillage in the semi-arid region of Australia, *Soil Till. Res.,* 34:105–113

Chan, K.Y. and A.S. Hodgson. 1984. Moisture regimes of a cracking clay used for cotton production, *Rev. Rural Sci.,* 5:176–180.

Chaney, K. and R.S. Swift. 1984. The influence of organic matter on aggregate stability in some British soils, *J. Soil Sci.,* 35:223–230.

Chaney, K. and R.S. Swift. 1986a. Studies on aggregate stability. I. Re-formation of soil aggregates, *J. Soil Sci.,* 37:329–335

Chaney, K. and R.S. Swift. 1986b. Studies on aggregate stability. II. The effect of humic substances on the stability of re-formed soil aggregates, *J. Soil Sci.,* 37:337–343.

Chantigny, M.H., D.A. Angers, D. Prévost, L.P. Vézina, and F.P. Chalifour. 1997. Soil aggregation, and fungal and bacterial biomass under annual and perennial cropping systems, *Soil Sci. Soc. Am. J.,* 61:262–267.

Chartres, C.J. and J.D. Fitzgerald. 1990. Properties of siliceous cements in some Australian soils and saprolites, p.199–205, *in* L.A. Douglas (ed.), Soil micromorphology, Elsevier Science Publishers, Amsterdam, Netherlands.

Chen, Y., A. Banin, and A. Borochovitch. 1983. Effect of potassium on soil structure in relation to hydraulic conductivity, *Geoderma,* 30:135–47.

Chenu, C. 1989. Influence of a fungal polysaccharide, scleroglucan, on clay microstructures, *Soil Biol. Biochem.,* 21:299–305.

Chenu, C. and J. Guérif. 1991. Mechanical strength of clay minerals as influenced by an adsorbed polysaccharide, *Soil Sci. Soc. Am. J.,* 55:1076–1080.

Chepil, W.S. and N.P. Woodruff. 1963. The physics of wind erosion and its control, *Adv. Agron.,* 15:211–302.

Cheshire, M.V., G.P. Sparling, and C.M. Mundie. 1983. Effect of periodate treatment of soil on carbohydrate constituents and soil aggregation, *J. Soil Sci.,* 34:105–112.

Churchman, G.J. and K.R. Tate. 1986. Effect of slaughterhouse effluent and water irrigation upon aggregation in seasonally dry New Zealand soil under pasture, *Aust. J. Soil Res.,* 24:505–516.

Churchman, G.J., J.O. Skjemstad, and J.M. Oades. 1993. Influence of clay minerals and organic matter on effects of sodicity on soils, *Aust. J. Soil Res.,* 31:779–800.

Clapp, C.E. and W.W. Emerson. 1965. The effect of periodate oxidation on the strength of soil crumbs, *Soil Sci. Soc. Am. Proc.,* 29:127–134.

Clapp, C.E., R.J. Davis, and S.H. Waugaman. 1962. The effect of rhizobial polysaccharides on aggregate stability, *Soil Sci. Soc. Am. Proc.,* 26:466–469.

Collis-George, N. and B.S. Figueroa. 1984. The use of high energy moisture characteristic to assess soil stability, *Aust. J. Soil Res.,* 22:349–56.

Combeau, A. 1965. Variations saisonnières de la stabilité structurale du sol en région tempérée. Comparaison avec la zone tropicale, Cah. ORSTOM Sér. Pédol. 3:123–135.

Concaret, J. 1967. Etude des mécanismes de la destruction des agrégats de terre au contact de solutions aqueuses, *Ann. Agron.,* 18:99–144.

Crawford, J.W. 1994. The relationship between structure and the hydraulic conductivity of soil, *Eur. J. Soil Sci.,* 45:493–502.

Crawford, J.W., N. Matsui, and I.M. Young. 1995. The relation between the moisture release curve and the structure of soil, *Eur. J. Soil Sci.,* 46:369–375.

Czeratzki, W. and H. Frese. 1958. Importance of water in formation of soil structure. Hwy. Res. Bd. Spec. Rep. 40:200–211, Highway Research Board, Washington, DC.

da Silva, A. and B.D. Kay. 1997a. Estimating the least limiting water range of soils from properties and management, *Soil Sci. Soc. Am. J.,* 61:877–883.

da Silva, A. and B.D. Kay. 1997b. The sensitivity of shoot growth of corn to the least limiting water range of soils, *Plant Soil,* 184:323–329.

da Silva, A. and B.D. Kay. 1997c. Effect of soil water content variation on the least limiting water range, *Soil Sci. Soc. Am. J.,* 61:994–888.

Dalal, R.C. and B.J. Bridge. 1996. Aggregation and organic matter storage in sub-humid and semi-arid soils, p. 263–307, *in* M.R. Carter and B.A. Stewart (eds.), Structure and organic matter storage in agricultural soils, CRC Press, Boca Raton, FL.

Danielson, R.E. and L.P. Sutherland. 1982. Porosity, p. 443–461, *in* A. Klute (ed.), Methods of soil analysis. Part 1. Physical and mineralogical methods, *Soil Sci. Soc. Am.,* Madison, WI.

Davies, P. 1985. Influence of organic matter content, moisture status and time after reworking on soil shear strength, *J. Soil Sci.,* 36:299–306.

Degens, B.P. 1997. Macro-aggregation of soils by biological bonding and binding mechanisms and the factors affecting these: a review, *Aust. J. Soil Res.,* 35:431–459.

Degens, B.P., G.P. Sparling, and L.K. Abbott. 1996. Increasing the length of hyphae in a sandy soil increases the amount of water-stable aggregates, *Appl. Soil Ecol.,* 3:149–159.

Derdour, H., D.A. Angers, and M.R. Laverdière. 1993. Caractérisation de l'space poral d'un sol argileux: Effets des ses constituants et du travail du sol, *Can. J. Soil Sci.,* 73:299–307.

Dexter, A.R. 1978. Tunnelling of soil by earthworms, *Soil Biol. Biochem.,* 10:447–449.

Dexter, A.R. 1988. Advances in the characterization of soil structure, *Soil Till. Res.,* 11:199–238.

Dexter, A.R., D. Hein, and J.S. Hewitt. 1982. Macro-structure of the surface layer of a self-mulching clay in relation to cereal stubble management, *Soil Till. Res.,* 2:251–64.

Dinel, H., P.E.M. Lévesque, and G.R. Mehuys. 1991. Effects of beeswax, a naturally occurring source of long-chain aliphatic compounds on the aggregate stability of lacustrine silty clay, *Soil Sci.,* 151:228–239.

Dinel, H., P.E.M. Lévesque, P. Jambu, and D. Righi. 1992. Microbial activity and long-chain aliphatics in the formation of stable soil aggregates, *Soil Sci. Soc. Am. J.,* 56:1455–1463.

Dorioz, J.M., M. Robert, and C. Chenu. 1993. The role of roots, fungi and bacteria on clay particle organization. An experimental approach, *Geoderma,* 56:179–194.

Dormaar, J.F. and R.C. Foster. 1991. Nascent aggregates in the rhizosphere of perennial ryegrass (*Lolium perenne* L.), *Can. J. Soil Sci.,* 71:465–474.

Dormaar, J.F., C.W. Lindwall, and G.C. Kosub. 1988. Effectiveness of manure and commercial fertilizer in restoring productivity of an artificially eroded dark brown chernozemic soil under dryland conditions, *Can. J. Soil Sci.,* 68:669–679.

Douglas, J.T. 1986. Effects of season and management on the vane shear strength of a clay topsoil, *J. Soil Sci.,* 37:669–679.

Douglas, J.T., M.G. Jarvis, K.R. Howse, and M.J. Goss. 1986. Structure of a silty soil in relation to management, *J. Soil Sci.,* 37:137–51.

Dufey, J.E., H. Halen, and R. Frankart. 1986. Evolution de la stabilité structurale du sol sous l'influence des racines de trèfle (*Trifolium pratense* L.) et de ray-grass (*Lolium multiflorum* Lmk.). Observations pendant et après culture, *Agronomie,* 6:811–817.

Edwards, C.A. and J.R. Lofty. 1978. The influence of arthropods and earthworms upon root growth of direct drilled cereals, *J. Appl. Ecol.,* 15:789–795.

Edwards, W.M., L.D. Norton, and C.E. Redmond. 1988. Characterizing macropores that affect infiltration into nontilled soil, *Soil Sci. Soc. Am. J.,* 52:483–487.

Ehlers, W. 1975. Observation on earthworm channels and infiltration on tilled and untilled loess soils, *Soil Sci.,* 119:242–249.

Ehlers, W. 1977. Measurement and calculation of hydraulic conductivity in horizons of tilled and untilled loess-derived soil, Germany, *Geoderma,* 19:293–306.

Ehlers, W., O. Wendroth, and F. de Mol. 1995. Characterizing pore organization by soil physical parameters, p. 257–75, *in* K.H. Hartge and B.A. Stewart (eds.), Soil structure: Its development and function, Lewis Publishers, Boca Raton, FL.

Elliott, E.T. 1986. Aggregate structure and carbon, nitrogen and phosphorus in native and cultivated soils, *Soil Sci. Soc. Am. J.,* 50:627–633.

Elliott, E.T., R.V. Anderson, D.C. Coleman, and C.V. Cole. 1980. Habitable pore space and microbial trophic interactions, *Oikos,* 35:327–335.

Elustondo, J., D.A Angers, M.R. Laverdière, and A. N'Dayegamiye. 1990. Étude comparative de l'agrégation et de la matière organique associée aux fractions granulométriques de sept sols sous culture de maïs ou en prairie, *Can. J. Soil Sci.,* 70:395–402.

Emerson, W.W. 1977. Physical properties and structure, p. 78–104, *in* J.S. Russel and E.L. Greacen (eds.), Soil factors in crop production in a semi-arid environment, Queensland University Press, Brisbane, Australia.

Emerson, W.W. 1983. Inter-particle bonding, p. 477–98, *in* Soils: An Australian viewpoint. Academic Press, London, UK.

Emerson, W.W., R.C. Foster, J.M. Tisdall, and D. Weissmann. 1994. Carbon content and bulk density of an irrigated Natrixeralf in relation to tree root growth and orchard management, *Aust. J. Soil Res.,* 32:939–951.

Emerson, W.W. 1995. Water retention, organic C and soil texture, *Aust. J. Soil Res.,* 33:241–251.

Fausey, N.R. and R. Lal. 1989. Drainage-tillage effects on Crosby-Kokomo soil association in Ohio, *Soil Tech.,* 2:359–70.

Fausey, N.R. and R. Lal. 1992. Drainage-tillage effects on a Crosby-Kokomo soil association in Ohio, 3. Organic matter content and chemical properties, *Soil Tech.,* 5:1–12.

Feller, C., A. Albrecht, and D. Tessier. 1996. Aggregation and organic matter storage in kaolinitc and smectitic tropical soils, p. 309–359, *in* M.R. Carter and B.A. Stewart (eds.), Structure and organic matter storage in agricultural soils, CRC Press, Boca Raton, FL.

Forster, S.M. 1990. The role of microorganisms in aggregate formation and soil stabilization: types of aggregation, *Arid Soil Res. Rehab.,* 4:85–98.

Foster, R.C. 1981. Polysaccharides in soil fabrics, *Science,* 214:665–667.

Foster, R.C. 1988. Microenvironments of soil microorganisms, *Biol. Fert. Soils,* 6:189–203.

Foster, R.C. and A.D. Rovira. 1976. Ultrastructure of wheat rhizosphere, *New Phytol.,* 76:343–352.

Fox, W.E. 1964. Cracking characteristics and field capacity in a swelling soil, *Soil Sci.,* 98:413.

Franzmeier, D.P., C.J. Chartres, and J.T. Wood. 1996. Hardsetting soils in southeast Australia: Landscape and profile processes, *Soil Sci. Soc. Am. J.,* 60:1178–87.

Gardner, E.A., R.J. Shaw, G.D. Smith, and K.J. Coughlan. 1984. Plant available water capacity: Concept, measurement and prediction, p.164–175, *in* J.W. McGarity, E.H. Hoult and H.B. So (eds.), The properties and utilization of cracking clay soils, University of New England, Armidale, Australia.

Garnier-Sillam, E., F. Toutain, and J. Renoux. 1988. Comparaison de l'influence de deux termitières (humivore et champignonniste) sur la stabilité structurale des sols forestiers tropicaux, *Pedopedologia,* 32:89–97.

Gifford, R.O. and D.F. Thran. 1974. Bonding mechanisms for soil crusts: Part II. Strength of silica cementation, *AZ Agric. Res. Sta. Tech. Bull.,* 214:28–32.

Giovannini, G. and P. Sequi. 1976. Iron and aluminum as cementing substances of soil aggregates: Changes in stability of soil aggregates following extraction of iron and aluminum by acetylacetone in a non-polar solvent, *J. Soil Sci.,* 27:148–153.

Giovannini, S. Lucchesi, and S. Cervelli. 1983. Water-repellent substances and aggregate stability in hydrophobic soil, *Soil Sci.,* 135:110–113.

Golchin, A., J.M. Oades, J.O. Skjemstad, and P. Clarke. 1994, Soil structure and carbon cycling, *Aust. J. Soil Res.,* 32:1043–1068.

Golchin, A., J.A. Baldock, and J.M. Oades. 1998. A model linking organic matter decomposition, chemistry and aggregate dynamics, p.245–266, *in* R. Lal et al. (eds.), Soil processes and the carbon cycle, CRC Press, Boca Raton, FL.

Goldberg, S. 1989. Interaction of aluminum and iron oxides and clay minerals and their effect on soil physical properties: A review, *Commun. Soil Sci. Plant Anal.,* 20:1181–207.

Gollany, H.T., T.E. Schumacher, P.D. Evanson, M.J. Lindstrom, and G.D. Lemme. 1991. Aggregate stability of an eroded and desurfaced Typic Haplustoll, *Soil Sci. Soc. Am. J.,* 55:811–816.

Goss, M.J., W. Ehlers, F.R. Boone, I. White, and K.R. Howse. 1984. Effects of soil management practice on soil physical conditions affecting root growth, *J. Agric. Eng. Res.,* 30:131–140.

Gouzou, L., G. Burtin, R. Philippy, F. Bartoli, and T. Heulin. 1993. Effect of inoculation with Bacillus polymyxa on soil aggregation in the wheat rhizosphere: preliminary examination, *Geoderma,* 56:479–491.

Greenland, D.J., G.R. Lindstrom, and J.P. Quirk. 1962. Organic materials which stabilize natural soil aggregates, *Soil Sci. Soc. Am. Proc.,* 26:366–371.

Greenland, D.J., D. Rimmer, and D. Payne. 1975. Determination of the structural stability class of English and Welsh soils using a water coherence test, *Soil Sci.,* 26:294–303.

Gregorich, E.G., C.F. Drury, B.H. Ellert, and B.C. Liang. 1997. Fertilization effects on physically protected light fraction organic matter, *Soil Sci. Soc. Am. J.,* 61:482–484.

Grevers, M.C.J. and E. de Jong. 1990. The characterization of soil macroporosity of a clay soil under ten grasses using image analysis, *Can. J. Soil Sci.,* 70:93–103.

Groenevelt, P.H. and B.D. Kay. 1981. On pressure distribution and effective stress in unsaturated soils, *Can. J. Soil Sci.,* 61:431–443.

Guckert, A., T. Chone, and F. Jacquin. 1975. Microflore et stabilité structurale du sol, *Rev. Ecol. Biol. Sol.,* 12:211–223.

Guérif, J. 1979. Rôle de la matière organique sur le comportement d'un sol au compactage. II. Matières organiques libres et liées, *Ann. Agron.,* 30:469–480.

Gupta, R.P. 1986. Criteria for physical rating of soils in relation to crop production, Proc. XIII Int. Cong, Soil Sci., Hamburg. 2:69–70.

Gupta, S.C., E.C. Schneider, W.E. Larson, and A. Hadas. 1987. Influence of corn residue on compression and compaction behavior of soils, *Soil Sci. Soc. Am. J.,* 51:207–212.

Gupta, V.V.S.R. and J.J. Germida. 1988. Distribution of microbial biomass and its activity in different soil aggregate size classes as affected by cultivation, *Soil Biol. Biochem.,* 20:777–786.

Guidi, G., G. Poggio, and G. Petruzelli. 1985. The porosity of soil aggregates from bulk soil and soil adhering to roots, *Plant Soil,* 87:311–314.

Gusli, S., A. Cass, D.A. MacLeod, and, P.S. Blackwell. 1994. Structural collapse and strength of some Australian soils in relation to hardsetting: l. Structural collapse on wetting and draining, *Eur. J. Soil Sci.,* 45:15–21.

Hadas, A. and D. Russo. 1974. Water uptake by seeds as affected by water stress, capillary conductivity, and seed-soil contact. II Analysis of experimental data, *Agron. J.,* 66:647–652.

Häkansson, I. 1990. A method for characterizing the state of compactness of the plough layer, *Soil Till. Res.,* 16:105–120.

Hardan, A. and A.N. Al-Ani. 1978. Improvement of soil structure by using date and sugar beet waste products, p. 305–308, *in* W.W. Emerson, R.D. Bond and A.R. Dexter (eds.), Modification of soil structure, A Wiley-Interscience Publication, Brisbane, Australia.

Harris, R.F., G. Chesters, and O.N. Allen. 1966. Dynamics of soil aggregation, *Adv. Agron.,* 18:107–169.

Hassink, J. 1992. Effects of soil texture and structure on C and N mineralization in grassland soils, *Biol. Fertil. Soils,* 14:126–134.

Haynes, R.J. and G.S. Francis. 1993. Changes in microbial biomass C, soil carbohydrates and aggregate stability induced by growth of selected crop and forage species under field conditions, *J. Soil Sci.,* 44:665–675.

Haynes, R.J. and R.S. Swift. 1990. Stability of soil aggregates in relation to organic constituents and soil water content, *J. Soil Sci.,* 41:73–83.

Haynes, R.J. and M.H. Beare. 1996. Aggregation and organic matter storage in meso-thermal, humid soils, p. 213–262, *in* M.R. Carter and B.A. Stewart (eds.), Structure and organic matter storage in agricultural soils, CRC Press, Boca Raton, FL.

Haynes, R.J., R.S. Swift, and R.C. Stephen. 1991. Influence of mixed cropping rotations (pasture-arable) on organic matter content, water-stable aggregation and clod porosity in a group of soils, *Soil Till. Res.,* 19:77–87.

Heard, J.R., E.J. Kladivko, and J.V. Mannering. 1988. Soil macroporosity, hydraulic conductivity and air permeability of silty soils under long-term conservation tillage in Indiana, *Soil Till. Res.,* 11:1–18.

Hénin, S., G. Monnier, and A. Combeau. 1958. Méthode pour l'étude de la stabilité struturale des sols, *Ann. Agron.,* 9:73–92.

Hodgson, A.S. and D.A. MacLeod. 1989a. Oxygen flux, air-filled porosity and bulk density as indices of vertisol structure, *Soil Sci. Soc. Am. J.,* 53:540–543.

Hodgson, A.S. and D.A. MacLeod. 1989b. Use of oxygen flux density to estimate critical air-filled porosity of a vertisol, *Soil Sci. Soc. Am. J.,* 53:355–61.

Hodgson, J.M. 1978. Soil sampling and soil description, Clarendon Press, Oxford, UK.

Hole, F.D. 1981. Effects of animals on soil, *Geoderma,* 25:75–112.

Horn, R. and A.R. Dexter. 1989. Dynamics of soil aggregation in a desert loess, *Soil Till. Res.,* 13:253–266.

Horn, R., H. Taubner, M. Wuttke, and T. Baumgartl. 1994. Soil physical properties related to soil structure, *Soil Till. Res.,* 30:187–216.

Hudson, B.D. 1994. Soil organic matter and available water capacity, *J. Soil Water Conser,* 49:189–194.

Hundal, S.S., G.O. Schwab, and G.S. Taylor. 1976. Drainage system effects on physical properties of a lakebed clay soil, *Soil Sci. Soc. Am. J.,* 40:300–305.

Ingles, O.G., 1965. The shatter test as an index of the strength of soil aggregates, p. 284–302, *in* C.J. Osborn (ed.), Proc. Tewksbury Symp. Fract. Faculty of Engineering, University of Melbourne, Melbourne, Australia.

Jakobsen, B. and A.R. Dexter. 1988. Influence of biopores on root growth, water uptake and grain yield of wheat. Predictions from a computer model, *Biol. Fert. Soils,* 6:315–321.

Jastrow, J.D. 1987. Changes in soil aggregation associated with tallgrass prairie restoration, *Am. J. Bot.,* 74:1656–1664.

Jastrow, J.D. 1996. Soil aggregate formation and the accrual of particulate and mineral-associated organic matter, *Soil Biol. Biochem.,* 45:665–676.

Juma, N.G. 1993. Interrelationships between soil structure/texture, soil biota/soil organic matter and crop production, *Geoderma,* 57:3–30.

Karlen, D.L., N.C. Wollenhaupt, D.C. Erbach, E.C. Berry, J.B. Swan, N.S. Eash, and J.L. Jordahl. 1994. Crop residue effects on soil quality following 10-years of no-till corn, *Soil Till. Res.,* 31:149–167.

Kay, B.D. 1990. Rates of change of soil structure under different cropping systems, *Adv. Soil Sci.,* 12:1–52.

Kay, B.D. 1998. Soil structure and organic C: A review, p. 169–197, *in* R. Lal et al. (eds.), Soil processes and the carbon cycle, CRC Press, Boca Raton, FL.

Kay, B.D. and A.R. Dexter. 1990. Influence of aggregate diameter, surface area and antecedent water content on the dispersibility of clay, *Can. J. Soil Sci.,* 70:655–671.

Kay, B.D., C.D. Grant, and P.H. Groenevelt. 1985. Significance of ground freezing on soil bulk density under zero tillage, *Soil Sci. Soc. Am. J.,* 49:973–978.

Kay, B.D., V. Rasiah, and E. Perfect. 1994a. Structural aspects of soil resiliency, p. 449–468, *in* D.J. Greenland and I. Szabolcs (eds.), Soil resilience and sustainable land use, CAB International, London, UK.

Kay, B.D., A.R. Dexter, V. Rasiah, and C.D. Grant. 1994b. Weather, cropping practices and sampling depth effects on tensile strength and aggregate stability, *Soil Till. Res.,* 32:135–148.

Kay, B.D., A.P. da Silva, and J.A. Baldock. 1997. Sensitivity of soil structure to changes in organic C content: predictions using pedotransfer functions, *Can. J. Soil Sci.,* 77:655–667.

Kemper W.D. and E.J. Koch. 1966. Aggregate stability of soils from the western portions of the United States and Canada. USDA Tech. Bull. 1355.

Kemper, W.D. and R.C. Rosenau. 1986. Aggregate stability and size distribution, p. 425–442, *in* A.L. Page (ed.), Methods of soil analysis. Part 1. Physical and mineralogical methods, *Soil Sci. Soc. Am.,* Madison, WI.

Kemper, W.D., R. Rosenau, and S. Nelson. 1985. Gas displacement and aggregate stability of soils, *Soil Sci. Soc. Am. J.,* 49:25–28.

Kemper, W.D., M.S. Bullock, and A.R. Dexter. 1989. Soil cohesion changes, p. 81–95, *in* W.E. Larson, G.R. Blake, R.R. Allmaras, W.B. Voorhees, and S.C. Gupta (eds.), Mechanics and related processes in structured agricultural soils, Kluwer Academic Publishers, Boston, MA.

Ketcheson, J.W. and E.G. Beauchamp. 1978. Effects of corn stover, manure, and nitrogen on soil properties and crop yield, *Agron. J.,* 70:792–797.

Killham, K., M. Amato, and J.N. Ladd. 1993. Effect of substrate location in soil and soil pore-water regime on carbon turnover, *Soil Biol. Biochem.,* 25:57–62.

Kleinfelder, D., S. Swanson, G. Norris, and W. Clary. 1992. Unconfined compressive strength of some streambank soils with herbaceous roots, *Soil Sci. Soc. Am. J.,* 56:1920–1925.

Klute, A. 1982. Tillage effects on the hydraulic properties of soil: A review, p. 29–43, *in* P. Unger and D.M. van Doren (eds.), Predicting tillage effects on soil physical properties and processes, *Soil Sci. Soc. Am.,* Madison, WI.

Kwaad, F.J.P.M. and H.J. Mucher. 1994. Degradation of soil structure by welding — A micromorphological study, *Catena,* 23:253–68.

Ladd, J.N., R.C. Foster, and J.M. Oades. 1996. Soil structure and biological activity, p. 23–78, *in* G. Stotzky and J.M. Bollag (eds.), *Soil Biochemistry,* Marcel Dekker Inc., NY.

Lafond, J., D.A. Angers, and M.R. Laverdière. 1992. Compression characteristics of a clay soil as influenced by crops and sampling dates, *Soil Till. Res.,* 22:233–241.

Lal, R. and N.R. Fausey. 1993. Drainage and tillage effects on a Crosby-Kokomo soil association in Ohio. 4. Soil physical properties, *Soil Tech.,* 6:123–35.

Lal, R., A.A. Mahboubi, and N.R. Fausey. 1994. Long-term tillage and rotation effects on properties of a central Ohio soil, *Soil Sci. Soc. Am. J.,* 57:517–522.

Larson, W.E., S.C. Gupta, and R.A. Useche. 1980. Compression of soils from eight soil orders, *Soil Sci. Soc. Am. J.,* 44:450–457.

le Bissonnais, Y. 1988. Comportement d'agrégats terreux soumis à l'action de l'eau: Analyse des mécanismes de désagrégation, *Agronomie,* 8:915–924.

le Bissonnais, Y. 1996. Aggregate stability and assessment of soil crustability and erodibility: I. Theory and methodology, *Eur. J. Soil Sci.,* 47:425–437.

le Bissonnais, Y. and D. Arrouays. 1997. Aggregate stability and assessment of soil crustabiltiy and erodibility: II. Application to humic loamy soils with various organic C contents, *Eur. J. Soil Sci.,* 48:39–48.

Lee, K.E. and R.C. Foster. 1991. Soil fauna and soil structure, *Aust. J. Soil Res.,* 29:745–775.

Letey, J. 1985. Relationship between physical properties and crop productions, *Adv. Soil Sci.,* 1:277–294.

Low, A.J. 1955. Improvements in the structural state of soil under leys, *J. Soil Sci.,* 6:177–199.

Lynch, J.M. and E. Bragg. 1985. Microorganisms and soil aggregate stability, *Adv. Soil Sci.,* 2:133–171.

MacRae, R.J. and G.R. Mehuys. 1987. Effects of green manuring in rotation with corn on the physical properties of two Quebec soils, *Biol. Agric. Hortic.,* 4:257–270.

Manrique, L.A. and C.A. Jones. 1991. Bulk density of soils in relation to soil physical and chemical properties, *Soil Sci. Soc. Am. J.,* 55:476–481.

Mapa, R.B., R.E. Green, and L. Santo. 1986. Temporal variability of soil hydraulic properties and wetting and drying subsequent to tillage, *Soil Sci. Soc. Am. J.,* 50:1133–38.

Marinissen, J.C.Y. and A.R. Dexter. 1990. Mechanisms of stabilization in earthworm casts and artificial casts, *Biol. Fertil. Soils.,* 9:163–167.

Marinissen, J.C.Y., E. Nijhuis, and N. van Breemen. 1996. Clay dispersibility in moist earthworm casts of different soils, *Appl. Soil Ecol.,* 4:83–92.

Marshall, T.J. and J.P. Quirk. 1950. Stability of structural aggregates of dry soil, *Aust. J. Agric. Sci.,* 1:266–275.

Martin, J.P. and S.A. Waksman. 1940. Influence of microorganisms on soil aggregation and erosion, *Soil Sci.,* 50:29–47.

Materechera, S.A., A.R. Dexter, and A.M. Alston. 1991. Penetration of very strong soils by seedling roots different plant species, *Plant Soil,* 135:31–41.

Materechera, S.A., A.R. Dexter, and A.M. Alston. 1992. Formation of aggregates by plant roots in homogenised soils, *Plant Soil,* 142:69–79.

Mazurak, A.P., L. Chesnin, and A. Amin Thijeel. 1977. Effects of beef cattle manure on water-stability of soil aggregates, *Soil Sci. Soc. Am. J.,* 41:613–615.

McBride, R.A. and G.C. Watson. 1990. An investigation of the re-expansion of unsaturated, structured soils during cyclic static loading, *Soil Till. Res.,* 17:241–53.

McDonald, D.C. and R. Julian. 1966. Quantitative estimation of soil total porosity and macro-porosity as part of the pedological description, *NZ J. Agric. Res.,* 8:927–946.

McGarry, D. and K.J. Smith. 1988. Indices of residual shrinkage to quantify the comparative effects of zero and mechanical tillage on a Vertisol, *Aust. J. Soil Res.,* 26:543–548.

McKeague, J.A., C. Wang, and G.C. Topp. 1982. Estimating saturated hydraulic conductivity from soil morphology, *Soil Sci. Soc. Am. J.,* 46:1239–1244.

McKeague, J.A., C.A. Fox, J.A. Stone, and R. Protz. 1987. Effects of cropping system on structure of Brookston clay loam in long-term experimental plots at Woodslee, Ontario, *Can. J. Soil Sci.,* 67:571–84.

McKenzie, D.C., P.J. Hulme, T.S. Abbott, D.A. MacLeod, and A. Cass. 1988. Vertisol structure dynamics following an irrigation of cotton, as influenced by prior rotation crops, p. 33–37, *in* K.J. Coughlan and P.N. Truong (eds.), Effects of management practices on soil physical properties, Proc. Natl. Worksh., Toowoomba, Queensland Department Primary Industries, Brisbane, Australia.

Meek, B.D., E.A. Rechel, L.M. Carter, and W.R. DeTar. 1989. Changes in infiltration under alfalfa as influences by time and wheel traffic, *Soil Sci. Soc. Am. J.,* 53:238–241.

Meek, B.D., W.R. De Tar, D. Rolph, E.R. Rechel, and L.M. Carter. 1990. Infiltration rate as affected by an alfalfa and no-till cotton cropping system, *Soil Sci. Soc. Am. J.,* 54:505–508.

Metting, B. 1987. Dynamics of wet and dry aggregate stability from a three-year microalgal soil conditioning experiment in the field, *Soil Sci.,* 143:139–143.

Metzger, L., D. Levanon, and U. Mingelgrin. 1987. The effect of sewage sludge on soil structural stability: microbiological aspects, *Soil Sci. Soc. Am. J.,* 51:346–351.

Miller, R.M. and J.D. Jastrow. 1990. Hierarchy of root and mycorrhizal fungal interactions with soil aggregation, *Soil Biol. Biochem.*, 5:579–584.

Mitchell, A.R. and M.T. van Genuchten. 1992. Shrinkage of bare and cultivated soil, *Soil Sci. Soc. Am. J.*, 56:1036–1042.

Mitchell, A.R., T.R. Ellsworth, and B.D. Meek. 1995. Effect of root systems on preferential flow in swelling soil, *Commun. Soil Sci. Plant Anal.*, 26:2655–2666.

Molope, M.B., I.C. Grieve, and E.R. Page. 1987. Contributions of fungi and bacteria to aggregate stability of cultivated soils, *J. Soil Sci.*, 38:71–77.

Monnier, G. 1965. Action des matières organiques sur la stabilité structurale des sols, *Ann. Agron.*, 16:327–400.

Monreal, C.M., M. Schnitzer, H.-R. Schulten, C.A. Campbell, and D.W. Anderson. 1995. Soil organic structures in macro- and microaggregates of a cultivated brown chernozem, *Soil Biol. Biochem.*, 27:845–853.

Moran, C.J., A.B. McBratney, and A.J. Koppi. 1989. A rapid method for analysis of soil macropore structure. I. Specimen preparation and digital binary image production, *Soil Sci. Soc. Am. J.*, 53:921–28.

Morel, J.L., L. Habib, S. Plantureux, and A. Guckert. 1991. Influence of maize root mucilage on soil aggregate stability, *Plant Soil*, 136:111–119.

N'Dayegamiye, A. and D.A. Angers. 1990. Effets de l'apport prolongé de fumier de bovins sur quelques propriétés physiques et biologiques d'un loam limoneux Neubois sous culture de maïs, *Can. J. Soil Sci.*, 70:259–262.

Norfleet, M.L. and A.D. Karathanasis. 1996. Some physical and chemical factors contributing to fragipan strength in Kentucky soils, *Geoderma*, 71:289–301.

Oades, J.M. 1984. Soil organic matter and structural stability mechanisms and implications for management, *Plant Soil*, 76:319–337.

Oades, J.M. 1993. The role of biology in the formation, stabilization and degradation of soil structure, *Geoderma*, 56:377–400.

Oades, J.M. and A.G. Waters. 1991. Aggregate hierarchy in soils, *Aust. J. Soil Res.*, 29:815–828.

Pagliai, M., G. Guidi, M. La Marca, M. Giachetti, and G. Lucanante. 1981. Effects of sewage sludges and compost on porosity and aggregation, *J. Envir. Qual.*, 10:556–561.

Panabokke, C.R. and J.P. Quirk. 1957. Effect of initial water content on the stability of soil aggregates in water, *Soil Sci.*, 83:185–189.

Pawluk, S. 1988. Freeze-thaw effects on granular structure reorganization for soil materials of varying texture and moisture content, *Can. J. Soil Sci.*, 68:485–94.

Perfect, E., B.D. Kay, W.K.P. van Loon, R.W. Sheard, and T. Pojasok. 1990a. Factors influencing soil structural stability within a growing season, *Soil Sci. Soc. Am. J.*, 54:173–179.

Perfect, E., W.K.P. van Loon, B.D. Kay, and P.H. Groenevelt. 1990b. Influence of ice segregation and solutes on soil structural stability, *Can. J. Soil Sci.*, 70:571–581.

Perfect, E., B.D. Kay, W.K.P. van Loon, R.W. Sheard, and T. Pojasok. 1990c. Rates of change in soil structural stability under forages and corn, *Soil Sci. Soc. Am. J.*, 54:179–186.

Perfect, E., B.D. Kay, J.A. Ferguson, A.P. da Silva, and K.A. Denholm. 1993. Comparison of functions for characterizing the dry aggregate size distribution of tilled soil, *Soil Till. Res.*, 28:123–139.

Piccolo, A. and J.S.C. Mbagwu. 1994. Humic substances and surfactants effects on the stability of two tropical soils, *Soil Sci. Soc. Am. J.*, 58:950–955.

Pojasok, T. and B.D. Kay. 1990. Effect of root exudates from corn and bromegrass on soil structural stability, *Can. J. Soil Sci.*, 70:351–362.

Puget, P., C. Chenu, and J. Balesdent. 1995. Total and young organic matter distributions in aggregates of silty cultivated soils, *Eur. J. Soil Sci.*, 46:449–459.

Quirk, J.P. and R.S. Murray. 1991. Towards a model for soil structural behavior, *Aust. J. Soil Res.*, 29:828–867.

Radcliffe, D.E., R.L. Clark, and M.E. Sumner. 1986. Effect of gypsum and deep-rooting perennials on subsoil mechanical impedance, *Soil Sci. Soc. Am. J.*, 50:1566–1570.

Raimbault, B.A. and T.J. Vyn. 1991. Crop rotation and tillage effects on corn growth and soil structural stability, *Agron. J.*, 83:979–985.

Rasiah, V. and B.D. Kay. 1995. Characterizing rate of wetting: Impact on structural destabilization, *Soil Sci.*, 160:176–182.

Rasiah, V., B.D. Kay, and T. Martin. 1992. Variation of structural stability with water content: Influence of selected soil properties, *Soil Sci. Soc. Am. J.*, 56:1604–1609.

Ratliff, L., J.T. Ritchie, and D.K. Cassel. 1983. Field-measured limits of soil water availability as related to laboratory-measured properties, *Soil Sci. Soc. Am. J.,* 47:770–775.

Rawitz, E., A. Hadas, H. Etkin, and M. Margolin. 1994. Short-term variations of soil physical properties as a function of the amounts and C/N ratio of decomposing cotton residues. II. Soil compressibility, water retention and hydraulic conductivity, *Soil Till. Res.,* 32:199–212.

Reeve, R.C. 1953. A method for determining the stability of soil structure based upon air and water permeability measurements, *Soil Sci. Soc. Am. Proc.,* 17:324–29.

Reid, J.B. and M.G. Goss. 1981. Effect of living roots of different plant species on the aggregate stability of two arable soils, *J. Soil Sci.,* 32:521–541.

Reid, J.B. and M.G. Goss. 1982. Interactions between soil drying due to plant water use and decreases in aggregate stability caused by maize roots, *J. Soil Sci.,* 33:47–53.

Rengasamy, P. and M.E. Sumner, 1998, *in* M.E. Sumner and R. Naidu (eds.) Sodic soils, Oxford University Press, New York, NY.

Rengasamy, P., R.S.B. Greene, and G.W. Ford. 1984. The role of clay fraction in the particle arrangement and stability of soil aggregates: A review, *Clay Res.,* 3:53–67.

Rengasamy, P., R. Beech, T.A. Naidu, K.Y. Chan, and C. Chartres. 1993. Rupture strength as related to dispersive potential in Australian soils, *Catena Suppl.,* 24:65–75.

Richardson, S.J. 1976. Effect of artificial weathering cycles on the structural stability of a dispersed silt soil, *J. Soil Sci.,* 27:287–294.

Roberson, E.B., S. Sarig, and M.K. Firestone. 1991. Cover crop management of polysaccharide-mediated aggregation in an orchard soil, *Soil Sci. Soc. Am. J.,* 55:734–739.

Rovira, A.D. and E.L. Greacen. 1957. The effect of aggregate disruption on enzyme activity of microorganisms in the soil, *Aust. J. Agric. Res.,* 8:659–673.

Rowell, D.L. and P.J. Dillon. 1972. Migration and aggregation of Na and Ca clays by the freezing of dispersed and flocculated suspensions, *J. Soil Sci.,* 23:442–447.

Russell, E.W. 1973. Soil conditions and plant growth, Longman Group, London, UK.

Schjønning, P. 1992. Size distribution of dispersed and aggregated particles and of soil pores in 12 Danish soils, Acta Agric. Scand. Sec. B, *Soil Plant Sci.,* 42:26–33.

Schjønning, P., B.T. Christensen, and B. Carstensen. 1994. Physical and chemical properties of a sandy loam receiving animal manure, mineral fertilizer or no fertilizer for 90 years, *Eur. J. Soil Sci.,* 45:257–68.

Schneider, E.C. and S.C. Gupta. 1985. Corn emergence as influenced by soil temperature, matric potential, and aggregate size distribution, *Soil Sci. Soc. Am. J.,* 49:415–422.

Scott Blair, G.W. 1937. Compressibility curves as a quantitative measure of soil tilth, *J. Agric. Sci.,* 27:541–56.

Seygold, C.A. 1994. Polyacrylamide review: Soil conditioning and environmental fate, *Commun, Soil Sci. Plant Anal.,* 25:2171–2185.

Shanmuganathan, R.T. and J.M. Oades. 1982. Modification of soil physical properties by manipulating the net surface charge on colloids through addition of Fe (III) polycations, *J. Soil Sci.,* 33:451–65.

Shanmuganathan, R.T. and J.M. Oades. 1983. Influence of anions on dispersion and physical properties of the A horizon of a red-brown earth, *Geoderma,* 29:257–277.

Shipitalo, M.J. and R. Protz. 1988. Factors influencing the dispersibility of clay in worm casts, *Soil Sci. Soc. Am. J.,* 52:764–769.

Skidmore, E.L. and J.B. Layton. 1992. Dry-aggregate stability as influenced by selected soil properties, *Soil Sci. Soc. Am. J.,* 56:557–561.

So, H.B. and L.A.G. Aylmore. 1993. How do sodic soils behave? The effects of sodicity on soil physical behavior, *Aust. J. Soil Res.,* 31:761–777.

Soane, B.D. 1975. Studies on some physical properties in relation to cultivation and traffic, p. 160–183, *in* Soil physical conditions and crop production, MN Agric. Food Fish. Tech. Bull. 29.

Soane, B.D. 1990. The role of organic matter in soil compactibility: A review of some practical aspects, *Soil Till. Res.,* 16:179–201.

Soil Science Society of America. 1997. Glossary of soil science terms, *Soil Sci. Soc. Am.,* Madison. WI.

Soil Survey Staff. 1975. Soil Taxonomy. USDA Handb. 436. U.S. Government Printing Office, Washington, D.C.

Sommerfeldt, T.G. and C. Chang. 1985. Changes in soil properties under annual applications of feedlot manure and different tillage practices, *Soil Sci. Soc. Am. J.,* 49:983–987.

Stengel, P. 1979. Utilisation de l'analyse des systèmes de porosité pour la caractérisation de l'etat physique du sol *in situ*, *Ann. Agron.,* 30:27–51.

Stewart, B.A. 1975. Soil conditioners. Soils Sci. Soc. Am. Spec. Pub. 7, *Soil Sci. Soc. Am.,* Madison, WI.

Stone J.A. and W.E. Larson. 1980. Rebound of five unidimensionally compressed unsaturated granular soils, *Soil Sci. Soc. Am. J.,* 44:819–822.

Stone, J.A. and B.R. Buttery. 1989. Nine forages and the aggregation of a clay loam soil, *Can. J. Soil Sci.,* 69:165–169.

Stypa, M., A. Nunez-Barrios, D.A. Barry, and M.H. Miller. 1987. Effects of subsoil bulk density, nutrient availability and soil moisture on corn root growth in the field, *Can. J. Soil Sci.,* 67:293–308.

Sutton, J.C. and B.R. Sheppard. 1976. Aggregation of sand dune soil by endomycorrhizal fungi, *Can. J. Bot.,* 54:326–333.

Taylor, H.M., G.M. Roberson, and J.J. Parker Jr. 1966. Soil strength-root penetration relations for medium to coarse-textured soil materials, *Soil Sci.,* 102:18–22.

Thomas, R.S., S. Dakessian, R.N. Ames, M.S. Brown, and G.J. Bethlenfalvay. 1986. Aggregation of a silty loam soil by mycorrhizal onion roots, *Soil Sci. Soc. Am. J.,* 50:1494–1499.

Thomasson, A.J. 1978. Towards an objective classification of soil structure, *J. Soil Sci.,* 29:38–46.

Thomasson, A.J. and A.D. Carter. 1992. Current and future uses of the U.K. soil water retention data set, p. 355–359, *in* M. Th. van Genuchten, F.J. Leij, and L.J. Lund (eds.), Proc. Int. Worksh. Indir. Meth. Estim. Hydr. Prop. Unsat. Soils, USDA/ARS/University of California, Riverside, CA.

Tiessen, H. and J.W.B. Stewart. 1988. Light and electron microscopy of stained microaggregates: the role of organic matter and microbes in soil aggregation, *Biogeochem.,* 5:312–322.

Tisdall, J.M. and J.M. Oades. 1979. Stabilization of soil aggregates by the root systems of ryegrass, *Aust. J. Soil Res.,* 17:429–441.

Tisdall, J.M. and J.M. Oades. 1982. Organic matter and water-stable aggregates, *J. Soil Sci.,* 33:141–163.

Tisdall, J.M., B. Cockroft, and N.C. Uren. 1978. The stability of soil aggregates as affected by organic materials, microbial activity and physical disruption, *Aust. J. Soil Res.,* 16:9–17.

Tisdall, J.M., S.E. Smith, and P. Rengasamy. 1997. Aggregation of soil by fungal hyphae, *Aust. J. Soil Res.,* 35:55–60.

Topp, G.C., Y.T. Galganov, K.C. Wires, and J.L.B. Culley. 1994. Non-limiting water range (NLWR): An approach for assessing soil structure. Soil Qual. Eval. Prog. Tech. Rep. 2. Centre for Land and Biological Resources Research, *Agr. Can.,* Ottawa, Canada.

Unger, P.W. 1996. Soil bulk density, penetration resistance, and hydraulic conductivity under controlled traffic conditions, *Soil Till. Res.,* 37:67–75.

Utomo, W.H. and A.R. Dexter. 1981a. Tilth mellowing, *J. Soil Sci.,* 32:187–201.

Utomo, W.H. and A.R. Dexter. 1981b. Soil friability, *J. Soil Sci.,* 32:203–213.

van Lanen, H.A.J., G.J. Reinds, O.H. Boersma, and J. Bouma. 1992. Impact of soil management systems on soil structure and physical properties in a clay loam soil, and the simulated effects on water deficits, soil aeration and workability, *Soil Till. Res.,* 23:203–20.

van Veen, J.A. and P.J. Kuikman. 1990. Soil structural aspects of decomposition of organic matter by micro-organisms, *Biogeochem.,* 11:213–233.

Veihmeyer, F.J. and A.H. Hendrickson. 1927. Soil moisture conditions in relation to plant growth, *Plant Physiol.,* 2:71–82.

Visser, S.A. and M. Caillier. 1988. Observations on the dispersion and aggregation of clays by humic substances, I. Dispersive effects of humic acids, *Geoderma,* 42:331–337.

Voorhees, W.B. 1983. Relative effectiveness of tillage and natural forces in alleviating wheel-induced soil compaction, *Soil Sci. Soc. Am. J.,* 47:129–133.

Voorhees, W.B. and M.J. Lindstrom. 1984. Long-term effects of tillage method on soil tilth independent of wheel traffic compaction, *Soil Sci. Soc. Am. J.,* 48:152–156.

Waldron, L.J. and S. Dakessian. 1982. Effect of grass, legume, and tree roots on soil shearing resistance, *Soil Sci. Soc. Am. J.,* 46:894–899.

Wang, C., J.A. McKeague, and G.C. Topp. 1985. Comparison of estimated and measured horizontal Ksat values, *Can. J. Soil Sci.,* 65:707–715.

Warner, G.S., J.L. Neiber, I.D. Moore, and R.A. Giese. 1989. Characterizing macropores in soil by computed tomography, *Soil Sci. Soc. Am. J.,* 53:653–660.

Waters, A.G. and J.M. Oades. 1991. Organic matter in water-stable aggregates, p. 163–174, *in* W.S. Wilson (ed.), Advances in soil organic matter research. The impact on agriculture and the environment, *Royal Soc. Chem.,* Cambridge, UK.

Watts, C.W., A.R. Dexter, E. Dumitru, and J. Arvidsson. 1996a. An assessment of the vulnerability of soil structure to destabilization during tillage. 1. A laboratory test, *Soil Till. Res.,* 37:161–74.

Watts, C.W., A.R. Dexter, and D.J. Longstaff. 1996b. An assessment of the vulnerability of soil structure to destabilization during tillage. 2. Field trials, *Soil Till. Res.,* 37:175–90.

Webber, L.R. 1965. Soil polysaccharides and aggregation in crop sequences, *Soil Sci. Soc. Am. Proc.,* 29:39–42.

Wenke, J.F. and C.D. Grant. 1994. The indexing of self-mulching behavior in soils, *Aust. J. Soil Res.,* 32:201–211.

White, R.E. 1985. The influence of macropores on the transport of dissolved and suspended matter through soil, *Adv. Soil Sci.,* 3:94–120.

White, W.M. 1993. Dry aggregate distribution, p. 659–662, *in* M.R. Carter (ed.), Manual on soil sampling and methods of analysis, CRC Press, Boca Raton, FL.

Williams, J., R.E. Prebble, W.T. Williams, and C.T. Hignett. 1983. The influence of texture, structure and clay mineralogy on the soil moisture characteristic, *Aust. J. Soil Res.,* 21:15–32.

Willis, W. 1955. Freezing and thawing, and wetting and drying in soils treated with organic chemicals, *Soil Sci. Soc. Am. Proc.,* 19:263–267.

Woodruff, N.P. and F.H. Siddoway. 1965. A wind erosion equation, *Soil Sci. Soc. Am. Proc.,* 29:602–608.

Wright, S.F and A. Upadhyaya. 1996. Extraction of an abundant and unusual protein from soil and comparison with hyphal protein of arbuscular mycorrhizal fungi, *Soil Sci.,* 161:575–586.

Wu, L. and J.A. Vomocil. 1992. Predicting the soil water characteristic from the aggregate-size distribution, p. 139–45, *in* M.Th. van Genuchten, F.J. Leij, and L.J. Lund (eds). Proc. Int. Worksh. Indir. Meth. Estim. Hydr. Prop. Unsat. Soils. USDA/ARS/University of California, Riverside, CA.

Xu, X., J.L. Nieber, and S.C. Gupta. 1992. Compaction effect on the gas diffusion coefficient in soils, *Soil Sci. Soc. Am. J.,* 56:1743–50.

Yoder, R.E. 1936. A direct method of aggregate analysis of soils and a study of the physical nature of erosion losses, *J. Am. Soc. Agron.,* 28:337–351.

Young, I. and C.E. Mullins. 1991. Water-suspensible solids and structural stability, *Soil Till. Res.,* 19:89–94.

Zausig, J., W. Stepniewski, and R. Horn. 1993. Oxygen concentration and redox potential gradients in unsaturated model soil aggregates, *Soil Sci. Soc. Am. J.,* 57:908–16.

# 8 Soil Gas Movement in Unsaturated Systems

*Bridget R. Scanlon, Jean Phillippe Nicot, and Joel W. Massmann*

## 8.1 GENERAL CONCEPTS RELATED TO GAS MOVEMENT

An understanding of gas transport in unsaturated media is important for evaluation of soil aeration or movement of $O_2$ from the atmosphere to the soil. Soil aeration is critical for plant root growth because roots generally cannot get enough $O_2$ from leaves. Evaluation of gas movement is also important for estimating transport of volatile and semivolatile organic compounds from contaminated sites through the unsaturated zone to the groundwater. The use of soil venting, or soil vapor extraction, as a technique for remediating contaminated sites has resulted in increased interest in gas transport in the unsaturated zone (Rathfelder et al., 1995). Migration of gases from landfills, such as methane formed by decomposition of organic material, is important in many areas (Moore et al., 1982; Thibodeaux et al., 1982). Soil gas composition has also been used as a tool for mineral and petroleum exploration and for mapping organic contaminant plumes. An understanding of gas transport is important for evaluating movement of volatile radionuclides, such as $^3H$, $^{14}C$ and Rd from radioactive waste disposal facilities. The adverse health effects of radon and its decay products have led to evaluation of transport in soils and into buildings (Nazaroff, 1992). A thorough understanding of gas transport is required to evaluate these issues.

Gas in the unsaturated zone is generally moist air, but it has higher $CO_2$ concentrations than atmospheric air because of plant root respiration and microbial degradation of organic compounds. Oxygen concentrations are generally inversely related to $CO_2$ concentrations because processes producing $CO_2$ generally deplete $O_2$ levels. Contaminated sites may have gas compositions that differ markedly from atmospheric air, depending on the type of contaminants.

This chapter will focus primarily on processes of gas transport in unsaturated media. The following issues will be evaluated:

1. How does gas move through the unsaturated zone?
2. How does one evaluate single gas and multicomponent gas transport?
3. How does one measure or estimate the various parameters required to quantify gas flow?
4. How does one numerically simulate gas flow?

Although chemical reactions are discussed in another part of this book, the model equations developed in this chapter need to be incorporated into them. The reader may find the discussion of water flow in unsaturated media in Chapters 3 and 4 helpful in understanding many of the concepts in this chapter.

### 8.1.1 GAS CONTENT

Unsaturated media consist generally of at least three phases: solid, liquid and gas. In some cases, a separate nonaqueous liquid phase may exist if the system is contaminated by organic compounds. In most cases the pore space is only partly filled with gas. The volumetric gas content ($\theta_G$) is defined as:

$$\theta_G = V_G / V_T \qquad [8.1]$$

where $V_G$ (L$^3$) is the volume of the gas and $V_T$ (L$^3$) is the total volume of the sample. This definition is similar to that used for volumetric water content in unsaturated-zone hydrology. In many cases, the volumetric gas content is referred to as the gas porosity. The saturation with respect to the gas phase ($S_G$) is:

$$S_G = V_G / V_v \qquad [8.2]$$

where $V_v$ (L$^3$) is the volume of voids or pores. Saturation values range from 0 to 1. Volumetric gas content and gas saturation are related as follows:

$$\theta_G = \phi S_G \qquad [8.3]$$

where $\phi$ is porosity ($V_p / V_T$). If only two fluids, gas and water, are in the system, the volumetric gas and water contents sum to the porosity. Therefore, volumetric gas content can be calculated if the volumetric water content and porosity are known. Volumetric water content can be measured using procedures described in Gardner (1986).

In unsaturated systems, water is the wetting and gas is the nonwetting phase. Therefore, water wets the solids and is in direct contact with them, whereas gas is generally separated from the solid phase by the water phase. Water fills the smaller pores, whereas gas is restricted to the larger pores.

## 8.1.2 DIFFERENCES BETWEEN GAS AND WATER PROPERTIES

Gas and water properties differ greatly and can be compared as follows: (1) gas density is much lower than liquid density. The density of air varies with composition but ranges generally from 1 to 1.5 kg m$^{-3}$, whereas the density of water is close to 1,000 kg m$^{-3}$ (Table 8.1); (2) although water is generally considered incompressible, the incompressibility assumption is not always valid for gas flow. Gas density depends on gas pressure, which results in a nonlinear flow equation; (3) at solid surfaces, water is assumed to have zero velocity whereas gas velocities are generally nonzero, and such velocities result in slip flow or the Klinkenberg effect (Figure 8.1); (4) because the dynamic viscosity of air is ~ 50 times lower than that of water, significant air flow can occur at much smaller pressure gradients. Gas viscosity increases with temperature, whereas water viscosity decreases with temperature; (5) air conductivities are generally an order of magnitude less than water or hydraulic conductivities in the same material because of density and viscosity differences between the fluids; (6) because gas molecular diffusion coefficients are about four orders of magnitude greater than those of water, gas diffusive fluxes are generally much greater than those of water.

## 8.1.3 TERMINOLOGY

Gas refers to a phase that may be single or multicomponent. Gas flux can be expressed in a variety of ways, such as volume (L$^3$ L$^{-2}$ t$^{-1}$), mass (M L$^{-2}$ t$^{-1}$), molar (mol L$^{-2}$ t$^{-1}$) or molecular (number of molecules L$^{-2}$ t$^{-1}$) fluxes. If a gas is incompressible, (density constant), then volume and mass are conservative, whereas if a gas is compressible only mass is conservative. The mean mass flux is weighted by the mass of each molecule. In the mean molar or molecular flux all molecules behave identically and contribute equally to the mean. The distinction between mass and molar flux will be clarified in Section 8.1.5.4. Parameters important in describing gas transport mechanisms include mean free path ($\lambda$), which is the average distance a gas molecule travels before colliding with another gas molecule; pore size ($\lambda_p$), which is equivalent to the average distance between soil particles in dry porous media; and particle size ($r_p$). Equimolar component gases are

**TABLE 8.1**
**Variations in density and viscosity with temperature for water, water vapor, and air. Moist air corresponds to a vapor pressure of 4.6 mm at 0°C to 261 mm at 100°C.**

| T (°C) | [1]Density water (kg m$^{-3}$) | [2]Density water vapor (kg m$^{-3}$) | [3]Density moist air (kg m$^{-3}$) | [4]Viscosity water (Pa s) | [5]Viscosity water vapor (Pa s) | [6]Viscosity dry air (Pa s) |
|---|---|---|---|---|---|---|
| 0 | 1000 | 0.005 | 1.290 | 1.79E-03 | 8.03E-06 | 1.73E-05 |
| 10 | 1000 | 0.009 | 1.242 | 1.31E-03 | 8.44E-06 | 1.77E-05 |
| 20 | 998 | 0.017 | 1.194 | 1.00E-03 | 8.85E-06 | 1.81E-05 |
| 30 | 996 | 0.030 | 1.146 | 7.97E-04 | 9.25E-06 | 1.85E-05 |
| 40 | 992 | 0.051 | 1.097 | 6.53E-04 | 9.66E-06 | 1.89E-05 |
| 50 | 988 | 0.083 | 1.043 | 5.48E-04 | 1.01E-05 | 1.93E-05 |
| 60 | 983 | 0.130 | 0.981 | 4.70E-04 | 1.05E-05 | 1.97E-05 |
| 70 | 978 | 0.198 | 0.909 | 4.10E-04 | 1.09E-05 | 2.01E-05 |
| 80 | 971 | 0.293 | 0.823 | 3.62E-04 | 1.13E-05 | 2.05E-05 |
| 90 | 965 | 0.423 | 0.718 | 3.24E-04 | 1.17E-05 | 2.09E-05 |
| 100 | 959 | 0.598 | 0.588 | 2.93E-04 | 1.21E-05 | 2.13E-05 |

[1,3,4]Weast, 1986; [2]Childs and Malstaff (1982), in Hampton (1989); [5,6]White and Oostrom (1996). Vapor pressure of moist air ranges from 4.6 mm at 0°C to 761 mm at 100°C.

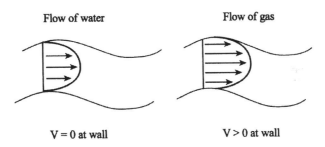

**FIGURE 8.1** Schematic demonstrating zero velocities at pore walls of liquid flow and nonzero velocities of gas flow.

components that have the same molecular weight, such as $N_2$ and CO, whereas nonequimolar component gases have different molecular weights. The terms *concentration* (M L$^{-3}$) and *partial pressure* (M L$^{-1}$t$^{-2}$) refer to individual gas components. The partial pressure of gas component $i$ is equal to the mole fraction times the total pressure and can be related to concentration through the ideal gas law for ideal gases. Advective flux refers to bulk gas phase and occurs in response to unbalanced mechanical driving forces acting on the phase as a whole.

### 8.1.4 Mass Transfer versus Mass Transport

Most of this chapter deals with transport of gas and associated chemicals through the unsaturated zone. Mass transfer refers to transfer of mass or partitioning of mass among gas, liquid and solid phases. Partitioning of chemicals into other phases retards their transport in the gas phase and can be described by:

$$C_T = \rho_b C_{ad} + \theta_l C_l + \theta_G C_G \qquad [8.4]$$

where $C_T$ is the total mass concentration (M L$^{-3}$ soil); $C_G$, $C_l$, and $C_{ad}$ are the mass concentrations in the gas and liquid phases and adsorbed on the solid phase; and $\rho_b$ is the bulk density (M L$^{-3}$). The relationship between gas and water mass concentrations can be described by Henry's law:

$$C_G = K_H C_l \qquad\qquad [8.5]$$

where $K_H$ is the dimensionless Henry's law constant. Many types of isotherms describe the adsorption onto the solid phase, the simplest being the linear adsorption isotherm:

$$C_{ad} = K_d C_l \qquad\qquad [8.6]$$

where $K_d$ (L$^3$ M$^{-1}$) is the distribution coefficient. The linear relationship generally applies to low polarity compounds (Brusseau, 1994).

   If linear relationships are valid, total concentration can be written in terms of gas concentration as follows (Jury et al., 1991):

$$C_T = (\rho_b K_d / K_H + \theta_l / K_H + \theta_G)C_G = B_G C_G \qquad\qquad [8.7]$$

where $B_G$ is the bulk gas phase partition coefficient (Charbeneau and Daniel, 1992). In some studies this partitioning behavior is used to quantify the amount of a liquid phase in the system. For example, transport of gas tracers that partition into water or nonaqueous phase liquids is compared with transport of those that do not partition (conservative) into that phase to determine the amount of water or organic compound in the system (Jin et al., 1995).

### 8.1.5 MECHANISMS OF GAS TRANSPORT

Many readers may be more familiar with transport mechanisms in the liquid than in the gas phase. Primary mechanisms of solute transport in the liquid phase include advection (movement with the bulk fluid) and hydrodynamic dispersion (mechanical dispersion and molecular diffusion) (Freeze and Cherry, 1979). Mechanical dispersion, resulting from variations in fluid velocity at the pore scale, is the product of dispersivity and advective velocity. Transport in the gas phase may also be described by advection and dispersion. Although some studies have found mechanical dispersion or velocity-dependent dispersion to be important for chemical transport in the gas phase (Rolston et al., 1969; Auer et al., 1996), in most cases mechanical dispersion is ignored because gas velocities are generally too small and the effects of diffusion are generally much greater than dispersion in the gas phase. Molecular diffusion coefficients are approximately four orders of magnitude greater in the gas than in the liquid phase.

   Diffusive transport in the liquid phase is described by molecular diffusion. Traditionally, diffusive transport in the gas phase has also been described by molecular diffusion. Diffusion in the gas phase may, however, be much more complicated and may include Knudsen, molecular and nonequimolar diffusion. Surface diffusion of adsorbed gases, generally not significant, is not discussed in this chapter. Pressure diffusion results from the separation of gases of different molecular weights under a pressure gradient and causes diffusion of heavier (lighter) molecules toward regions of higher (lower) pressure (Amali and Ralston, 1993). Pressure diffusion is generally negligible at depths of less than 100 m, which include most unsaturated sections (Amali and Rolston, 1993). Although temperature gradients in unsaturated media are generally too low to result in significant diffusion except at the land surface, thermal diffusion is important for water vapor transport (Section 8.6.3).

### 8.1.5.1 Advective Flux

If a total pressure gradient exists in a soil as a result of external forces such as atmospheric pumping (Section 8.2), gases will flow from points of higher to those of lower pressure. It has been shown that relatively small gradients in total pressure can result in advective gas fluxes that are much larger than diffusive gas fluxes (Alzaydi and Moore, 1978; Thorstenson and Pollock, 1989; Massmann and Farrier, 1992). The driving force for flow is the total pressure gradient, and the resistance to flow is caused by viscosity of the gas. Other terms for advective flux are pressure driven or viscous flux. Under a total pressure gradient, advection is dominant when the mean free path of the gas molecules ($\lambda$) is much less than the pore radius ($\lambda_p$) and the particle radius ($r_p$) ($\lambda \ll \lambda_p$ and $\lambda \ll r_p$), resulting in intermolecular collisions being dominant relative to collisions between gas molecules and the pore walls (Cunningham and Williams, 1980). Because the mean free path is inversely proportional to the mean pressure, low mean free paths relative to pore size may occur in dry, coarse-grained media and/or under high mean pressure. A pressure gradient is required to maintain advective flux because of the velocity reduction near the pore walls caused by gas molecules rebounding from collisions with the pore walls. In unsaturated media the pore walls will generally consist of the gas–liquid interface. Advective flux is also termed nonsegregative or nonseparative because bulk flow as a result of a pressure gradient does not segregate the gas into individual components. The viscous flux of gas component $i$ is proportional to its mole fraction ($x_i$) in the mixture:

$$N_i^V = x_i N^V \qquad [8.8]$$

where $N_i^V$ is the molar viscous flux (mol L$^{-2}$ t$^{-1}$) of gas i and $N^V$ is the total molar viscous flux. As the mean pressure and/or as the pore size decrease as a result of decreasing grain size or gas saturation (increasing water saturation), a continuum results, from advective or viscous flux to viscous slip flux (Klinkenberg effect) to Knudsen diffusive flux. Viscous slip flux occurs when the mean free path of the gas molecules becomes approximately the same as the pore radius and results from nonzero velocity at the pore wall. Stonestrom and Rubin (1989) and Detty (1992) found that errors resulting from ignoring viscous slip flux in dry, coarse-grained porous media (sands) were less than 7%.

### 8.1.5.2 Knudsen Diffusive Flux

Knudsen flux occurs when the gas mean free path is much greater than the pore radius ($\lambda \gg \lambda_p$), Knudsen number ($K_n = \lambda/\lambda_p$) = 10 corresponding to the Knudsen regime (Alzaydi, 1975; Abu-El-Sha'r, 1993). Molecule–wall collisions dominate over molecule–molecule collisions. The term *free molecule flux* is also used to describe Knudsen flux because rebounding molecules do not collide with other gas molecules. The Knudsen diffusive flux depends on the molecular weights and temperatures of gases and the radius of the pores. It is not influenced by the presence of other species of gas and is described as follows:

$$N_i^k = -D_i^k \nabla C_i \qquad [8.9]$$

where $N_i^k$ is the Knudsen molar flux of component $i$ (mol L$^{-2}$ t$^{-1}$), $D_i^k$ is the effective Knudsen diffusivity (L$^2$ t$^{-1}$), and $C_i$ is the molar concentration of gas i (mol L$^{-3}$) (Cunningham and Williams, 1980). $D_i^k$ is defined as:

$$D_i^k = Q_p \left( \frac{RT}{M_i} \right)^{0.5} \qquad [8.10]$$

where $Q_P$ is the Knudsen radius (L), $R$ is the ideal gas constant (M L$^2$ t$^{-2}$ T$^{-1}$ mol$^{-1}$), $T$ is temperature (T), and $M_i$ is the molecular weight of gas i (M mol$^{-1}$).

### 8.1.5.3  Molecular Diffusive Flux

Molecular diffusion is the only type of transport mechanism that occurs under isothermal, isobaric conditions when equimolar pairs of gases (e.g., $N_2$, CO) counterdiffuse in pores whose size is much greater than that of the mean free path of the gas molecules. In this case molecule–molecule collisions dominate over molecule–wall collisions. The molecular diffusive flux depends on gas molecular weights and temperatures in the pore space and is unaffected by the physical nature of the pore walls. Because molecular diffusion results in segregation of the different component gases, it is termed *segregative*. The molar diffusive flux of component i resulting from molecular diffusion $J_{iM}^m$, in a binary gas mixture under isothermal, isobaric conditions is proportional to the concentration gradient and is described by Mason and Malinauskas (1983):

$$J_{iM}^m = -D_{ij}^e \nabla n_{iM} \qquad [8.11]$$

where $D_{ij}^e$ is the effective molecular diffusion coefficient (L$^2$ t$^{-1}$) and $n_{iM}$ is the molar concentration of gas i (mol L$^{-3}$) (Abu-El-Sha'r, 1993). The effective diffusion coefficient in porous media is calculated from the free air diffusion coefficient as follows:

$$D_{ij}^e = \tau \theta_G D_{ij} \qquad [8.12]$$

where $\tau$ is the tortuosity, $\theta_G$ is the gas content and $D_{ij}$ is the free air diffusion coefficient (L$^2$ t$^{-1}$). Volumetric gas content ($\theta_G$) accounts for reduced cross-sectional area in porous media relative to free air, and tortuosity ($\tau$) accounts for increased path length. A variety of equations have been developed to calculate tortuosity. The most commonly used equations are those of Penman (1940a, b) and Millington and Quirk (1961) (Table 8.2). The effective diffusion coefficient decreases with increased water content in unsaturated systems; the rate of decrease is low at low water contents because gas is in the large and water in the small pore spaces. The effective diffusion coefficient decreases sharply as soils become saturated with water because the large pores become occluded.

### 8.1.5.4  Bulk Diffusive Flux

Bulk diffusion includes molecular and nonequimolar diffusion. Nonequimolar diffusion occurs when gas components have different molecular weights. According to the kinetic theory of gases, gas molecules have the same kinetic energy in an isothermal, isobaric system; therefore, lighter gas molecules have higher velocities than heavier gas molecules. In a binary mixture of nonequimolar gases, the more rapid diffusion of the lighter gas molecules results in a pressure gradient. Such pressure gradients should be distinguished from external pressure gradients that result in advective flux. The flux resulting from the buildup of pressure is diffusive and is called the *nonequimolar flux* or the *diffusive slip flux* (Cunningham and Williams, 1980). Because the nonequimolar flux does not result in separation of gas components (nonsegregative), the molar bulk diffusive flux includes the segregative molecular diffusive flux and the nonsegregative nonequimolar diffusive flux:

$$N_i^D = J_{iM}^m + x_i \sum_{j=1}^{v} N_j^D \qquad [8.13]$$

**TABLE 8.2**
**Models for estimating tortuosity, $\theta_G$ is the gas content, and $\phi$ is the porosity; a more complete listing is provided in Abu-El-Sha'r (1993).**

| Tortuosity ($\tau$) times $\theta G$ | Reference | Comments |
|---|---|---|
| $0.66\phi$ | Penman (1940a, b) | experimental |
| $\left(\theta_G\right)^{3/2}$ | Marshall (1958) | pseudotheoretical |
| $\left(\theta_G\right)^{4/3}$ | Millington (1959) | pseudotheoretical |
| $\theta_G^{-10/3} / \phi^2$ | Millington and Quirk (1961) | theoretical |
| $(1-S_w)^2 *(\phi-\theta)^{2x}$ | Millington and Shearer (1970) | theoretical |
| $\theta_G^{7/3}$ | Lai et al. (1976) | experimental |
| $0.435\phi$ | Abu-El-Sha'r Abriola (1997) | experimental |

where $N_i^D$ is the bulk molar diffusive flux of component $i$ (mol L$^{-2}$ t$^{-1}$), $J_{iM}^m$ is the molar diffusive flux of component $i$ resulting from molecular diffusion, $x_i$ is the mole fraction of component $i$, $\nu$ is the number of gas components, and the second term on the right $x_i \sum_{j=1}^{\nu} N_j^D$ is the nonequimolar flux.

It is important to distinguish between diffusive gas flux with respect to average velocity of the gas molecules (molar velocity) and that with respect to velocity of the center of mass of the gas (mass velocity) because the two are not necessarily the same (Cunningham and Williams, 1980). In nonequimolar gas mixtures, because the lighter gas molecules diffuse more rapidly, the molar gas flux moves in the direction in which the lighter gas molecules are moving, whereas the center of mass of the gas moves in the direction in which the heavier molecules diffuse. The molar and mass fluxes can therefore move in opposite directions.

## 8.1.6 GAS TRANSPORT MODELS

Darcy's law is used to model advective or viscous gas transport, and traditionally Fick's law has been used to model molecular diffusion. The Stefan-Maxwell equations and the dusty gas model can also be used to simulate diffusive gas flux. If there is a total pressure gradient in a system, then the diffusive fluxes are calculated relative to the bulk molar flux. The total molar flux is calculated by adding the molar viscous or advective flux to the bulk molar diffusive flux:

$$N_i^T = N_i^V + N_i^D \qquad [8.14]$$

where $N_i^T$ is the total gas molar flux, $N_i^V$ is the molar viscous flux and $N_i^D$ is the molar diffusive flux of component $i$. For analysis of transient gas flux, the constitutive flux equations are incorporated into the conservation of mass. If the mass density of the fluid and the diffusion coefficients are assumed constant, then the transport equation can be written as:

$$\frac{\partial(Cx_i)}{\partial t} + \nabla \cdot \left(N_i^T\right) = 0 \qquad [8.15]$$

where $C$ is the molar concentration of the gas (mol L$^{-3}$) and $x_i$ is the mole fraction of component $i$ (Massmann and Farrier, 1992). This equation is strictly valid for relatively dilute gas mixtures or for vapors with molecular weights close to the average molecular weight of air (Bird et al., 1960).

### 8.1.6.1  Darcy's Law

Darcy's law is used to describe advective gas transport. The simplest form of Darcy's law is as follows:

$$J_G = -\frac{k_G}{\mu_G}\nabla P \qquad [8.16]$$

where $J_G$ is the volumetric flux (L$^3$ L$^{-2}$ t$^{-1}$), $k_G$ is the permeability (L$^2$), $\mu_G$ is the dynamic gas viscosity (M L$^{-1}$ t$^{-1}$), and $\nabla P$ is the applied pressure gradient. Gravity gradients, assumed negligible here, are treated in Section 8.2.1. The ideal gas law can be used to convert the volumetric gas flux ($J_G$) to the molar gas flux ($N^V$):

$$PV = nRT, \qquad \rho_G = \frac{nM}{V} = \frac{PM}{RT} \qquad [8.17]$$

where $n$ is the number of moles of the gas, $M$ is the molar mass of the gas, $R$ is the ideal gas constant (M L$^2$ t$^{-2}$ T$^{-1}$ mol$^{-1}$) and $T$ is temperature (K). The molar viscous or advective flux is:

$$N^V = -\frac{P}{RT}\frac{k_G}{\mu_G}\nabla P \qquad [8.18]$$

The validity of Darcy's law can be evaluated by plotting the gas flux for a single component against the pressure gradient. The relationship should be linear with a zero intercept if Darcy's law is valid.

### 8.1.6.2  Fick's Law

Fick's law is an empirical expression originally developed to describe molecular diffusion of solutes in the liquid phase. Fick's law is generally used to describe molecular diffusion of gas $i$ in gas $j$ (Bird et al., 1960; Jaynes and Rogowski, 1983):

$$J_{iM}^{m} = -D_{ij}C\nabla x_i \qquad [8.19]$$

where $C$ is the total molar concentration (mol L$^{-3}$, constant in an isothermal, isobaric system) and $x_i$ is the mole fraction of gas and $D_{ij} = D_{ji}$ (Bird et al., 1960). Fick's law is strictly applicable to molecular diffusion of equimolar gases in isothermal, isobaric systems. Fick's law excludes the effects of Knudsen diffusion and nonequimolar diffusion. Fick's law can predict the flux of only one component. The adequacy of Fick's law will be discussed after the other models have been described.

### 8.1.6.3  Dusty Gas Model

In contrast to Fick's law, which is empirical, the dusty gas model is based on the full Chapman Enskog kinetic theory of gases. Application of the kinetic theory of gases is only possible by considering the dust particles (porous medium) as giant molecules that constitute another component in the gas phase. The multicomponent equations, which are based on the dusty gas model originally proposed by Mason et al. (1967), are presented by Satterfield and Cadle (1968), Cunningham and

Williams (1980), Mason and Malinauskas (1983), and Thorstenson and Pollock (1989). The equations for the dusty gas model of binary and multicomponent gas diffusion are (Cunningham and Williams, 1980):

$$\frac{x_1 N_2^D - x_2 N_1^D}{D_{12}^e} - \frac{N_1^D}{D_1^k} = \frac{\nabla P_1}{RT}$$

$$\sum_{\substack{i=1 \\ i \neq j}}^{\nu} \frac{x_i N_j^D - x_j N_i^D}{D_{ij}^e} - \frac{N_i^D}{D_i^k} = \frac{\nabla P_i}{RT} \qquad [8.20]$$

$$\begin{array}{ccc} \textit{molecular} & \textit{Knudsen} & \textit{P gradient} \\ \textit{diffusion} & \textit{diffusion} & \textit{term} \end{array}$$

where $x_i$ is the mole fraction of component $i$, $N_i^D$ is the bulk molar diffusive flux of gas component $i$ (mol $L^{-2}$ $t^{-1}$), $D_{ij}^e$ is the effective molecular diffusivity ($L^2$ $t^{-1}$), $D_i^k$ is the effective Knudsen diffusivity ($L^2$ $t^{-1}$), and $P_i$ is the partial pressure of component $i$ (M $L^{-1}t^{-2}$). Although one generally writes equations in terms of a flux of a component being proportional to a gradient, multicomponent gas equations are simplest when written in terms of a gradient of gas component $i$ being proportional to the fluxes of all other components (Mason and Malinauskas, 1983). Derivation of the above form of these equations for a binary gas mixture is shown in Section 8.7. Knudsen diffusion is important when the second term in the dusty gas model is larger than the first term. This requires that $D_i^k \ll D_{ij}^e$ which appears counterintuitive (Thorstenson and Pollock, 1989), but which can be understood if one considers the reciprocal of $D_i^k$ as a resistance. Knudsen diffusion is thus important when $1/D_i^k$ is large and $D_i^k$ is small. Although there is a gradient in the natural log of temperature in the dusty gas model (Thorstenson and Pollock, 1989), gradients in temperature in the subsurface are generally small and gradients in the natural log of temperature are generally negligible. It has therefore been omitted in Equation [8.20]. The primary assumptions of the dusty gas model are that the dust particles are spherical and that there are no external forces on the gas. The dusty gas model can predict the flux of all components in a gas mixture. The system of equations can be solved analytically (Jackson, 1977) or numerically (Massmann and Farrier, 1992). Many analyses of the dusty gas model assume steady-state flow (Thorstenson and Pollock, 1989; Abu-El-Sha'r, 1993).

When both partial and total pressure gradients are important for gas flux, the combined effects of advection and diffusion need to be considered. Knudsen diffusion, which is proportional to the total mole fraction, provides the link between advection and diffusion. Because the dusty gas model is the only model that includes Knudsen diffusion, it is the only model that can theoretically link advection and diffusion through the Knudsen diffusion term. The dusty gas model for combined advection and diffusion is as follows (Thorstenson and Pollock, 1989):

$$\sum_{\substack{j \neq 1 \\ j \neq i}}^{\nu} \frac{x_i N_j^T - x_j N_i^T}{D_{ij}^e} - \frac{N_i^T}{D_i^k} = \frac{P \nabla x_i}{RT} + \left(1 + \frac{kP}{D_i^k \mu_G}\right) \frac{x_i \nabla P}{RT} \qquad [8.21]$$

where $x_i$ is the mole fraction of component $i$, $N_i^T$ is the total molar gas flux of component $i$ relative to a fixed coordinate system (mol $L^{-2}$, $t^{-1}$), $D_{ij}^e$ is the effective molecular diffusion coefficient of component $i$ in $j$ ($L^2$ $t^{-1}$), $D_i^K$ is the effective Knudsen diffusion coefficient of component $i$ ($L^2$ $t^{-1}$), $P$ is pressure (M $L^{-1}$ $t^{-2}$), $R$ is the ideal gas constant, $T$ is absolute temperature, $k$ is permeability ($L^2$) and $\mu_G$ is gas viscosity (M $L^{-1}$ $t^{-1}$).

#### 8.1.6.4  Stefan-Maxwell Equation

The Stefan-Maxwell equations, which can predict the fluxes of all but one gas component in a multicomponent gas mixture, can be obtained from the dusty gas model by assuming negligible Knudsen diffusion (no molecule particle collisions):

$$\sum_{\substack{j=1 \\ j \neq i}}^{v} \frac{x_i N_j^D - x_j N_i^D}{D_{ij}^e} = \frac{\nabla P_i}{RT} \qquad [8.22]$$

## 8.2  TRANSPORT OF A HOMOGENEOUS GAS

Transport of a homogeneous gas in dry, coarse-grained media can be described by advection. Because such a gas can be considered as a single component, the only type of diffusion possible is Knudsen diffusion. Single-component gases in dry, coarse-grained media are dominated by advective or viscous flux because Knudsen diffusion is negligible in such systems. This analysis of gas transport is appropriate when gas velocities are high. The single fluid, nonreactive, noncompositional approximation is appropriate only when one is interested in the bulk flow of a homogeneous gas.

Natural advective gas transport can occur in response to barometric pressure fluctuations, wind effects, water-table fluctuations, density effects, and can also be induced by injection or extraction, as in soil vapor extraction systems. Barometric pressure fluctuations consist of (1) diurnal fluctuations due to thermal and gravitational effects, which are on the order of a few millibars, and (2) longer term fluctuations that result from regional scale weather patterns, which are on the order of tens of millibars within a few hours of when a high or low pressure front moves through. The penetration depth of barometric pressure fluctuations increases with the thickness of the unsaturated zone and with the permeability of the medium. Because highs are balanced by lows, the net transport of the gas may be negligible, except in fractured media, where contaminants may migrate large distances (Nilson et al., 1991). Smaller scale fluctuations, such as gusts and lulls related to wind, may be important in fractured media (Weeks, 1993). Water-table fluctuations, resulting in changes in the gas volume, can produce advective flow; however, advective fluxes as a result of water-table fluctuations are considered important only if the rate of rise or decline of the water-table is rapid, the permeability of the material is high and the water table is shallow.

### 8.2.1  GOVERNING EQUATIONS

Darcy's law, which was developed for water flow in saturated media, can also be applied to single-phase gas flow. Darcy's law is an empirical expression and the general form is:

$$J_G = -\frac{k_G}{\mu_G}\left(\nabla P + \rho_G g \nabla z\right) \qquad [8.23]$$

where $J_G$ is the volumetric flux density ($L^3 L^{-2} t^{-1}$), $k_G$ is gas permeability ($L^2$), $\rho_G$ is gas density ($M L^{-3}$), $g$ is gravitational acceleration ($L t^{-2}$), $\mu_G$ is gas viscosity ($M L^{-1} t^{-1}$), $P$ is pressure ($M L^{-1} t^{-2}$), and $z$ is elevation ($L$). The negative sign in Equation [8.23] is required because flow occurs in the direction of decreasing pressure. The first term in parentheses is the driving force due to pressure and the second term is the driving force due to gravity. Generally gas density is a function of pressure and composition, and the above form of Darcy's law is the only valid form in these cases. The pressure gradient provides the main driving force for advective gas transport, and the resistance to flow results from the gas viscosity. Small pressure gradients can result in substantial advective gas fluxes because the resistance to flow is small. The small flow resistance is attributed to the negligible viscosities of gases relative to those of liquids and to the gas being present in the largest pores. If the components of a gas all have the same molecular mass, then gas density is independent of composition and is only a function of pressure.

Under static equilibrium and isothermal conditions, gas does not move. If the only forces operating on the gas are pressure and gravity, then these two forces must be balanced. The condition of zero motion results when:

$$\nabla P = -\rho_G g \nabla z \qquad\qquad [8.24a]$$

$$\frac{dP}{dz} = -\rho_G g \qquad\qquad [8.24b]$$

If gas density varies because of variable composition, the above equation cannot be simplified further. If the gas is barotropic (i.e., $\rho_G = \rho_G(P)$ only), however, the ideal gas law (Equation [8.17]) can be used to relate density and pressure when pressure variations are close to atmospheric. Equation [8.24a] can be solved through integration by inserting the ideal gas law:

$$\frac{RT}{gM}\int_{P_0}^{P}\frac{dP}{P} = -\int_{z_0}^{z}dz; \quad \frac{RT}{gM}\ln\left(\frac{P}{P_0}\right) = -z \qquad\qquad [8.25]$$

$$P = P_0 \exp\left(-\frac{Mg}{RT}z\right)$$

where $P$ is assumed to be equal to $P_0$ at $z = 0$. If $z = 1$ m, $T = 290$ $K$, $R$ is 8.314 $J$ mol$^{-1}$ K$^{-1}$, $M$ is 28.96 $g$ mol$^{-1}$ (air), and $g$ is 9.8 m s$^{-2}$, then $P = 0.99988$ $P_0$. Gas pressure therefore changes by 0.012% over a height of 1 m. Because the gas pressure changes are so small, pressure can generally be considered independent of height.

If gas density is only a function of pressure, (compressible gas), Equation [8.23] can be manipulated to obtain:

$$q_G = -\frac{k_G \rho_G}{\mu_G}\left(\frac{\nabla P}{\rho_G} + g\nabla z\right) = -\frac{k_G \rho_G}{\mu_G}\left(\nabla\int_{P_0}^{P}\frac{dP}{\rho_G} + g\nabla z\right) \qquad\qquad [8.26]$$

If gas density is independent of pressure, then the gas is incompressible:

$$\nabla\int_{P_0}^{P}\frac{dP}{\rho_G} = \nabla\left(\frac{P}{\rho_G}\right) \qquad\qquad [8.27]$$

$$q_G = -\frac{k_G \rho_G g}{\mu_G}\left(\nabla\left(\frac{P}{\rho_G g}\right) + \nabla z\right) = -\frac{k_G \rho_G g}{\mu_G}\left(\nabla h + \nabla z\right) = -\frac{k_G \rho_G g}{\mu_G}\left(\nabla H\right) \qquad\qquad [8.28]$$

where h is pressure head (L) and H is total head (L, pressure + gravitational head). Because gas densities are generally very low (density of air is approximately three orders of magnitude less than water density), the gravitational term in Equation [8.23] ($\rho_G g \nabla z$) is generally $< 1\%$ of the pressure term and is ignored in most cases, as shown in Equation [8.16]. For transient gas flow, Darcy's law (ignoring gravity, Equation [8.16]), substituted into the conservation of mass equation, results in:

$$\frac{\partial(\rho_G \theta_G)}{\partial t} = -\nabla \cdot \left( -\frac{\rho_G k_G}{\mu_G} \nabla P \right)$$

[8.29]

Equation [8.29] is nonlinear because gas density depends on gas pressure and can be linearized in terms of $P^2$ by assuming ideal gas behavior (Equation [8.17]), which results in (Section 8.7):

$$\frac{\partial P^2}{\partial t} \approx \frac{k_G P_0}{\phi \mu_G} \nabla^2 P^2 \approx \alpha \nabla^2 P^2$$

[8.30]

where $k_G P_0 / \phi \mu_G$ is the pneumatic diffusivity ($\alpha$). If pressure variations are small, another linear approximation to Equation [8.29] can be obtained:

$$\frac{\partial P}{\partial t} = \frac{k_G P_0}{\phi \mu_G} \nabla^2 P$$

[8.31]

Derivations of Equations [8.30] and [8.31] are given in Section 8.7.

## 8.2.2 GAS PERMEABILITY

Gas permeability describes the ability of the unsaturated zone to conduct gas. Permeability ($k$) should be simply a function of the porous medium if the fluid does not react with the solid. Gas permeability ($k_G$; $L^2$) is related to gas conductivity ($K_G$) as follows:

$$K_G = k_G \frac{\rho_G g}{\mu_G}$$

[8.32]

Some groups use the term *gas permeability* for $K_G$, which is called *gas conductivity*, and use the term *intrinsic permeability* for $k_G$, which is called *gas permeability*. Equation [8.32] shows that gas conductivity increases with gas density and decreases with gas viscosity.

The above material describes single-phase gas flow. In unsaturated media, the pore space generally contains at least two fluids, gas and liquid. Darcy's law is applied to each fluid in the system, which assumes that there is no interaction between the fluids (Dullien, 1979). Because the cross-sectional area available for flow of each fluid is less than if the system were saturated with a single fluid, the permeability with respect to that fluid is also less. The gas permeability decreases as the gas saturation decreases. The relative permeability ($k_{rG}$) is a function of the gas saturation ($S_G$) and is defined as the permeability of the unsaturated medium at a particular gas saturation ($k_G(S_G)$) divided by the permeability at 100% saturation ($k_G$):

$$k_{rG}(S_G) = k_G(S_G) / k_G$$

[8.33]

Relative permeability varies with (1) fluid saturation, (2) whether the fluid is wetting or nonwetting, and (3) whether the system is wetting or drying (hysteresis).

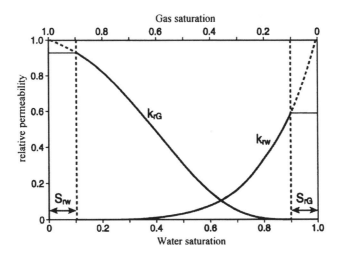

**FIGURE 8.2** Schematic relative permeabilities with respect to gas and water saturation.

The relative permeability of the gas phase is greater than that of the liquid phase at low to moderate liquid saturations because the gas generally occurs in larger pores (Demond and Roberts, 1987). Relative permeabilities do not sum to one. This has been attributed to flow pathways traversed by two fluid phases being more tortuous than those traversed by a single phase (Scheidegger, 1974) or to pores with static menisci that cannot result in flow (Demond and Roberts, 1987). Zero relative permeabilities correspond to nonzero fluid saturations (Figure 8.2). For example, in systems initially water saturated that begin to drain, gas will not begin to flow until a minimum gas saturation has been reached, the residual gas saturation ($S_{rG}$; Figure 8.2). The gas zero relative permeability region corresponds to trapped gas and disruption of gas connectivity caused by water blockages (Stonestrom and Rubin, 1989). Liquid-phase permeability also exhibits a zero-permeability region that corresponds to residual liquid or water saturation (Chapter 3). In natural systems, the zone of residual water saturation corresponds to a zone of constant gas relative permeability because gas content does not increase. The dashed line in Figure 8.2 in this zone corresponds with further increases in gas relative permeability ($k_{rG}$), which is related to increased gas content if the soil is oven dried and the water content is reduced to zero. A similar effect occurs with the water relative permeability ($k_{rw}$), where the dashed line shows increased water relative permeability corresponding to vacuum saturation or saturation of the sample under back pressure. Gas permeability is hysteretic at low water content, which indicates that gas permeability is not a unique function of gas saturation but depends on saturation history (i.e., whether the system is drying or wetting). At a given saturation, gas permeability is generally greater for wetting than for drying (Stonestrom and Rubin, 1989).

Various expressions have been developed to relate relative permeability to gas saturation (Table 8.3). Many of the expressions were developed to estimate relative permeabilities with respect to water; the corresponding expressions for gas relative permeability were obtained by replacing the effective liquid saturation ($S_e$) by $1-S_e$, where $S_e$ is:

$$S_e = \frac{S - S_r}{1 - S_r} \qquad [8.34]$$

where $S_r$ is the residual water saturation (Brooks and Corey, 1964).

Darcy's law can be written as:

$$J_G = -\frac{k_{rG}(S_G)k_G}{\mu_G}(\nabla P) = -\frac{k_G(S_G)}{\mu_G}\nabla P \qquad [8.35]$$

**TABLE 8.3**
**Expressions relating relative gas permeability and saturation. $S_e$ is the effective saturation with respect to water; $S_e = (S_w - S_{rw})/(1 - S_{rw})$, $S_{rw}$ is the residual wetting phase saturation (water in water-air system), $\lambda$ is the bubbling pressure, and $m$ and $n$ are fitting parameters.**

| | | |
|---|---|---|
| Brooks and Corey (1964) | $k_{rG} = \left(1 - S_e\right)^2\left(1 - S_e^{\frac{2+\lambda}{\lambda}}\right)$ | |
| Corey (1954) | $k_{rG} = (1 - S_e)^2(1 - S_e^2)$ | |
| Falta et al. (1989) | $k_{rG} = (1 - S_e)^3$ | |
| van Genuchten (1980); Mualem (1976) | $k_{rG} = \left(1 - S_e^{0.5}\right)\left\{1 - \left[1 - \left(1 - S_e\right)^{1/m}\right]^m\right\}^2$ | $m = 1 - 2/n$ <br> $0 < m < 1$ |
| van Genuchten (1980); Burdine (1953) | $k_{rG} = \left(1 - S_e\right)^2\left\{1 - \left[1 - \left(1 - S_e\right)^{1/m}\right]^m\right\}$ | $m = 1 - 2/n$ <br> $0 < m < 1;\ n > 2$ |

The velocity of the gas particles ($V_G$) (L t$^{-1}$) can be calculated from the volumetric flux density by dividing by the volumetric gas content:

$$V_G = J_G / \theta_G \qquad [8.36]$$

### 8.2.3 DEVIATIONS FROM DARCY'S LAW

Non-Darcian effects such as viscous slip flow and inertial flow can occur under many circumstances. As mentioned earlier, gas flow differs from liquid flow in that the velocity of the gas molecules is nonzero at the pore walls (Figure 8.1) and is called *slip velocity* (Dullien, 1979). Slip flow, or the Klinkenberg effect, results in underestimation of gas flux by Darcy's law. Because of slip flow, permeability depends on pressure and the flow equation is nonlinear (Collins, 1961). Klinkenberg (1941) evaluated the relationship between slip-enhanced, or apparent, permeability $k_G$ (L$^2$) and the permeability at infinite pressure $k$ (L$^2$) when the gas behaves like a liquid:

$$k_G = k\left[1 + \frac{4c\lambda}{\overline{\lambda}_p}\right] \qquad [8.37]$$

where $c$ is a constant characteristic of the porous medium, $\lambda$ is the mean free path of the gas at the mean pressure, and $\overline{\lambda}_p$ is the mean pore radius (Klinkenberg, 1941). Because the mean free path is inversely proportional to the mean pressure in the system, the equation can be modified to:

$$k_G = k\left(1 + \frac{b_i}{\overline{P}}\right) \qquad [8.38]$$

where $\overline{P}$ is the mean pressure and $b_i$ (M L$^{-1}$ t$^{-2}$) is a constant (slip parameter) that depends on the porous medium and the gas $i$ used in the measurement (Klinkenberg, 1941). These two equations show that gas slippage, or slip-enhanced permeability, is enhanced when the mean pressure is low. As the mean pressure increases, gas permeability decreases and approaches liquid permeability.

Detty (1992) found that the slip parameter is not only a function of the reciprocal mean pressure as indicated by Klinkenberg (1941), but also a function of the pressure gradient and saturation. Slip correction factors measured in unconsolidated sands are generally fairly low (1 to 6%) (Stonestrom and Rubin, 1989; Detty, 1992) and resemble those measured in consolidated sands (Estes and Fulton, 1956).

Non-Darcian behavior can also occur at high flow velocities as a result of inertial effects. According to Darcy's law, the flux is linearly proportional to the pressure gradient. At high flow velocities, the relationship becomes nonlinear and inertial effects result in fluxes lower than those predicted by Darcy's law. Forcheimer (1901) modified Darcy's law for high velocities:

$$\nabla P = \frac{\mu_G J_G}{k_G} + \beta \rho_G J_G{}^2 \qquad [8.39]$$

where P is pressure, $\mu_G$ is gas viscosity, $J_G$ is volumetric gas flux, and $\beta$ is the inertial flow factor ($L^{-1}$) (Detty, 1992). The second-order term results from kinetic energy losses from high-velocity flows. There is a spectrum from viscous (linear–laminar or Darcy flow) to inertial flow (fully turbulent) with a visco-inertial regime in between. Laminar flow is not restricted to the region where Darcy's law is valid, but extends into the visco-inertial regime where nonlinear–laminar flow occurs ($q \sim \nabla P^n$) (Detty, 1992). Detty (1992) indicated that deviations from viscous flow can be significant. The flow rates and pressure gradients that result in visco-inertial flow vary with water saturation.

## 8.3   MULTICOMPONENT GAS TRANSPORT

In isobaric systems, gas transport occurs by diffusion, whereas in nonisobaric systems gas transport occurs by advection (Darcy's law) and diffusion. A variety of models are available to describe multicomponent gas diffusion. Traditionally, gas diffusion has been described by Fick's law. As noted earlier, Fick's law is valid strictly for isothermal, isobaric and equimolar countercurrent diffusion of a binary gas mixture. Theoretical analysis by Jaynes and Rogowski (1983) shows that the diffusion coefficient of Fick's law is only a function of the porous medium under certain conditions: equimolar countercurrent diffusion in a binary gas mixture, diffusion of a dilute component gas in a multicomponent gas mixture, and diffusion in a ternary system when one gas component is stagnant (zero flux, no sources or sinks, no reactions). Studies by Leffelaar (1987) indicate that if binary gas diffusion coefficients differ by more than a factor of 2, then Fick's law is invalid. Amali and Rolston (1993) also added that the total mole fraction of the diffusing gas components needs to be considered and should be low in order for Fick's law to be applicable.

If the flux of each gas component depends on the flux of the other gas components, then Fick's law no longer applies. In some cases, the calculated Fick's law flux is not only different in magnitude relative to multicomponent molecular diffusive flux, but opposite in direction also, as a result of molecular diffusion against a concentration gradient (Thorstenson and Pollock, 1989). Because Fick's law predicts the diffusion of only one component, variations in concentration of other components are attributed to other processes. Baehr and Bruell (1990) showed that physical displacement of naturally occurring gases such as $O_2$ by organic vapors and evaporative advective fluxes can be incorrectly attributed to aerobic microbial degradation or other sink terms when Fick's law is used. Unlike Fick's law, which can only consider binary gas mixtures, the Stefan-Maxwell equations or the dusty gas model can evaluate multicomponent gas transport. For multicomponent analysis, the transport of one component depends upon the transport of all other components that are present in the gas mixture. It is impossible to separate fully the effects of diffusion from those of advection in multicomponent gas systems (Thorstenson and Pollock, 1989). As noted earlier, Stefan-Maxwell equations are valid strictly for isobaric systems where Knudsen diffusion is negligible.

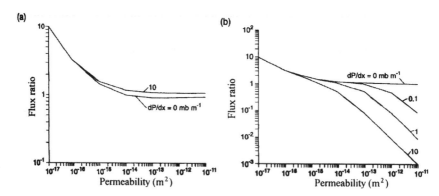

**FIGURE 8.3** (a) Ratios of TCE fluxes calculated using the single-component equation (Fick's law for diffusion) divided by fluxes calculated using the dusty gas model; (b) ratios of TCE fluxes calculated using the Stefan-Maxwell equation divided by fluxes calculated using the dusty gas model (Massmann and Farrier, 1992).

The validity of the single-component advection–diffusion (Fick's law) equation for simulating gas phase transport depends on the pore sizes and permeability of the porous media, the relative concentrations of the gas components that are involved, and their molecular weights, viscosities and pressure gradients. Massmann and Farrier (1992) compared fluxes using the single-component advection–diffusion equation with those calculated using the multicomponent Stefan-Maxwell equation and dusty gas model. The comparisons were developed for transport conditions similar to what might be observed in volatile organic compounds (VOCs) in the near surface vadose zone. For total pressure gradients ranging from 100 to 1,000 Pa m$^{-1}$ (1 to 10 mbar m$^{-1}$), the single-component advection–diffusion (Fick's law) equation significantly overestimates fluxes for toluene and trichloroethylene (TCE) in soils having permeabilities on the order of 10$^{-16}$ to 10$^{-17}$ m$^2$ (Figure 8.3a). These permeabilities might correspond to unweathered clays, glacial tills or very fine silts (Table 8.4). The Stefan-Maxwell equations also overestimate fluxes in this permeability range because Knudsen diffusion is not included in Fick's law or in the Stefan-Maxwell equations (Figure 8.3b). The flux predicted by the single-component advection–diffusion equation is within a factor of 2 of that predicted by the dusty gas model in materials having permeabilities greater than 10$^{-14}$ m$^2$. This permeability corresponds to a relatively dry silty-sand material (Table 8.4). The Stefan-Maxwell equation underestimates fluxes in high-permeability materials for moderate total pressure gradients (0.1 to 1 mbar m$^{-1}$; 10 to 100 Pa m$^{-1}$) (Figure 8.3b). Massmann and Farrier (1992) also illustrated how multicomponent equations may result in situations in which diffusion can occur in a direction opposite that of the partial-pressure gradients. In multicomponent equations, the diffusive flux of one component depends on the diffusive fluxes of all other components in the system. For example, a large partial-pressure gradient for TCE can cause the diffusive flux of N$_2$ and O$_2$ to occur in a direction opposite to their partial pressure gradients. In general, single component equations will be more valid for dilute gases having molecular weights similar to those of other species in the gas mixture.

Abriola et al. (1992) evaluated advective and diffusive fluxes in a system comprising two components. They reported that under an applied total pressure gradient of 0.05 mbar m$^{-1}$ (5 Pa m$^{-1}$), the single and multicomponent models give virtually indistinguishable results for transport predictions in a sandy soil. For transport in a low permeability material, such as a clay under a total pressure gradient of 0.05 mbar m$^{-1}$, diffusive fluxes dominate, and single- and multicomponent equations give different transport predictions. Multicomponent gas experiments that evaluated transport of methane and TCE in air were examined by Abu-El-Sha'r (1993). The system was evaluated as either a binary (considering N$_2$ and O$_2$ in air as a single component with methane or TCE as the other component) or a ternary system. Results of the experiments showed that model predictions based on Fick's law, the Stefan-Maxwell equations and the dusty gas model were

**TABLE 8.4**
**Typical values of permeability for different types of sediments and rocks.**
(From Terzaghi and Peck, 1968, John Wiley & Sons, New York.)

$10^{-7}$ $10^{-8}$ $10^{-9}$ $10^{-10}$ $10^{-11}$ $10^{-12}$ $10^{-13}$ $10^{-14}$ $10^{-15}$ $10^{-16}$ $10^{-17}$ $10^{-18}$ $10^{-19}$ $10^{-20}$ $m^2$

| Clean gravel | Clean sands, clean sand and gravel mixtures | Very fine sands, silts, mixtures of sand, silt, and clay, stratified clay deposits, etc. | Homogeneous clays below zone of weathering |
|---|---|---|---|

(From Freeze and Cherry, 1979, *Groundwater*, 604, Table 2.2, p. 29, Prentice-Hall, Englewood Cliffs, NJ.)

$10^{-7}$ $10^{-8}$ $10^{-9}$ $10^{-10}$ $10^{-11}$ $10^{-12}$ $10^{-13}$ $10^{-14}$ $10^{-15}$ $10^{-16}$ $10^{-17}$ $10^{-18}$ $10^{-19}$ $10^{-20}$ $m^2$

Gravel

Clean sand and silty sand

Silt, loess

Glacial till

Unweathered marine clay

Shale

indistinguishable in sand samples, whereas the dusty gas model provided slightly better predictions in kaolinite. All three models can therefore predict diffusional gas fluxes under isobaric conditions when transport is dominated by molecular diffusion. However, under pressure gradients, deviations of measured and predicted values of gas fluxes increased with increasing total pressure gradients. All three models underestimated gas fluxes under nonisobaric conditions. Deviations from measured fluxes were lowest for the dusty gas model, increased for the Stefan-Maxwell equations, and were greatest for Fick's law. Theoretical analysis by Thorstenson and Pollock (1989) indicates that multicomponent analysis using the dusty gas model is required for evaluation of stagnant gases such as $N_2$ or Ar in the air. Analysis of transient gas flux on the basis of column experiments using benzene and TCE shows that Fick's law and the Stefan-Maxwell equations give similar results in the transient phase of the experiment, but that Fick's law underestimates measured fluxes by as much as 10% in the steady-state phase. The Stefan-Maxwell equations gave much better results (Amali et al., 1996).

These studies were used to develop Table 8.5, which generally outlines which models are most applicable under which pressure, permeability and concentration conditions. For low-permeability material, the dusty gas model is required because it incorporates Knudsen diffusion. For high-permeability material under isobaric conditions, Fick's law can be used if the diffusing gas component has a low concentration, whereas the Stefan-Maxwell equation is required if concentration is high because of the interdependence of flux. The Stefan-Maxwell equations or the dusty gas model can be used when Fick's law applies. Similarly, the dusty gas model can also be used when the Stefan-Maxwell equations are valid. Under nonisobaric conditions, Darcy's law is combined with a diffusion model. The single-component advection diffusion (Fick's law) model can be used in low concentration situations in high-permeability material, whereas the dusty gas model is required in high concentration situations. The Stefan-Maxwell equations generally do not apply

**TABLE 8.5**
**Most appropriate model for describing gas flux under different pressure, permeability and concentration conditions. Pressure gradient refers to the existence of an external pressure gradient, concentration refers to concentration of the diffusing gas component, dusty gas model\* includes Darcy's law to describe advective gas flow.**

| Pressure Gradient | Permeability | Low Concentration | High Concentration |
|---|---|---|---|
| Isobaric | low | dusty gas model | dusty gas model |
| Isobaric | high | Fick's law | Stefan-Maxwell |
| Nonisobaric | low | dusty gas model* | dusty gas model* |
| Nonisobaric | high | advection diffusion | dusty gas model |

under nonisobaric conditions (Massmann and Farrier, 1992; pressure gradients $> 10$ Pa m$^{-1}$, Figure 8.3b).

### 8.3.1 DENSITY-DRIVEN ADVECTION

Density-driven advective transport, a specific type of multicomponent gas transport, is important for high molecular-weight compounds that have high vapor densities, such as dense volatile organic contaminants (Falta et al., 1989; Mendoza and Frind, 1990; Mendoza and McAlary, 1990). Important factors controlling density-driven advective flow include saturated vapor density, molecular mass of the chemical, and gas phase permeability. Vapor densities are maximized when the air mixture is saturated by the vapor phase of the chemical. Such high saturations are restricted to areas close to the free phase. Numerical simulations indicate that density-driven advective flow in homogeneous media is important at the high permeabilities ($> 10^{-11}$ m$^2$) typical of sands and gravels (Falta et al., 1989). The presence of fractures can greatly enhance density-driven advective flow.

An order of magnitude estimate of the advective gas flux resulting from density variations may be obtained from (Falta et al., 1989):

$$J_G = -\frac{k_G g}{\mu_G} \nabla (\rho - \rho_0)$$

[8.40]

Equation [8.40] describes steady-state downward flux of fluid of density $\rho$ through a stagnant fluid of density $\rho_0$ in material of permeability $k_G$ and ignores diffusion and phase partitioning. An estimate of the density potential of the gas mixture ($\rho$) with respect to air ($\rho_0$) can be obtained from relative vapor density ($\rho_{rv}$) (Mendoza and Frind, 1990):

$$\rho_{rv} = \frac{\rho}{\rho_0} \frac{x_c M_c + (1 - x_c) M_a}{M_a}$$

[8.41]

where $x_c$ is the fractional molar concentration of the compound and $M_c$ and $M_a$ are the molecular weights of the compound and air, respectively. Relative vapor density of the source depends on the vapor pressure and molecular weight of the compound (Mendoza and Frind, 1990). Density-driven advection occurs generally in areas contaminated by high-molecular-weight volatile organic contaminants that have high vapor densities.

## 8.4  METHODS

### 8.4.1  ESTIMATED PARAMETERS

Because theoretically permeability should be independent of fluid used, gas and liquid permeabilities should be the same. In fine-grained media or under small mean pressures, gas permeability is greater than liquid permeability because of gas slippage (Klinkenberg effect). Permeability can be corrected for gas slippage using Equation [8.38]. If gas slippage is negligible, as in coarse media having high mean pressures, gas conductivity can be estimated from hydraulic conductivity by:

$$K_G = K_w \left( \frac{\rho_G}{\rho_w} \right) \left( \frac{\mu_w}{\mu_G} \right) \qquad [8.42]$$

Air conductivities are about one-tenth of hydraulic conductivities because the air density is approximately three orders of magnitude less than water density and air viscosity is approximately 50 times less than water viscosity (Table 8.1). The estimated gas conductivities (Equation [8.42]) assume that pores are completely filled with gas. Typical values of permeability for different sediment textures are found in Freeze and Cherry (1979) and Terzaghi and Peck (1968) (Table 8.4), and permeability varies over 13 orders of magnitude from clay to gravel.

A rough estimate of gas permeability in gas saturated systems can be obtained from particle size data:

$$k = C\left( D_{10} \right)^2 \qquad \text{(Hazen, 1911)} \qquad [8.43]$$

$$k = 1,250(D_{15})^2 \qquad \text{(Massmann, 1989)} \qquad [8.44]$$

where $C$ is a dimensionless shape factor and $D_{10}$ and $D_{15}$ correspond to the grain diameters at which 10 or 15% of the particles by weight are finer. Other estimates of exponents include 1.65 and 1.85 instead of 2 in the Hazen formula (Shepherd, 1989).

The Klinkenberg $b$ parameter can be estimated according to the following empirical equation developed by Heid et al. (1950) for air-dry consolidated media (standard correction for the Klinkenberg effect by the American Petroleum Industry):

$$b_{air} = \left( 3.98 \times 10^{-5} \right) k_\infty^{-0.39} \qquad [8.45]$$

where $b$ is in atmospheres and $k_\infty$ is in cm$^2$. The Klinkenberg $b$ parameter for any gas ($b_i$) can be estimated from that for air ($b_{air}$), developed by Heid et al. (1950) according to the following (Thorstenson and Pollock, 1989):

$$b_i = \left( \mu_i / \mu_{air} \right) \left( M_{air} / M_i \right)^{1/2} b_{air} \qquad [8.46]$$

The Knudsen diffusion coefficient is related to the Klinkenberg effect because both are related to the ratio of the mean free path of the gas molecules to the pore radius. Thorstenson and Pollock (1989) showed how the Knudsen diffusion coefficient ($D_i^k$) can be estimated from the true permeability and the Klinkenberg $b$ parameter for gas $i$:

$$D_i^k = k b_i / \mu_i \qquad [8.47]$$

Gas conductivities and permeabilities vary with gas content. Estimates of gas conductivity at different gas saturations are provided by equations listed in Table 8.3.

## 8.4.2 LABORATORY TECHNIQUES

A variety of laboratory techniques are available for measuring parameters related to gas movement, such as gas pressure and permeability. Gas pressure can be measured with U-tube manometers containing different fluids, such as water or mercury. Such manometers generally measure differential pressure because one end of the tube is inserted into soil or rock and the other is exposed to the atmosphere. Manometers can be inclined to increase sensitivity. Pressure transducers are used widely to measure absolute or differential gas pressure. The operational range of these instruments varies, and their precision is generally a percentage of the full-scale measurement range. Manufacturers include Setra (Acton, Massachusetts) and Microswitch Honeywell (Freeport, Illinois). Because transducers are subject to drift, they have to be calibrated periodically. Data can be recorded automatically in a data logger.

Gas permeability can be measured in the laboratory on undisturbed or repacked cores. Repacking should be done only on samples having low clay content. Because repacking changes structure, gas transport parameters are affected. The more structured the soil, the bigger the potential change. Clay soils tend to be more structured. Permeability measurements include determination of flow rate of each phase under an applied pressure gradient and measurement of saturation in unsaturated systems (Scheidegger, 1974). Various methods used to measure air and water permeabilities include those in which both phases move and are measured at the same time and those in which the permeability of one fluid phase is measured while the other phase remains stationary. Steady-state methods that hold the wetting phase stationary are used most widely (Corey, 1986). Tempe cells (Soilmoisture Equipment Corp., Santa Barbara, California) used for measuring water retention functions can be adapted as permeameters and include a sample holder with ceramic end plates. Alternatively, if sample shrinkage is expected, flexible wall permeameters can be used to minimize air flow along the annulus between the sample and the holder. Sharp et al. (1994) described an electronic minipermeameter for measuring gas permeability in the laboratory. Gas is injected at a constant pressure and the steady-state flow is measured by electronic mass flow transducers. Permeabilities can be measured over a wide range ($10^{-15}$ to $10^{-8}$ m$^2$). The Hassler method (Hassler, 1944) controls the capillary pressure at both ends of a soil core by means of capillary barriers and measures air and water relative permeabilities at the same time under a pressure gradient. The capillary barriers separate wetting and nonwetting phases. The Hassler method was used by Stonestrom (1987) and by Springer (1993). Other procedures for laboratory measurement of gas permeability were described in Springer et al. (1995) and Detty (1992). Darcy's law is used to estimate gas permeability:

$$k_G(S_G) = -\left(J_G \mu_G\right) / (dP / dz)$$ [8.48]

Additional equipment required for gas permeability measurements includes a pressure transducer or manometer, a flow sensor such as a soap film bubble meter, and a temperature sensor such as a thermistor. Gas permeabilities are measured at different gas saturations or water contents. If the sample is initially saturated with water, a minimum pressure must be reached (air entry pressure) before a continuous gas phase is achieved. Gas permeability increases as water content decreases.

The Klinkenberg $b$ parameter can be estimated by plotting gas permeabilities measured at different mean pressures ($\bar{P}$) as a function of the reciprocal mean pressure at which the tests were performed (Equation [8.38]). Rearranging Equation [8.38] results in:

$$k_G = k\left(1 + b_i / \bar{P}\right) = k + kb_i(1 / \bar{P})$$ [8.49]

Therefore, the slope is the product of $k$ times $b_i$ and the intercept is the true permeability.

Gas diffusivities can be measured in open, semi-open and closed systems (Abu-El-Sha'r, 1993). An open system involves component gases flowing past the edges of the soil. Pressure gradients and absolute pressures can be controlled by regulating the flow rate of the component gases. A semi-open system is generally termed a Stefan tube that consists of the diffusing substance in liquid form at the base and either free air (if measuring free-air diffusivity) or the porous medium (if measuring effective diffusivity of the porous medium) at the top. Closed systems generally consist of two chambers connected by a capillary or chamber filled with the porous medium (Glauz and Rolston, 1989). This system can be used in the study of noxious gases. Semi-open systems are generally used in hydrology to measure effective binary diffusion coefficients. Fick's law is generally used to analyze these experiments. Although most studies in the past used nonequimolar pairs of gases, none of these studies considered nonequimolar diffusion. Experiments conducted by Abu-El-Sha'r (1993) were the first to use an equimolar pair of gases ($N_2$ and CO) to determine effective binary molecular diffusion coefficients. Details of various procedures for measuring molecular diffusion coefficients are outlined in Rolston (1986).

Single-gas experiments are used to measure the Knudsen diffusion coefficient by applying the dusty gas model (Abu-El-Sha'r, 1993):

$$N_i^T = -\left(\frac{D_i^k}{RT} + \frac{\bar{P}k}{\mu_G RT}\right)\nabla P \qquad [8.50]$$

Rearranging Equation [8.50] results in:

$$N_i^T = -\left(\frac{D_i^k \mu_G}{\bar{P}} + k\right)\frac{\bar{P}}{\mu_G RT}\nabla P \qquad [8.51]$$

Plotting $N_i^T LRT/\Delta P$ versus $\bar{P}$ should result in a straight line with an intercept of $D_i^k$ and a slope of $k/\mu_G$, where $L$ is the length of the column in the experiment (Abu-El-Sha'r, 1993).

### 8.4.3 FIELD TECHNIQUES

#### 8.4.3.1 Estimation of Gas Permeability for Advective Gas Flow

Advective transport of gases depends on gas permeability and pressure gradient. Gas permeability can be estimated from (1) analysis of atmospheric pumping data, (2) pneumatic tests, or (3) measurements by air minipermeameters. Comparison of gas permeabilities from laboratory and field indicates that field derived estimates of gas permeabilities generally exceed laboratory derived estimates by as much as orders of magnitude (Weeks, 1978; Edwards, 1994). These differences in permeability are attributed primarily to the increase in scale from laboratory to field measurements and inclusion of macropores, fractures and heterogeneities in field measurements. Field permeability measurements in low permeability media include the effects of viscous slip and Knudsen diffusion.

#### 8.4.3.2 Analysis of Atmospheric Pumping Data

Comparison of temporal variations in gas pressure, monitored at different depths in the unsaturated zone, with atmospheric pressure fluctuations at the surface can be used to determine minimum vertical air permeability between land surface and monitoring depth (Weeks, 1978; Nilson et al., 1991). Differential pressure transducers are used to monitor gas pressures in the unsaturated zone. Gas ports generally consist of screened intervals in boreholes of varying diameter. Flexible tubing (Cu or nylon) connects the gas port at depth with a differential pressure transducer at the surface.

One port of the transducer is left open to the atmosphere. Atmospheric pressure is monitored at the surface by a barometer.

Data analysis consists of expressing the variations in atmospheric pressure as time harmonic functions. The attenuation of surface waves at different depths in the unsaturated zone provides information on how well or poorly unsaturated sections are connected to the surface. The accuracy of the results increases with the amplitude of the surface signals. Pressure fluctuations resulting from irregular weather variations change by as much as 20 to 30 mbar (2,000 to 3,000 Pa) during a 24-h period (Massmann and Farrier, 1992).

If the surface pressure (upper boundary condition) is assumed to vary harmonically with time, and the water table or a low-permeability layer acts as a no-flow boundary, an analytical solution can be derived (Carslaw and Jaeger, 1959, in Nilson et al., 1991). Pneumatic diffusivity can be estimated graphically by means of the amplitude ratio. The ratio of the amplitude at a certain depth $z$ compared with the amplitude at the surface can be obtained graphically or by time-series analysis (Rojstaczer and Tunks, 1995). Air permeability is estimated from the pneumatic diffusivity by dividing by the volumetric air content.

---

**EXAMPLE 8.1**   Estimation of gas conductivity using subsurface attenuation of barometric pressure fluctuations

The use of barometric pressure fluctuations to estimate subsurface gas conductivity is demonstrated with data from the Hanford Nuclear Reservation in the Columbia Plateau of south-central Washington state. The depth to the water table at the site is 66 m. The stratigraphy in the unsaturated zone consists of about 38 m of permeable sand and gravel, underlain by an 8-m thick caliche layer, and a 20-m thick gravel layer to the water table.

The average barometric pressure at the Hanford Nuclear Reservation is approximately 100,2144 Pa. Fluctuations in barometric pressure cause vertical pressure gradients in subsurface gases, as shown in Diagram 8.1.1. These data are from a stainless-steel, 15-cm diameter well with a single-screened interval from 52.5 to 56.6 m which is beneath the caliche layer. Pressure differentials, defined as subsurface pressure minus barometric pressure, are on the order of 1,000 Pa in this deeper layer.

Pressure attenuation and phase shift at a given frequency can be used to infer subsurface vertical gas conductivity. The calculated conductivity reflects the lowest conductivity in the section, which is the caliche layer in this section. A low-pass filter and a Fourier transform were applied to isolate diurnal and cyclonic periods (Diagrams 8.1.2 and 8.1.3). A 24-h period results in an attenuation of 0.24 and a 205-h period results in an attenuation of 0.70.

$$\frac{P - P_0}{\Delta P}$$

Nilson et al. (1991) described how to calculate the gas conductivity from the amplitude ratio $((P - P_0) / \Delta P)$ assuming the medium is homogeneous and gas flow can be described by Equation [8.31] and that the surface pressure varies harmonically as $P = P_0 + \Delta P e^{i\omega t}$:

$$\frac{P - P_0}{\Delta P} = \frac{\cosh \lambda \sqrt{i}\left(1 - \dfrac{z}{L}\right)}{\cosh \lambda \sqrt{i}} e^{i\omega t}$$

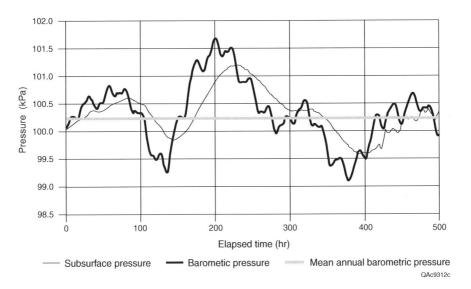

**DIAGRAM 8.1.1** Surface and subsurface pressure variations. Reprinted by permission of Ground Water. Copyright 1999.

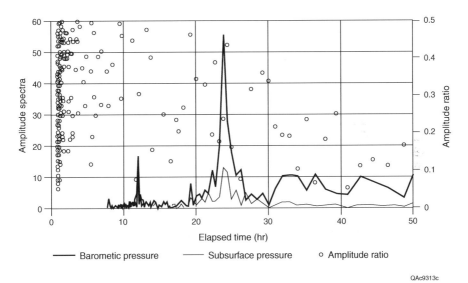

**DIAGRAM 8.1.2** Pressure variations in the frequency domain (daily pressure variations).

where $\lambda = L\sqrt{\dfrac{\omega}{\alpha}} = \left[\dfrac{2\pi}{\alpha T / L^2}\right]^{1/2}$, $\sqrt{i} = (1+i)/\sqrt{2}$, z/L is the fractional depth, $\alpha$ is the pneumatic diffusivity $(k_G P_0)/(\phi\mu_G)$, $\phi$ is the volumetric gas content, $\mu_G$ is the gas viscosity, $T = 2\pi/\omega$ is the period of the pressure disturbance, and $\omega$ is the angular frequency. The caliche layer is the likely source of attenuation and barometric pressure boundary conditions are applied to the top of this layer. Therefore, the fractional depth is based on the caliche layer:

$$z/L = (52.5 - 38)/(66 - 38) = 0.518$$

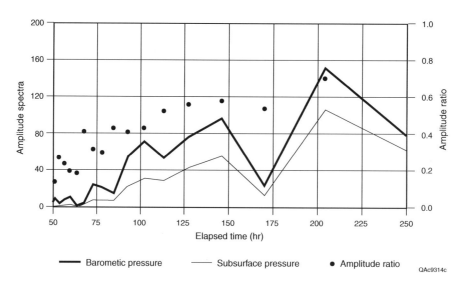

DIAGRAM 8.1.3 Pressure variations in the frequency domain (cyclonic pressure variations).

The amplitude ratio and the fractional depth are used to calculate $\lambda$ which is then used to calculate $\alpha$. Once $\alpha$ has been determined, the gas permeability can be calculated by inserting a value for the volumetric gas content (0.15 for the caliche layer). The gas conductivity can then be calculated according to Equation [8.32]. The diurnal signal results in an amplitude ratio of 0.24 and a conductivity of $0.73 \times 10^{-5}$ cm s$^{-1}$, whereas the cyclonic signal (205 hr) results in an amplitude ratio of 0.70 and a gas conductivity of $0.41 \times 10^{-5}$ cm s$^{-1}$. These values are similar to that calculated by calibration of a numerical model ($8 \times 10^{-6}$ cm s$^{-1}$ (Ellerd et al., 1999).

### 8.4.3.3 Pneumatic Tests

Pneumatic tests are also used widely to evaluate gas permeability in the unsaturated zone. In these tests, air is either injected into or extracted from a well and pressure is monitored in gas ports installed at different depths in surrounding monitoring wells (Figure 8.4). A reversible air pump is used to inject or extract air. Most analyses of pneumatic tests assume that the gas content ($\theta_G$) is constant over time, i.e., that no redistribution of water occurs during the test. To evaluate this assumption, injection or extraction tests should be conducted at several different rates. If results from the different rates are similar, the assumption of constant gas content is valid. The tests can be conducted in horizontal or vertical wells.

A variety of techniques are available for analyzing pneumatic tests. The initial transient phase of the test or the steady-state portion of the test can be analyzed. If transient data are available, volumetric gas content can also be estimated. Analysis of pneumatic tests resembles the inverse problem in well hydraulics, where permeabilities are estimated from pressure data. Solutions for estimating gas permeability differ in terms of the boundary conditions that are assumed at the ground surface (such as unconfined, leaky confined and confined) and the method of solution. The lower boundary is generally assumed to be the water table or an impermeable layer. All solutions assume radial flow to a vertical well.

As discussed previously, the gas flow equation is nonlinear because of the pressure dependence of the density, viscosity and permeability (Klinkenberg effect). In many cases the pressure dependence of the density is approximated by ideal gas behavior (Equation [8.17]). Under low to moderate pressures and pressure gradients typical of unsaturated media, pressure dependence of the viscosity

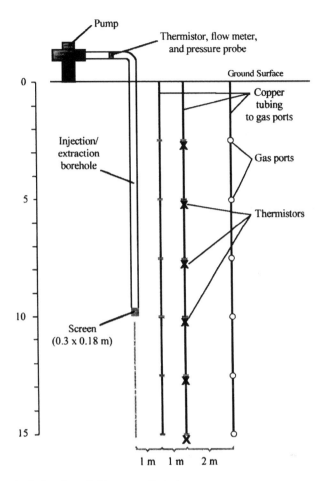

**FIGURE 8.4** Schematic design for a field pneumatic test.

can be neglected. In most analyses, the Klinkenberg effect is also neglected (Massmann, 1989; Baehr and Hult, 1991). If pressure variations are assumed to be small, the transient gas flow Equation [8.52] can be written in a form similar to that of the groundwater flow Equation [8.53]:

$$\frac{\theta_G \mu_G}{P_0} \frac{\partial P}{\partial t} = \nabla(k_G \nabla P) \qquad [8.52]$$

$$S_s \frac{\partial h_w}{\partial t} = \nabla(K_w \nabla h_w) \qquad [8.53]$$

where $S_s$ is specific storage, $h_w$ is hydraulic head (L) and $K_w$ is hydraulic conductivity (L $t^{-1}$).

A summary of various techniques for evaluating field-scale pneumatic tests is provided in Table 8.6. Massmann (1989) used many techniques developed for groundwater hydraulics to analyze transient gas tests and used a modified Theis solution for systems with no leakage from the ground surface and a modified Hantush solution for systems with leakage. The computer software AQTE-SOLV (Duffield and Rumbaugh, 1989) was used to estimate the parameters according to the Marquadt nonlinear least squares technique. Johnson et al. (1990a) developed an analytical model to evaluate transient (1-D) radial flow in a homogeneous, isotropic flow field. McWhorter (1990) developed an analytical model to analyze transient flow in a 1-D, radially symmetrical flow field

**TABLE 8.6**
**Various approaches for evaluating field scale pneumatic tests (modified from Massmann and Madden, 1994).**

| Approach | Transient or steady state | No. of dimensions | Layered or homogeneous | Anisotropic or isotropic | Data |
|---|---|---|---|---|---|
| Massmann (1989) | transient | 1-D | layered | anisotropic | data |
| Johnson et al. (1990a,b) | transient | 1-D | homog. | isotropic | data |
| McWhorter (1990) | transient | 1-D | homog. | isotropic | data |
| Baehr and Hult (1991) | steady state | 2-D | layered | anisotropic | data |
| Shan et al. (1992) | steady state | 2-D | homog. | anisotropic | no data |
| Croise and Kaleris (1992) | steady state | 2-D | layered | anisotropic | data |
| Massmann and Madden (1994) | transient | 2-D | layered | isotropic | data* |

\* includes horizontal wells; all other studies do not include horizontal wells.

that includes the nonlinear effects of compressible flow. He found that the nonlinear effects resulting from the pressure dependence of permeability were negligible for pressures within $\pm 20\%$ of $P_0$.

The transient phase of gas tests is generally short (approximately seconds to hours) (Edwards, 1994), and it is sometimes difficult to collect reliable data. Many studies analyze the steady-state portion of the pneumatic test. Baehr and Hult (1991) developed analytical solutions for steady-state, 2-D, axisymmetric gas flow to a partially screened well for an open system and a leaky confined system. Analysis of the leaky confined system required a Hantush type leakage term. A computer code (AIR2D) is available that includes these analytical solutions (Joss and Baehr, 1997). Falta (1995) developed analytical solutions for steady-state and transient gas pressure and stream function fields using parallel horizontal injection and extraction wells where the ground surface is open to the atmosphere. Shan et al. (1992) also developed an analytical solution for steady-state flow in homogeneous and anisotropic media that includes the effects of leakage. A constant pressure upper boundary is used that assumes that the system is completely open. Horizontal and vertical gas permeabilities are estimated using type curves. Kaluarachchi (1995) developed an analytical solution for 2-D, axisymmetric flow with anisotropic gas permeability that includes the Klinkenberg effect. Errors resulting from neglecting the Klinkenberg effect are highest in low permeability materials and near the well. Edwards (1994) used a numerical model of steady-state, radial and vertical, anisotropic, heterogeneous compressible flow with an optimization routine to estimate gas permeabilities. All tests reached steady state before 5 s (Edwards, 1994). Results of this analysis indicate that vertical heterogeneities were significant. Variations in gas permeabilities with depth were therefore attributed to increases in water content with depth. Test data can also be analyzed by means of groundwater flow models such as MODFLOW with preprocessing (Joss and Baehr, 1995), which is discussed in Section 8.5.

### 8.4.3.4 Air Minipermeameters

A minipermeameter is a device for measuring gas flow that is used to determine permeability in the field at a localized scale. Measurements are made rapidly and are nondestructive. A mechanically based field minipermeameter was described by Goggin et al. (1988). Compressed $N_2$ is injected at a constant pressure through a tip pressed against the measurement surface. The steady-state flow rate is measured by a series of rotameters, and the gas injection pressure is measured at the tip seal. The flow rate at a particular injection pressure is calibrated against measurements on core plugs of known permeability.

### 8.4.3.5 Evaluation of Diffusive Transport Parameters

Knudsen diffusion coefficients are measured in the laboratory using single gas experiments (Section 8.4.2), while the effective molecular diffusion coefficient can be determined from field experiments (Raney, 1949; McIntyre and Philip, 1964; Lai et al., 1976; Rolston, 1986). The simplest method for measuring gas diffusivity in near surface soils involves inserting a tube into the soil surface and supplying gas from a chamber over the inserted tube (McIntyre and Philip, 1964; Rolston, 1986). Gas diffusivity can be calculated by measuring initial gas concentration in the soil and chamber and gas concentrations in the chamber at different times. An independent measurement of volumetric gas content in the soil is also required. If gas samplers are installed at different depths in the soil and the gas concentration gradient is calculated, gas flux can be estimated by using information on the diffusivity and gas concentration gradient in conjunction with Fick's law. Kreamer et al. (1988) used atmospheric fluorocarbons to determine field tortuosities. A permeation device was used that slowly released fluorocarbon gases, the concentrations of which were monitored for several days. Nicot (1995) used instantaneous release of a tracer and continuous monitoring of gas concentration to determine field tortuosities. The data resulting from the tracer tests can be evaluated by analytical or numerical methods.

Carslaw and Jaeger (1959) presented a derivation of an analytic solution for diffusion in an isotropic, homogeneous and infinitely porous medium. The partial differential equation reduces to:

$$\frac{\partial C}{\partial t} = D_e \frac{\partial^2 C}{\partial r^2} + D_e \frac{2}{r} \frac{\partial C}{\partial r} \qquad [8.54]$$

where $C$ is the concentration (M L$^{-3}$), $D_e$ is the effective diffusion coefficient, and $r$ is the radial distance (L). The solution to the above equation, based on instantaneous release of mass M at a point source, is:

$$C = \frac{M}{8(\pi D_e t)^{3/2}} \exp\left(-\frac{r^2}{4D_e t}\right) \qquad [8.55]$$

where $M$ is the molecular mass (M). The inverse problem is solved by estimating effective diffusion coefficients from the measured concentration data, a procedure similar to the one described for permeability estimation.

## 8.5 APPLIED NUMERICAL MODELING

### 8.5.1 SINGLE PHASE FLOW

A summary of the various types of codes available for simulating gas flow is provided in Table 8.7. Most numerical simulations of water flow in unsaturated media have generally ignored the gas phase by assuming that the gas is at atmospheric pressure. The equation that is solved is the Richards equation, which describes a single-phase (liquid), single-component (water) system. The Richards approximation is generally valid for most cases of unsaturated flow (Chapters 3 and 4).

A variety of numerical models have been developed to simulate gas flow in unsaturated systems. An important consideration in choosing a code for evaluating gas transport is the assumptions of each code. Groundwater flow models can be used to simulate gas flow in cases where the vapor behaves as an ideal gas and where pressure fluctuations are small and gas content is constant (no water redistribution) (Massmann, 1989). Such assumptions are generally valid for vapor extraction remediation systems. The code most widely used to simulate groundwater flow is MODFLOW

**TABLE 8.7**
**Summary of types of codes available to simulate gas flow.**

| Code | Dim. | No. Phases | Components | Energy Balance | Porous/Fractured Systems |
|------|------|------------|------------|----------------|--------------------------|
| AIR3D | 3 | 1 | gas | no | porous |
| BREATH | 1 | 2 | water | yes | porous |
| SPLaSHWaTr | 1 | 2 | water | yes | porous |
| UNSAT-H | 1 | 2 | water | yes | porous |
| Princeton Code | 2 | 2 | water and air | no | porous |
| FEHM | 3 | 2 | water and air | yes | porous/fractured |
| TOUGH | 3 | 2 | water and air | yes | porous/fractured |
| STOMP | 3 | 3 | water, air, NAPL | yes | porous/fractured |

(McDonald and Harbaugh, 1988). Joss and Baehr (1995) developed a sequence of computer codes (AIR3D) adapted from MODFLOW to simulate gas flow in the unsaturated zone. The codes can be used to simulate 3-D air flow in a heterogeneous, anisotropic system including induced air flow in dry wells or trenches. Pre- and postprocessors are included. AIR3D can also be used to simulate natural advective air transport in response to barometric pressure fluctuations in shallow, unsaturated systems when gravity and temperature gradients can be neglected. AIR3D transforms the air flow Equation [8.52] into a form similar to the groundwater flow Equation [8.53] that is solved by MODFLOW. Air compressibility is approximated by the ideal gas law. The simulations can be conducted (1) in a calibration mode to evaluate parameters such as permeability from pneumatic tests or (2) in a predictive mode.

### 8.5.2 TWO-PHASE FLOW (WATER AND AIR)

Codes have also been developed to simulate two-phase (liquid and gas), two-component (water and air) systems. In situations where the air phase retards infiltration of the water phase, a two-phase code is required (Touma and Vauclin, 1986). Two-phase codes are also required to simulate transport of volatile organic compounds. Various approaches have been developed to simulate two-phase flow. In the petroleum industry, Buckley and Leverett (1942) developed an approach that excluded the effects of capillary pressure and gravity. The basic theory of Buckley and Leverett (1942) was extended by Morel-Seytoux (1973) and Vauclin (1989). If the fluids are considered incompressible, the 1-D pressure equation reduces to (Celia and Binning, 1992):

$$q_w + q_a = constant \qquad [8.56]$$

where $q_w$ is the water flux and $q_a$ is the air flux. Only one equation, therefore, has to be solved. The fractional flow equation was solved by Morel-Seytoux and Billica (1985) using a finite difference approximation. Because for 2-D flow both the pressure and saturation equations must be solved, the problem no longer reduces to one equation.

Celia and Binning (1992) developed a finite element, 1-D code that considers dynamic coupling of the gas and water phases. The code is inherently mass conservative because the mixed formulation is used: temporal differentiation in terms of water content and spatial differentiation in terms of pressure head. The code was used to simulate laboratory two-phase flow experiments conducted by Touma and Vauclin (1986). The results of these simulations were compared with those achieved by means of a finite element, two-phase flow code developed by Kaluarachchi and Parker (1989) that uses the traditional h-based formulation and a finite difference code that was based on fractional flow theory used by Morel-Seytoux and Billica (1985). The simulations were used to evaluate the conditions under which water flow is significantly altered by air flow. Infiltration experiments in a bounded column that resulted in ponded conditions showed significantly altered water flow because the air could not readily escape. An important insight gained from these simulations was that the reduction in water velocity resulted from a reduction in hydraulic conductivity with increased air content, rather than from a buildup of air pressure ahead of the wetting front (Celia and Binning, 1992).

More general codes have been developed by the National Laboratories to simulate multiphase flow and transport. TOUGH (Transport of Unsaturated Groundwater and Heat) is a 3-D code that simulates nonisothermal flow and transport of two fluid phases (liquid and gas) and two components (water and air) (Pruess, 1987). Subsequent upgrades include ATOUGH, VTOUGH, CTOUGH, ITOUGH (Finsterle and Pruess, 1995) and TOUGH2 (Pruess, 1991). A separate module has been developed for simulating transport of volatile contaminants (T2VOC). FEHM (Finite Element transport of Heat and Mass) is a 3-D, nonisothermal code that simulates two-phase flow and transport of multiple components (Zyvoloski et al., 1996). Both TOUGH and FEHM are used for simulation of flow and transport in variably saturated fractured systems at Yucca Mountain, Nevada, the proposed high-level radioactive waste disposal site. STOMP (Subsurface Transport Over Multiple Phases) simulates nonisothermal flow and transport in porous media (White and Oostrom, 1996). STOMP was developed specifically to evaluate remediation of sites contaminated by organics and includes a separate NAPL phase. Both TOUGH and STOMP use an integrated finite difference method to solve mass- and energy-balance equations.

There has been considerable interest in simulating flow in fractured systems because the proposed high-level nuclear waste facility will be located in fractured tuff. Mathematical models for simulating flow in fractured media can be subdivided into continuum and discrete fracture models (Rosenberg et al., 1994; NRC, 1996). In addition, models can be deterministic or stochastic. Continuum models can be further subdivided into equivalent continuum, dual porosity and dual permeability models (Rosenberg et al., 1994). In equivalent continuum models the system is described by a single equivalent porous medium with average hydraulic properties. This approach is considered valid if transport between fractures and matrix is rapid relative to transport within fractures (Pruess et al., 1990). Dual porosity models consider the fractures and the matrix as two interacting media with interaction between the fractures and the matrix restricted to a local scale (Pruess and Narasimhan, 1985). There is no continuous flow in the matrix. Dual porosity models are generally used to simulate transient flow and transport in saturated zones. Dual permeability models allow continuous flow in either the fractures or the matrix and an interaction term is used to describe coupling between the two media. The dual permeability formulation is generally used to simulate transient flow and transport in unsaturated systems. Dual porosity and dual permeability models treat fractures as high permeability porous media (Rosenberg et al., 1994). Discrete fracture network models assume that the matrix is not involved in flow and transport and that fluid flow can be predicted from information on the geometry of the fractures and the permeability of individual fractures (NRC, 1996). Stochastic simulation is generally used to produce several realizations of the fracture system.

## 8.6  APPLICATIONS OF GAS TRANSPORT THEORY

### 8.6.1  Soil Vapor Extraction

A detailed review of various issues related to soil vapor extraction is provided in Rathfelder et al. (1995). Soil vapor extraction has become the most common innovative technology for treating subsurface soils contaminated by volatile and semivolatile organic compounds (U.S. EPA, 1992). This popularity is partly due to its low cost relative to other available technologies, especially when contamination occurs relatively deep below the ground surface. Vapor extraction systems are also attractive because mitigation is completed *in situ,* reducing the exposure of onsite workers and the offsite public to chemical contaminants. Vapor extraction also offers flexibility in terms of installation and operation. This flexibility allows systems to be installed at locations crowded by existing structures, roadways and other facilities and to be readily adjusted during the course of remediation to improve mass removal efficiency.

Vapor extraction involves two major processes: mass transfer and mass transport. Mass transfer is the movement from one phase to another at a particular location. Volatile and semivolatile compounds will enter the vapor phase by desorption from the soil particles through volatilization from the soil water and by evaporation from nonaqueous phase liquids (NAPLs), such as gasoline or liquid solvents. The rate at which this mass transfer occurs depends on subsurface temperature, humidity and pressure; the properties of the contaminant, including vapor pressure and solubility; and the sorptive properties of the soil.

The second major process involved in vapor extraction is mass transport, the movement of vapor from one location to another. This transport, which is primarily due to advection, is caused by pressure gradients that are developed by using extraction wells or trenches. In some instances, mass transport is enhanced through passive or active injection wells and trenches and through low permeability soil covers and cutoff walls. The rate at which mass transport occurs is a function of the distribution of soil permeabilities, soil moisture content and pressure gradients induced through the extraction and injection systems. In highly heterogeneous systems, flow may be concentrated in high permeability layers, with flow bypassing low permeability material.

The applicability of vapor extraction at a site depends on the volatility of the contaminants and the ability to generate advective gas flow through the subsurface (Hutzler et al., 1989, Johnson et al., 1995). High volatility contaminants and uniformly high permeability soils are optimal for successful vapor extraction.

The time required to clean a particular site by means of vapor extraction depends on the amount and distribution of contaminants in the subsurface and the rates at which mass transfer and mass transport occur. It is often assumed in vapor extraction applications that mass transfer is faster than mass transport, so that local equilibrium conditions exist in terms of vapor concentrations (Baehr et al., 1989; Hayden et al., 1994). Under conditions of local equilibrium, the uncertainty in cleanup time is due primarily to uncertainties in the amount and distribution of subsurface contaminants and in the amount and distribution of air flow induced by the extraction system. The distribution of both contaminants and air flow is controlled, at least partly, by the spatial distribution of soil permeability near the spill or leak. In other cases, laboratory and field experiments indicate that mass transfer from pore water to the gas phase is rate limited (Gierke et al., 1992; McClellan and Gillham, 1992). The rate limitation has been attributed to intra-aggregate or intraparticle diffusion or bypassing of low permeability zones (Gierke et al., 1992; Travis and McGinnis, 1992).

Vapor extraction systems are most often used to treat soils that have been contaminated from leaks or spills of NAPLs that occur near the ground surface. The NAPL that is released from the spill migrates through the unsaturated soil as a separate phase. As the NAPL moves through the soil, portions are trapped within the pores by capillary forces. The amount of NAPL that is trapped is termed *residual saturation*. This residual saturation depends upon the volume and rate of the NAPL release, characteristics of the soil, and properties of the contaminant. In typical situations, between 5 and 50% of the pore space may be filled by this residual saturation (Mercer and Cohen, 1990). The spatial distribution of the residual NAPL will depend upon heterogeneities in the soil column. In general, lower permeability soils tend to trap more of the NAPL than higher permeability materials (Pfannkuch, 1983; Hoag and Marley, 1986; Schwille et al., 1988). After the contaminant has entered the subsurface, it will become partitioned among the NAPL, solid, aqueous and vapor phases. In relatively fresh spills, the major fraction of the contaminant will remain in the NAPL form.

Operation of a vapor extraction system at the spill location will induce air flow through subsurface soils. The spatial distribution of air flow in the subsurface will depend on the distribution of the soil permeability: the flow will tend to occur through higher permeability channels. As the air flows through the subsurface, it will become saturated with contaminant vapors that evaporate from the residual NAPL. After the contaminant has volatilized, it will migrate in the vapor phase to the extraction wells.

The design and operation of vapor extraction systems are described in a variety of texts, including Hutzler et al. (1988), Baehr et al. (1989), Johnson et al. (1990a, b), Pedersen and Curtis (1991) and the U.S. EPA (1991). The simplest design consists of a single vapor extraction well. In most cases, several extraction wells are used. The radius of influence of the extraction wells is often used to design the optimal distance between extraction wells (Pedersen and Curtis, 1991). Horizontal wells are used when contamination is restricted to the shallow subsurface (Connor, 1988; Hutzler et al., 1989; Pedersen and Curtis, 1991). Air injection and passive vents may be used to increase gas flow. Many studies have shown that surface seals can greatly increase the radius of influence of the extraction well and increase the efficiency of vapor extraction systems (Rathfelder et al., 1991; 1995). Modeling of soil vapor extraction systems was reviewed by Rathfelder et al. (1995) and described in references including Massmann (1989), Gierke et al. (1992), Benson et al. (1993), and Falta et al. (1993).

---

**EXAMPLE 8.2**   Using barometric pressures to help clean contaminated soils

An example of the use of passive vapor extraction to help clean up contaminated sites is provided by the cleanup of carbon tetrachloride at the Hanford Nuclear Reservation in the Columbia Plateau of south-central Washington State that is described in Example 8.1. Barometric pressure fluctuations at this site are shown in Diagram 8.1.1.

Diagram 8.2.1 shows data that describe vapor flow and carbon tetrachloride concentration. These data were collected at the same well as those shown in Diagram 8.1.1. The data from approximately 20 to 80 h correspond to a period in which air is flowing into the formation. Approximately 400 $m^3$ of air entered the subsurface during this 60-h period. The concentration of carbon tetrachloride in this air is essentially zero because the air entering the well is atmospheric.

At approximately 80 h the barometric pressure decreased, and the direction of airflow was reversed. Air flowed out of the well for approximately 40 h, and approximately 500 $m^3$ of air exited the subsurface during this period. Although air began to flow out of the well at approximately 80 h, there was a 6-h period where carbon tetrachloride concentration within this air remained essentially zero. The 6-h lag before the carbon tetrachloride concentration began to increase was the result of relatively clean air that had entered the subsurface during the previous period of air inflow. Approximately 27 g of carbon tetrachloride was removed from the subsurface during the outflow event shown in Diagram 8.2.1.

Without a check valve, atmospheric air will flow back into the subsurface through an extraction well when barometric pressure is greater than the subsurface gas pressure. Although this does not directly compromise the integrity of the system, the introduction of ambient air to the soil may result in lower concentrations of contaminants being drawn from the soil when the flow direction reverses. This in turn will reduce mass removal rates. In typical applications, these check valves will open and close at pressure differentials of less than 50 Pa. Rohay (1996) gave examples of several designs that have been used for check valves, including use of a table tennis ball or solenoid valves that open and close in response to an electrical signal controlled by differential or absolute pressures. Simulations described in Ellerd et al. (1999) indicate that use of check valves can increase the extraction of subsurface gas by 270% at the Hanford site.

Surface covers are intended to reduce vertical flow of atmospheric air into the subsurface and to enhance horizontal flow toward the extraction well by preventing short-circuiting adjacent to extraction wells. These surface seals may also reduce the dilution of contaminant

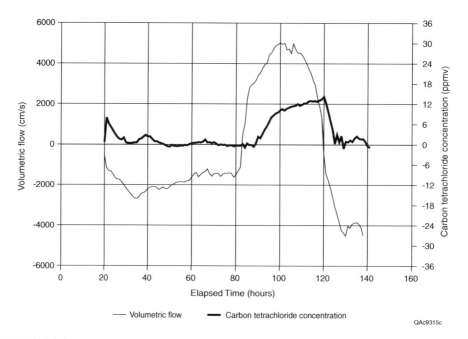

QAc9315c

**DIAGRAM 8.2.1**

concentrations by preventing clean air from entering the subsurface. Surface seals would be particularly important for relatively shallow systems with permeable soils near the ground surface. By covering the ground surface with an impermeable cover, the differential between surface and subsurface gas pressure will increase for a given fluctuation in barometric pressure. Numerical simulations of covers of different radii at the Hanford site showed that volumetric flow rates more than doubled with covers of up to 90 m radius.

**EXAMPLE 8.3**   Case study: optimal vapor velocities for active soil vapor extraction

Most soil vapor extraction systems are operated as active systems in which pressure gradients and airflow are induced through extraction wells using mechanical blowers or pumps. The intent of active vapor extraction systems is to induce air flow through the contaminated soils. The distribution of this air flow can be conceptualized as a set of flow tubes that begin at the ground surface (or at an injection well or trench) and end at the extraction point (Diagram 8.3.1b). Clean air that is drawn into each flow tube from the atmosphere will cause NAPL to volatilize within the subsurface. If there is NAPL within the flow tube, contaminant vapors will enter the air. The maximum concentration within the vapor phase, $C_S$ (M L$^{-3}$) is given by the following expression:

$$C_S = M_i P_i / RT \tag{1}$$

where $M_i$ is the molecular weight of gas $i$ (M mol$^{-1}$), $P_i$ is the vapor pressure for gas $i$ (M L$^{-1}$ t$^{-2}$), $R$ is the ideal gas constant (M L$^2$ t$^{-2}$ T$^{-1}$ mol$^{-1}$), and $T$ is temperature (T). Active systems should be designed based on velocities rather than pressure gradients. The optimal velocity is that which would result in saturated concentrations of volatile organics in the gas

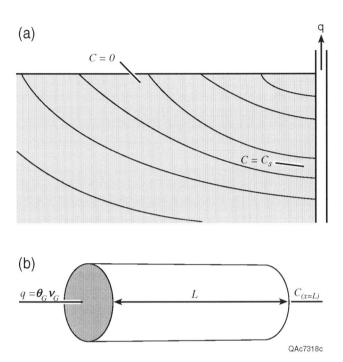

**DIAGRAM 8.3.1** (a) Schematic of active vapor extraction demonstrating change in vapor concentrations as vapor moves from land surface to extraction well. (b) Detailed image of single flow tube.

phase at the well. Velocities greater than this optimal velocity would result in greater clean-up costs because more contaminated air would have to be treated. Alternatively, less-than-optimal velocities would result in longer clean-up times. Clean up progresses from the land surface to the well. The last area to be cleaned is adjacent to the well because the gas becomes saturated near the well; velocities increase near the well.

An estimate of the optimal velocity can be obtained from the following analysis that considers a one-dimensional flow tube (Figure 1a). The actual concentration in the vapor phase will approach the saturated concentration, $C_S$, as the air velocity within a flow tube decreases. For a flow tube with one-dimensional flow and uniform distribution of NAPL, the ratio of the effluent vapor concentration measured at the end of the tube ($C_{(x=L)}$) to the saturated vapor concentration (equilibrium concentration, $C_S$) is given by the following expression (Hines and Maddox, 1985):

$$\frac{C_{(x=L)}}{C^S} = 1 - \exp(-M_T^{*}aL / \theta_G v_G) \qquad [2]$$

$M_T^{*}$ is the mass transfer coefficient (L t$^{-1}$), $a$ is the contact area between liquid and gas per unit volume (L$^2$ L$^{-3}$), $\theta_G$ is the volumetric gas content, and $v_G$ is the gas velocity. From this equation one can estimate the velocity required to reach saturation near the well. For most applications, the mass transfer coefficient and the contact area are lumped together into an effective mass transfer coefficient with units of t$^{-1}$.

Laboratory column experiments can be conducted to estimate the effective mass transfer coefficient, $M_T^{*} a$, by measuring the vapor concentration as a function of flow velocity. If it is assumed that the laboratory-derived transfer coefficient is representative of field-scale conditions, then Equation [2] can be used to select a velocity in the field based upon the lengths of the flow tubes that are developed with the vapor extraction system. The field velocities can be approximated by calculating the flux using Darcy's law based on head measurements at two points and dividing the flux by the volumetric gas content. Gas velocities should be used as the design parameter for active soil vapor extraction systems. Because permeabilities can be highly variable, pressure decreases are not that important. Optimal velocities in one flow tube may not be optimal in another. The pumping rates should be controlled to achieve maximum saturations at the well.

### 8.6.2 Radon Transport

Radon transport has created considerable interest because of the radiological health hazard (lung cancer) associated with the decay products of radon. Radium radioactively decays to radon, and the greatest doses of radiation result from $^{222}$Rn. Many of the issues related to radon transport were reviewed in Nazaroff (1992). Recent evaluations of contaminated sites have focused on risk assessment and suggest that one of the most critical exposures to radon and volatile organic contaminants is from movement of soil gas into buildings. Much research is therefore currently being conducted on gas movement from soil into buildings.

The typical range of radium content in surficial sediments in the United States is 10 to 100 Bq kg$^{-1}$ (Bq, becquerel is the SI unit for activity, which corresponds to the number of atoms needed to yield a radioactive decay rate of one per second) (Nazaroff, 1992). Much higher concentrations of radon occur near uranium mines and mill tailings. Radon is also emitted from low-level radioactive waste disposal sites. The half-life of $^{222}$Rn is 3.8 days. Radon partitions among solid, liquid and gas phases. Diffusive transport is sufficient to account for radon migration from unsaturated material into the atmosphere. Radon fluxes estimated on the basis of diffusion are consistent with experimental measurement (Nazaroff, 1992). The diffusion coefficient of $^{222}$Rn in air is $1.2 \times 10^{-5}$ m$^2$ s$^{-1}$ (Hirst and Harrison, 1939). The effective diffusion coefficient, including the effect of reduced cross-sectional area and increased path length (tortuosity), is about a factor of four smaller than the free air diffusion coefficient in fairly dry soils (Nazaroff, 1992). Effective diffusivities in saturated systems are about four orders of magnitude less than those in dry soils (Tanner, 1964). This difference has been used in the design of engineered barriers for uranium mill tailings to minimize upward movement of radon into the atmosphere. Thick clay layers (0.6 to 2 m) are used as radon barriers because they retain much more water than does sand and thus reduce radon diffusion. Desiccation of clay, however, results in cracking and development of preferred pathways for gas transport. Typical diffusive fluxes for radon estimated by Nazaroff (1992) average 0.022 Bq m$^{-2}$ s$^{-1}$, which is the same average as diffusive fluxes in Australian soils estimated by Schery et al. (1989).

Advection is generally considered more important than diffusion for radon transport into buildings because concrete slabs provide a barrier to diffusion and ventilation causes a pressure differential between indoor and outdoor air that drives advective gas transport. Advective gas transport, described by Darcy's law, varies with permeability and pressure gradient. Because permeability generally decreases with grain size, Nazaroff and Sextro (1989) suggested that diffusion is dominant in low permeability materials and advection dominant in high permeability materials, the cutoff being $\sim 10^{-11}$ m$^2$. Gas permeability is generally anisotropic because sediment layering results in higher permeabilities horizontally than vertically. Openings in the substructures of buildings provide pathways for advective radon transport into buildings. The average indoor $^{222}$Rn concentration in buildings is 50 Bq m$^{-3}$; in soil, $\sim$ 30 Bq m$^{-3}$ (Nazaroff, 1992). Pressure differentials across buildings may result from ventilation caused by open windows and doors, fans, heating and air conditioning. Heating generally increases radon entry into buildings; air conditioning reduces it.

The construction of building substructures can have important implications for advective gas transport. Concrete slabs are commonly underlain by gravel layers. Penetrations at floor wall joints and perimeter drain systems allow gas transport. Temperature differences, barometric pressure fluctuations, wind and rain cause pressure differentials (Nazaroff, 1992).

Diffusive fluxes can account for about 10% of mean radon concentration in single family dwellings (Nero et al., 1986). Nazaroff (1992) presented several different paradigms for radon entry into dwellings. In the base case, transport of radon from the soil into buildings is attributed primarily to advective airflow driven by weather and ventilation. An alternative paradigm would be soil with strongly varying permeability, in which case diffusion would be important for transporting gas from high- to low-permeability regions, such as gravel seams or cracked clays. Another alternative is a high-permeability material at the interface between the building and the soil, such as a gravel layer or a thin air gap. In this case, molecular diffusion from the native low-permeability soil to the high-permeability gravel is important, and advection alone would result in much slower transport of radon.

### 8.6.3 Water Vapor Diffusion

Analysis of water vapor transport in natural systems can be done by approximating soil gas as a two-component system containing water vapor and air. Diffusion of water vapor is important in near-surface sediments where evaporation is occurring and in arid systems where water contents are extremely low. The simple theory of water vapor diffusion assumes that water vapor behaves like other gases. The flux of water vapor (water in the gas phase) is:

$$q_G^w = -\theta_G \tau D_G^w \nabla \rho_v \qquad [8.57]$$

where $D_G^w$ is the binary diffusion coefficient for water vapor in air and $\rho_v$ is the water vapor density. The Kelvin equation can be used to express water vapor density in terms of temperature and matric potential head:

$$\rho_v = \rho_v^0 \cdot RH = \rho_v^0 \exp\frac{h_m V_w}{\rho_w RT} \qquad [8.58]$$

where $\rho_v^0$ is saturated vapor density, $RH$ is relative humidity, $h_m$ is matric potential (Pa), $V_w$ is the molar volume of water ($1.8 \times 10^{-5}$ m$^3$ mol$^{-1}$), $R$ is the gas constant and $T$ is Kelvin temperature. Laboratory experiments showed that water vapor fluxes estimated by Equation [8.58] underestimated measured water vapor fluxes by as much as one order of magnitude. Philip and de Vries (1957) attributed the discrepancy between laboratory measured and estimated water vapor fluxes to (1) liquid island enhancements and (2) increased temperature gradients in the air phase. Liquid island enhancement refers to the fact that when the liquid phase is discontinuous, liquid islands act as short circuits for thermal vapor diffusion. Water vapor condenses on one side of liquid water and evaporates on the other side. This results in an increased cross-sectional area for diffusion from the volumetric gas content to the sum of the volumetric gas and liquid contents (porosity) when the water phase is discontinuous. Temperature gradients in the gas phase are much higher than the average temperature gradients in porous media because thermal conductivity in the gas phase is much lower than that in the liquid and solid phases:

$$q_G^w = -\theta_G \tau_G D_G^w \frac{\partial \rho_g^w}{\partial \psi}\bigg|_T \nabla \psi - f \frac{\nabla T_G}{\nabla T} \tau_G D_G^w \frac{\partial \rho_v}{\partial T}\bigg|_\psi \nabla T \qquad [8.59]$$

where $\nabla T_G$ is the average temperature gradient in the gas phase, $\nabla T$ is the average bulk temperature gradient in the system, and $f$ is a correction factor for liquid island enhancement and is equal to the porosity when the liquid phase is discontinuous (Milly and Eagleson, 1980). Thermal vapor flux, resulting from variations in saturated vapor pressure with temperature, is generally considered much more important than isothermal vapor flux. A temperature difference of 1°C at 20°C results in a greater difference in vapor density ($1.04 \times 10^{-3}$ kg m$^{-3}$) than does a 1.5 MPa difference in matric potentials from $-0.01$ to $-1.5$ MPa ($0.17 \times 10^{-3}$ kg m$^{-3}$) (Hanks, 1992). The temperature dependence of saturated vapor pressure is given in Table 8.1. Temperature gradients are generally high in near surface sediments, and nonisothermal vapor diffusion is important.

One-dimensional liquid and vapor transport in arid and semiarid systems has been simulated by two-phase (liquid and gas), single-component (water), nonisothermal codes (BREATH, SPLaSH-WaTr, UNSAT-H) (Stothoff, 1995; Milly, 1982; Fayer and Jones, 1990). Advective gas transport is not included. In many applications of these codes, liquid and vapor flow is simulated in response to atmospheric forcing functions. Mass and energy equations are solved sequentially, and the resultant tridiagonal equations are readily solved by the efficient Thomas algorithm. These simulators solve the continuity equation for water:

$$\frac{\partial}{\partial t}\left(\rho_l^w \theta_l + \rho_G^v \theta_G\right) = -\nabla \cdot \left(\rho_l^w q_l + \rho_G^v q_v\right) \tag{8.60}$$

where $\rho_l^w$ is the density of water in the liquid phase and $\rho_G^v$ is the density of water vapor in the gas phase. Darcy's law is used to solve the liquid water flux, and the theory of Philip and de Vries (1957) is used to solve the vapor diffusive flux. The energy balance equation is also solved in these codes. Numerical simulations of liquid and vapor flux provide insights on flow processes in the unsaturated zone. Simulations in response to 1 yr of atmospheric forcing in the Chihuahuan Desert of Texas showed net downward water flux in response to seasonally varying temperature gradients that was consistent with isotopic tracer data (Scanlon and Milly, 1994). Simulations at Yucca Mountain, Nevada, evaluated the impact of hydraulic properties on infiltration (Stothoff, 1997). For low permeabilities in this system, vapor transport was dominant. The simulations demonstrated the importance of the alluvial cover on fractured bedrock in decreasing downward water flux. The UNSAT-H code was used to evaluate the performance of engineered barriers, including a capillary barrier at the Hanford site (Fayer et al., 1992). Hysteresis was found to be important in reproducing drainage measured through the capillary barrier.

### 8.6.4 PREFERENTIAL FLOW

Preferential flow refers to nonuniform movement of a fluid. In most cases preferential flow has been examined with respect to water flow. Preferential flow can occur in macropores that are noncapillary size pores, such as desiccation cracks in clays, worm holes and root tubules in soils, and fractures in rocks. Because macropores drain rapidly they are generally dry, and preferential flow of gases should therefore be much greater than that of liquids.

There has been much interest in fractured systems in recent years because the proposed high level radioactive waste disposal facility is located in fractured tuff. Fractures allow rapid transport of contaminants in unsaturated systems. The cubic law is used to describe fracture flow, which is proportional to the fracture aperture cubed for a given head gradient. Laminar flow between smooth parallel plates is given by:

$$Q = A q_G = (bw)\frac{b^2}{12\mu_G}\frac{\partial P}{\partial z} = \frac{b^3 w}{12\mu_G}\frac{\partial P}{\partial z} \tag{8.61}$$

where $Q$ is the volumetric flow rate (L$^3$ t$^{-1}$), $b$ is aperture opening (L), $w$ is the width of the fracture perpendicular to the direction of flow (L), $bw$ is the cross-sectional area (L$^2$) and $P$ is the hydraulic head (Pa) (Snow, 1968). Fractures are important because they account for most of the permeability in the system, whereas the matrix is important because it accounts for most of the porosity. Nilson et al. (1991) evaluated contaminant gas transport in fractured permeable systems. For example, a system that has 1 mm wide fractures ($\delta_f$) separated by 1 m slabs of matrix ($\delta_m$) with a matrix porosity of 0.10 ($\phi_m$) will have a ratio of capacitance volumes ($V$) as follows (Nilson et al., 1991):

$$\frac{V_m}{V_f} = \frac{\phi_m \delta_m}{\delta_f} \sim 100 \qquad [8.62]$$

Most mass flow occurs through the fractures as shown by the following (Nilson et al., 1991):

$$\frac{Q_f}{Q_m} = \frac{A_f q_f}{A_m q_m} \sim \frac{\delta_f}{\delta_m} \frac{\left(\delta_f^2 / 12\right)}{k_m} \sim \frac{1}{12} \frac{\delta_f^3}{\delta_m k_m} \sim 10^5 \qquad [8.63]$$

where $Q$ is the mass flow, $q$ is the mass flux density, and $k_m$ is the gas permeability of the matrix ($10^{-15}$ m$^2$). Even with high matrix permeability ($\sim 10^{-12}$ m$^2$) and narrow fractures ($10^{-4}$ m), about 99% of the flow will occur in the fractures.

In fractured, permeable media, advective fluxes resulting from barometric pressure fluctuations may be orders of magnitude greater than diffusive fluxes and could result in upward movement of contaminated gases into the atmosphere (Nilson et al., 1991). Theoretical analyses by Nilson et al. (1991) showed that in homogeneous media, a differential barometric pressure fluctuation of 7% ($\Delta P/P_0$) would result in the upward movement of an interface between contaminated and pure gas of 20 m if the interface were 300 m deep. In contrast, in fractured systems the volume expansion of the gas is the same as that in porous media; however, gas expansion occurs primarily in the fractures and the interface moves as much as two orders of magnitude higher in the system. The conceptual model developed for upward movement of contaminants involves upward gas movement in the fractures as a result of barometric pressure fluctuations and lateral diffusion from the fractures into the surrounding matrix, which holds the contaminants at that level until the next barometric pressure fluctuation. This is a racheting mechanism for transporting gases upward. A gas tracer experiment conducted by Nilson et al. (1992) confirmed the results of earlier theoretical analyses. These experiments showed upward movement of gas tracers in a period of months from a spherical cavity (depth ~300 m) created by underground nuclear tests at the Nevada Test Site. This rapid upward movement of gases was attributed to the effects of barometric pumping in fractured tuff.

Weeks (1987; 1993) also conducted detailed studies of flow in fractured tuff at Yucca Mountain. In areas of steep topography such as at Yucca Mountain, temperature and density-driven topographic effects result in continuous exhalation of air through open boreholes at the mountain crest in the winter, as cold dry air from the flanks of the mountain replaces warm moist air within the rock/borehole system (Weeks, 1987). Wind also results in air discharge from the boreholes that is about 60% of that resulting from temperature induced density differences (Weeks, 1993). Open boreholes greatly enhance the advective air flow at this site; numerical simulations indicate that water fluxes resulting from advective air flow under natural conditions (0.04 mm yr$^{-1}$) are five orders of magnitude less than those found in the borehole (Kipp, 1987) and similar in magnitude to estimated vapor fluxes as a result of the geothermal gradient (0.025 to 0.05 mm yr$^{-1}$; Montazer et al., 1985). These processes could cause drying of fractured rock uplands and could expedite the release of gases from underground repositories to the atmosphere (Weeks, 1993).

## UNITS

Pressure:        1 atm. = 760.0 mm Hg = 101.325 kPa = 1.01 bars = 10.2 m H$_2$0
Permeability:   1 darcy = 9.87 × 10$^{-13}$ m$^2$
Viscosity:        1 Pa s = 1,000 cp

## 8.7 DERIVATION OF EQUATIONS

Derivation of Equation [8.30]:

$$\frac{\partial \rho_G \phi}{\partial t} = -\nabla \cdot \left( -\frac{\rho_G k_G}{\mu_G} \nabla P \right) \qquad [8.64; 8.29]$$

Assuming $\phi$ is constant and writing gas density in terms of pressure (Equation [8.17]):

$$\phi \frac{\partial \dfrac{PM}{RT}}{\partial t} = \nabla \cdot \left( \frac{\dfrac{PM}{RT} k_G}{\mu_G} \nabla P \right) \qquad [8.65]$$

The term $M/RT$ is constant and can be canceled:

$$\frac{M}{RT} \phi \frac{\partial P}{\partial t} = \frac{M}{RT} \nabla \cdot \left( \frac{P k_G}{\mu_G} \nabla P \right) \qquad [8.66]$$

Assuming permeability and viscosity are constant:

$$\phi 2P \frac{\partial P}{\partial t} = \frac{k_G}{\mu_G} 2P \nabla \cdot (P \nabla P) \qquad [8.67]$$

$$\frac{\partial P^2}{\partial t} \approx \frac{k_G P_0}{\phi \mu_G} \nabla^2 P^2 \qquad [8.68; 8.30]$$

Equation [8.69] is Equation [8.30] in Section 8.2.2. The derivation assumes single-phase flow. Similar expressions can be developed for two-phase flow (air and water) by using the volumetric gas content ($\theta_G$) instead of porosity.

Equation [8.31] is derived by assuming that the pressure fluctuations are small, an assumption satisfied in many cases in the unsaturated zone:

$$P = P_0 + \Delta P$$
$$P^2 = P_0^2 + 2\Delta P P_0 + (\Delta P)^2 \qquad [8.69]$$

The first term on the right of Equation [8.70] is a constant; therefore, its derivative is zero. The third term is negligible and is neglected. Equation [8.30] can be converted to Equation [8.31] using the following:

$$\frac{\partial P^2}{\partial t} = 2P_0 \frac{\partial P}{\partial t} \qquad [8.70]$$

$$\nabla^2 P^2 = 2P_0 \nabla^2 P$$

$$\frac{\partial P}{\partial t} = \alpha \nabla^2 P \qquad [8.71; 8.31]$$

Derivation of the form of equations (Section 8.1.6.3) for a binary gas mixture is shown in the following. Diffusive transport of nonequimolar gases in an isobaric system mixture is described by (Cunningham and Williams, 1980):

$$N_i^D = J_{iM}^m + N_i^N \qquad [8.72]$$

The molar diffusive flux of component $i$ ( $N_i^D$) results from the molecular diffusive flux ( $J_{iM}^m$) and the nonequimolar diffusive flux ( $N_i^N$) of gas component $i$. Because the nonequimolar diffusive flux is nonsegregative, the contribution of each species is proportional to its mole fraction $x_i$ and Equation [8.72] can be reformulated:

$$N_i^D = J_{iM}^m + x_i N^N \qquad [8.73]$$

$$\sum_{i=1}^{v} N_i^D = \sum_{i=1}^{v} J_{iM}^m + \left( \sum_{i=1}^{v} x_i \right) N^N \qquad [8.74]$$

Because $\sum_{i=1}^{v} x_i = 1$, it can be proved that $\sum_{i=1}^{v} J_{iM}^m = 0$; Equation [8.74] can be rewritten as:

$$\sum_{i=1}^{v} N_i^D = N^N \qquad [8.75]$$

If Equation [8.73] is applied to the two components A and B of an isobaric, binary mixture, Equation [8.74] becomes:

$$N_A^D = -D_{AB} \nabla n_{AM} + x_A (N_A^D + N_B^D) \qquad [8.76]$$

$$N_B^D = -D_{BA} \nabla n_{BM} + x_B (N_A^D + N_B^D) \qquad [8.77]$$

where $n_{AM}$ and $n_{BM}$ are the number of moles of gases A and B. Because the system is isobaric, the total flux is zero (otherwise a pressure gradient would result), and Equations [8.76] and [8.77] reduce to Fick's law applied to the total diffusive flux. This is the only case when Fick's law is strictly valid. In deriving Equations [8.76] and [8.77], the segregative diffusive flux is assumed to follow a gradient law:

$$J_{AM}^m = -D_{AB} \nabla n_{AM} \qquad [8.78]$$

The following two equations result:

$$N_A^D = -D_{AB} \nabla n_{AM} + x_A (N_A^D + N_B^D) \qquad [8.79]$$

$$N_B^D = -D_{BA} \nabla n_{BM} + x_B (N_A^D + N_B^D) \qquad [8.80]$$

Multiplying Equation [8.79] with $x_B$ and [8.80] with $x_A$ results in:

$$x_B N_A^D = -x_B D_{AB} \nabla n_{AM} + x_A x_B (N_A^D + N_B^D) \qquad [8.81]$$

$$x_A N_B^D = -x_A D_{BA} \nabla n_{BM} + x_A x_B (N_A^D + N_B^D) \qquad [8.82]$$

Subtracting Equation [8.82] from Equation [8.81] results in:

$$x_B N_A^D - x_A N_B^D = -D_{AB}(x_B \nabla n_{AM} - x_A \nabla n_{BM})$$  [8.83]

Because $n_{AM} + n_{BM} = n =$ constant and $x_A + x_B = 1$, Equation [8.83] can be rewritten as:

$$x_B N_A^D - x_A N_B^D = -D_{AB}\big((1 - x_A)\nabla n_{AM} - x_A(\nabla n_{BM})\big)$$

$$x_B N_A^D - x_A N_B^D = -D_{AB}\big(\nabla n_{AM} - x_A \nabla(n_{AM} + n_{BM})\big)$$  [8.84]

$$x_B N_A^D - x_A N_B^D = -D_{AB} \nabla n_{AM}$$

According to the ideal gas law $P_A = \dfrac{n_{AM}}{RT}$; therefore:

$$\frac{x_B N_A^D - x_A N_B^D}{D_{AB}} = -\frac{\nabla P_A}{RT}$$  [8.85]

## REFERENCES

Abriola, L.M., C.S. Fen, and H.W. Reeves. 1992. Numerical simulation of unsteady organic vapor transport in porous media using the dusty gas model, 195–202, *in Proc. of the Int. Assoc. Hydrol. Conf. on Subsurface Contamination by Immiscible Fluids*, A.A. Balkema, Rotterdam, Netherlands.

Abu-El-Sha'r, W.Y. 1993. Experimental Assessment of Multicomponent Gas Transport Mechanisms in Subsurface Systems, Ph.D. dissertation, 75, Univ. of Mich., Ann Arbor.

Abu-El-Sha'r, W.Y. and L.M. Abriola. 1997. Experimental Assessment of Gas Transport Mechanisms in Natural Porous Media: Parameter Evaluation, *Water Resour. Res.,* 33:505–516.

Alzaydi, A.A. 1975. Flow of Gases through Porous Media, 172 pp., Ohio State University, Columbus.

Alzaydi, A.A. and C.A. Moore. 1978. Combined pressure and diffusional transition region flow of gases in porous media, *AICHE J.,* 24(1):35–43.

Amali, S. and D.E. Rolston. 1993. Theoretical investigation of multicomponent volatile organic vapor diffusion: steady-state fluxes, *J. Environ. Qual.,* 22:825–831.

Amali, S., D.E. Rolston, and T. Yamaguchi. 1996. Transient multicomponent gas-phase transport of volatile organic chemicals in porous media, *J. Environ. Qual.,* 25:1041–1047.

Auer, L.H., N.D. Rosenberg, K.H. Birdsell, and E.M. Whitney. 1996. The effects of barometric pumping on contaminant transport, *J. Contam. Hydrol.,* 24:145–166.

Baehr, A.L. and C.J. Bruell. 1990. Application of the Stefan-Maxwell equations to determine limitations of Fick's law when modeling organic vapor transport in sand columns, *Water Resour. Res.,* 26:1155–1163.

Baehr, A.L., G.E. Hoag, and M.C. Marley. 1989. Removing volatile contaminants from the unsaturated zone by inducing advective air-phase transport, *J. Contam. Hydrol.,* 4:1–26.

Baehr, A.L. and M.F. Hult. 1991. Evaluation of unsaturated zone air permeability through pneumatic tests, *Water Resour. Res.,* 27:2605–2617.

Benson, D.A., D. Huntley, and P.C. Johnson. 1993. Modeling vapor extraction and general transport in the presence of NAPL mixtures and nonideal conditions, *Ground Water,* 31(3):437–445.

Bird, R.B., W.E. Stewart, and E.N. Lightfoot. 1960. Transport Phenomena, 780 pp., John Wiley, NY.

Brooks, R.H. and A.T. Corey. 1964. Hydraulic Properties of Porous Media, 27 pp., Colorado State Univ. Hydrology Paper No. 3.

Brusseau, M.L. 1994. Transport of reactive contaminants in heterogeneous porous media, *Rev. Geophys.,* 32:285–313.

Buckley, S.E. and M.C. Leverett. 1942. Mechanisms of fluid displacement in sands, *Trans. Am. Inst. Mining Metallurgical Engineers,* 146:107–116.

Burdine, N.T. 1953. Relative permeability calculations from pore size distribution data, *Petr. Trans., AIME,* 198:71–78.

Carslaw, H.S. and J.C. Jaeger. 1959. The Conduction of Heat in Solids, 510 pp., Oxford Univ. Press, London.

Celia, M.A. and P. Binning. 1992. A mass conservative numerical solution for two-phase flow in porous media with application to unsaturated flow, *Water Resour. Res.,* 28:2819–2828.

Charbeneau, R.J. and D.E. Daniel. 1992. Contaminant transport in unsaturated flow, 15.1–15.54, *in Handbook of Hydrology,* D.R. Maidment, (ed.), McGraw-Hill, Inc., NY.

Childs, S.W. and G. Malstaff. 1982. Heat and Mass Transfer in Unsaturated Porous Media, 173, Pacific Northwest Laboratory.

Collins, R.E. 1961. Flow of Fluids through Porous Media, 270 pp., van Nostrand-Reinhold, Princeton, NJ.

Connor, J.R. 1988. Case study of soil venting, *Pollution Engr.,* 20(7):74–78.

Corey, A.T. 1954. The interrelation between gas and oil relative permeabilities, *Producer's Monthly,* 19(1):38–44.

Corey, A.T. 1986. Air permeability, 1121–1136, *in* A. Klute (ed.), *Methods of Soil Analysis, Am. Soc. Agron.,* Madison, Wisconsin.

Croise, J. and V. Kaleris. 1992. Field measurements and numerical simulations of pressure drop during air stripping in the vadose zone, *in* K.U. Weger (ed.), *Subsurface Contamination by Immiscible Fluids,* A.A. Balkema, Rotterdam.

Cunningham, R.E. and R.J.J. Williams. 1980. Diffusion of Gases and Porous Media, 275 pp., Plenum, NY.

Demond, A.H. and P.V. Roberts. 1987. An examination of relative permeability relations for two-phase flow in porous media, *Water Resour. Bull.,* 23(4):617–628.

Detty, T.E. 1992. Determination of Air and Water Relative Permeability Relationships for Selected Unconsolidated Porous Materials, Ph.D. dissertation, 194 pp., Univ. Ariz., Tucson.

Duffield, G.M. and J.O. Rumbaugh, 1989. AQTESOLV aquifer test solver documentation; version 1.00, Geraghty and Miller Modeling Group, Reston, VA.

Dullien, F.A.L. 1979. Porous Media: Fluid Transport and Pore Structure, 396 pp., Academic Press, NY.

Edwards, K.B. 1994. Air permeability from pneumatic tests in oxidized till, *J. Environ. Eng.,* 120(2):329–346.

Ellerd, M., J.W. Massmann, D.P. Schwaegler, and V.J. Rohay. 1999. Enhancements for passive vapor extraction: the Hanford study, *Ground Water,* Vol. 37, 427–437.

Estes, R.K. and P.F. Fulton. 1956. Gas slippage and permeability measurements, *Trans. Am. Inst. Min. Metall. Pet. Eng.,* 207:338–342.

Falta, R.W. 1995. Analytical solutions for gas flow due to gas injection and extraction from horizontal wells, *Ground Water,* 33(2):235–246.

Falta, R.W., I. Javandel, K. Pruess, and P.A. Witherspoon. 1989. Density-driven flow of gas in the unsaturated zone due to the evaporation of volatile organic compounds, *Water Resour. Res.,* 25:2159–2169.

Falta, R.W., K. Pruess, and D.A. Chesnut. 1993. Modeling advective contaminant transport during soil vapor extraction, *Ground Water,* 31(6):1011–1020.

Fayer, M.J. and T.L. Jones. 1990. UNSAT-H Version 2.0: Unsaturated soil water and heat flow model, Pac. Northwest Lab., Richland, WA.

Fayer, M.J., M.L. Rockhold, and M.D. Campbell. 1992. Hydrologic modeling of protective barriers: comparison of field data and simulation results, *Soil Sci. Soc. Am. J.,* 56:690–700.

Finsterle, S. and K. Pruess. 1995. Solving the estimation-identification problem in two-phase flow modeling, *Water Resour. Res.,* 31(4):913–924.

Forcheimer, P. 1901. Wasserbewegung durch boden, *Zeitschrift Des Verines Deutch Ing,* 49:1736–1749.

Freeze, R.A. and J.A. Cherry. 1979. *Groundwater,* 604, Prentice Hall, NJ.

Gardner, W.H. 1986. Water content, 493–545, in A. Klute (ed.), *Methods of Soil Analysis, Part 1, Physical and Mineralogical Methods, Monograph 9,* Am. Soc. Agron., Madison, Wisconsin.

Gierke, J.S., N.J. Hutzler, and D.B. McKenzie. 1992. Vapor transport in unsaturated soil columns: Implications for vapor extraction, *Water Resour. Res.,* 28(2):323–335.

Glauz, R.D. and D.E. Rolston. 1989. Optimal design of two-chamber, gas diffusion cells, *Soil Sci. Soc. Am. J.,* 53:1619–1624.

Goggin, D.J., R.L. Thrasher, and L.W. Lake. 1988. A theoretical and experimental analysis of minipermeameter response including gas slippage and high velocity flow effects, *In Situ,* 12:79–116.

Hampton, D. 1989. Coupled heat and fluid flow in saturated-unsaturated compressible porous media, 293 pp., Colorado State University, Fort Collins, CO.

Hanks, R.J., 1992. Applied Soil Physics, 176 pp., Springer-Verlag, NY.

Hassler, G.L. 1944. Method and apparatus for permeability measurements, Patent Number 2,345,935 April 2, 1944, U.S. Patent Office, Washington, DC.

Hayden, N.J., T.C. Voice, M.D. Annable, and R.B. Wallace. 1994. Change in gasoline constituent mass transfer during soil venting, *J. Environ. Eng.*, 120:1598–1614.

Hazen, A. 1911. Discussion: dams on sand foundations, *Trans. Am. Soc. Civ. Eng.*, 73:199.

Heid, J.G., J.J. McMahon, R.F. Nielsen, and S.T. Yuster. 1950. Study of the permeability of rocks to homogeneous fluids, *in A.P.I. Drilling and Production Practice*, 230, A.P.I.

Hirst, W. and G.E. Harrison. 1939. The diffusion of radon gas mixtures, *Proc. R. Soc. London, Ser. A*, 169:573–586.

Hines, A.L. and R.N. Maddox. 1985. *Mass Transfer: Fundamentals and Applications*, 529 pp., Prentice Hall, Englewood Cliffs, NJ.

Hoag, G.E. and M.C. Marley. 1986. Gasoline residual saturation in unsaturated uniform aquifer materials, *J. Environ. Eng.*, 112(3):586–604.

Hutzler, N.J., B.E. Murphy, and J.S. Gierke. 1989. State of technology review: soil vapor extraction systems, *Final Report to U.S. EPA*, 36 pp., Hazardous Waste Engineering Research Laboratory.

Jackson, R. 1977. Transport in Porous Catalysts, Elsevier Science Publ. Co., NY.

Jaynes, D.B. and A.S. Rogowski. 1983. Applicability of Fick's law to gas diffusion, *Soil Sci. Soc. Am. J.*, 47:425–430.

Jin, M., M. Delshad, B. Dwarakanath, D.C. McKinney, G.A. Pope, K. Sepehrnoori, C. Tilburg, and R.E. Jackson. 1995. Partitioning tracer test for detection, estimation, and remediation performance assessment of subsurface nonaqueous phase liquids, *Water Resour. Res.*, 31:1201–1211.

Johnson, P.C., A.L. Baehr, R.A. Brown, R.E. Hinchee, and G.E. Hoag. 1995. Innovative Site Remediation Technology: Vol. 8 – Vacuum Vapor Extraction, American Academy of Environmental Engineers, Annapolis.

Johnson, P.C., M.W. Kemblowski, and J.D. Colthart. 1990a. Quantitative analysis for the cleanup of hydrocarbon-contaminated soils by *in situ* soil venting, *Ground Water*, 28(3):413–429.

Johnson, P.C., C.C. Stanley, M.W. Kemblowski, D.L. Byers, and J.D. Colthart. 1990b. A practical approach to the design, operation, and monitoring of *in situ* soil-venting systems, *Ground Water Monit. Rev.*, 10(2):159–178.

Joss, C.J. and A.L. Baehr. 1995. Documentation of AIR3D, an adaptation of the ground water flow code MODFLOW to simulate three-dimensional air flow in the unsaturated zone, *in U.S. Geol. Surv. Open File Rept. 94-533*, 154 pp., U.S. Geol. Survey.

Joss, C.J. and A.L. Baehr 1997. AIR2D — A computer code to simulate two-dimensional, radially symmetric airflow in the unsaturated zone, *in U.S. Geological Survey Open File Report 97-588*, 106, U.S. Geological Survey.

Jury, W.A., W.R. Gardner, and W.H. Gardner. 1991. Soil Physics, 328 pp., John Wiley & Sons, Inc., NY.

Kaluarachchi, J.J. 1995. Analytical solution to two-dimensional axisymmetric gas flow with Klinkenberg effect, *J. Envir. Eng.*, 121(5):417–420.

Kaluarachchi, J.J. and J.C. Parker. 1989. An efficient finite element method for modeling multiphase flow, *Water Resources Research*, 25(1):43–54.

Kipp, K.L.J. 1987. Effect of topography on gas flow in unsaturated fractured rock: Numerical simulation, 171–176, *in* D.D. Evans and T.J. Nicholson (eds.), *Flow and Transport Through Unsaturated Fractured Rock*, Geophys. Monog. 9, Am. Geophys. Union, Washington, DC.

Klinkenberg, L.J. 1941. The permeability of porous media to liquids and gases, 200 pp., A.P.I. Drilling and Production Practice.

Kreamer, D.K., E.P. Weeks, and G.M. Thompson. 1988. A field technique to measure the tortuosity and sorption-affected porosity for gaseous diffusion of materials in the unsaturated zone with experimental results from near Barnwell, South Carolina, *Water Resour. Res.*, 24:331–341.

Lai, S.H., J.M. Tiedje, and A.E. Erickson. 1976. *In situ* measurement of gas diffusion coefficient in soils, *Soil Sci. Soc. Am. J.*, 40:3–6.

Leffelaar, P.A. 1987. Dynamic simulation of multinary diffusion problems related to soil, *Soil Sci.*, 143:79–91.

Marshall, T.J. 1958. A relation between permeability and size distribution of pores, *J. Soil Sci.*, 9:1–8.

Mason, E.A. and A.P. Malinauskas. 1983. Gas transport in porous media: the Dusty-Gas Model, *in Chem. Engr. Monogr.*, Vol. 17, 194, Elsevier, NY.

Mason, E.A., A.P. Malinauskas, and R.B. Evans. 1967. Flow and diffusion of gases in porous media, *J. Chem. Phys.,* 46:3199–3126.

Massmann, J.W. 1989. Applying groundwater flow models in vapor extraction system design, *J. Env. Engin.,* 115:129–149.

Massmann, J. and D.F. Farrier. 1992. Effects of atmospheric pressures on gas transport in the vadose zone, *Water Resour. Res.,* 28:777–791.

Massmann, J.W. and M. Madden. 1994. Estimating air conductivity and porosity from vadose-zone pumping tests, *J. Env. Eng.,* 120(2):313–328.

McClellan, R.D. and R.W. Gillham. 1992. Vapour extraction of trichloroethylene under controlled conditions at the Borden site, *in Subsurface Contamination by Immiscible Fluids,* 89–96, A.A. Balkema, Rotterdam.

McDonald, M.G. and A.W. Harbaugh. 1988. A modular three-dimensional finite-difference ground-water flow model, *in Techniques of Water Resources Investigations,* 576 pp., USGS.

McIntyre, D.S. and J.R. Philip. 1964. A field method for measurement of gas diffusion into soils, *Aust. J. Soil Res.,* 2:133–145.

McWhorter, D.B. 1990. Unsteady radial flow of gas in the vadose zone, *J. Contam. Hydrol.,* 5:297–314.

Mendoza, C.A. and E.O. Frind. 1990. Advective-dispersive transport of dense organic vapors in the unsaturated zone: Sensitivity analysis, *Water Resour. Res.,* 26(3):388–398.

Mendoza, C.A. and T.A. McAlary. 1990. Modeling of ground-water contamination caused by organic solvent vapors, *Ground Water,* 28(2):199–206.

Mercer, J.W. and R.M. Cohen. 1990. A review of immiscible fluids in the subsurface: properties, models, characterization, and remediation, *J. Contam. Hydrol.,* 6(2):107–163.

Millington, R.J. 1959. Gas diffusion in porous media, *Science,* 130:100–102.

Millington, R.J. and J.P. Quirk. 1961. Permeability of porous solids, *Faraday Soc. Trans.,* 57(7):1200–1207.

Millington, R.J. and R.C. Shearer. 1970. Diffusion in aggregated porous media, *Soil Sci.,* 3(6):372–378.

Milly, P.C.D. 1982. Moisture and heat transport in hysteretic, inhomogeneous porous media: a matric head-based formulation and a numerical model, *Water Resour. Res.,* 18:489–498.

Milly, P.C.D. and P.S. Eagleson. 1980. The coupled transport of water and heat in a vertical soil column under atmospheric excitation, 234, Ralph M. Parsons Lab., Mass. Inst. Technol.

Montazer, P., E.P. Weeks, F. Thamir, S.N. Yard, and P.B. Hofrichter. 1985. Monitoring the vadose zone in fractured tuff, Yucca Mountain, Nevada, 439–469, *in Proc. on Characterization and Monitoring of the Vadose (Unsaturated) Zone,* National Water Well Association, Denver.

Moore, C., I.S. Rai, and J. Lynch. 1982. Computer design of landfill methane migration control, *J. Env. Eng. Div. Am. Soc. Civ. Eng.,* 108:89–107.

Morel-Seytoux, H.J. 1973. Two-phase flow in porous media, *Adv. Hydrosci.,* 9:119–202.

Morel-Seytoux, H.J. and J.A. Billica. 1985. A two-phase numerical model for prediction of infiltration: Application to a semi-infinite column, *Water Resour. Res.,* 21(4):607–615.

Mualem, Y. 1976. A new model for predicting the hydraulic conductivity of unsaturated porous media, *Water Resour. Res.,* 12(3):513–521.

Muskat, M. 1946. Flow through Porous Media, 72 pp., McGraw Hill, NY.

Nazaroff, W.W. 1992. Radon transport from soil to air, *Rev. Geophys.,* 30:137–160.

Nazaroff, W.W. and R.G. Sextro. 1989. Technique for measuring the indoor $^{222}$Rn source potential of soil, *Environ. Sci. Technol.,* 23:451–458.

Nero, A.V., A.J. Gadgil, W.W. Nazaroff, and K.L. Revzan. 1986. Distribution of airborne radon-222 concentrations in U.S. homes, *Science,* 234:992–997.

Nicot, J.P. 1995. Characterization of gas transport in a playa subsurface Pantex Site, Amarillo, Texas, Master's thesis, 178 pp., The University Texas at Austin.

Nilson, R.H., W.B. McKinnis, P.L. Lagus, J.R. Hearst, N.R. Burkhard, and C.F. Smith. 1992. Field measurements of tracer gas transport induced by barometric pumping, 710–716, *in Proc. of the Third International Conference of High Level Radioactive Waste Management,* Am. Nucl. Soc.

Nilson, R.H., E.W. Peterson, K.H. Lie, N.R. Burkard, and J.R. Hearst. 1991. Atmospheric pumping: a mechanism causing vertical transport of contaminated gases through fractured permeable media, *J. Geophys. Res.,* 96(B13):21933–21948.

NRC. 1996. Rock Fractures and Fluid Flow Contemporary Understanding and Applications, 551 pp., National Academy Press, Washington, DC.

Pederson, T.A. and J.T. Curtis. 1991. Soil Vapor Extraction Technology — Reference Handbook, Office of Research and Development.

Penman, H.L. 1940a. Gas and vapor movements in the soil. I. The diffusion of vapors through porous solids, *J. Agr. Sci.,* 30:437–462.

Penman, H.L. 1940b. Gas and vapor movements in the soil. II. The diffusion of carbon dioxide through porous solids, *J. Agr. Sci.,* 30:570–581.

Pfannnkuch, H. 1983. Hydrocarbon spills, their retention in the subsurface and propagation into shallow aquifers, *Office Water Resources Technology,* 51 pp., Office of Water Resources Technology, Washington, DC.

Philip, J.R. and D.A. de Vries. 1957. Moisture movement in porous materials under temperature gradients, *Trans. AGU,* 38:222–232.

Pruess, K. 1987. TOUGH User's Guide, *NUREG/CR-4645,* 78, Nuclear Regulatory Commission, Washington, DC.

Pruess, K. 1991. TOUGH2 — A general-purpose numerical simulator for multiphase fluid and heat flow, Lawrence Berkeley Lab., Berkeley, CA.

Pruess, K. and T.N. Narasimhan, 1985. A practical method for modeling fluid and heat flow in fractured porous media, *Soc. Petrol. Engin. J.,* 25:14–26.

Pruess, K., J.S.Y. Wang, and Y.W. Tsang. 1990. On thermohydrologic conditions near high-level nuclear wastes emplaced in partially saturated fractured tuff, 1, simulation studies with explicit consideration of fracture effects, *Water Resour. Res.,* 26:1235–1248.

Raney, W.A. 1949. Field measurement of oxygen diffusion through soil, *Soil Sci. Soc. Am. Proc.,* 14:61–65.

Rathfelder, K., J.R. Lang, and L.M. Abriola. 1995. Soil vapor extraction and bioventing: Applications, limitations, and future research directions, *U.S. Natl. Rep. Int. Union Geod. Geophys. 1991–1994, Rev. Geophys.,* 33:1067–1081.

Rathfelder, K., W.W.-G. Yeh, and D. Mackay. 1991. Mathematical simulation of soil vapor extraction systems: model development and numerical examples, *J. Contam. Hydrol.,* 8:263–297.

Rohay, V.J. 1996. Field tests of passive soil vapor extraction at the Hanford Site, Washington, BHI-00766, Rev. 0, Bechtel Hanford, Inc., Richland, WA.

Rojstaczer, S. and J.P. Tunks. 1995. Field-based determination of air diffusivity using soil-air and atmospheric pressure time series, *Water Resour. Res.,* 12:3337–3343.

Rolston, D.E.. 1986. Gas diffusivity, 1089–1102, *in* A. Klute (ed.), *Methods of Soil Analysis,* Vol. 1, Am. Soc. Agron., Madison, Wisconsin.

Rolston, D.E., D. Kirkham, and D.R. Nielsen. 1969. Miscible displacement of gases through soils columns, *Soil Sci. Soc. Am. Proc.,* 33:488–492.

Rosenberg, N.D., Soll, W.E., and Zyvoloski, G.A. 1994. Microscale and macroscale modeling of flow in unsaturated fractured rock: Am. Geophys. Union, Chapman Conf. on Aqueous Phase and Multiphase Transport in Fractured Rock, 22 pp., Sept. 12–15, Burlington, VT.

Satterfield, C.N. and P.J. Cadle. 1968. Diffusion and flow in commercial catalysts at pressure levels about atmospheric, *J. Ind. Eng. Chem. Fundamentals,* 7:202.

Scanlon, B.R. and P.C.D. Milly. 1994. Water and heat fluxes in desert soils 2. Numerical simulations, *Water Resour. Res.,* 30:721–733.

Scheidegger, A.E. 1974. The Physics of Flow through Porous Media, 353 pp., Univ. Toronto Press, Toronto, 3rd ed.

Schery, S.D., S. Whittlestone, K.P. Hart, and S.E. Hill. 1989. The flux of radon and thoron from Australian soils, *J. Geophys. Res.,* 94:8567–8576.

Schwille, F., W. Bertsch, R. Linke, W. Reif, and S. Zauter. 1988. Dense Chlorinated Solvents in Porous and Fractured Media: Model Experiments, Lewis Publishers, Chelsea, MI.

Shan, C., R.W. Falta, and I. Javandel. 1992. Analytical solutions for steady state gas flow to a soil vapor extraction well, *Water Resour. Res.,* 28:1105–1120.

Sharp, J.M., L. Fu, P. Cortez, and E. Wheeler. 1994. An electronic minipermeameter for use in the field and laboratory, *Ground Water,* 32(1):41–46.

Shepherd, R.G. 1989. Correlations of permeability and grain size, *Ground Water,* 27(5):633–638.

Snow, D.T. 1968. Rock fracture spacings, openings, and porosities, *J. Soil Mech.,* 94:73–91.

Springer, D.S. 1993. Determining the Air Permeability of Porous Materials as a Function of a Variable Water Content under Controlled Laboratory Conditions, 71 pp., Univ. Calif., Santa Barbara.

Springer, D.S., S.J. Cullen, and L.G. Everett. 1995. Laboratory studies on air permeability, 217–247, *in* L.G. Wilson, L.G. Everett and S.J. Cullen (eds.), *Handbook of Vadose Zone Characterization and Monitoring*, Lewis Publishers.

Stonestrom, D.A. 1987. Co-Determination and Comparisons of Hysteresis-Affected, Parametric Functions of Unsaturated Flow: Water-Content Dependence of Matric Pressure, Air Trapping, and Fluid Permeabilities in a Non-Swelling Soil, 292 pp., Stanford University.

Stonestrom, D.A. and J. Rubin. 1989. Air permeability and trapped-air content in two soils, *Water Resour. Res.*, 25(9):1959–1969.

Stothoff, S.A. 1995. BREATH Version 1.1 — coupled flow and energy transport in porous media, Simulator description and user guide, U.S. Nucl. Reg. Comm.

Stothoff, S.A. 1997. Sensitivity of long-term bare soil infiltration simulations to hydraulic properties in an arid environment, *Water Resour. Res.*, 33:547–558.

Tanner, A.B. 1964. Radon migration in the ground: A review, 161–190, *in* J.A.S. Adams and W.M. Lowder (eds.), *Natural Radiation Environment*, Chicago Press.

Terzaghi, K. and R.B. Peck. 1968. Soil Mechanics in Engineering Practice, 729 pp., John Wiley & Sons, New York.

Thibodeaux, L.J., C. Springer, and L.M. Riley. 1982. Models and mechanisms for the vapor phase emission of hazardous chemicals from landfills, *J. Hazard. Mater.*, 7:63–74.

Thorstenson, D.C. and D.W. Pollock. 1989. Gas transport in unsaturated porous media, the adequacy of Fick's law, *Rev. Geophys.*, 27:61–78.

Touma, J. and M. Vauclin. 1986. Experimental and numerical analysis of two-phase infiltration in a partially saturated soil, *Transp. Porous Media*, 1:22–55.

Travis, C.C. and J.M. Macinnis. 1992. Vapor extraction of organics from subsurface soils — Is it effective?, *Env. Sci. Technol.*, 26:1885–1887.

U.S. EPA. 1991. Soil Vapor Extraction Technology Reference Handbook.

U.S. EPA. 1992. A Technology Assessment of Soil Vapor Extraction and Air Sparging, Office of Research and Development, Washington, DC.

van Genuchten, M.T. 1980. A closed-form equation for predicting the hydraulic conductivity of unsaturated soils, *Soil Sci. Soc. Am. J.*, 44:892–898.

Vauclin, M. 1989. Flow of water and air in soils, 53–91, *in* Theoretical and experimental aspects, *Unsaturated Flow in Hydrologic Modelling, Theory and Practice*, H.J. Morel-Seytoux (ed.), Kluwer Academic, Dordrecht, Netherlands.

Weast, R.C. 1986. CRC Handbook of Chemistry and Physics, CRC Press, Boca Raton, FL.

Weeks, E.P. 1978. Field determination of vertical permeability to air in the unsaturated zone, 41, *in U.S. G.S. Prof. Paper*, USGS.

Weeks, E.P. 1987. Effect of topography on gas flow in unsaturated fractured rock: concepts and observations, 165–170, *in* D.D. Evans and T.J. Nicholson (eds.), *Flow and Transport through Unsaturated Fractured Rock*, Geophy. Monograph 9, Am. Geophys. Union, Washington, DC, 1987.

Weeks, E.P. 1993. Does the wind blow through Yucca Mountain, 45–53, *in Proc. of Workshop V: Flow and Transport through Unsaturated Fractured Rock Related to High-Level Radioactive Waste Disposal*, D.D. Evans and T.J. Nicholson (eds.), U.S. Nucl. Reg. Comm., NUREG CP-0040.

White, M.D. and M. Oostrom. 1996. STOMP Subsurface Transport Over Multiple Phases Theory Guide, variously paginated, Pacific Northwest National Laboratory.

Zyvoloski, G.A., B.A. Robinson, Z.V. Dash, and L.L. Trease. 1996. Models and Methods Summary for the FEHM Application, 65 pp., Los Alamos National Lab., Los Alamos, NM.

# 9 Soil Spatial Variability

## D. J. Mulla and Alex B. McBratney

## 9.1 VARIABILITY IN SOIL PROPERTIES FROM SOIL CLASSIFICATION

A key feature of soil is the variation with depth in soil properties. Soil is formed as a result of the influences of climate, plants and time acting on geologic parent material in different landscape positions. Soils are uniquely different from geologic parent material such as loess, glacial till or sedimentary rock because soils develop horizonation, in which each horizon has a distinct set of characteristic and diagnostic soil properties. Horizons may differ in organic matter content, color, structure, texture, pH, base saturation, cation exchange capacity, bulk density and water holding capacity, as well as many other soil physical and chemical properties.

Variation in soil properties also occurs across the landscape and in response to regional variations in climate and parent material. Broad differences in soil types at the soil order level are identified based on horizon characteristics, such as thickness and organic C content of the A horizon, base saturation of the B horizon, or the presence of a pale colored, extensively leached subsurface E horizon. Differences in soil horizonation have been extensively described and classified, giving rise to the 12 major soil orders.

Changes in soil properties at the order level of classification generally, but not always, take place over relatively large distances, often across a significant climatic and/or orographic gradient. Within fields and across shorter distances, soil properties also vary significantly, even across locations that involve only one soil order. This variation has also been extensively described and classified, giving rise to the soil series level of classification. In the United States, the variability in soil properties at the series mapping unit level is depicted using County Soil Series maps. Although a given field or farm may be mapped as having from 5 to 10 or more soil mapping units, it is rare for these units to represent more than a few soil orders.

The variability in soil properties at the soil series level is often caused by small changes in topography that affect the transport and storage of water across and within the soil profile. Hillslopes are typically divided into five landscape positions which are closely related to patterns in water transport and storage, which strongly influence soil development. These landscape positions are the summit, shoulder, backslope, footslope and toeslope positions. The soil variation across a hillslope is usually referred to as a catena, which is an association of soil mapping units that are closely linked to hillslope position.

One of the most common soil catenas in the midwestern United States is the Clarion-Nicollet-Webster catena. These are all silt loam textured soils with a thick organic C rich A horizon (mollic epipedon) formed from calcareous glacial till parent material. The Clarion soil is typically well drained and occurs at the summit and shoulder positions. The Nicollet is moderately to somewhat poorly to well drained and occurs at the backslope and footslope positions. The Webster is a poorly drained soil which occurs at the toeslope position. Thus, variations in soil properties at the scale of a hillslope are often controlled by landscape position and water flow behavior.

### 9.1.1 VARIABILITY WITHIN SOIL MAPPING UNITS

It would be wrong to give the impression that soil variability at the field/scale can be completely described by soil mapping units because there is considerable variation in soil properties that are not accounted for by soil mapping unit classification. Two types of variation can be identified. (1) There is soil variabilty within soil mapping units that is caused by mapping and classification error (Burrough, 1991). The scale of most soil series maps in the United States is 1:24,000. At this scale, up to 40% of the region within a soil mapping unit can consist of dissimilar inclusions, which are soils that are not classified the same as the dominant soil in the mapping unit. Errors in mapping may occur anywhere within the soil mapping unit, whether near the boundary with other mapping units or in the center of the mapping unit. Often, these errors appear as lighter- or darker-colored regions within a mapping unit when viewed in an aerial photograph of bare soil (Thompson and Robert, 1995). (2) There is variation in soil properties within soil mapping units due to the effects of human management activities. For example, the application of animal manure to soil results in significant increases of soil nutrient levels and soil electrical conductivity relative to unmanured soil. Since the soil survey does not classify soils based on nutrient content or electrical conductivity, these human-induced variations would not be diagnostic criteria used in soil classification. Alternatively, the deposition of wind-blown ash from an industrial site may bring with it trace metals that contaminate soil in an isolated area. A large variety of other human activities can alter or disturb the natural characteristics of soils, including landfill disposal, industrial discharge of solvents, mining and logging operations, tillage and drainage of agricultural fields, application of fertilizers and pesticides, rotation of crops, irrigation practices, and installation of septic tanks.

Thus, it is evident that soil maps published in the County Soil Surveys are not sufficient alone for describing the detailed patterns in variation of soil properties that occur across hillslopes, within fields or parcels of land, and across regions. A significant amount of research has been conducted on methods for sampling, characterizing and representing soil variability at a finer resolution than is given by soil mapping units (Di et al.,1989; Rogowski and Wolf, 1994).

## 9.2 CLASSICAL MEASURES OF VARIABILITY

Soil variability can be estimated using either continuous or discrete sampling. In continuous sampling, measurements of a selected soil property are obtained at all locations in a field. This is usually accomplished through analysis of a satellite or aerial image, or through ground collection of soil data by noninvasive remote sensing techniques. An example of the latter is sensing of soil electrical conductivity with the Geonics EM-38 electromagnetic induction sensor. Continuous sampling results in data for all locations of a field, with no need for interpolation between measurements or emphasis on soil sampling design strategies.

Discrete sampling is usually accomplished by collecting soil samples at predetermined locations and depths using destructive sampling techniques. Unlike continuous sampling, only a subset of the sample population is observed in discrete sampling. If enough samples have been collected, the characteristics of the sample population can be inferred using statistical techniques.

### 9.2.1 CLASSICAL MEASURES OF CENTRAL TENDENCY AND SPREAD

Two key characteristics of sample populations are measures of central tendency and measures of data variation (Snedecor and Cochran, 1980; Upchurch and Edmonds, 1991). Measures of central tendency include the mean, median and mode. The mean ($\bar{z}$) is the arithmetic sample average from:

$$\bar{z} = \frac{1}{N} \sum_{i=1}^{N} z_i$$

[9.1]

where $N$ is the number of sampled locations, and $z_i$ is the measured value at the $i$th location. At the median, half the sampled population has a value greater than the median, and half has a value less than the median. When the data for a sampled population are ranked by value, the mode is the value that occurs with the greatest frequency.

The spread of sample values around the mean is an important measure of variability in the sample population and is computed as the standard deviation ($s$), which is the square root of the sample variance ($s^2$). The variance is estimated from:

$$s^2 = \frac{1}{N} \sum_{i=1}^{N} \left[ z_i - \bar{z} \right]^2 \qquad [9.2]$$

where $N$ is the number of measured values $z_i$. For a normal frequency distribution, approximately 95% of the population will have a value of the mean plus or minus 2 standard deviations.

Other measures of spreading about the mean include the range, quartile range and coefficient of variation. As the value of the range, standard deviation and coefficient of variation increases, so too does population variability. The range is the difference between the maximum and minimum values, while the quartile range is the difference between the first and third quartile values. The coefficient of variation (CV) gives a normalized measure of spreading about the mean, and is estimated using:

$$CV = \frac{s}{\bar{z}} \times 100\% \qquad [9.3]$$

Properties with larger CV values are more variable than those with smaller CV values.

Wilding (1985) has described a classification scheme for identifying the extent of variability for soil properties based on their CV values in which CV values of 0–15, 16–35 and > 36 indicate little, moderate and high variability, respectively. Typical ranges of CV values from published studies of soil properties (Jury, 1986; Jury et al., 1987; Beven et al., 1993; Wollehaupt et al., 1997) are given in Table 9.1. Properties such as soil pH and porosity are among the least variable, while those pertaining to water or solute transport are among the most variable.

In classical statistics, a sample population is characterized by frequency distributions which are parameterized using the mean and standard deviation. A frequency distribution is the number of observations occurring in several class intervals between the minimum and maximum values of the ample population. Frequency distributions may either be normally or non-normally distributed, being symmetrical and asymmetrical about the mean, respectively. In a normal frequency distribution, the mean, median and mode are all equal whereas in a non-normal distribution, they are not. One of the most common non-normal distributions is the log normal distribution in which the median is larger than the mode but smaller than the mean. For log normal frequency distributions, the mean is heavily influenced by a few large measurement values; hence, the median is a more accurate measure of central tendency than the mean.

Asymmetrical frequency distributions can often be transformed to normal distributions by taking the natural logarithm of the data. Examples of properties where a log transform produced normally distributed data include studies of soil hydraulic properties (Vieira et al., 1988; Wierenga et al., 1991; Russo and Bouton, 1992), soil nutrient content (Wade et al., 1996) and trace metal soil concentrations (Markus and McBratney, 1996). The mean ($m$) and variance ($s$) of the distribution for the log transformed data can be used to estimate the arithmetic mean of the untransformed data using the expression (Webster and Oliver, 1990):

$$\bar{z} = \exp\left( m + 0.5s^2 \right) \qquad [9.4]$$

**TABLE 9.1**
**Ranges in values for the coefficient of variation (CV)**
**of selected soil and crop properties**

| Property | CV (%) | Magnitude of Variability |
|---|---|---|
| pH | 2–15 | Low |
| Porosity | 7–11 | Low |
| Bulk Density | 3–26 | Low to Moderate |
| Crop Yield | 8–29 | Low to Moderate |
| %Sand | 3–37 | Low to Moderate |
| 0.01 MPa Water Content | 4–20 | Low to Moderate |
| Pesticide Adsorption Coeff. | 12–31 | Moderate |
| Organic Matter Content | 21–41 | Moderate to High |
| 1.5 MPa Water Content | 14–45 | Moderate to High |
| %Clay | 16–53 | Moderate to High |
| Soil Nitrate-N | 28–58 | Moderate to High |
| Soil Water Infiltration Rate | 23–97 | Moderate to High |
| Soil Available Potassium | 39–157 | High |
| Soil Available Phosphorus | 39–157 | High |
| Soil Electrical Conductivity | 91–263 | High |
| Saturated Hydraulic Conductivity | 48–352 | High |
| Solute Saturated Velocity | 78–194 | High |
| Solute Dispersion Coeff. | 79–178 | High |
| Solute Dispersivity | 78–539 | High |

The spread of the normal frequency distribution about the mean is an important measure of variability in the sample population. As an example, consider the frequency distributions for two artificial data sets generated with mean values of 50, and a standard deviation of either 10 (Figure 9.1A) or 5 (Figure 9.1B). A sampled population with more variability (Figure 9.1A) will have a wider spread in the frequency distribution than a population with less variability (Figure 9.1B). The frequency distribution for a sampled population with a larger standard deviation appears relatively flat and spread out. In contrast, a sampled population with little variability will have a narrow spread in the frequency distribution, and most of the sampled values will be close to the mean and median.

The frequency distribution characterizes the mean and spread of sample values about the mean. Comparison of two frequency distributions for the same soil property (for instance, soil moisture content) allows comparison of the mean values for each sampled population, as well as comparisons about the relative amount of sample variation observed. Comparisons of the mean values for the normal frequency distributions from two samples of the same property are often achieved using t-tests (Snedecor and Cochran, 1980; Upchurch and Edmonds, 1991).

---

**EXAMPLE 9.1**   The mean and standard deviation for percent clay in a 100 ha field were found to be 25% and 5%, respectively. In contrast, the mean and standard deviation for saturated hydraulic conductivity in the same field were found to be 1 cm/hr and 1.5 cm/hr, respectively. Which of these properties has the largest magnitude of variability within the field?

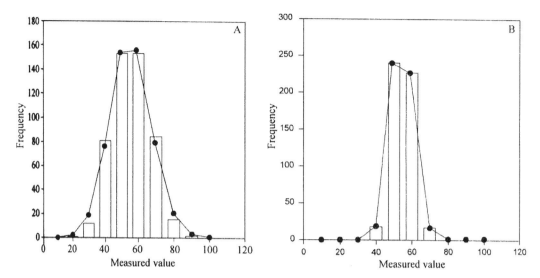

**FIGURE 9.1** (A) Normal frequency plot (●) and sample frequency distribution (bars) for a population with a mean of 50 and a standard deviation of 10; (B) normal frequency plot (●) and sample frequency distribution (bars) for a population with a mean of 50 and a standard deviation of 5.

## 9.3 GEOSTATISTICS

The frequency distribution is somewhat limited in its ability to describe variability of a sampled population because it does not provide any information about the spatial correlation between soil samples at a given location. Most often, soil properties do not occur across the landscape in a random fashion. Typically, the values for a soil property from samples taken at close spacings will be similar or spatially correlated (Oliver, 1987). Soil properties from large sample spacings will typically be dissimilar and spatially uncorrelated.

The important differences between spatially correlated and randomly distributed data can be illustrated using the following example. Suppose soil surface samples are taken from a large field, and the organic C content of each sample is measured. One would expect samples collected close to one another to have similar organic C contents, while samples taken farther apart would be less similar. Figure 9.2A shows the measured patterns in surface organic C content for a study site in Minnesota from 234 samples collected on a regular grid at spacings of $30 \times 45$ m. For comparison, a random artificial data set was generated having the same mean and standard deviation as the measured organic C contents (Figure 9.2B). Comparison of Figures 9.2A and 9.2B shows the great differences in appearance of the spatial patterns for organic C content when spatial correlation is absent. Notice that the randomly distributed values for organic C content appear much patchier than the spatially correlated values, which show larger spatial groupings where either large or small values for organic C occur. This is an illustration of spatial dependence, which manifests itself as a greater similarity between closely spaced than widely spaced samples.

Thus, in addition to knowing what the mean and standard deviation of a sample population are, one is often interested in knowing the spatial correlation structure of the population. The spatial structure of a population can be estimated using approaches developed in geostatistics (Journel and Huijbregts, 1978; Isaaks and Srivastava, 1989; Cressie, 1991). Geostatistics is a branch of applied statistics that quantifies the spatial dependence and spatial structure of a measured property and, in turn, uses that spatial structure to predict values of the property at unsampled locations. These two steps typically involve spatial modeling (variography) and spatial interpolation (kriging).

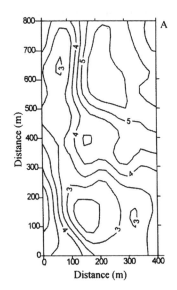
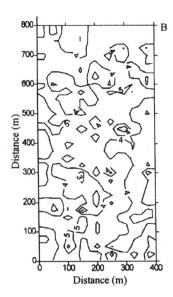

**FIGURE 9.2** (A) Spatially correlated patterns in measured soil surface organic carbon content (%) from a field in southern Minnesota; (B) Random distribution of soil surface organic carbon contents (%) having the same mean and standard deviation as the data presented in Figure 9.2A.

In the jargon of geostatistics, a regionalized variable is a property such as soil moisture or hydraulic conductivity which can be sampled. This regionalized variable has a value at every point within the region to be studied, and the collection of regionalized variables at all sampled points is known as a random function. One realization of this random function is a regionalized variable $Z(x)$, having data values $z(x_i)$, or simply $z_i$, at sampling locations $x_i$. There can be many realizations of the regionalized variable if the region is sampled many different times, with each realization being a subsample of the sample population.

### 9.3.1 VARIOGRAPHY

Spatial dependence can be quantified and modeled using the semivariogram (Burgess and Webster, 1980a). The semivariogram $\gamma(h)$ is estimated using the equation:

$$\hat{\gamma}(h) = \frac{1}{2n(h)} \sum_{i=1}^{n(h)} \left[z_i - z_{i+h}\right]^2 \qquad [9.5]$$

where $h$ is the separation distance between locations $x_i$ and $x_{i+h}$, $z_i$ and $z_{i+h}$ are the measured values for the regionalized variable at locations $x_i$ or $x_{i+h}$, and $n(h)$ is the number of pairs at any separation distance $h$.

The semivariogram is theoretically related to the autocorrelation function [r($h$)] by the expression (Burgess and Webster, 1980a; Vauclin et al., 1982):

$$\gamma(h) = s^2\left[1 - \rho(h)\right] \qquad [9.6]$$

where $s^2$ is the population variance. This relationship applies best to data which have no trend, meaning that the population mean and variance are constant over the sampled area (Rossi et al., 1992). Based on this relationship, the semivariogram should theoretically be equal to the population variance when the separation distance is very large. At very small separation distances, the semivariogram approaches a value of zero. Thus, the semivariogram is a quantitative measure of how the variance between sampled points is reduced as separation distance decreases.

Often, the separation distances are normalized by dividing separation distance by the smallest separation distance between locations in the sampling design. The resulting normalized separation distance is termed a lag or a lag separation distance ($h$) in geostatistical jargon. Lags are integers taking on values 1, 2, 3, ... etc., whereas separation distances (for example, 30, 60, 90, ..., etc.) are the actual averaged distances between pairs.

If pairwise squared differences are averaged without regard to direction, the semivariogram is considered to be isotropic and omnidirectional. When pairs of a given separation distance are considered as a function of direction, the semivariogram is considered to be directional. There are two types of anisotropy (Webster and Oliver, 1990): zonal where the sill varies with direction, and geometric where the range varies with direction. Anisotropic forms of the semivariogram can also be estimated using explicitly two-dimensional equations (Mulla, 1988a; Webster and Oliver, 1990), rather than simply computing a collection of one-dimensional semivariograms in various directions, and modeling their direction dependence (Nash et al., 1988).

---

**EXAMPLE 9.2** Assuming that good spatial structure exists for a soil property (i.e., nonrandom spatial patterns), qualitatively draw the relationship described by Equation [9.6] by plotting the semivariogram and the autocorrelation function of the soil property on the y-axis versus separation distance between samples on the x-axis.

---

### 9.3.1.1 Binning

A key consideration in the estimation of semivariograms is deciding how to group pairs when calculating the squared differences at varying separation distances. In geostatistical jargon, this is a decision about bin distance, or simply binning procedures. The term bin refers to a group of pairs having a narrow range of separation distances. For example, consider a situation where samples are collected on a regular grid at spacings of $30 \times 50$ m, and an omnidirectional semivariogram is to be estimated. The separation distances between pairs of points are 30, 50, 58.3, 60, 78.1, 90, 100,..., etc. It is inefficient to estimate the semivariogram at each of these separation distances because there is likely to be little gain in useful information by having separate values for the semivariogram at closely spaced separation distances (e.g., 58.3 and 60). Therefore, one can average values for the semivariogram in bins which represent a narrow range of separation distances, i.e., 55–60 inclusive.

To illustrate binning procedures, various ways of grouping pairs by separation distance from the previous example will be investigated. As a first step, one might estimate the semivariogram using the bins 48.7, 71.7, 104.4, ..., etc. Note that this involves averaging the data from pairs at separation distances of 30, 50, and 58.3 m, and averaging data from pairs separated by 60, 78.1, and 90 m. Much of the information about variability at short separation distances is lost using this approach. A second approach could involve bins with pairs separated by an average distance of 30, 56.6, 82.1, 108.3, ..., etc. Clearly, there is more information about short-range variability in this approach, but still considerable averaging at larger separation distances. A third approach which produces even less averaging of pairs with dissimilar separation distances is to use bins of 30, 55.4, 60, 78.1, 90, 102.9, ..., etc. Clearly this approach provides more detail about spatial variability at separation distances representative of the actual sampling pattern but fewer data points are available in each bin.

The previous example on binning of pairs shows that the best choice of bins on a regular grid is one that best represents the actual separation distances between sampled points with a minimum of averaging, especially at short separation distances. The only reason to use an approach that involves more spatial averaging is when there is an insufficient number of pairs (e.g., less than 30–50 pairs) at a given separation distance. In this case, binning provides a way to increase the number of sample pairs upon which a semivariogram is estimated.

Some general guidelines for binning procedures are as follows. For omnidirectional semivariograms, use bins that include the closest separation distances in the sample design so that the maximum amount of information is obtained about the semivariogram at small separation distances. Avoid making bins so narrow that the number of pairs in any bin is less than 30. For directional semivariograms, it is necessary to have bins for both separation distance and direction (Webster and Oliver, 1990). As an example, if semivariograms are to be estimated in the four principal directions, the directional bins could be $0° \pm 22.5°$, $90° \pm 22.5°$, $180° \pm 22.5°$, and $270° \pm 22.5°$. Narrower directional bins could also be selected, but wider ones should probably be avoided, since they would lead to a loss of information about directional effects.

---

**EXAMPLE 9.3**  Consider a sampling strategy involving a regular grid at spacings of 15 m × 25 m. What are the five smallest separation distances between pairs of sample points on this regular grid, considering all possible directions for pairs of samples? How would you propose to group sample pairs into bins for this sampling strategy?

---

### 9.3.1.2  Modeling the Semivariogram

The semivariogram can be modeled using any of several authorized models (Journel and Huijbregts, 1978; Oliver, 1987; Isaaks and Srivastava, 1989) that are commonly fitted to the semivariogram data using nonlinear least squares with either uniform weighting of points or weighting that favors data at small over large separation distances (Cressie, 1985). The most common models are linear, spherical and exponential models. The linear model (Figure 9.3A) is given by:

$$\gamma(h) = C_0 + mh \quad h < a \qquad [9.7a]$$

and

$$\gamma(h) = C_0 + C_1 \quad \text{for} \ h \geq a \qquad [9.7b]$$

The spherical model (Figure 9.3B) is given by:

$$\gamma(h) = C_0 + 0.5C_1 \left[ \left( \frac{3h}{a} \right) - \left( \frac{h}{a} \right)^3 \right] \quad \text{for} \ h < a \qquad [9.8a]$$

and

$$\gamma(h) = C_0 + C_1 \quad \text{for} \ h \geq a \qquad [9.8b]$$

The exponential model (Figure 9.3C) is given by:

$$\gamma(h) = C_0 + C_1 \left[ 1 - \exp\left( \frac{-h}{r} \right) \right] \qquad [9.9]$$

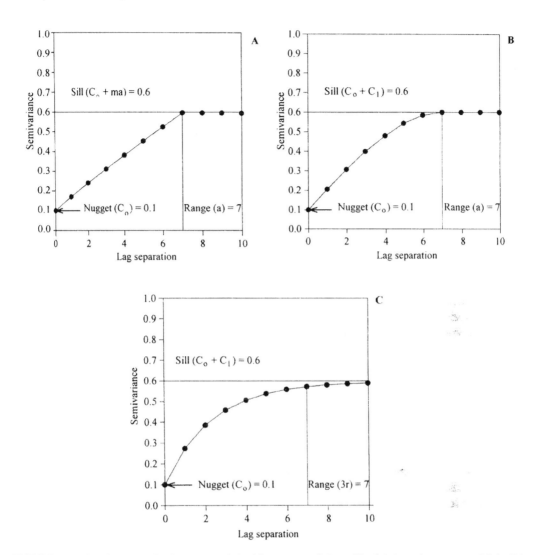

**FIGURE 9.3** (A) Linear semivariogram model with a range of 7, a sill of 0.6, and a nugget of 0.1; (B) Spherical semivariogram model with a range of 7, a sill of 0.6, and a nugget of 0.1; (C) Exponential semivariogram model with a range of 7, a sill of 0.6, and a nugget of 0.1.

The semivariogram model and its parameters provide a quantitative expression of spatial structure for the measured property. In these models, the fitted parameters have the following definitions (Journel and Huijbregts, 1978; Oliver, 1987; Oliver and Webster, 1991). The nugget parameter ($C_0$) is a measure of the amount of variance due to errors in sampling, measurement and other unexplained sources of variance. The sum of parameters $C_0$ and $C_1$ is the sill, theoretically equal to the variance of the sampled population at large separation distances if the data have no trend. If the nugget parameter is about equal to the sample variance, it is an indication that the sampled property has very little spatial structure or varies randomly. In this case, the semivariogram is best described using a linear semivariogram with a slope of zero, often referred to as a pure nugget effect model (Oliver, 1987; Isaaks and Srivastava, 1989). Parameter $a$ is the range. For the linear and spherical models, the range ($a$) is the distance at which samples become spatially independent and uncorrelated with one another. For the exponential model, the range is operationally equivalent to the value $3r$. Parameter m of the linear model expresses the rate of change in variance as separation distance increases, in other words, the slope of the linear model.

**TABLE 9.2**
**Variation in the range parameter for semivariogram models of selected soil and crop properties.**

| Property | Range (m) | Spatial Dependence |
|---|---|---|
| Saturated Hydraulic Conductivity | 1–34 | short range |
| %Sand | 5–40 | short range |
| Saturated Water Content | 14–76 | short to moderate range |
| Soil pH | 20–260 | short to long range |
| Crop Yield | 70–700 | moderate to long range |
| Soil Nitrate-N | 40–275 | moderate to long range |
| Soil Available Potassium | 75–428 | moderate to long range |
| Soil Available Phosphorus | 68–260 | moderate to long range |
| Organic Matter Content | 112–250 | long range |

Probably the most important of the semivariogram parameters for decisions concerning the spacing between sample locations is the range ($a$). At separation distances greater than the range, sampled points are no longer spatially correlated, which has great implications for sampling design. If a region is being sampled in order to understand the spatial pattern of a given property, it is advisable that the sampling design use separation distances that are, at most, no greater than the value for the range parameter of the semivariogram. It is preferable that the sample spacing be from ¼ to ½ of the range (Flatman and Yfantis, 1984). If samples are separated by distances greater than the range, there is no spatial dependence between locations. It is inappropriate to use geostatistics in such a situation.

Table 9.2 summarizes published values for the range of semivariogram models for several measured soil and agronomic properties (Jury, 1986; Warrick et al., 1986; Wollenhaupt et al., 1997; McBratney and Pringle, 1997).

When fitting data to semivariogram models, there are a few important guidelines to follow. First, the number of sample data values used in estimating the semivariogram must not be < 30 (Journel and Huijbregts, 1978), while some authors recommend at least 100–200 sampling locations for accurate estimation of the semivariogram (Oliver and Webster, 1991). For regularly spaced sample designs, the maximum number of pairs in the semivariogram calculation at the smallest separation distance is always equal to $N-1$, where $N$ is the number of sample data points. Thus, it follows that the number of sample pairs used in the estimation of the semivariogram at any separation distance should not be < 30. Second, the semivariogram model should not be fitted to semivariogram data at separation distances greater than ½ of the largest dimension of the study area (Burrough, 1991). This is because the pairs of measured data at large separation distances are representative of the variance structure at the edges of the field, not the majority of the samples.

---

**EXAMPLE 9.4**    Consider the following data for soil water content (kg/kg) collected on two transects in a field at sample spacings of 20 m. Estimate the sample mean, standard deviation and semivariance for each of these two transects separately. Discuss the differences between the two transects in terms of the average water content, magnitude of variability, and extent of spatial structure present.

| Transect #1: | 0.05 | 0.10 | 0.15 | 0.20 | 0.25 | 0.20 | 0.15 | 0.10 | 0.05 |
|---|---|---|---|---|---|---|---|---|---|
| Transect #2: | 0.10 | 0.25 | 0.15 | 0.10 | 0.20 | 0.05 | 0.15 | 0.05 | 0.20 |

---

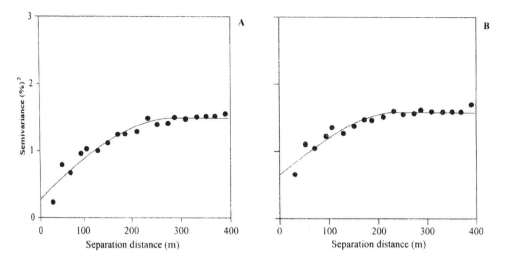

**FIGURE 9.4** (A) Spherical semivariogram model for measured soil surface organic carbon contents displayed in Figure 9.2A; (B) Spherical semivariogram model for measured soil surface organic carbon contents from Figure 9.2A with 5 added influential observations.

### 9.3.1.3 Influential Observations

When there are influential observations in the measured data, estimating the semivariogram can be problematic leading to significant increases in the nugget and sill values for the semivariogram. In the worst case, a few outliers can obscure all spatial structure in the semivariogram and make it appear as if the measured data are spatially independent (Rossi et al., 1992).

The following example shows the effect of a few outliers on semivariogram modeling. A study site in Minnesota was sampled at 234 locations and analyzed for organic C content. Spatial structure from the resulting data set obeyed a spherical semivariogram (Equation [9.8] and Figure 9.4A) with a nugget of ~ 0.12, a range of ~ 260 m and a sill of ~ 1.4. For illustration, five outliers were inserted into the data set and the semivariogram was recomputed. The resulting semivariogram was significantly affected by the five outliers (Figure 9.4B), with the main effects being an increase in the nugget to ~ 0.7, and a decrease in the range to ~ 210 m. The overall effect of the outliers was to increase the perceived short-range variability, and to decrease the range of spatial dependence.

Influential observations can be removed using several methods. One of the best is to remove influential observations when computing the semivariogram using the method of Hawkins (1980). Another good approach is to estimate the semivariogram using a robust resistant rather than the classical semivariogram estimator (Cressie and Hawkins, 1980; Birrell et al., 1996).

### 9.3.2 Interpolation by Kriging

Kriging is a general term describing a geostatistical approach for interpolation at unsampled locations. There are several types of univariate kriging methods including punctual (Burgess and Webster, 1980a), indicator (Journel, 1986), disjunctive (Yates et al., 1986a,b), universal (Webster and Burgess, 1980) and block kriging (Burgess and Webster, 1980b). A multivariate form of kriging is known as cokriging (Vauclin et al., 1983). Conditional simulation has also been applied to kriging (Hoeksma and Kitanidis, 1985; Warrick et al., 1986; Gutjahr, 1991). The methods used for punctual kriging cokriging, block kriging and indicator kriging will be described in this chapter.

The science of kriging was developed for the mining industry by Matheron (1963), and first applied in the mining industry by Krige (1966); it was applied to soil science by Burgess and Webster (1980a,b) and Webster and Burgess (1980) in Europe and in the United States by Vauclin et al. (1982, 1983) and Vieira et al. (1983).

No other linear interpolation method provides less bias in predictions than kriging, which is known as a best linear unbiased predictor (BLUP) (Burgess and Webster, 1980a). This is because the interpolated or kriged values are computed from equations that minimize the variance of the estimated value. In fact, when interpolating at a location where a measurement exists, kriging will always generate a value equal to the measured value. For this reason, kriging is often loosely described as being a type of linear regression in which the regression line always passes through every one of the measured data points.

Several studies have compared kriging and classical methods for interpolation, such as inverse distance weighting and cubic splines (Dubrule, 1984; Laslett et al., 1987). As a general rule of thumb, kriging methods are equivalent or superior to classical methods when the data to be interpolated have well-developed spatial structure, have a semivariogram without a significant nugget effect, and are sampled at spacings less than the range of the semivariogram in clusters or at irregular spacings. Inverse distance weighting tends to be more suitable for use with data having short-range variability (Cooke et al., 1993) than with data having long-range spatial dependence (Gotway et al., 1996). Spline-based interpolation is perhaps better than kriging only in situations where an abrupt change in measured soil property values occurs across a short distance (Voltz and Webster, 1990).

A key conclusion from the comparisons by Gotway et al. (1996) is that interpolation accuracy is more dependent on the adequacy of the sampling design than on the type of interpolator used. When intensive sampling is conducted on a regular grid, there may be only small differences in interpolation with kriging, inverse distance weighting or cubic splines. In view of this, Warrick et al. (1988) suggested that kriging is the best choice for an interpolator, because it is the only method that allows the variance of an interpolated point to be estimated. Whelan et al. (1996) suggest that inverse distance weighting interpolators will outperform kriging interpolators for small sample sizes collected at moderate intensity.

### 9.3.2.1 Stationarity

The kriging interpolator uses information from the semivariogram model about spatial structure of the measured property. In order to use kriging methods, one very important condition must be met, which is the intrinsic hypothesis (Warrick et al., 1986). It states that the mean of the measured property is stationary and the semivariogram at any separation distance is finite and they do not depend on location. This means that the expectation of the squared differences between values depends only on separation distance and not on location. If these conditions are not met, the measured property is said to be nonstationary, and kriging is not appropriate without removing the causes of nonstationarity in the data.

Nonstationarity can result from three types of problems. The first is a long-range systematic change in the mean with location, often referred to as trend. This type of effect can be removed by fitting a regression model to the data and removing the long-range trend. Alternatively, mean or median polish can be used to remove trend (Cressie, 1991). The latter method works best when the trend is aligned with the rows or columns of the sample design. The second is a short-range stochastic change in the mean with location, often referred to as drift. This is much more problematic, and the primary method for removing drift is to use universal kriging. The third type of nonstationarity involves a change in variance with location. Universal kriging can also be used to remove this effect.

### 9.3.2.2 Punctual Kriging

Punctual kriging is essentially a linear interpolator, which sums up weighted values for measurements at locations neighboring the unsampled location. The expression used in kriging is:

$$z_0 = \sum_{i=1}^{N} \lambda_i z_i \qquad [9.10]$$

where $z_0$ is the interpolated value, $N$ is the number of points neighboring the interpolated location, $z_i$ are the measured values at neighboring locations, and $\lambda_i$ is the kriging weighting factor for each of the neighboring measured values. The weighting factors must sum to unity so that the expected value of the interpolated points minus the measured points is zero. To further illustrate this concept, if interpolation occurs at a location where a measured value exists, the weighting factor is unity for the measured point at the location to be interpolated, and the weighting factors for all neighboring locations are zero.

### 9.3.2.3   Neighborhood Search Strategy

Identifying the $N$ locations that neighbor the unsampled location is an important step in the kriging process. There are a few important guidelines concerning the search for neighboring locations around an unsampled location. First, for omnidirectional kriging, the search radius should be set so that at least 6–8 locations, but not more than 16–24 locations, are identified as neighboring locations (Wollenhaupt et al., 1997). With fewer than 6–8 locations, the matrix solver may not be able to converge to a solution. The search radius should not exceed the range of the semivariogram, and is typically less than ½ the range. For systematic grid sample designs, the search radius typically describes a circle which extends to and includes the closest four neighbors as well as the next four neighbors along a diagonal from the unsampled location. If the measured data are not regularly spaced, or are clustered, it may be difficult to use a fixed search radius. Some unsampled locations may have too few nearest neighbors in order to include at least eight. In this case, it is best to use a flexible search radius so that each unsampled location has approximately the same number of neighbors. Second, the neighbors should be evenly distributed in all directions outward from the unsampled location, if possible. This avoids low weighting factors that result from data redundancies. A quadrant or octant search strategy can be used to find neighbors that are located in the four or eight principal directions away from the unsampled location (Webster and Oliver, 1990). Third, using a large search radius may give rise to a large number of neighbors and many small or negative weights. In extreme cases, this can lead to a negative value at the unsampled location which can be avoided by placing an upper limit on either the search radius or the maximum number of neighbors.

For randomly oriented or clustered sample designs, the search radius should be set to a value slightly larger than the average spacing between data points (Wollenhaupt et al., 1997). If anisotropic kriging is to be used, the search neighborhood is usually elliptical in shape. The longest axis of the search ellipse corresponds to the direction in which the range of the semivariogram is greatest, while the shortest axis corresponds to the direction in which the range is smallest.

### 9.3.2.4   Estimation Variance

The kriging technique has one major advantage over all other interpolation methods. Only with kriging is it possible to estimate the variance of the interpolated values. The estimation variance depends upon only two factors. The first is the spatial structure for the measured property as shown in the semivariogram, and the second is the weighting factors for neighbors of the interpolated location. These weighting factors depend, in turn, upon the arrangement of sampled locations around the unsampled location. The estimation variance is calculated using the expression:

$$\sigma^2\left(z_o\right) = \mu + \sum_{i=1}^{N} \lambda_i \gamma_{io} \qquad [9.11]$$

where $\sigma^2$ is the estimation variance, $\mu$ is the Lagrangian multiplier and $\gamma_{io}$ is the semivariance at the separation distance between the locations for measured ($z_i$) and interpolated values ($z_o$).

If the sample population can be assumed to follow a normal distribution, the kriging estimation variance can be used to estimate the confidence interval for prediction at the unsampled location (McBratney and Pringle, 1997). The 95% confidence interval about the mean is estimated from the expression given by Birrell et al. (1996):

$$\left( \bar{z} \pm 1.96 \frac{\sigma_o}{\sqrt{N}} \right)$$ [9.12]

where $\sigma_o$ is the maximum estimation variance obtained at all unsampled locations.

As shown by Burrough (1991), the estimation variance is affected primarily by two factors. The first of these is the form and shape of the semivariogram. As the nugget effect increases, so too does the estimation variance. In many cases, the nugget effect can be minimized by sampling at close spacings and by compositing or bulking samples. The optimum number of bulked samples depends upon the CV for the measured property and the shape of its semivariogram. A cost-effective and optimum number of composite samples may be anywhere from 4–16 samples (Webster and Burgess, 1984; Burrough, 1991; Oliver et al., 1997). The second factor is the number and spatial arrangement of neighboring locations. As separation distance increases and the number of neighboring locations decreases, the estimation variance at an unsampled location increases. Generally, the arrangement of sampled neighbors which minimizes estimation variance is along a regular square grid or an equilateral triangle (Webster and Burgess, 1984; Burrough, 1991).

### 9.3.2.5 Solving for Weighting Factors

The key step in kriging is to estimate the n weighting factors for locations that neighbor the unsampled location where interpolation is to occur. This is accomplished by solving a set of $N+1$ simultaneous equations with $N+1$ unknowns (the $N$ weighting factors and the undetermined Lagrangian multiplier). These equations can be written in matrix notation using:

$$[F][L] = [B]$$ [9.13]

where $[F]$ is an $N+1$ by $N+1$ matrix of semivariogram values between measured locations, $[L]$ is an $N+1$ by 1 matrix of weighting factors and the Lagrangian multiplier, and $[B]$ is an $N+1$ by 1 matrix of semivariogram values between the interpolated location and its neighboring locations. The full mathematical details for each matrix are given below:

$$[F] = \begin{bmatrix} 0 & \gamma_{12} & \gamma_{13} & \cdots & \gamma_{1n} & 1 \\ \gamma_{21} & 0 & \gamma_{23} & \cdots & \gamma_{2n} & 1 \\ \vdots & & & & \vdots & 1 \\ \gamma_{n1} & \gamma_{n2} & \gamma_{n3} & \cdots & 0 & 1 \\ 1 & 1 & 1 & \cdots & 1 & 0 \end{bmatrix}$$ [9.14]

$$[L] = \begin{bmatrix} \lambda_1 \\ \lambda_2 \\ \vdots \\ \lambda_n \\ \mu \end{bmatrix}$$ [9.15]

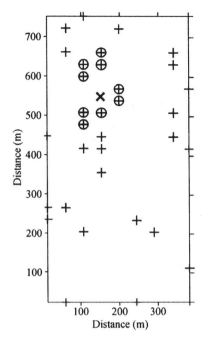

**FIGURE 9.5** Sample locations of alachlor Freundlich adsorption partition coefficient measurements (+) from a field in southern Minnesota. Sample locations denoted with circles are further described in Table 9.3. Location denoted with the symbol X is an unsampled location.

$$[B] = \begin{bmatrix} \gamma_{o1} \\ \gamma_{o2} \\ \vdots \\ \gamma_{on} \\ 1 \end{bmatrix}$$

[9.16]

### 9.3.3 Example: Spatial Variability of Alachlor Sorption

Soil surface samples were collected from a field in south central Minnesota at 35 locations (Figure 9.5). The Freundlich adsorption partition coefficients ($K_f$) were obtained from experiments conducted in the laboratory on each sample. A subset of these $K_f$ values (shown with circles in Figure 9.5) are given in Table 9.3 to illustrate the estimation of the semivariogram, and for an example of interpolation by kriging. The interpolated value is to be estimated at location o, which has the x,y coordinates 150,550.

#### 9.3.3.1 Semivariogram Example

The semivariogram for this subset of alachlor adsorption data is estimated from Equation [9.5] as follows. First, one estimates the value of the semivariogram at the closest separation distance (also known as the first lag distance) which is 30.5 m. As is evident from Table 9.3 and Figure 9.5 for this subset of the data, there are four pairs of data separated by 30.5 m:

**TABLE 9.3**
**Location and value for alachlor Freundlich adsorption**
**partition coefficients (K$_f$) from a field in southern Minnesota**

| Location No. | X Distance (m) | Y Distance (m) | K$_f$ Value (mL g$^{-1}$) |
|---|---|---|---|
| 1 | 106.7 | 478.5 | 6.4 |
| 2 | 106.7 | 509.0 | 7.6 |
| 3 | 106.7 | 600.5 | 5.3 |
| 4 | 106.7 | 630.9 | 5.9 |
| 5 | 152.4 | 509.0 | 12.2 |
| 6 | 152.4 | 630.9 | 5.9 |
| 7 | 152.4 | 661.4 | 5.5 |
| 8 | 198.1 | 539.5 | 9.9 |
| 9 | 198.1 | 570.0 | 6.8 |
| 10 | 150.0 | 550.0 | unsampled |

$$\gamma(h = 30.5) = 0.5\left(\frac{1}{4}\right)\left[(7.6 - 6.4)^2 + (5.9 - 5.3)^2 + (5.5 - 5.9)^2 + (6.8 - 9.9)^2\right]$$

$$= \left(\frac{1}{8}\right)(11.57) = 1.45$$

[9.17]

The semivariogram for the second lag is estimated from all pairs of measurements separated by either 54.9 m or 45.7 m. These two distances correspond to the distance between sample locations 1 and 5 or 2 and 5, respectively. There are a total of six pairs of data with these separation distances, for an average separation distance of 51.9 m. The semivariogram at the second lag (h = 51.9 m) is given by:

$$\gamma(h = 51.9) = 0.5\left(\frac{1}{6}\right)\left[(12.2 - 6.4)^2 + (12.2 - 7.6)^2 + (5.9 - 5.3)^2 + (5.9 - 5.9)^2\right] +$$

$$(5.5 - 5.9)^2 + (9.9 - 12.2)^2\right] = \left(\frac{1}{12}\right)(60.61) = 5.05$$

[9.18]

The semivariogram for the full set of alachlor $K_f$ values is shown in Figure 9.6 and was best described using a linear model (Equation [9.7]). Notice that the semivariogram values for the first and second lag distances in the full model are similar to those estimated using the partial dataset.

### 9.3.3.2  Kriging Example

For the interpolation of measured $K_f$ values at the unsampled location (150, 550), all measured values within a search radius of 100 m were considered. From Figure 9.5, it is evident that there are eight (not nine) neighbors within a distance of 100 m from the unsampled location. Using the linear semivariogram model shown in Figure 9.6, the semivariance values for the [F] and [B] matrices of Equations [9.14] and [9.16] were estimated. These values are shown in Equation [9.19] in the first and last matrix. The lower left portion of the [F] matrix is left blank for convenience, but since the matrix is symmetrical, the values omitted are the same as those

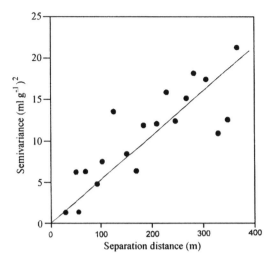

**FIGURE 9.6** Linear semivariogram model for alachlor Freundlich adsorption partition coefficients from a field in southern Minnesota.

across the diagonal at an equivalent position. After inversion of the [F] matrix, values were determined for the weighting factors and Lagrangian multiplier. These solutions are shown in the second matrix (the [L] matrix):

$$
\begin{bmatrix}
0 & 1.6 & 6.4 & 8.0 & 2.9 & 8.3 & 5.7 & 6.8 & 1 \\
 & 0 & 4.8 & 6.4 & 2.4 & 6.8 & 5.0 & 5.7 & 1 \\
 & & 0 & 1.6 & 5.3 & 2.9 & 5.7 & 5.0 & 1 \\
 & & & 0 & 6.8 & 2.4 & 6.8 & 5.7 & 1 \\
 & & & & 0 & 6.4 & 2.9 & 4.0 & 1 \\
 & & & & & 0 & 5.3 & 4.0 & 1 \\
 & & & & & & 0 & 1.6 & 1 \\
 & & & & & & & 0 & 1 \\
1 & 1 & 1 & 1 & 1 & 1 & 1 & 1 & 0
\end{bmatrix}
\times
\begin{bmatrix}
-0.08 \\ 0.19 \\ 0.22 \\ -0.04 \\ 0.33 \\ 0.06 \\ 0.14 \\ 0.18 \\ -0.49
\end{bmatrix}
=
\begin{bmatrix}
4.37 \\ 3.12 \\ 3.48 \\ 4.80 \\ 2.15 \\ 4.23 \\ 2.57 \\ 2.72 \\ 1
\end{bmatrix}
\qquad [9.19]
$$

Note that the sum of the weighting factors in the [L] matrix is one, as required. Also note that the heaviest weight (0.33) is for location 5, with moderately large weights for locations 2 (0.19), 3 (0.22), 8 (0.14) and 9 (0.18). This makes sense because location 5 is closest to the unsampled location.

Note that in spite of being farther from the unsampled location, the weights for locations 2 and 3 are each larger than the weights for locations 8 and 9. This is because locations 8 and 9 are very close to one another, and so their values are somewhat redundant. Rather than give each location a large weighting factor, the kriging approach accounts for the redundancy in closely spaced values. The combined weights for locations 8 and 9 is 0.32, about the same as the weighting factor for location 5. Because measured values at locations 2 and 3 are in different directions from the unsampled location, there is little redundancy in the measured values, and the weighting factors are not corrected for redundancy.

Note also that neighboring locations 1, 4 and 6 have very small or even negative weighting factors for two reasons. The first is that the latter neighboring locations are farther from the unsampled location than the other neighbors and so their spatial correlation with the unsampled location is smaller. Second, the low weighting results from a screening effect. The path from each of the low weighted neighbors to the unsampled location is nearly blocked by a closer neighbor. This makes the information contained in the screened out neighbors redundant, so the weighting factors are adjusted downward.

Now that the weighting factors have been obtained, one can use them to estimate the alachlor $K_f$ value at the unsampled location. The interpolated value (8.85 mL g$^{-1}$) is obtained using Equation [9.10] as follows:

$$8.85 = (-0.08)6.4 + (0.19)7.6 + (0.22)5.3 + (-0.04)5.9 +$$
$$(0.33)12.2 + (0.06)5.9 + (0.14)9.9 + (0.18)6.8 \qquad [9.20]$$

Note that the interpolated value (8.85) is higher than the average value of 7.5 for the eight neighboring locations. In averaging, each of the measured values is assigned an equal weighting factor, which is ⅛ in this example. In kriging, the weighting factors depend upon the distance between the measured value and the unsampled location, as well as upon the arrangement of the sampled location relative to all other sampled locations.

The weighting factors can also be used to obtain the estimation variance for the unsampled location. The estimation variance [2.12 (mL g$^{-1}$)2] is obtained as follows:

$$2.12 = (0.49) + (-0.08)4.4 + (0.19)3.1 + (0.22)3.5 + (-0.04)4.8 +$$
$$(0.33)2.1 + (0.06)4.2 + (0.14)2.6 + (0.18)2.7 \qquad [9.21]$$

Of course, this is a very simplified example of estimating the semivariogram and using kriging for interpolation. In reality, a computer program would use the procedures described above at hundreds of unsampled locations, rather than just one.

A primary objective of kriging is to produce a set of interpolated values from which spatial patterns of the measured property can be mapped. To illustrate this, the alachlor $K_f$ values were interpolated throughout the Minnesota study site using all 35 measured values and kriging on a regular grid at spacings of 25 × 25 m. The resulting map for $K_f$ values (Figure 9.7A) shows a region of relatively large sorption centered near location (250,200). In addition, a region with relatively weak sorption is centered near location (200,650). This map, produced using 35 measured $K_f$ values, shows the usefulness of information produced by kriging. One could use the map as an aid to precision application of the alachlor herbicide by applying lower rates of herbicide in the regions where sorption is relatively weak, and higher rates in the regions where sorption is relatively strong. Or one could use the maps to estimate spatial patterns in leaching and runoff losses of the herbicide using a fate and transport model.

The estimation variance is also useful in this example, because it shows the uncertainty in interpolated values at various locations for the study site. In this case, there are a few regions (175, 125) and (400,300) which are undersampled as shown by larger estimation variances (Figure 9.7B). The uncertainty of interpolated values in regions with larger estimation variances is greater than that in regions with smaller estimation variances. Additional samples could be collected in regions with large estimation variances to improve the accuracy of prediction.

### 9.3.4 BLOCK KRIGING

As the previous section shows, punctual kriging is a method for estimating the value of a property at an unsampled point location. There are times when it is more appropriate to estimate the average value of a property in a block or small region around an unsampled point location. For example, data from the Landsat Thematic Mapper satellite are typically available at a spatial resolution of

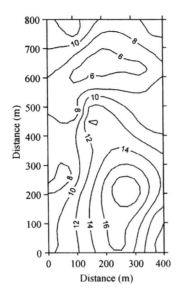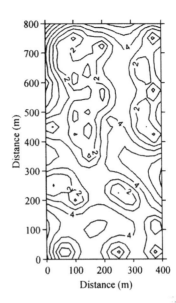

**FIGURE 9.7** (A) Kriged alachlor Freundlich adsorption partition coefficients for a field in southern Minnesota; (B) estimation variances from kriging of alachlor Freundlich adsorption partition coefficients for a field in southern Minnesota.

30 m. If data are missing from some sections of the image due to interference from clouds, it would be useful to be able to estimate the average value for the missing data in the unsampled blocks. Alternatively, one may wish to estimate the average value of a soil property for mapping units of the soil survey, or the effective permeability in a particular stratum of the subsurface geology (Journel et al., 1986). The geostatistical approach for estimating the average value of a property in an unsampled block is called block kriging (Burgess and Webster, 1980b).

Block kriging is analogous to punctual kriging in many ways. The average value of a block, $z_o(B)$ is estimated from a linear weighting of measured values inside and outside of the block using the expression (Webster and Oliver, 1991):

$$z_o(B) = \sum_{i=1}^{N} \lambda_i z_i \qquad [9.22]$$

where the weighting factors sum to unity. The weighting factors are estimated by solving an equation similar to Equation [9.13], except that the semivariograms in this equation are not estimated between points, but between points and blocks, or within a block.

The average semivariogram between a sampled point and the block is estimated using:

$$\overline{\gamma}_{iB} = \frac{1}{B} \int_B \gamma(x_i, x) dx \qquad [9.23]$$

where $B$ is the area of the block, $x_i$ is the measured point and $x$ is a point inside the block. Similarly, the average semivariogram within the block is estimated using:

$$\overline{\gamma}_{BB} = \frac{1}{B^2} \int_B \int_B \gamma(x, x') dx dx' \qquad [9.24]$$

where both $x$ and $x'$ are points within the block. The integrations required to estimate semivariograms for block kriging can be obtained using analytical functions (Clark, 1979; Webster and Burgess, 1984) based on the type of semivariogram and the geometry of the block, or by direct numerical integration.

Finally, the estimation variance of a block $\sigma^2(B)$ is obtained using the expression:

$$\sigma^2(B) = \sum_{i=1}^{N} \lambda_i \overline{\gamma_{iB}} + \mu - \overline{\gamma_{BB}} \qquad [9.25]$$

where $\mu$ is the Lagrangian multiplier. One of the advantages of block kriging is that the estimation variances are typically much smaller than those obtained from punctual kriging.

For normally distributed data, there is no need to estimate block-averaged semivariograms using analytical functions in order to use block kriging. A simpler approach is to use punctual kriging with a fine-resolution grid spacing, and then to average the interpolated values within blocks of a specified size and geometry. The average value of a block is equivalent to the value obtained by averaging all of the point-kriged values generated on a dense grid within the block (Isaaks and Srivastava, 1989).

### 9.3.5 COKRIGING

Cokriging is an interpolation technique that uses information about the spatial patterns of two different, but spatially correlated properties to interpolate only one of the properties (Vauclin et al., 1983). Typically, cokriging is used to map the property that is more difficult or expensive to measure ($z_2$) based on its spatial dependence on a property that is easier or less expensive to measure ($z_1$). An example of this is the interpolation by cokriging of sparsely sampled soil test P levels using intensively sampled soil organic C content values (Bhatti et al., 1991; Mulla, 1997). Another example might be the interpolation by cokriging of sparsely sampled soil moisture content levels (Yates and Warrick, 1987; Mulla, 1988b; Stein et al., 1988).

Cokriging requires that the semivariogram models $[\gamma_1(h), \gamma_2(h)]$ be estimated for both of the measured properties. In addition, cokriging requires estimation of the cross-semivariogram model $[\gamma_{12}(h)]$ describing spatial dependence between the two measured properties. This cannot be done unless measurements of the more intensively sampled property are available at each of the locations where the sparsely sampled property is measured.

The cross-semivariogram is estimated using the following expression:

$$\gamma_{12}(h) = \frac{1}{2n(h)} \sum_{i=1}^{n(h)} \left[ z_{1i} - z_{1,i+h} \right] \left[ z_{2i} - z_{2,i+h} \right] \qquad [9.26]$$

where $z_{1i}$ and $z_{1,i+h}$ are the measured values for property 1 at locations $i$ and $i+h$, respectively, and $z_{2i}$ and $z_{2,i+h}$ are the measured values for property 2 at locations $i$ and $i+h$, respectively. Zhang et al. (1992) have proposed a method for estimating pseudo cross-semivariograms using data that are sampled at nearly the same, but not identical, locations.

Interpolation with cokriging involves an approach that is similar to that for kriging, with terms for weighting factors of both measured properties. Interpolated values for the second property ($z_{2o}$, the sparsely measured property) are obtained using the expression:

$$z_{20} = \sum_{i=1}^{N_1} \lambda_{1i} z_{1i} + \sum_{i=1}^{N_2} \lambda_{2i} z_{2i} \qquad [9.27]$$

where $\lambda_{1i}$ and $\lambda_{2j}$ are the weighting factors for property 1 at location i and property 2 at location $j$, respectively, and $z_{1i}$ and $z_{2j}$ are the measured values for property 1 at location $i$ and property 2 at location $j$, respectively. Thus, the interpolated value is simply a linear combination of the measured values for both properties at locations that neighbor the unsampled location.

As in kriging, the cokriging prediction is a best linear unbiased predictor. This is ensured by requiring that the weighting factors for property 1 sum to zero, and the weighting factors for property 2 sum to unity.

The cokriging technique provides an estimate for the variance of the cokriging prediction at all unsampled locations. At locations where a measured value exists for the cokriged property, the estimation variance is zero, and the interpolated value equals the measured value. The expression for the cokriging estimation variance ($\sigma_o^2$) is given by:

$$\sigma_o^2 = \mu_2 + \sum_{i=1}^{N_1} \lambda_{1i} \gamma_{12,io} + \sum_{j=1}^{N_2} \lambda_{2j} \gamma_{22,jo} \qquad [9.28]$$

where $\mu_2$ is the Lagrangian multiplier for property 2, $\lambda_{1i}$ is the weighting factor for the $i$th location of property 1, $\lambda_{2j}$ is the weighting factor for the $j$th location of property 2, $\gamma_{12,io}$ is the cross-semivariance at a separation distance between the $i$th location and the unsampled location, and $\gamma_{22,jo}$ is the semivariance for property 2 at a separation distance between the $j$th location and the unsampled location.

### 9.3.6 INDICATOR KRIGING

Punctual kriging, block kriging and cokriging are all useful for estimating the value of a property at an unsampled location. The arithmetic average of all interpolated values is an estimate of the mean. Oftentimes, rather than estimating the mean, it is necessary to estimate the proportion of values at a site which is above or below some critical value. For example, the extent of cleanup and remediation required at a site contaminated with trace metals depends upon the proportion of the site where the soil concentrations of trace metals exceed statutory limits. The kriging methods discussed up to this point are unsuitable for making such estimates.

Indicator kriging is an approach for estimating the proportion of values that fall within specified class intervals or the proportion of values that are below a threshold level (Journel, 1986; Isaaks and Srivastava, 1989). The basic approaches in indicator kriging are: (1) to transform measured values into indicator variables, (2) to estimate the semivariogram for the indicator transformed values, (3) to use simple kriging to interpolate the indicator transformed values across the study site, and (4) to compute the average of the kriged indicator values. This average is the proportion of values that is less than the threshold.

The indicator transform ($I_i$) is simply defined as follows (Rossi et al., 1992):

$$I_i = \begin{cases} 1 & if \quad z_i \leq k \\ 0 & if \quad z_i > k \end{cases} \qquad [9.29]$$

where $k$ is the specified threshold cutoff value. Indicator values can be averaged over the site using a series of different threshold cutoff values if there is some uncertainty about the most appropriate value to use for the threshold. For each threshold value specified, a new semivariogram model should be estimated. However, it has often been found that the semivariogram estimated using the population median as a threshold is similar to the semivariograms estimated using other threshold values (Isaaks and Srivastava, 1989).

## 9.4 SAMPLING DESIGN

The variability and spatial structure of a sample population are of great importance in developing a valid soil sampling design. Soil sampling design is concerned with developing a statistically rigorous and unbiased strategy for collecting soil samples (Brown, 1993; Carter, 1993). The main considerations in sampling design include determining the optimal number of samples to collect, and determining the spatial arrangement of the samples to be collected (Wollenhaupt et al., 1997). A rigorous sampling design ensures that sampling points are representative of the region studied (ASTM, 1997a), meaning that the mean of the sampled points is a very good estimate for the mean of the population. It also minimizes the bias or systematic error caused by human judgment. Sample designs may also be developed to provide quality control and quality assurance, involving replicate measurements and split samples (ASTM, 1997b). Finally, the sampling design should provide the basis for accurate identification of spatial patterns in the measured property.

The optimal number of samples to collect depends upon the variability of the sample population, the desired level of accuracy in estimating the population mean, the confidence interval desired for estimation of the population mean, considerations for cost of sampling and sample analysis, and availability of labor or equipment. When samples are expected to obey a normal distribution and are uncorrelated, the following formula (Snedecor and Cochran, 1980; Warrick et al., 1986; Burrough, 1991; ASTM, 1997a) is often used to estimate the number ($N$) of samples required to achieve a desired precision in estimating the mean value for the property studied:

$$N = \frac{\left(t^2 s^2\right)}{d^2} \qquad [9.30]$$

where $t$ is the tabulated value of student's $t$ for a two-sided confidence interval at a given probability level, $s$ is a preliminary estimate for the standard deviation of the population, and $d$ is the deviation desired between the population and measured means. In theory, since $N$ is unknown, the proper number of degrees of freedom for estimating $t$ in Equation [9.30] is also unknown. Skopp et al. (1995) have provided an iterative approach for successively estimating an initial value for $N$, its corresponding degrees of freedom ($t$) and the true value of $N$.

In practice, one often assumes that the sample size ($N$) is large enough that the degrees of freedom can be obtained for a large sample size. Consider an example in which one computes the number of samples required to estimate the mean within a 95% confidence interval. The number of samples required is $(1.96)^2(s^2/d^2)$. For this example, the number of samples needed to estimate the mean for the two sample populations whose frequency distributions were plotted in Figures 9.1A and B will be computed. The two populations had means of 50 and standard deviations of 10 and 5, respectively. If the desired deviation cannot exceed $d = 1$, one would have to collect 384 and 96 samples to characterize the means of the first and second populations, respectively. If the desired deviation cannot exceed $d = 3$, one would have to collect 43 and 11 samples, respectively. Finally, if the desired deviation is $d = 10$, only 4 and 1 sample would be required. Clearly, as the desired deviation from the mean decreases, the number of samples required increases dramatically, although the rate of increase depends on the standard deviation of the population.

For sample populations in which spatial correlation is expected, the formula for estimating the number of samples is modified to take spatial correlation into account. Generally, when spatial correlation is present, the amount of information in any sample is diminished, and a greater number of samples are required to estimate the mean. This increase in the number of samples needed can be estimated by computing the equivalent number of independent observations giving the same population variance. In general, the number of equivalent independent observations is much less than the number of correlated samples. The following formula shows how to estimate the equivalent coefficient (r):

$$N' = N \Big/ \Big[ 1 + 2\{\rho/(1-\rho)\} \; \{1 - (1/N)\} - 2\{\rho/(1-\rho)\}^2 \, (1-\rho^N)/N \Big] \qquad [9.31]$$

For spatially correlated sample populations, the value of $N'$ is always smaller than $N$. Thus, the true number of samples needed to estimate the mean is $N^2/N'$. Another consequence of correlation is that the confidence intervals computed for independent samples are too narrow for correlated samples.

Other considerations for soil sampling, in addition to the number of samples, include sampling depth, sampling time, sample volume or support, and compositing (Wollenhaupt et al., 1997). The proper sampling depth is dependent on many factors, including the type of property measured, the type of equipment used to collect samples, the type and depth of tillage (Kitchen et al., 1990), the application of broadcast or banded fertilizer (Mahler, 1990; Tyler and Howard, 1991) and the condition of the soil (wet, dry, compacted, etc.). In some cases, it is desirable to sample at depths corresponding to distinct soil horizons. When soil horizons are very thick, this may not be practical. The proper timing of sample collection is particularly important when measuring a temporally variable soil property such as soil $NO_3$-N (Lockman and Molloy, 1984), soil moisture content (Bertuzzi et al., 1994) or soil hydraulic conductivity.

The proper sample support or volume depends to some extent on the small scale variability of the property (Johnson et al., 1990; Lame and Defize, 1993; Starr et al., 1995). For instance, undisturbed cores used to estimate saturated soil hydraulic conductivity should be large enough to represent all of the structural heterogeneities (especially macropores) that are likely to affect infiltration. A rigorous approach for determining the optimum sample support from the shape of the semivariogram was developed by Zhang et al. (1990). van Wesenbeeck and Kachanoski (1991) used the range of the semivariogram for solute travel times to estimate the optimum support size for studies of spatial variability in solute leaching. Kamgar et al. (1993) determined the optimum size of field plots and the optimum number of neutron probe access tubes in each plot using semivariograms in a study of water storage; a theoretical relationship between plot size and sample variance developed by Smith (1938) was used. Gilbert and Doctor (1985) developed an approach for determining optimum sample volume with consideration for not only the variance-volume relationship, but also the cost of sampling and analysis at different sample volumes.

Composite disturbed samples are often collected from a small region and mixed when undesirable small-scale spatial patterns would obscure broad-scale spatial patterns. The optimum number of composite samples collected depends on the variance of the measured property, and its spatial structure (Webster and Burgess, 1984; Giesler and Lundstrom, 1993). Typically, a series of from four to eight subsamples are collected and composited from an area having a radius no larger than 5 m (Oliver et al., 1997).

To avoid bias in sampling, a statistically rigorous arrangement of sample locations is needed. Several strategies for determining sample locations have been developed, including judgment, simple random, stratified random, systematic, stratified systematic unaligned, targeted, adaptive and geostatistical sampling. Each of these approaches will be described below.

## 9.4.1 JUDGMENTAL SAMPLING

This is an approach used when there is little interest in statistical rigor. The objective of the sampling program may be to evaluate soil properties at several locations where some type of problem is thought to exist. This problem could, for example, be low crop yield, soil contamination, poor drainage or severe erosion. The decisions about where to sample in this approach are often based on visual evidence of a change in soil properties, which could include changes in color, soil moisture, vegetation height and color, or salinity.

Judgmental sampling is not statistically rigorous. Sample numbers are often not large enough for accurate estimation of the population mean. Because the choice of sampling locations is based on personal experience, judgment and opinion, there is almost always bias in the sample design. For example, the objective of judgmental sampling may be to diagnose the reason for low crop yields at several locations in the field. Samples could be collected only from those areas with low yields, and analyzed for soil nutrient availability or permeability. The samples collected are not representative of the whole field.

Even more problematic, there is judgment involved in determining what constitutes a low yielding portion of the field. Two people may have very different criteria for deciding what portions of the field are low yielding, and hence, their sampling programs may be quite different. If one of the two people makes a living from selling fertilizer, his or her judgment about where to sample may be biased by a desire to sell more fertilizer. Thus, judgment sampling is subject to human bias.

### 9.4.2 SIMPLE RANDOM SAMPLING

Bias can be avoided by using statistically rigorous methods to identify sampling locations. Random selection of sampling locations avoids bias. In this approach, a random number generator is used to select locations for sampling (Figure 9.8A). All locations have an equal probability of being selected, and successive selections for sampling locations are completely independent of the previous selections.

It may not be easy to actually sample from the locations selected by the random number generator. This is because there may be features in the field, such as a tree, a pond, a fence or a building, which were not accounted for when selecting sample locations. In such cases, the desired sampling location may be replaced by another nearby one which is selected by randomly selecting a direction and distance to the new point. Unless sophisticated position-locator equipment is available (GPS), it may also prove difficult to find the randomly selected location with any great degree of accuracy. Finally, the random sampling design often has an uneven or clustered appearance, and large areas of the field may be unsampled. This leads to a large degree of uncertainty about the soil properties in the unsampled regions.

### 9.4.3 STRATIFIED RANDOM SAMPLING

Stratified random sampling is a refinement of simple random sampling in which the area is divided into cells or strata that can be regularly or irregularly shaped. The most common types of strata include grid cells that are square in shape, or irregular cells that correspond to soil mapping units. If square cells are chosen, there is some judgment involved in specifying the dimension of the cells, which can lead to bias.

Sampling locations within each cell are selected at random, using a random number generator (Figure 9.8B). This reduces, but may not eliminate, the unevenness and clustering that often occur in simple random sampling. The number of sampling locations within each cell is preset based upon an evaluation of the total number of samples needed to accurately define the population mean.

Stratification of the study area by soil type may be most useful when the soil variation studied is strongly linked to soil mapping units. This may be especially true when studying a soil morphological property such as depth of the A horizon. When studying soil properties that are strongly influenced by soil management, this approach for stratifying sample locations may not lead to improvements in understanding sample variation. An example of this might be the study of soil P availability, in which the major source of variability is due to previous management effects rather than soil morphology and mapping unit delineations.

### 9.4.4 SYSTEMATIC SAMPLING

To avoid clustered sampling and difficulties associated with finding randomly spaced locations in the field, many sampling designs use systematic sampling (Wollenhaupt et al., 1994). In this approach, sample locations occur either at the center of regularly shaped grid cells or at the

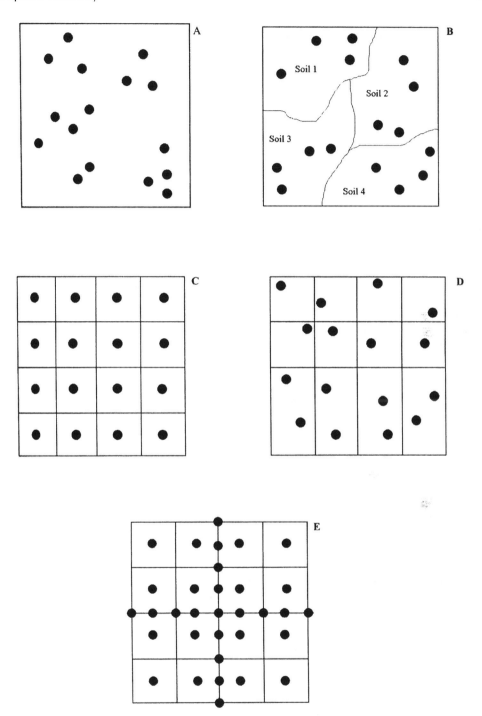

**FIGURE 9.8** (A) Example of a simple random sampling design; (B) example of a stratified random sampling design; (C) example of a systematic sampling design; (D) example of a sampling design for random sampling within cells; (E) example of a geostatistical sampling design.

intersection of the grid lines. When samples are located at the center of cells, the resulting values are thought to be representative of the soil within the grid cell (Figure 9.8C). When they are located at the intersections of grid lines, all of the values along the edge of the grid cell can be averaged to obtain a representative value for the grid cell.

Grid cells can be of varying size and shape. The most common shapes are square, rectangular, hexagonal and triangular grid cells. Triangular grid cells are widely considered to be more efficient than the other methods. Rectangular grid cells may be used when there is some reason to believe that the spatial variability in the sampled property exhibits anisotropy due to topographic, tillage or other types of influences. The square grid cell is probably the most widely used approach in systematic sampling, and is often simply referred to as grid sampling.

One reason for the popularity of the square grid cell sampling design is the ease in finding or surveying sampling locations. On flat fields, two parallel rows of sampling locations can be surveyed on adjacent transects. Then, starting at the midpoint of these two rows, two more columns can be surveyed in a direction perpendicular to the first two rows. Stakes or flags can be driven at each sampling location along the four transects. All other sampling locations can then be found by sighting toward the two rows and the two columns until the two flags at the right sampling distance in those transects are both perfectly lined up with your location. Alternatively, global positioning satellites, distance measuring devices and dead reckoning can be used to find sampling locations.

The major disadvantage of systematic sampling is that the sample rows may be perfectly aligned with soil or management features that vary systematically (Wollenhaupt et al., 1997). For instance, sample rows may be alternately aligned with irrigation furrows and then hills, or with tillage rows and then wheel tracks. In either case, the analysis of spatial patterns in the direction of the columns might show periodicity. To avoid this bias, random sampling may be done within cells (Figure 9.8D).

### 9.4.5 STRATIFIED, SYSTEMATIC, UNALIGNED SAMPLING

The bias introduced by the systematic sampling strategy can be overcome by reducing the alignment of sampling locations in the columns and rows. Stratified systematic unaligned sampling involves stratifying the study area into regular-sized cells which may be square, rectangular, triangular or hexagonal. Each cell is further subdivided into many smaller cells of the same shape. The approach for choosing sampling locations within each larger cell is described by Wollenhaupt et al. (1997). The stratified, systematic, unaligned sampling strategy combines the best features of the systematic and stratified random sampling designs. Finding sample locations in the field is relatively easy, yet the random element of the design overcomes problems associated with regular features caused by row spacing and tillage.

### 9.4.6 TARGETED OR DIRECTED SAMPLING

There are two major types of variation in soil properties, namely, those that can be seen and those that cannot. An example of the former is the spatial variation in soil color as a result of patterns in soil surface organic C or moisture contents; an example of the latter is the spatial variation in soil nutrient availability caused by a management history of spatially varying cropping patterns, fertilizer or manure applications.

A statistically rigorous sampling design is appropriate for the characterization of spatial patterns arising from unseen sources of variability such as soil nutrient availability. If, however, spatial patterns in the field arise from both unseen and visually obvious sources of variability, it may be advisable to supplement the statistically rigorous sampling design with some targeted or directed samples.

Targeted or directed samples are taken at locations where there is some visual evidence for a change in value of the measured property. For instance, an aerial photo taken prior to sampling (Bhatti et al., 1991) or during the early stages of crop growth (Ferguson et al., 1996) may show color variations within the study site. A map of variations in crop yield may show small regions with either very poor or very good yields. Other sources of prior information about the study site may also be used to design a directed sampling strategy, including soil, digital elevation, ground penetrating radar and electromagnetic induction maps (Pocknee et al., 1996). In these cases, a few sample locations may be added to the statistically rigorous sampling design strategy to find out how the measured soil property changes in the targeted region. This approach is particularly useful when the objective of the sampling program is to identify and characterize regions of the field that are distinctly different from the majority of the field, as is needed for Precision Farming (Mulla, 1991, 1993; Francis and Schepers, 1997).

### 9.4.7 ADAPTIVE SAMPLING

Some soil properties and many biological properties such as weeds and pests often are distributed in clusters. Sampling locations are much more likely to have a weed or pest infestation if weeds or pests were or were not observed at a nearby sampling location. If this is the case, the efficiency for any one of the sampling designs discussed previously can be improved by adaptive sampling (Thompson, 1997). In adaptive sampling, a statistically rigorous sampling design is first selected. During the field sampling program, if a weed or pest infestation is observed at one of the sample locations, then another nearby location is also sampled. If that location also shows a weed or pest, then another nearby point is selected for sampling. This continues until weeds or pests are no longer observed in the neighborhood. The advantage of this approach is that it allows the spatial extent of the cluster for weed or pest infestation to be more competely delineated than in the case of using just the initial sample design.

### 9.4.8 GEOSTATISTICAL SAMPLING

In geostatistical sampling there are two primary concerns. The first concern is systematic sampling for the purposes of accurate interpolation by kriging to produce spatial pattern maps. A regular grid with square, triangular or hexagonal elements is most often used to achieve this objective (Webster and Burgess, 1984; Burrough, 1991; Wollenhaupt et al., 1997). For a regular grid sampling program, the most efficient placement of sample locations is in the center of each grid cell (Webster and Burgess, 1984; Burrough, 1991). Sample spacings for these grid cells should be less than ½ of the range for the semivariogram (Flatman and Yfantis, 1984). The second concern is sampling for accurate estimation of the semivariogram (Russo, 1984; Russo and Jury, 1988). The semivariogram is useful as a tool for modeling spatial structure in a measured soil property, and is important for accurate interpolation by kriging. A typical approach for accurate estimation of the semivariogram is to supplement the systematic sampling grid with one or more transects consisting of closely spaced sample locations (Figure 9.8E). This helps define the shape of the semivariogram at small separation distances.

One of the dilemmas in designing geostatistically based sampling schemes is that the design depends upon the semivariogram, yet the semivariogram is often unknown until the study site is sampled. There are two approaches for solving this dilemma. The first is to use pre-existing information about the range of the semivariogram for the soil property of interest, whether this information is from a study site, a nearby site, or a site in the same region. Once an estimate for the range is obtained, it can be used to set the approximate sample spacings in the sampling design of interest. The second approach is used if no pre-existing information about the semivariogram can be obtained. In this case, it is necessary to conduct a preliminary sampling survey of the study site along several transects (Flatman and Yfantis, 1984).

If the semivariogram is known, it is a powerful tool which can be used to evaluate various sampling strategies before any samples are collected (McBratney et al., 1981; Rouhani, 1985; Burrough, 1991). The first type of evaluation is the determination of optimum spacing for sample locations (Rouhani, 1985; Warrick et al., 1986; Oliver and Webster, 1987; Burrough, 1991; Gutjahr, 1991). This is achieved by computing kriging estimation variances for a wide range of sample spacings and arrangements. The estimation variances can be computed even when no measurements are available at sample locations, because the estimation variance depends only upon the semivariogram and separation distances between potential sampled and unsampled locations. The second type of evaluation is determination of the optimum number of composite or bulked samples to collect at each sample location (Webster and Burgess, 1984; Oliver et al., 1997). Significant reductions in estimation error are possible by mixing samples taken from small blocks around a sampled location and analyzing the composite sample.

---

**EXAMPLE 9.5**   The main elements of a good sampling strategy include the proper number of samples, the proper arrangement of samples and the proper spacing between samples. Propose a good sampling strategy for the soil organic carbon content values shown in Figure 9.2 using information provided in Table 9.1, Figure 9.4A, Equation [9.30] and Figure 9.8. Assume that the mean value for soil organic carbon content is 4%.

---

## REFERENCES

ASTM. 1997a. Standard guide for general planning of waste sampling, p. 460–469, *in* J. Azara, N.C. Baldini, E. Barszcewski, L. Bernhardt, E.L. Gutman, J.G. Kramer, C.M. Leinweber, V.A. Mayer, P.A. McGee, T.J. Sandler, and R.F. Wilhelm (eds.), Standards on Environmental Sampling, 2nd ed., American Society for Testing Materials, West Conshohocken, PA.

ASTM. 1997b. Standard guide for sampling strategies for heterogeneous wastes, p. 521–537, *in* J. Azara, N.C. Baldini, E. Barszcewski, L. Bernhardt, E.L. Gutman, J.G. Kramer, C.M. Leinweber, V.A. Mayer, P.A. McGee, T.J. Sandler, and R.F. Wilhelm (eds.), Standards on Environmental Sampling, 2nd ed., American Society for Testing Materials, West Conshohocken, PA.

Bertuzzi, P., L. Bruckler, D. Bay, and A. Chanzy. 1994. Sampling strategies for soil water content to estimate evapotranspiration, *Irrig. Sci.,* 14:105–115.

Beven, K.J., D.E. Henderson, and A.D. Reeves. 1993. Dispersion parameters for undisturbed partially saturated soil, *J. Hydrol.,* 143:19–43.

Bhatti, A.U., D.J. Mulla, and B.E. Frazier. 1991. Estimation of soil properties and wheat yields on complex eroded hills using geostatistics and thematic mapper images, *Remote Sens. Environ.,* 37:181–191.

Birrell, S.J., K.A. Sudduth, and N.R. Kitchen. 1996. Nutrient mapping implications of short-range variability, p. 207–216, *in* P.C. Robert, R.H. Rust, and W.E. Larson (eds.), Precision Agriculture, *Am. Soc. Agron.,* Madison, WI.

Brown, A.J. 1993. A review of soil sampling for chemical analysis, *Aust. J. Exp. Agric.,* 33:983–1006.

Burgess, T.M. and R. Webster. 1980a. Optimal interpolation and isarithmic mapping of soil properties. I. The semi-variogram and punctual kriging, *J. Soil Sci.,* 31:315–331.

Burgess, T.M. and R. Webster. 1980b. Optimal interpolation and isarithmic mapping of soil properties. II. Block kriging, *J. Soil Sci.,* 31:333–341.

Burrough, P.A. 1991. Sampling designs for quantifying map unit composition, p. 89–125, *in* M.J. Mausbach and L.P. Wilding (eds.), Spatial Variabilities in Soils and Landforms, *Soil Sci., Soc. Am.,* Spec. Pub. 28, Madison, WI.

Carter, M.R. 1993. Soil sampling and methods of analysis, Lewis Publishers, Boca Raton, FL.

Clark, I. 1979. Practical geostatistics, Appl. Sci. Publ., London, UK.

Cooke, R.A., S. Mostaghimi, and J.B Campbell. 1993. Assessment of methods for interpolating steady-state infiltrability, Trans. ASAE, 36:1333–1341.

Cressie, N. 1985. Fitting variogram models by weighted least squares, *Math. Geol.,* 17:563–586.

Cressie, N. 1991. Statistics for spatial data, Wiley Interscience, NY.

Cressie, N. and D.M. Hawkins. 1980. Robust estimation of the variogram, *J. Inter. Assoc. Math. Geol.,* 12:115–125.

Di, H.J., B.B. Trangmar, and R.A. Kemp. 1989. Use of geostatistics in designing sampling strategies for soil survey, *Soil Sci. Soc. Am. J.,* 53:1163–1167.

Dubrule, O. 1984. Comparing splines and kriging, *Computers Geosci.,* 10:327–338.

Ferguson, R.B., C.A. Gotway, G.W. Hergert, and T.A. Peterson. 1996. Soil sampling for site-specific nitrogen management, p. 13–22, *in* P.C. Robert, R.H. Rust, and W.E. Larson (eds.) Precision Agriculture. Am. Soc. Agron., Madison, WI.

Flatman, G.T. and A.A. Yfantis. 1984. Geostatistical strategy for soil sampling: The survey and the census, *Environ. Monit. Assess.*, 4:335:349.

Francis, D.D. and J.S. Schepers. 1997. Selective soil sampling for site-specific nutrient management, p. 119–126, *in* J.V. Stafford (ed.), Precision Agriculture '97. Vol. 1: Spatial variability in soil and crop. BIOS Sci. Publ. Ltd., Oxford, UK.

Giesler, R. and U. Lundstrom. 1993. Soil solution chemistry: effects of bulking soil samples, *Soil Sci. Soc. Am. J.*, 57:1283–1288.

Gilbert, R.O. and P.D. Doctor. 1985. Determining the number and size of soil aliquots for assessing particulate contaminant concentrations, *J. Environ. Qual.*, 14:286–292.

Gotway, C.A., R.B. Ferguson, G.W. Hergert, and T.A. Peterson. 1996. Comparison of kriging and inverse-distance methods for mapping soil parameters, *Soil Sci. Soc. Am. J.*, 60:1237–1247.

Gutjahr, A. 1991. Geostatistics for sampling designs and analysis, ACS Symp. Ser. 465:48–90.

Hawkins, D.M. 1980. Identification of outliers, Chapman and Hall, London, UK.

Hoeksma, R.J. and P.K. Kitanidis. 1985. Comparison of Gaussian conditional mean and kriging estimation in the geostatistical solution of the inverse problem, *Water Resour. Res.*, 21:825–836.

Isaaks, E.H. and R.M. Srivastava. 1989. An introduction to applied geostatistics, Oxford University Press, NY.

Johnson, C.E., A.H. Johnson, and T.G. Huntington. 1990. Sample size requirements for the determination of changes in soil nutrient pools, *Soil Sci.*, 150:637–644.

Journel, A.G. 1986. Constrained interpolation and qualitative information — The soft kriging approach, *Math. Geol.*, 18:269–286.

Journel, A.G. and C.J. Huijbregts. 1978. Mining geostatistics, Academic Press, London, UK.

Journel, A.G., C. Deutsch, and A.J. Desbarats. 1986. Power averaging for block effective permeability, *Soc. Petrol. Eng.*, Pap. SPE 15128.

Jury, W.A. 1986. Spatial variability of soil properties, p.245–269, *in* S.C. Hern and S.M. Melancon (eds.), Vadose Zone Modeling of Organic Pollutants, Lewis Publishers, Chelsea, MI.

Jury, W.A., D. Russo, G. Sposito, and H. Elabd. 1987. The spatial variability of water and solute transport properties in unsaturated soil: I. Analysis of property variation and spatial structure with statistical models, Hilgardia, 55:1–32.

Kamgar, A., J.W. Hopmans, W.W. Wallender, and O. Wendroth. 1993. Plot size and sample number for neutron probe measurements in small field trials, *Soil Sci.*, 156:213–224.

Kitchen, N.R., J.L. Havlin, and D.G. Westfall. 1990. Soil sampling under no-till banded phosphorus, *Soil Sci. Soc. Am. J.*, 54:1661–1665.

Krige, D.G. 1966. Two dimensional weighted moving average trend surfaces for ore-evaluation, *J.S. Afr. Inst. Min. Metall.*, 66:13–38.

Laslett, G.M., A.B. McBratney, P.J. Pahl, and M.F. Hutchinson. 1987. Comparison of several spatial prediction methods for soil pH, *J. Soil Sci.*, 38:325–341.

Lame, F.P.J. and P.R. Defize. 1993. Sampling of contaminated soil: Sampling error in relation to sample size and segregation, *Environ. Sci. Technol.*, 27:2035–2044.

Lockman, R.B. and M.G. Molloy. 1984. Seasonal variations in soil test results, *Commun. Soil Sci. Plant Anal.*, 15:741–757.

Mahler, R.L. 1990. Soil sampling fields that have received banded fertilizer applications, *Commun. Soil Sci. Plant Anal.*, 21:1793–1802.

Markus, J.A. and A.B. McBratney. 1996. An urban soil study: Heavy metals in Glebe, Australia, *Aust. J. Soil Res.*, 34:453–465.

Matheron, G. 1963. Principles of geostatistics, *Econ. Geol.*, 58:146–1266.

McBratney, A.B., R. Webster, and T.M. Burgess. 1981. The design of optimal sampling schemes for local estimation and mapping of regionalized variables. I. Theory and method, *Comp. Geosci.*, 7:331–334.

McBratney A.B. and M.J. Pringle. 1997. Spatial variability in soil: Implications for precision agriculture, p. 3–31, *in* J.V. Stafford (ed.), Precision agriculture '97. Volume 1: Spatial variability in soil and crop, BIOS Sci. Publ. Ltd., Oxford, UK.

Mulla, D.J. 1988a. Using geostatistics and spectral analysis to study spatial patterns in the topography of southeastern Washington State, U.S.A., *Earth Surf. Proc. Land.*, 13:389–405.

Mulla, D.J. 1988b. Estimating spatial patterns in water content, matric suction, and hydraulic conductivity, *Soil Sci. Soc. Am. J.*, 52:1547–1553.

Mulla, D.J. 1991. Using geostatistics and GIS to manage spatial patterns in soil fertility, p. 336–345, *in* G. Kranzler (ed.), Automated Agriculture for the 21st Century, *Am. Soc. Agric. Eng.*, St. Joseph, MI.

Mulla, D.J. 1993. Mapping and managing spatial patterns in soil fertility and crop yield, p. 15–26, *in* P.C. Robert, R.H. Rust, and W.E. Larson (eds.), Site Specific Crop Management, *Am. Soc. Agron.*, Madison, WI.

Mulla, D.J. 1997. Geostatistics, remote sensing and precision farming, p. 100–114, *in* J.V. Lake, G.R. Bock, and J.A. Goode (eds.), Precision agriculture: Spatial and Temporal Variability of Environmental Quality, John Wiley and Sons, Chichester, UK.

Nash, M.H., L.A. Daugherty, A. Gutjahr, P.J. Wierenga, and S.A. Nance. 1988. Horizontal and vertical kriging of soil properties along a transect in Southern New Mexico, *Soil Sci. Soc. Am. J.*, 52:1086–1090.

Oliver, M.A. 1987. Geostatistics and its application to soil science, *Soil Use Manag.*, 3:8–20.

Oliver, M.A. and R. Webster. 1987. The elucidation of soil pattern in the Wyre Forest of the West Midlands, England. II. Spatial distribution, *J. Soil Sci.*, 38:293–307.

Oliver, M.A. and R. Webster. 1991. How geostatistics can help you, *Soil Use Manag.*, 7:206–217.

Oliver, M.A., Z. Frogbrook, R. Webster, and C.J. Dawson. 1997. A rational strategy for determining the number of cores for bulked sampling of soil, p. 155–162, *in* J.V. Stafford (ed.), Precision Agriculture '97. Volume 1: Spatial variability in soil and crop, BIOS Sci. Publ. Ltd., Oxford, UK.

Pocknee, S., B.C. Boydell, H.M. Green, D.J. Waters, and C.K. Kvien. 1996. Directed soil sampling, p. 159–168, *in* P.C. Robert, R.H. Rust, and W.E. Larson (eds.), Precision Agriculture, *Am. Soc. Agron.*, Madison, WI.

Rogowski, A.S. and J.K. Wolf. 1994. Incorporating variability into soil map unit delineations, *Soil Sci. Soc. Am. J.*, 58:163–174.

Rossi, R.E., D.J. Mulla, A.G. Journel, and E.H. Franz. 1992. Geostatistical tools for modeling and interpreting ecological spatial dependence, *Ecol. Mono.*, 62:277–314.

Rouhani, S. 1985. Variance reduction analysis, *Water Resour. Res.*, 21:837–846.

Russo, D. 1984. Design of an optimal sampling network for estimating the variogram, *Soil Sci. Soc. Am. J.*, 48:708–716.

Russo, D. and M. Bouton. 1992. Statistical analysis of spatial variability in unsaturated flow parameters, *Water Resour. Res.*, 28:1911–1925.

Russo, D. and W.A. Jury. 1988. Effect of the sampling network on estimates of the covariance function of stationary fields, *Soil Sci. Soc. Am. J.*, 52:1228–1234.

Skopp, J., S.D. Kachman, and G.W. Hergert. 1995. Comparison of procedures for estimating sample numbers, *Commun. Soil Sci. Plant Anal.*, 26:2559–2568.

Smith, H.F. 1938. An empirical law describing heterogeneity in the yields of agricultural crops, *J. Agri. Sci.*, 28:1–23.

Snedecor, G.W. and W.G. Cochran. 1980. Statistical methods, 7th ed., Iowa State University Press, Ames, IA.

Starr, J.L., T.B. Parkin, and J.J. Meisinger. 1995. Influence of sample size on chemical and physical soil measurements, *Soil Sci. Soc. Am. J.*, 59:713–719.

Stein, A., W. van Dooremolen, J. Bouma, and A.K. Bregt. 1988. Cokriging point data on moisture deficit, *Soil Sci. Soc. Am. J.*, 52:1418–1423.

Thompson, S.K. 1997. Spatial sampling, p. 161–168, *in* J.V. Lake, G.R. Bock, and J.A. Goode (eds.), Precision Agriculture: Spatial and temporal variability of environmental quality, John Wiley and Sons, Chichester, UK.

Thompson, W.H. and P.C. Robert. 1995. Evaluation of mapping strategies for variable rate applications, p. 303–323, *in* P.C. Robert, R.H. Rust, and W.E. Larson (eds.), Site-specific management for agricultural systems, *Am. Soc. Agron.*, Madison, WI.

Tyler, D.D. and D.D. Howard. 1991. Soil sampling patterns for assessing no-tillage fertilization techniques, *J. Fert. Issues*, 8:52–56.

Upchurch, D.R. and W.J. Edmonds. 1991. Statistical procedures for specific objectives, p. 49–71, *in* M.J. Mausbach and L.P. Wilding (eds.), Spatial Variabilities of Soils and Landforms, *Soil Sci. Soc. Am.*, Spec. Publ. 28, *Soil Sci. Soc. Am.*, Madison, WI.

van Wesenbeeck, I.J. and R.G. Kachanoski. 1991. Spatial scale dependence of *in situ* solute transport, *Soil Sci. Soc. Am. J.*, 55:3–7.

Vauclin, M., S.R. Vieira, R. Bernard, and J.L. Hatfield. 1982. Spatial variability of surface temperature along two transects of a bare soil, *Water Resour. Res.*, 18:1677–1686.

Vauclin, M., S.R. Vieira, G. Vachaud, and D.R. Nielsen. 1983. The use of cokriging with limited field soil observations, *Soil Sci. Soc. Am. J.,* 47:175–284.

Vieira, S.R., J.L. Hatfield, D.R. Nielsen, and J.W. Biggar. 1983. Geostatistical theory and application to variability of some agronomical properties, *Hilgardia,* 51:1–75.

Vieira, S.R., W.D. Reynolds, and G.C. Topp. 1988. Spatial variability of hydraulic properties in a highly structured clay soil, p. 471–483, *in* P.J. Wierenga and D. Bachelet (eds.), International conference and workshop on the validation of flow and transport models for the unsaturated zone, NM State University, Las Cruces, NM.

Voltz, M. and R. Webster. 1990. A comparison of kriging, cubic splines and classification for predicting soil properties from sample information, *J. Soil Sci.,* 41:473–490.

Wade, S.D., I.D.L. Foster, and S.M.J. Baban. 1996. The spatial variability of soil nitrates in arable land and pasture landscapes: Implications for the development of geographical information system models of nitrate leaching, *Soil Use Manag.,* 12:95–101.

Warrick, A.W., D.E. Myers, and D.R. Nielsen. 1986. Geostatistical methods applied to soil science, p. 53–82, *in* A. Klute (ed.), Methods of Soil Analysis. Part 1. 2nd ed., *Soil Sci. Soc. Am.,* Madison, WI.

Warrick, A.W., R. Zhang, M.K. El-Harris, and D.E. Myers. 1988. Direct comparisons between kriging and other interpolators, p. 505–510, *in* P.J. Wierenga and D. Bachelet (eds.), International conference and workshop on the validation of flow and transport models for the unsaturated zone, NM State University, Las Cruces, NM.

Webster, R. and T.M. Burgess. 1980. Optimal interpolation and isarithmic mapping of soil properties. III. Changing drift and universal kriging, *J. Soil Sci.,* 31:505–524.

Webster, R. and T.M. Burgess. 1984. Sampling and bulking strategies for estimating soil properties in small regions, *J. Soil Sci.,* 35:127–140.

Webster, R. and M.A. Oliver. 1990. Statistical methods in soil and land resource survey, Oxford University Press, NY.

Whelan, B.M., A.B. McBratney, and R.A. Viscarra Rossel. 1996. Spatial prediction for precision agriculture, p. 331–342, *in* P.C. Robert, R.H. Rust, and W.E. Larson (eds.), Precision agriculture, *Am. Soc. Agron.,* Madison, WI.

Wierenga, P.J., R.G. Hills, and D.B. Hudson. 1991. The Las Cruces trench site: Characterization, experimental results, and one-dimensional flow predictions, *Water Resour. Res.,* 27:2695–2705.

Wilding, L.P. 1985. Spatial variability: its documentation, accomodation and implication to soil surveys, p. 166–194, *in* D.R. Nielsen and J. Bouma (eds.), Soil Spatial Variability, Pudoc, Wageningen, Netherlands.

Wollenhaupt, N.C., D.J. Mulla, and C.A. Gotway Crawford. 1997. Soil sampling and interpolation techniques for mapping spatial variability of soil properties, p. 19–53, *in* F.J. Pierce and E.J. Sadler (eds.), The State of Site Specific Management for Agriculture, *Am. Soc. Agron.,* Madison, WI.

Wollenhaupt, N.C., R.P. Wolkowski, and M.K. Clayton. 1994. Mapping soil test phosphorus and potassium for variable-rate fertilizer application, *J. Prod. Agric.,* 7:441–448.

Yates, S.R., A.W. Warrick, and D.E. Myers. 1986a. Disjunctive kriging. 1. Overview of estimation and conditional probability, *Water Resour. Res.,* 22:615–621.

Yates, S.R., A.W. Warrick, and D.E. Myers. 1986b. Disjunctive kriging. 2. Examples, *Water Resour. Res.,* 22:623–630.

Yates, S.R. and A.W. Warrick. 1987. Estimating soil water content using cokriging, *Soil Sci. Soc. Am. J.,* 51:23–30.

Zhang, R., A.W. Warrick, and D.E. Myers. 1990. Variance as a function of sample support size, *Math. Geol.,* 22:107–121.

Zhang, R., D.E. Myers, and A.W. Warrick. 1992. Estimation of the spacial distribution of soil chemicals using pseudo-cross-variograms, *Soil Sci. Soc. Am. J.,* 56:1444–1452.

# Index